LOW-METALLICITY STAR FORMATION:
FROM THE FIRST STARS TO DWARF GALAXIES

IAU SYMPOSIUM No. 255

COVER ILLUSTRATION: The first stars, dwarf galaxies in the Local Universe, and the "Antico Castello sul Mare", Rapallo.

The upper panel is a collage of simulations of the First Stars (Tom Abel and WMAP Team), together with *Hubble Space Telescope* images of two nearby low-metallicity dwarf galaxies (Alessandra Aloisi, Monica Tosi, and the ESA, NASA, Hubble Heritage Teams); the collage was composed by Roberto Baglioni (Universitá di Firenze).

The lower panel is the "Antico Castello sul Mare" located at Rapallo (Genova), Italy, near the Symposium venue at the Clarisse Convent. The photograph is reproduced with the permission of the Tipografia Canessa, Rapallo.

INTERNATIONAL ASTRONOMICAL UNION

UNION ASTRONOMIQUE INTERNATIONALE

LOW-METALLICITY STAR FORMATION: FROM THE FIRST STARS TO DWARF GALAXIES

PROCEEDINGS OF THE 255th SYMPOSIUM OF THE INTERNATIONAL ASTRONOMICAL UNION HELD IN RAPALLO (GENOVA), ITALY JUNE 16–20, 2008

Edited by

LESLIE K. HUNT

INAF-Istituto di Radioastronomia, Firenze, Italy

SUZANNE C. MADDEN

CEA, Saclay, France

and

RAFFAELLA SCHNEIDER

INAF-Osservatorio Astrofisico di Arcetri, Firenze, Italy

CAMBRIDGE
UNIVERSITY PRESS

CAMBRIDGE UNIVERSITY PRESS
The Edinburgh Building, Cambridge CB2 8RU, United Kingdom
32 Avenue of the Americas, New York, NY 10013-2473, USA
477 Williamstown Road, Port Melbourne, VIC 3207, Australia
Ruiz de Alarcón 13, 28014 Madrid, Spain
Dock House, The Waterfront, Cape Town 8001, South Africa

First published 2008

Printed in the United Kingdom at the University Press, Cambridge

Typeset in System LaTeX 2_ε

A catalogue record for this book is available from the British Library

Library of Congress Cataloguing in Publication data

ISBN 9780 521 889865 hardback
ISSN 1743-9213

Table of Contents

Session I. Population III and metal-free star formation

Session II. Metal enrichment, chemical evolution, and feedback

Session III. Explosive events in low-metallicity environments

Session IV. Dust and gas as seeds for metal-poor star formation

Session V. Metal-poor IMFs, stellar evolution, and star-formation histories

Session VI. Low-metallicity star formation in the local Universe

Preface

Although low-mass metal-poor galaxies in the local universe have often been proposed as the "primordial building blocks" in the hierarchical scenario of structure formation, several lines of evidence suggest that this may not be true.

From the local perspective, the stellar populations and the interstellar medium (ISM) in low-metallicity star-forming dwarf galaxies are intimately related, but the importance of dust and molecular gas and their effect on low-metallicity star formation are only now beginning to be appreciated. The mechanisms of feedback and their influence on chemical evolution are still controversial, even though they control the observed abundance patterns in the metal-poor ISM.

At cosmological redshifts, recent evidence suggests that gamma-ray bursts (GRBs) may be hosted preferentially by metal-poor low-mass galaxies. GRBs may help achieve a better understanding of the physics of stellar evolution in the chemically unenriched environments of the high-redshift universe. Dust in high-redshift objects must have been created by supernova explosions, since there has not been sufficient time at high redshifts for dust production from intermediate-age and evolved stars. The metals synthesized and released in the surrounding gas need to be transported out of the galaxies in which they are produced into the low density regions of the intergalactic medium (IGM), where they are observed up to redshifts as high as $\sim$6. Although metal enrichment of the IGM has received great attention, both the nature of the sources and the epoch where most of the pollution took place remain highly debated.

The metal enrichment history of the gas within galaxies and in the diffuse IGM has a strong impact on structure formation and the nature of stellar populations. At the highest redshifts, chemical feedback might be responsible for the transition from very massive Population III stars to Population II stars spanning the conventional mass range. Theoretical models suggest that primordial star formation and evolution is strongly influenced by the peculiar environment in which it takes place through a combination of mechanical, chemical, and radiative feedback processes whose importance is only now emerging. These early epochs of star formation will be accessible to direct observations with JWST and ALMA. In view of these major breakthroughs, "stellar archeology" of the most metal-poor stars observed in the Galactic halo or in nearby dwarf satellites of the Milky Way, is already providing a powerful probe of high redshift star formation.

The idea for the Symposium came about at the Summer Solstice, June, 2006, during the annual picnic hosted by the IAP and the Observatoire de Paris. In addition to the importance of the science, it was decided that such topics were best discussed at the seaside, preferably during early summer. The aim of the Symposium was to foster an open exchange among different astronomical communities by placing in juxtaposition, for the first time, low-redshift observational astronomers, observational cosmologists, and theoreticians. We felt the general impact of IAU Symposia would help accomplish this by joining, in the context of metal abundance, the rapidly advancing fields of the ISM, stellar evolution, local star-formation scenarios, chemical evolution and feedback, and primordial star and structure formation.

Later that year, the SOC was formed, and following endorsement and sponsoring by IAU Commissions and Divisions and subsequent approval by the IAU Executive Committee in May 2007, we compiled the list of invited speakers and informed the community at large. We soon had almost 200 registered participants, with a final program comprising 5 invited reviews, 17 invited papers, 45 oral contributions, and 90 poster papers, with the participation of 166 astronomers from 24 countries.

IAUS 255 was marked by a special event, in addition to the scientific program. On the evening of 17 June, in the Clarisse Auditorium, the Province of Genova offered the Symposium participants a private concert (by Dario Bonuccelli, pianist) and poetry reading (by Rachele Ghersi, actress), in Italian with written English translation. The program, entitled "Starry Nights" (Notte delle Stelle), was compiled by Dr. Giorgio Devoto, the Cultural Attaché from the Province of Genova, and dedicated to poetry and music with astronomical themes. An extraordinary musical piece for piano entitled "Big Bang – Low Metallicity" was composed by Bonuccelli for the occasion of IAUS 255, and performed by him that night as a world premiere. Incredibly, it captured musically the essence of the primordial universe, and was received with a standing ovation.

We acknowledge with great pleasure the financial support of our sponsors listed on page xiii of these Proceedings, as well as the cultural support provided by the Province of Genova and the City of Rapallo. We would also like to thank Donatella Balbo and Martino Tassano of Rapallo, and Luca Fini and Francesco Tribioli of Firenze, who with good spirit and competence ensured the smooth execution of the oral program at the Clarisse Auditorium. Finally, without Emanuela Masini, the LOC chair, who was in perpetual motion to organize the many logistical details, this Symposium would never have happened.

It is our hope that these Proceedings will provide a nexus between near-field cosmology and the primordial Universe, and that star formation at low metallicity will continue to be a flourishing topic at all redshifts.

Leslie Hunt, Suzanne Madden & Raffaella Schneider,
SOC co-chairs, Firenze, Paris, September 20, 2008

THE ORGANIZING COMMITTEE

Scientific

Roger Chevalier (USA)
Andrea Ferrara (Italia)
Deidre Hunter (USA)
Suzanne Madden (co-chair, la France)
Francesca Matteucci (Italia)
Daniel Schaerer (la Suisse)
Evan Skillman (USA)

Eli Dwek (USA)
Leslie Hunt (co-chair, Italia)
Yuri Izotov (Ukraine)
André Maeder (la Suisse)
Sandra Savaglio (Deutschland)
Raffaella Schneider (Italia)
Eduardo Telles (Brasil)

Local

Roberto Baglioni
Marco Grossi
Emanuela Masini (chair)

Viviana Casasola
Leslie Hunt
Raffaella Schneider

Acknowledgements

The symposium is sponsored and supported by the IAU Divisions IV (Stars), VI (Interstellar Matter), and VIII (Galaxies & the Universe); by the IAU Commissions No. 28 (Galaxies), No. 34 (Interstellar Matter), No. 37 (Star Clusters and Associations), and No. 47 (Cosmology); and by INAF (Italy), CEA (France), the City of Rapallo (Genova, Italy), and the Province of Genova (Italy).

The Local Organizing Committee operated under the auspices of the
INAF.

Funding by the
International Astronomical Union,
INAF, CEA,
the Istituto di Radioastronomia, the Osservatorio Astrofisico di Arcetri,
the City of Rapallo, and the Province of Genova
is gratefully acknowledged.

Participants

Tom **Abel**, KIPAC, Stanford Linear Accelerator, Menlo Park, CA, USA — tabel@stanford.edu
Alessandra **Aloisi**, STScI/ESA, Baltimore, Maryland, USA — aloisi@stsci.edu
Hakim **Atek**, Institut d'Astrophysique de Paris, Paris, France — atek@iap.fr
Aycin **Aykutalp**, Kapteyn Astronomical Institute, Groningen, Netherlands — aykutalp@astro.rug.nl
David **Bahena**, Institute of Astronomy, Praha, Czech Republic — bahen@hotmail.com
Arturs **Barzdis**, Institute of Astronomy, University of Latvia, Riga, Latvia — arturs_lv@inbox.lv
Timothy **Beers**, Michigan State University / Dept. of Physics & Astronomy, E. Lansing, MI, USA — beers@pa.msu.edu
Edouard **Bernard**, Instituto de Astrofisica de Canarias, La Laguna, Tenerife, Spain — ebernard@iac.es
Simone **Bianchi**, INAF-IRA, Firenze, Italy — sbianchi@arcetri.astro.it
Alberto **Bolatto**, Dept. of Astronomy, University of Maryland, College Park, MD, USA — bolatto@astro.umd.edu
Jonathan **Braine**, Observatoire de Bordeaux, Universite de Bordeaux 1, Floirac, France — braine@obs.u-bordeaux1.fr
Maarten **Breddels**, Kapteyn Astronomical Institute, Groningen, Netherlands — breddels@astro.rug.nl
Elias **Brinks**, University of Hertfordshire, Hatfield, UK — e.brinks@herts.ac.uk
Daniel **Brito de Freitas**, Osservatorio Astrofisico di Arcetri - OAA, Firenze, Italy — brito@arcetri.astro.it
Volker **Bromm**, University of Texas, Dept. of Astronomy, Austin, TX, U.S.A. — vbromm@astro.as.utexas.edu
Ines **Brott**, Sterrenkundig Instituut Utrecht, Utrecht, The Netherlands — brott@astro.uu.nl
Francesco **Calura**, Dipartimento di Astronomia, Universita' di Trieste, Trieste, Italy — fcalura@oats.inaf.it
Lynn **Carlson**, Johns Hopkins University/STScI, Baltimore, MD, USA — carlson@stsci.edu
Gabriele **Cescutti**, Dept. of Astronomy, Trieste University, Trieste, Italy — cescutti@oats.inaf.it
Nicolas **Champavert**, CRAL - Observatoire de Lyon, Saint Genis Laval, France — nicolas.champavert@obs.univ-lyon1.fr
Roger **Chevalier**, University of Virginia, Dept. of Astronomy, Charlottesville, VA, USA — rac5x@virginia.edu
Ena **Choi**, Yonsei University, Seoul, South Korea — ena0choi@gmail.com
Michele **Cignoni**, INAF, Osservatorio Astronomico di Bologna, Bologna, Italy — michele.cignoni@unibo.it
Paul **Clark**, Institute for Theoretical Astrophysics, University of Heidelberg, Heidelberg, Germany — pcc@ita.uni-heidelberg.de
Sergio **Cristallo**, Osservatorio Astronomico di Teramo (INAF), Teramo, Italy — cristallo@oa-teramo.inaf.it
Robert **Cumming**, Dept of Astronomy, Stockholm University, Stockholm, Sweden — robert@astro.su.se
Gary **Da Costa**, Research School of Astronomy and Astrophysics, ANU, Weston, Australia — gdc@mso.anu.edu.au
Ignacio **de la Rosa**, Instituto de Astrofisica de Canarias, La Laguna, Spain — irosa@iac.es
Sven **De Rijcke**, Astronomical Observatory, Ghent University, Gent, Belgium — sven.derijcke@UGent.be
Miroslava **Dessauges-Zavadsky**, Geneva Observatory, Sauverny, Switzerland — miroslava.dessauges@obs.unige.ch
Sperello **di Serego Alighieri**, INAF - Osservatorio Astrofisico di Arcetri, Firenze, Italy — sperello@arcetri.astro.it
Andrey **Doroshkevich**, Astro Space Center FIAN, Moscow, Russia — dorr@asc.rssi.ru
Sylvia **Ekstrom**, Geneva Observatory, Sauverny, Switzerland — sylvia.ekstrom@obs.unige.ch
Ekta, National Centre for Radio Astrophysics - TIFR, Pune, India — ekta@ncra.tifr.res.in
Andrew **Ferrara**, SISSA/ISA, Trieste, Italy — ferrara@sissa.it
Giuliana **Fiorentino**, Kapteyn Astronomical Institute, University of Groningen, Groningen, the Netherlands — fiorentino@astro.rug.nl
Andreea **Font**, Institute for Computational Cosmology, Durham, Durham, UK — andreea.font@durham.ac.uk
Bi-Qing **For**, Dept. of Astronomy, University of Texas, Austin, Austin, TX, USA — biqing@astro.as.utexas.edu
Anna **Frebel**, McDonald Observatory, University of Texas at Austin, Austin, TX, USA — anna@astro.as.utexas.edu
Katherine **Freese**, Physics Dept., University of Michigan, Ann Arbor, MI, USA — ktfreese@umich.edu
Maud **Galametz**, CEA/DSM/IRFU/SAP, Gif sur Yvette, France — maud.galametz@cea.fr
Christa **Gall**, Niels Bohr Institute/Dark Cosmology Centre, Copenhagen, Denmark — christa@dark-cosmology.dk
Frederic **Galliano**, University of Maryland, College Park, MD, USA — galliano@astro.umd.edu
Cyril **Georgy**, Geneva Observatory / Geneva University, Geneva, Switzerland — Cyril.Georgy@obs.unige.ch
Simon **Glover**, Astrophysikalisches Institut Potsdam, Potsdam, Germany — sglover@aip.de
Claus **Goessl**, Uni-Sternwarte München, München, Germany — cag@usm.lmu.de
Denise **Goncalves**, Observatorio do Valongo - UFRJ, Rio de Janeiro, Brasil — denise@ov.ufrj.br
Dimitrios **Gouliermis**, Max Planck Institute for Astronomy, Heidelberg, Heidelberg, Germany — dgoulier@mpia.de
Pierre **Gratier**, Observatoire de Bordeaux, Floirac, France — pierre.gratier@mpia.de
Thomas **Greif**, Institut fuer theoretische Astrophysik, Heidelberg, Germany — tgreif@ita.uni-heidelberg.de
Natalia **Guseva**, Main Astronomical Observatory, Kyiv, Ukraine — guseva@mao.kiev.ua
Kenji **Hasegawa**, University of Tsukuba/ Center for Computational Sciences, Tsukuba, Japan — hasegawa@ccs.tsukuba.ac.jp
John **Hibbard**, NRAO, Charlottesville, VA, USA — jhibbard@nrao.edu
Ana **Hidalgo-Gamez**, Escuela Superior de Fisica y Matematicas-IPN, México City, México — ahidalgo@esfm.ipn.mx
Hiroyuki **Hirashita**, Institute of Astronomy and Astrophysics, Academia Sinica, Taipei, Taiwan — hirashita@asiaa.sinica.edu.tw
Raphael **Hirschi**, Keele University (Astrophysics Group), Keele, UK — r.hirschi@epsam.keele.ac.uk
Seyit **Hocuk**, Kapteyn Astronomical Institute, Groningen, Netherlands — seyit@astro.rug.nl
Michael **Hood**, University of California, Irvine, CA, USA — mahood@uci.edu
Adam **Hosford**, University of Hertfordshire, Hatfield, Hertfordshire, UK — a.hosford@herts.ac.uk
Leslie **Hunt**, INAF-Institute of Radioastronomy/Sezione Firenze, Firenze, Italy — hunt@arcetri.astro.it
Deidre **Hunter**, Lowell Observatory, Flagstaff, AZ, USA — dah@lowell.edu
Laura **Husti**, University of Torino, Torino, Italy — lotesileanu@yahoo.com
Fabio **Iocco**, INAF-Oss Astrofisico Arcetri, Firenze, Italia — iocco@arcetri.astro.it
Yuri **Izotov**, Main Astronomical Observatory, Kyiv, Ukraine — izotov@mao.kiev.ua
Bethan **James**, University College London / Physics & Astronomy , London, UK — bj@star.ucl.ac.uk
Luc **Jamet**, Instituto de Astronomia - UNAM, México, D.F. — ljamet@astroscu.unam.mx
Anne-Katharina **Jappsen**, Cardiff University, School of Physics and Astronomy, Cardiff, UK — jappsen@cita.utoronto.ca
Jennifer **Johnson**, Ohio State/Dept. of Astronomy, Columbus, OH, USA — jaj@astronomy.ohio-state.edu
Kelsey **Johnson**, University of Virginia/NRAO, Charlottesville, VA, USA — kej7a@virginia.edu
Jarrett **Johnson**, University of Texas at Austin , Austin, TX, USA — jljohnson@astro.as.utexas.edu
Yuko **Kakazu**, Institut d'Astrophysique de Paris, Paris, France — kakazu@iap.fr
Carolina **Kehrig**, University of Michigan/Dept. of Astronomy , Ann Arbor, MI, USA — kehrig@umich.edu
Jaime **Klapp**, Instituto Nacional de Investigaciones Nucleares, Salazar, México — klapp@nuclear.inin.mx
Uli **Klein**, AIfA, Univ. Bonn, Bonn, Germany — uklein@astro.uni-bonn.de
Ralf **Klessen**, Zentrum fuer Astronomie der Universitat Heidelberg, Heidelberg, Germany — rklessen@ita.uni-heidelberg.de
Mina **Koleva**, Observatoire de Lyon / University of Sofia, St Genis Laval, France — mina.koleva@obs.univ-lyon1.fr
Yutaka **Komiya**, Tohoku Uninersity /Graduate School of Science, Sendai, Japan — komiya@astr.tohoku.ac.jp
Pavel **Kroupa**, Argelander Institute for Astronomy, University of Bonn, Bonn, Germany — pavel@astro.uni-bonn.de
Jiri **Krticka**, Institute of Theoretical Physics and Astrophysics, Masaryk University, Brno, Czech Republic — krticka@physics.muni.cz
Gustavo **Lanfranchi**, Nucleo de Astrofisica Teorica - UNICSUL, São Paulo, SP, Brasil — gustavo.lanfranchi@unicsul.br
Ryan **Leaman**, University of Victoria / Dept. of Physics & Astronomy, Victoria, BC, Canada — rleaman@uvic.ca
Vianney **Lebouteiller**, Cornell University, Ithaca, NY, USA — vianney@isc.astro.cornell.edu
Henry **Lee**, Gemini Observatory (South), La Serena, Chile — hlee@gemini.edu
Claus **Leitherer**, Space Telescope Science Institute, Baltimore, MD, USA — leitherer@stsci.edu

Adam **Leroy**, Max Planck Institute for Astronomy, Heidelberg, Germany   leroy@mpia-hd.mpg.de
Emily **Levesque**, Institute for Astronomy, University of Hawaii, Honolulu, HI, USA   emsque@ifa.hawaii.edu
Ute **Lisenfeld**, Universidad Granada, Granada, Spain   ute@ugr.es
Sara **Lucatello**, INAF Padova and Excellente Cluster Universe,
    Garching b. München, Germany   sara.lucatello@oapd.inaf.it
Suzanne **Madden**, CEA/DSM/DAPNIA/Service d'Astrophysique, Gif-sur-Yvette, France   smadden@cea.fr
Umberto **Maio**, Max-Planck-Institut fuer Astrophysik, Garching bei München, Germany   maio@mpa-garching.mpg.de
Dmitry **Makarov**, Special Astrophysical Observatory, RAS, Karachai-Cirkassian Republic, Russia   dim@sao.ru
Lidia **Makarova**, Special Astrophysical Observatory, RAS, Karachai-Cirkassian Republic, Russia   lidia@sao.ru
Filippo **Mannucci**, INAF-IRA, Firenze, Firenze, Italy   filippo@arcetri.astro.it
Andrea **Marcolini**, University of Central Lancashire, Preston, Lancashire, UK   amarcolini@uclan.ac.uk
Brian **Marsteller**, University of California, Irvine, California, USA   bmarstel@uci.edu
Mariluz **Martin-Manjon**, Dep.to de Fisica Teórica, Universidad Autónoma de Madrid,
    Madrid, Spain   mariluz.martin@uam.es
Donald **Martins**, University of Alaska/Dept. of Physics and Astronomy, Anchorage, Alaska, USA   afdhm@uaa.alaska.edu
Thomas **Masseron**, The Ohio State University, Columbus, OH, USA   masseron@astronomy.ohio-state.edu
Kenta **Matsuoka**, Ehime university, Matsuyama, Japan   k.matsuoka@cosmos.phys.sci.ehime-u.ac.jp
Francesca **Matteucci**, Dipartimento di Astronomia Universita' di Trieste, Trieste, Italy   matteucc@oats.inaf.it
Andre **Milone**, University of Central Lancashire, Preston, Lancashire, UK   adcmilone@uclan.ac.uk
Matteo **Monelli**, Instituto de astrofisica de Canarias, La Laguna, Spain   monelli@iac.es
Lianne **Muijres**, Anton Pannekoek Instituut, Unversity of Amsterdam, Amsterdam, Netherlands   lmuijres@science.uva.nl
Tohru **Nagao**, Ehime University, Matsuyama, Japan   tohru@cosmos.phys.sci.ehime-u.ac.jp
Ken **Nomoto**, University of Tokyo, Physics and Mathematics, Kashiwa-city,
    Chiba, Japan   nomoto@astron.s.u-tokyo.ac.jp
John **Norris**, Mount Stromlo Observatory, Weston, Australia   jen@mso.anu.edu.au
Takaya **Nozawa**, Hokkaido University, Dept. of Cosmosciences, Sapporo, Japan   nozawa@kern.ep.sci.hokudai.ac.jp
Joana **Oliveira**, Astrophysics Group, Keele University, Staffordshire, UK   joana@astro.keele.ac.uk
Kazuyuki **Omukai**, NAOJ, Tokyo, Japan   omukai@th.nao.ac.jp
Juergen **Ott**, NRAO/California Institute of Technology, Pasadena, CA, USA   jott@nrao.edu
Francesco **Palla**, INAF-Osservatorio Astrofisico di Arcetri, Firenze, Italy   palla@arcetri.astro.it
Polychronis **Papaderos**, Instituto de Astrofisica de Andalucia , Granada, Spain   papaderos@iaa.es
Davis **Philip**, ISO and Union College, Schenectady, NY, USA   agdp@union.edu
Vinicius **Placco**, IAG/USP - Depto. de Astronomia, São Paulo, Brazil   vmplacco@astro.iag.usp.br
Simon **Pustilnik**, Special Astrophysical Observatory, Russian Academy of Sciences, Russia   sap@sao.ru
Anna **Raiter**, European Southern Observatory, Garching bei München, Germany   araiter@eso.org
Massimo **Ricotti**, University of Maryland, College Park, MD, USA   ricotti@astro.umd.edu
Carlos **Rodriguez-Rico**, Dep.to de Astronomia, Universidad de Guanajuato,
    Guanajuato, G.to, México   carlos@astro.ugto.mx
Ian **Roederer**, University of Texas at Austin, Austin, TX, USA   iur@astro.as.utexas.edu
David **Rosario**, UCO/Lick Observatory, Santa Cruz, CA, USA   rosario@ucolick.org
Elena **Sabbi**, STScI, Baltimore, MD, USA   sabbi@stsci.edu
Stefania **Salvadori**, SISSA/International School for Advanced Studies, Trieste, Italy   salvas@sissa.it
Ruben **Salvaterra**, Universitá di Milano-Bicocca, Milano, Italy   salvaterra@mib.infn.it
Marc **Sauvage**, CEA/DSM/DAPNIA/Service d'Astrophysique, Gif sur Yvette CEDEX, France   msauvage@cea.fr
Sandra **Savaglio**, Max-Planck-Institut fuer Extraterrestrische Physik, Garching bei München,
    Germany   savaglio@mpe.mpg.de
Daniel **Schaerer**, Geneva Observatory, Versoix, Switzerland   daniel.schaerer@obs.unige.ch
Dominik **Schleicher**, Institute of Theoretical Astrophysics / ZAH, Heidelberg, Germany   dschleic@ita.uni-heidelberg.de
Katharine **Schlesinger**, The Ohio State University Dept. of Astronomy, Columbus, OH, USA   kschlesinger@gmail.com
Raffaella **Schneider**, INAF/Osservatorio Astrofisico di Arcetri, Firenze, Italy   raffa@arcetri.astro.it
Regina **Schulte-Ladbeck**, University of Pittsburgh, Dept. of Physics and Astronomy ,
    Pittsburgh, PA, USA   rsl@pitt.edu
Andrew **Schurer**, SISSA / INAF, Osservatorio Astronomico di Padova, Trieste, Italy   schurer@sissa.it
William **Schuster**, Institute of Astronomy UNAM (México), San Diego, CA, USA   schuster@astrosen.unam.mx
Fernando **Selman**, European Southern Observatory, Santiago, Chile   fselman@eso.org
Zhengyi **Shao**, Shanghai Astronomical Observatory, Shanghai, China   zyshao@shao.ac.cn
Caroline **Simpson**, Florida International University, Miami, FL, USA   simpsonc@galaxy.fiu.edu
Evan **Skillman**, Astronomy Dept., University of Minnesota, Minneapolis, MN, USA   skillman@astro.umn.edu
Tammy **Smecker-Hane**, University of California, Irvine, CA, USA   tsmecker@uci.edu
Britton **Smith**, Center for Astrophysics & Space Astronomy, University of Colorado,
    Boulder, CO, USA   brittons@origins.colorado.edu
Jan **Snigula**, Max-Planck-Institut fuer Extraterrestrische Physik, Garching bei München,
    Germany   snigula@mpe.mpg.de
Marco **Spaans**, Kapteyn Astronomical Institute, Groningen, Netherlands   spaans@astro.rug.nl
Athena **Stacy**, University of Texas at Austin, Austin, TX, USA   minerva@astro.as.utexas.edu
Else **Starkenburg**, Kapteyn Astronomical Institute, Groningen, the Netherlands   else@astro.rug.nl
Grazyna **Stasinska**, LUTH, Observatoire de Paris, Meudon, France   grazyna.stasinska@obspm.fr
Takuma **Suda**, Hokkaido University, Sapporo,
    Hokkaido, Japan   suda@astro1.sci.hokudai.ac.jp
Eon-Chang **Sung**, Korea Astronomy and Space Science Insititute, Daejeon, South Korea   ecsung@kasi.re.kr
Jonathan **Tan**, University of Florida, Gainesville, FL, USA   jt@astro.ufl.edu
Eduardo **Telles**, University of Virginia/Observatório Nacional, Brazil, Charlottesville, VA, USA   etelles@virginia.edu
Ovidiu **Tesileanu**, Universita' degli Studi di Torino / Dip. Fisica Generale, Torino, Italy   ovidiu.tesileanu@ph.unito.it
Trinh **Thuan**, University of Virginia, Astronomy Dept., Charlottesville, VA, USA   txt@virginia.edu
Eline **Tolstoy**, Kapteyn Astronomical Institute, Groningen, Netherlands   etolstoy@astro.rug.nl
Nozomu **Tominaga**, National Astronomical Observatory, Mitaka, Tokyo, Japan   tominaga@astron.s.u-tokyo.ac.jp
Luca **Tornatore**, Universita di Trieste, Dipartimento di Astronomia, Trieste, Italy   tornatore@oats.inaf.it
Monica **Tosi**, INAF - Osservatorio Astronomico di Bologna, Bologna, Italy   monica.tosi@oabo.inaf.it
Takuji **Tsujimoto**, National Astronomical Observatory, Tokyo, Japan   taku.tsujimoto@nao.ac.jp
Brent **Tully**, Institute for Astronomy, University of Hawaii, Honolulu, HI, USA   tully@ifa.hawaii.edu
Masayuki **Umemura**, Center for Computational Sciences, University of Tsukuba,
    Tsukuba, Japan   umemura@ccs.tsukuba.ac.jp
Sander **Valcke**, Astronomical Observatory, University of Ghent, Gent, Belgium   Sander.Valcke@UGent.be
Rosa **Valiante**, Osservatorio Astrofisico di Arcetri, Firenze, Italia   valiante@arcetri.astro.it
Jacco **van Loon**, Astrophysics Group, Keele University, Staffordshire, UK   jacco@astro.keele.ac.uk
Anne **Verhamme**, Geneva Observatory, Versoix, Switzerland   anne.verhamme@obs.unige.ch
George **Wallerstein**, University of Washington, Seattle, Washington, USA   wall@astro.washington.edu
Daniel **Whalen**, Los Alamos National Laboratory, Los Alamos, NM, USA   dwhalen@lanl.gov
Taras **Yakobchuk**, Main Astronomical Observatory, Kyiv, Ukraine   yakobchuk@mao.kiev.ua
Sukyoung **Yi**, Yonsei University/Dept. of Astronomy, Seoul, Korea   yi@yonsei.ac.kr
Naoki **Yoshida**, Dept. of Physics, Nagoya University, Nagoya, Japan   nyoshida@a.phys.nagoya-u.ac.jp
Laimons **Zacs**, Faculty of Physics and Mathematics / University of Latvia, Riga, Latvia   zacs@latnet.lv

Life at the Conference

The City: Rapallo

The Venue: The Clarisse Convent

The Context: "Antico Castello sul Mare"

The Opening: Registration and Cultural Sponsor, Dr. Giorgio Devoto, Genoa

The Talks

The Breaks

The Free Afternoon

The Social Dinner

The Closing

Photo credits: Sperello di Serego Alighieri & Simone Bianchi

CONFERENCE PHOTOGRAPH

Photo credits: Jonathan Tan

Session I

Population III and metal-free star formation

Low-Metallicity Star Formation:
From the First Stars to Dwarf Galaxies
Proceedings IAU Symposium No. 255, 2008
L.K. Hunt, S. Madden & R. Schneider, eds.

Open questions in the study
of population III star formation

S. C. O. Glover[1,2]**, P. C. Clark**[2]**, T. H. Greif**[2]**, J. L. Johnson**[3]**,**
V. Bromm[3]**, R. S. Klessen**[2]**, & A. Stacy**[3]

[1]Astrophysikalisches Institut Potsdam, An der Sternwarte 16, 14482 Potsdam, Germany

[2]Institut für theoretische Astrophysik, Albert-Ueberle Strasse 2, 69120 Heidelberg, Germany

[3]Department of Astronomy, University of Texas, Austin, TX 78712

Abstract. The first stars were key drivers of early cosmic evolution. We review the main physical elements of the current consensus view, positing that the first stars were predominantly very massive. We continue with a discussion of important open questions that confront the standard model. Among them are uncertainties in the atomic and molecular physics of the hydrogen and helium gas, the multiplicity of stars that form in minihalos, and the possible existence of two separate modes of metal-free star formation.

Keywords. astrochemistry, stars: formation, galaxies: formation, cosmology: theory

1. Introduction

We have learnt a great deal over the last decade concerning the formation of the first stars, the so-called population III or Pop III. A consensus has emerged regarding the properties of the protogalaxies (or 'minihalos') in which the first stars formed and the major physical and chemical processes involved in their formation. Their masses remain uncertain, but are widely expected to be significantly larger than the characteristic mass for present-day star formation. Good summaries of the present state of the field can be found in Bromm & Larson (2004), Glover (2005) and Norman (2008). Nevertheless, there remain some important open questions. In this review, we discuss four of the most important of these issues and the efforts being made to resolve them. These issues are the impact of chemical rate coefficient uncertainties on the accuracy of our models of Pop III star formation; the perennial question of whether we have identified all of the important physical processes responsible for cooling the gas; the number of population III stars that form in each minihalo; and the question of whether population III actually consists of two sub-populations (Pop III.1 and Pop III.2) with different characteristic masses. Two further important issues – the question of what physical process terminates accretion onto the earliest protostars, and the impact of dark matter decay and annihilation on the formation of the first stars – are not discussed here as they are covered in detail elsewhere in these proceedings (see e.g. the contributions by Whalen, Tan, Freese & Iocco).

2. Population III star formation: the consensus view

In the ΛCDM model for cosmic structure formation, the first gravitationally bound structures to form are very small (e.g. $M \sim M_{\odot}$ if cold dark matter consists of neutralinos; Diemand, Moore & Stadel 2005) and consist purely of dark matter. Larger bound structures form through accretion and through the merger of these smaller objects, in a process known as hierarchical clustering. Once the mass of these gravitationally bound dark matter objects (conventionally referred to as dark matter halos) exceeds the cosmological Jeans mass, pressure forces can no longer prevent gas from falling into the potential wells created by the dark matter. Infalling gas is heated by adiabatic compression and by shocks, and in the absence of radiative cooling it will eventually reach a state of hydrostatic equilibrium, with a mean temperature given by the virial temperature of the halo, T_{vir}. The first dark matter halos to have masses greater than the cosmological Jeans mass have virial temperatures much smaller than the $\sim 10^4$ K temperature at which cooling from electronic excitation of atomic hydrogen becomes effective, and so the gas in these halos must rely on other, less effective forms of cooling.

It has long been recognized that the most important coolant at $T < 10^4$ K in primordial gas is molecular hydrogen, H_2 (Saslaw & Zipoy 1967; Peebles & Dicke 1968). In the absence of dust, this forms in the gas phase via the reaction chains

$$H + e^- \rightarrow H^- + \gamma, \tag{2.1}$$
$$H^- + H \rightarrow H_2 + e^-, \tag{2.2}$$

and

$$H + H^+ \rightarrow H_2^+ + \gamma, \tag{2.3}$$
$$H_2^+ + H \rightarrow H_2 + H^+, \tag{2.4}$$

with the H^- mechanism generally dominating. Gas-phase H_2 formation therefore relies on the presence of free electrons and protons, and the amount of H_2 that can be formed by these reactions is limited by the recombination of the gas. A number of studies have examined the conditions required in order for the gas to form enough H_2 to be able to cool within a Hubble time (see e.g. Haiman, Thoul, & Loeb 1996; Tegmark *et al.* 1997; Yoshida *et al.* 2003), with the consensus being that virial temperatures $T_{\mathrm{vir}} \gtrsim 1000$ K are required, corresponding to halo masses of a few times 10^5 M$_\odot$ or more.

The redshift at which the first halos of the required mass are formed depends to some extent of the definition of 'first'. The very first objects of this mass to form within a Hubble volume do so at redshifts $z \sim 50$–60 (Reed *et al.* 2005; Naoz, Noter & Barkana 2006), but are exceedingly rare; on the other hand, the first object to form within a reasonable local volume, say 1 comoving Mpc^3, does so at a somewhat lower redshift $z \sim 30$ (Glover 2005). For technical reasons, numerical simulations of population III star formation typically focus on the latter case.

If enough H_2 can be formed to efficiently cool the gas, then it will undergo gravitational collapse. Initially, this collapse occurs rapidly. The gas temperature drops as the density increases, and so the gas becomes increasingly gravitationally unstable. However, as the collapse proceeds, cooling from H_2 becomes steadily less efficient. The exponential fall-off in the H_2 cooling rate at low temperatures prevents it from cooling the gas below $T \sim 200$ K, while the approach of the rotational and vibrational level populations of H_2 to their local thermodynamic equilibrium (LTE) values at densities above $n_{\mathrm{cr}} \sim 10^4$ cm^{-3} renders H_2 cooling inefficient at higher densities. The gas therefore accumulates in a quasi-hydrostatic "loitering state" with a characteristic temperature $T_{\mathrm{char}} = 200$ K and characteristic density $n_{\mathrm{char}} = 10^4$ cm^{-3} (Abel, Bryan, & Norman 2002; Bromm, Coppi, & Larson 2002). Eventually, the amount of mass accumulated at this density exceeds the Bonnor-Ebert mass, which at this point is a few hundred M$_\odot$, and the gravitational collapse resumes. However, beyond this point H_2 cooling is unable to maintain the temperature at $T = 200$ K; it steadily reheats, evolving with an effective polytropic index of $\gamma_{\mathrm{eff}} = 1.1$ (Omukai & Nishi 1998).

At densities $n > 10^8$ cm^{-3}, three-body formation of H$_2$ via the reactions

$$H + H + H \rightarrow H_2 + H, \qquad (2.5)$$
$$H + H + H_2 \rightarrow H_2 + H_2, \qquad (2.6)$$

becomes effective and rapidly converts all of the hydrogen to molecular form (Palla, Salpeter, & Stahler 1983). Although the dramatic increase in the H$_2$ abundance boosts the H$_2$ cooling rate, the high gas density, the growing optical depth in the H$_2$ rotational and vibrational lines (see e.g. Omukai & Nishi 1998; Ripamonti *et al.* 2002; Ripamonti & Abel 2004) and the heat input from H$_2$ formation all combine to prevent the temperature from dropping significantly.

At densities $n \sim 10^{14}$ cm^{-3} and above, a second form of H$_2$ cooling becomes important: H$_2$ collision-induced emission (CIE). Although H$_2$ has no permanent dipole moment, the complexes formed in collisions of H$_2$ with H, H$_2$ or He can act as 'super-molecules', with non-zero dipole moments. Although these excited complexes last for only a very short time ($t_{coll} \sim 10^{-12}$ s; Ripamonti & Abel 2004), there is nevertheless always a small probability that a photon will be emitted. At very high densities, collisions occur frequently enough to make this emission a viable means of cooling the gas. However, this CIE-dominated phase lasts for only a short time before the gas becomes optically thick in the continuum, and so CIE cooling is unable to significantly reduce the temperature.

At even higher densities ($n > 10^{16}$ cm^{-3}), the gas becomes increasingly optically thick, and so radiative cooling becomes completely ineffective. However, collisional dissociation of H$_2$ acts as a heat sink, keeping the temperature evolution of the gas close to isothermal until most of the H$_2$ has been destroyed (Omukai *et al.* 2005; Turk, Abel & O'Shea 2008). This occurs by the time that the density reaches 10^{21} cm^{-3}, and beyond this point the evolution of the gas becomes adiabatic. This is the moment at which we first have something that we can identify as a true protostar.

At the moment that this protostar forms, its mass is less than 0.01 M$_\odot$, but it is surrounded by a dense, massive envelope of infalling gas with a mass of hundreds of solar masses. Accretion of this envelope is expected to occur at a rapid rate: a simple scaling argument suggests that the accretion rate should scale as $\dot{M} \propto c_s^3 \propto T^{3/2}$, and since the temperature of the gas is between 10–100 times larger than in local star forming regions, the expected accretion rates are orders of magnitude larger than the rate of order 10^{-5} M$_\odot$ yr^{-1} that is typical locally. Accretion rates have been estimated using a variety of techniques (see Glover 2005, section 4), and although the estimated rates differ somewhat, in every case one expects the star to be able to accrete more than 100 M$_\odot$ of gas within the Kelvin-Helmholtz relaxation time. Moreover, stellar feedback in the form of radiation-driven winds is expected to be far less effective in primordial gas than in metal-enriched gas (see e.g. Kudritzki 2002), and so there seems little to prevent the star from becoming massive.

Hence, population III stars are expected to be massive and short-lived, surviving for only a few million years. How they end their lives depends on both their mass and the speed at which they are rotating. Non-rotating population III stars with masses in the range 10–50 M$_\odot$ are expected to explode as either conventional type II supernovae or as hypernovae (Umeda & Nomoto 2002). For masses between 50 and 140 M$_\odot$ and above 260 M$_\odot$, direct collapse to form a black hole is expected, while non-rotating Pop III stars with masses in the range 140–260 M$_\odot$ are expected to end their lives as pair-instability supernovae (Heger & Woosley 2002). Rapid rotation changes this picture somewhat, by enhancing the effects of mass-loss, and by inducing mixing within the star. The latter effect leads to the star having a larger helium core at the end of its main sequence evolution, and so likely allows less massive stars to become pair instability supernovae (Ekström *et al.* 2008).

3. Open questions

3.1. *Are uncertainties in the chemical rate coefficients important?*

Our ability to construct accurate models of the chemical evolution of metal-free gas is constrained by the level of accuracy with which the rate coefficients of the key chemical reactions have been determined. This varies significantly depending on the reaction in question. Many of the most important reactions involved in the formation and destruction of H_2 have rate coefficients that have been determined to within an accuracy of the order of 10–20 % at typical protogalactic temperatures (Abel *et al.* 1997; Galli & Palla 1998). However, there are several important reactions whose rates are far more uncertain.

One example is the charge transfer reaction

$$H_2 + H^+ \rightarrow H_2^+ + H, \tag{3.1}$$

which is a major destruction pathway for H_2 in hot, ionized gas. Savin *et al.* (2004) present a new calculation of the rate of this reaction, and show that previous determinations, some of which remain widely used in the literature, differ by orders of magnitude. Fortunately, this process is unimportant in cold gas, owing to its large endothermicity, and so the large uncertainty in this reaction has little impact on the accuracy with which we can model the formation of the first stars. However, its impact on the formation of so-called Pop III.2 stars (discussed in more detail in §3.4 below) may be much larger and deserves further study.

Another source of uncertainty stems from the competition between two of the main destruction pathways for H^-, associative detachment (reaction 2.2 above) and mutual neutralization

$$H^- + H^+ \rightarrow H + H. \tag{3.2}$$

Glover, Savin & Jappsen (2006) surveyed the literature available on the rates of these reactions at low temperatures ($T < 10^4$ K), and showed that both were uncertain by an order of magnitude. They also examined the effects of this uncertainty on the chemistry, cooling and dynamics of the gas. In the conventional Pop III formation scenario, the fractional ionization of the gas is small enough to ensure that associative detachment always dominates over mutual neutralization, and so any uncertainty in the rate coefficients has almost no effect. In gas cooling from a highly ionized state, however, as in some of the Pop III.2 scenarios discussed in section 3.4 below, mutual neutralization dominates initially, with associative detachment becoming important only once the gas has recombined sufficiently. In this case, any uncertainty in the rates of these two reactions leads to an uncertainty in the H_2 formation rate, and in the final amount of H_2 formed. More recently, Glover & Abel (2008) have shown that this uncertainty affects the amount of HD that can form, and so also influences the minimum temperature that the collapsing gas can reach.

Fortunately, experimentalists have begun to address this source of uncertainty. Recent measurements of the mutual neutralization reaction rate at low temperatures by Xavier Urbain have reduced what was an order of magnitude uncertainty to something closer to a 50% uncertainty (X. Urbain, private communication). At the same time, an experiment designed to accurately measure the rate coefficient for the associative detachment reaction has been funded and is currently under construction (Bruhns *et al.* 2008; D. Savin, private communication), and so this source of uncertainty may also have been removed in a few years time.

Finally, and perhaps most importantly, a very large uncertainty exists in the rate of the three-body H_2 formation reaction. The current state of the literature regarding the rate of reaction 2.5 was surveyed in Glover (2008), who showed that in the temperature range $200 < T < 2000\,\mathrm{K}$ relevant for population III star formation, there is an uncertainty in the rate coefficient of two to three orders of magnitude. Moreover, there is no sign that this uncertainty is reducing: indeed, the two most recent determinations of the rate coefficient (by Abel, Bryan, & Norman 2002 and by Flower & Harris 2007) are the two with the greatest disagreement. The uncertainty in the rate of reaction 2.6 is harder to quantify, as there have been fewer studies made of this reaction. However, if we assume (following Jacobs, Giedt & Cohen 1967) that the rate of this reaction is 1/8th that of reaction 2.5, then the obvious implication is that this rate has a similarly large uncertainty.

The effects of the uncertainty in the three-body H_2 formation rate on the thermal evolution of the gas have been examined by Glover & Abel (2008) and Glover & Savin (2008) using highly simplified one-zone models. These studies find that the rate coefficient uncertainties lead to an uncertainty of approximately 50% in the temperature evolution of the gas in the density range $10^8 < n < 10^{13}\,\mathrm{cm}^{-3}$. The effect of this uncertainty on the dynamical evolution of the gas and in particular on the predicted protostellar accretion rates are not currently known, although work is currently under way to address this.

3.2. Are we including all of the significant coolants?

As previously noted, H_2 has long been recognized as the most important coolant in primordial gas at temperatures $T < 10^4\,\mathrm{K}$. At the same time, it is clear from our previous discussion that H_2 becomes increasingly ineffective as a coolant as we move to higher densities, owing to the low critical density at which its rotational and vibrational level populations reach their LTE values, and to the fact that at densities $n \gtrsim 10^{10}\,\mathrm{cm}^{-3}$, optical depth effects further suppress H_2 cooling.

The comparative ineffectiveness of H_2 as a coolant in high density metal-free gas has motivated various authors to examine the role that might be played by other coolants at high densities. Perhaps the best studied alternative coolant is hydrogen deuteride, HD. It has excited rotational and vibrational levels that have radiative lifetimes that are about a factor of 100 shorter than those of H_2, and so the HD cooling rate does not reach its LTE limit until $n \sim 10^6\,\mathrm{cm}^{-3}$. It is also a far more effective coolant than H_2 at low temperatures ($T \lesssim 200\,\mathrm{K}$; see e.g., Flower *et al.* 2000). This is due primarily to the fact that radiative transitions can occur between rotational levels with odd and even values of J, allowing cooling to occur through the $J = 1 \rightarrow 0$ transition. The corresponding odd $\leftrightarrow$ even transitions in the case of H_2 represent conversions from ortho-H_2 to para-H_2 or vice versa, and are highly forbidden. Furthermore, at low temperatures the ratio of HD to H_2 can be significantly enhanced with respect to the cosmological D:H ratio by chemical fractionation (see e.g., Glover 2008).

The role of HD cooling in early minihalos has been investigated by a number of authors. In the case of the earliest generation of minihalos, which form from very cold neutral gas that is never heated to more than a few thousand K during the course of the galaxy formation process, the importance of HD appears to be a function of the size and dynamical history of the minihalo (Ripamonti 2007; McGreer & Bryan 2008). In small minihalos ($M \lesssim 10^6\,\mathrm{M_\odot}$) that collapse in an unperturbed, relatively uniform fashion, H_2 cooling can lower the temperature of the gas enough to allow HD (which is strongly enhanced at low temperatures by chemical fractionation; Glover 2008) to take over and dominate the cooling. In larger minihalos ($M \gtrsim 10^6\,\mathrm{M_\odot}$) that have a more complex dynamical history,

H_2 cooling is unable to lower the temperature to the same extent, and so the gas never becomes cold enough for HD to dominate. In this case, it contributes no more than about 20–30% of the total cooling (Glover & Savin 2008) and does not appear to significantly affect the dynamics of the gas (Bromm, Coppi, & Larson 2002). HD cooling is also of great importance in situations where an increase in the fractional ionization of the gas has led to an increase in the H_2 fraction. In this situation, the gas often becomes cool enough for HD to dominate (see e.g., Nakamura & Umemura 2002; Nagakura & Omukai 2005; Johnson & Bromm 2006; Shchekinov & Vasiliev 2006). This scenario is discussed in more detail in §3.4 below.

Another molecule to have attracted considerable attention is lithium hydride, LiH. This molecule has a very large dipole moment, $\mu = 5.888$ debyes (Zemke & Stwalley 1980), and consequently its excited levels have very short radiative lifetimes. Therefore, despite the very low lithium abundance in primordial gas ($x_{Li} = 4.3 \times 10^{-10}$, by number; see Cyburt 2004), it was thought for a time that LiH would dominate the cooling at very high densities (see e.g., Lepp & Shull 1984). However, accurate quantal calculations of the rate of formation of LiH by radiative association (Dalgarno *et al.* 1996; Gianturco & Gori Giorgi 1996; Bennett *et al.* 2003)

$$\text{Li} + \text{H} \rightarrow \text{LiH} + \gamma, \tag{3.3}$$

have shown that the rate is much smaller than was initially assumed, while recent work by Defazio *et al.* (2005) has shown that the reaction

$$\text{LiH} + \text{H} \rightarrow \text{Li} + \text{H}_2, \tag{3.4}$$

has no activation energy and so will be an efficient destruction mechanism for LiH for as long as some atomic hydrogen remains in the gas. Consequently, the amount of lithium hydride present in the gas is predicted to be very small, even at very high densities, and so LiH cooling is no longer believed to be important (Mizusawa, Omukai, & Nishi 2005; Glover & Savin 2008).

Finally, molecular ions such as H_2^+, H_3^+ or HeH^+ provide another possible source of cooling in dense primordial gas. Early work on H_2^+ cooling in ionized primordial gas can be found in Suchkov & Shchekinov (1977, 1978), and its possible importance in hot, highly ionized conditions has recently been re-emphasized by Yoshida *et al.* (2007). However, it has a low critical density ($n_{cr} \sim 10^3$ cm^{-3}) and is also readily destroyed in collisions with atomic hydrogen

$$\text{H}_2^+ + \text{H} \rightarrow \text{H}_2 + \text{H}^+. \tag{3.5}$$

These factors make it unlikely to be an important coolant at high densities.

HeH^+ is a more promising candidate: it has a large dipole moment, a large cooling rate per molecule when in LTE, and hence a very large critical density (Engel *et al.* 2005). However, once again it is readily destroyed in collisions with atomic hydrogen

$$\text{HeH}^+ + \text{H} \rightarrow \text{H}_2^+ + \text{He}, \tag{3.6}$$

and so its abundance in high density gas is very small (Glover & Savin 2008). It therefore never becomes a significant coolant.

The last of these three molecular ions, H_3^+, is perhaps the most interesting. It has a large cooling rate per molecule when in LTE (Neale, Miller, & Tennyson 1996) and hence a large critical density (Glover & Savin 2006, 2008). Unlike H_2^+ and HeH^+, it is not readily destroyed by collisions with atomic hydrogen – the reaction

$$\text{H}_3^+ + \text{H} \rightarrow \text{H}_2^+ + \text{H}_2 \tag{3.7}$$

does occur, but must overcome a large energy barrier, and so proceeds slowly at temperatures $T < 1000$ K. Moreover, H_3^+ is known to be an important coolant in at least one astrophysical scenario, namely in the upper atmospheres of gas giants (Miller *et al.* 2000). Glover & Savin (2008) have examined in detail the role that H_3^+ plays in the cooling of primordial gas. They find that in most variations of the conventional Pop III.1 formation scenario, H_3^+ comes close to being an important coolant, but never quite succeeds. It contributes to the total cooling rate at densities $10^7 < n < 10^9$ cm^{-3} at the level of a few percent, making it the third most important coolant after H_2 and HD, but unimportant for the overall thermal evolution of the gas. However, Glover & Savin (2008) do identify one scenario in which H_3^+ can become the dominant coolant. If a significant ionization rate can be maintained at densities $n > 10^8$ cm^{-3}, for instance by cosmic rays or very hard X-rays, then ionization of H_2 to H_2^+ is quickly followed by the reaction

$$H_2^+ + H_2 \rightarrow H_3^+ + H \qquad (3.8)$$

resulting in the production of a large number of H_3^+ ions. In this scenario, the H_3^+ abundance can be maintained at a high enough level to allow H_3^+ to dominate the cooling rate. The required ionization rate is of the order of 10^{-17} s^{-1}. This is far larger than could be produced by plausible extragalactic sources, but is perhaps consistent with production by local sources. An interesting possibility in this context is that dark matter annihilation within the dense core may provide the necessary source of ionization.

3.3. *How many stars form per minihalo?*

High resolution AMR and SPH simulations of the formation of the first stars typically find that only a single collapsing protostellar core forms in each minihalo (see e.g. Abel, Bryan, & Norman 2002; Yoshida *et al.* 2006; O'Shea & Norman 2007). However, it is possible that this result is a consequence of the numerical methods used to simulate the gas, rather than of the gas physics. In order to properly resolve the gravitational collapse of the gas, it is necessary to ensure that the gravitational Jeans length is resolved with sufficient computational elements, a criterion that has been formalized by Truelove *et al.* (1997) for grid-based codes and by Bate & Burkert (1997) for SPH. If the gas remains close to isothermal during the collapse, then the Jeans length will continually decrease, as will the Courant timestep, the largest timestep on which the hydrodynamical evolution can be followed while still maintaining numerical stability. Therefore, once the gas reaches very high densities, the simulations can take only very small timesteps, and it is common practice in the numerical study of population III star formation to terminate the simulations at this point. However, this practice means that the simulations can follow the evolution of multiple collapsing objects only if the collapses are very closely synchronized in time. In reality, we know from the numerical study of local star formation that gravitational fragmentation is rarely so well synchronized. Typically, there is always some region with a higher density, or a lower angular momentum, that collapses first, with other objects forming only after a few local dynamical times. For example, the overall duration of star formation in nearby molecular clouds is found to be comparable to the global crossing time of the cloud (e.g. Elmegreen 2000; Mac Low & Klessen 2004) and exceeds the collapse timescale of individual stars by one to two orders of magnitude. If only the initial collapse is simulated, the formation of these other objects can be missed. To avoid this problem, it is common in numerical studies of local star formation to replace gas which has collapsed beyond the limiting resolution of the simulation with artifical sink particles. These particles possess the mass and linear momentum of the gas that they replace, and continue to interact with the surrounding gas via gravity. They are able to accrete additional infalling gas, provided that it is gravitationally bound

to them and comes within a preset accretion radius. However, they no longer feel the
effects of hydrodynamical pressure gradients, and the subsequent evolution of the gas
incorporated into them is not followed (Bate, Bonnell, & Price 1995). By replacing high
density, unresolved gas with sink particles, it becomes possible to follow the process of
gravitational fragmentation for many dynamical times.

Despite their wide usage in the study of present-day star formation, sink particles have
been used in only a few studies of primordial star formation. Bromm, Coppi, & Larson
(1999, 2002) used sink particles in their study of Pop III star formation, creating them
once the gas density exceeded $n_{\rm th} = 10^8$ cm^{-3}. They found that several massive clumps
formed in most of their simulations. The only case for which this was not true was for the
smallest mass halo they simulated ($M_{\rm tot} = 2 \times 10^5$ M$_\odot$), in which pressure forces would
be expected to have the greatest effect. Taken at face value, these results suggest that the
formation of several massive stars per minihalo could be a common outcome of population
III star formation. However, the initial conditions used in these simulations – specifically,
the adoption of a solid-body initial rotation profile – have been criticized on the grounds
that they are overly prone to fragmentation (Jappsen *et al.* 2007). The large, rotationally-
supported disks formed in the Bromm, Coppi, & Larson (1999, 2002) simulations are not
seen in simulations that start from self-consistent cosmological initial conditions (e.g.
Abel, Bryan, & Norman 2002; Yoshida *et al.* 2006), and so the fragmentation may also
not occur.

A second study to utilize sink particles was that of Bromm & Loeb (2004). They
adopted similar initial conditions to Bromm, Coppi, & Larson (2002), but used a numerical technique called particle splitting (Kitsionas & Whitworth 2002; Bromm & Loeb
2003) to allow them to follow the evolution of the first dense clump up to a much higher
density ($n_{\rm th} = 10^{12}$ cm^{-3}). They found no evidence for sub-fragmentation of this dense
clump on timescales $t \lesssim 10^4$ yr after the formation of the central sink particle.

More recently, Clark, Glover & Klessen (2008) used SPH with sink particles to simulate
the collapse of dense protostellar cores of various metallicities. Their simulations followed
collapse from an initial density of 5×10^5 cm^{-3} up to a density of $n_{\rm th} \sim 10^{17}$ cm^{-3}. The
highest resolution simulation used 25 million SPH particles to represent 500 M$_\odot$, and so
had a mass resolution of $M_{\rm res} = 2 \times 10^{-3}$ M$_\odot$. The thermodynamic evolution of the gas
was treated using a tabulated equation of state, based on the one-zone results of Omukai
et al. (2005). The initial rotational and turbulent energies were chosen to be consistent
with the results of previous studies of Pop III star formation, such as Abel, Bryan, &
Norman (2002). Although the primary focus of Clark *et al.*'s study was an examination
of the effects of metal enrichment, they also modelled a $Z = 0$ core for comparison.
They found that even in the primordial case, the core fragmented, forming of the order
of 20 sink particles. The mass function of these fragments was considerably flatter than
the present day IMF, implying that most of the mass was concentrated in the few most
massive fragments. Clark *et al.* also found that there was a delay of several local free-
fall times between the formation of the first and second sink particles, and that at the
time that the first sink particle formed, the radial profiles of mass density and specific
angular momentum were similar to those seen in previous high-resolution simulations
performed without sink particles. Slices through the densest structure at the time that
the first sink particle forms also show little sign of the impending fragmentation (see
e.g. Figure 1). These results support the view that simulations without sink particles (or
some comparable treatment of unresolved gas) run the risk of missing the formation of
all but the first protostar.

Careful analysis of the Omukai *et al.* (2005) equation of state for zero-metallicity gas
shows roughly isothermal behavior in the density range 10^{14} cm$^{-3} \leqslant n \leqslant 10^{16}$ cm^{-3}, i.e.

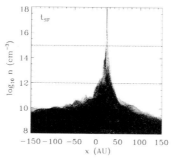

Figure 1. The densities of the SPH particles in the Clark, Glover & Klessen (2008) high resolution simulation, plotted as a function of their x-position at the moment that the first sink particle forms. Only the particles contained within the central 300 AU of the simulation are shown. At this point, there is little sign of the secondary fragmentation that will shortly occur.

just before the gas becomes optically thick and begins to heat up adiabatically. Conservation of angular momentum during this phase of the collapses leads to the build-up of a rotationally supported massive disk-like structure, which becomes gravitationally unstable and fragments. This is understandable, as isothermal disks are susceptible to gravitational instability (Bodenheimer 1995) once they have accumulated sufficient mass. Further, Goodwin, Whitworth & Ward-Thompson (2004a,b) show how even very low levels of turbulence can induce fragmentation. Since turbulence creates local anisotropies in the angular momentum on all scales, it can always provide some centrifugal support against gravitational collapse. This support can then provide a window in which fragmentation can occur. Moreover, the density at which this occurs is significantly higher than the maximum density resolved in the earlier Bromm & Loeb (2004) simulation, and we would therefore not expect the fragmentation to have been observed in that simulation.

However, the results of the Clark, Glover & Klessen (2008) simulations come with several caveats attached. The most important relates to the use of a tabulated equation of state to represent the thermodynamic evolution. In this approach, one is essentially assuming that as the density of the gas changes, the gas temperature changes instantaneously to reflect the behaviour prescribed by the equation of state. In reality, the gas temperature will adjust itself on a timescale t_{cool}, which in primordial gas is comparable to or greater than the dynamical timescale. This delay may damp out density fluctuations, thereby helping to suppress fragmentation. A second concern is that the Clark, Glover & Klessen (2008) simulations do not allow sink particles to merge, and so may overestimate the number of fragments that survive in reality. However, there are two reasons for believing that this is not a major concern. Firstly, the sink particle volume is considerably larger than the actual protostellar object in its center and so the cross section for two real protostars to collide is orders of magnitude smaller than the geometric cross section of the sink particles. Secondly, Clarke & Bonnell (2008) have demonstrated that the importance of collisions depends on the balance between shinkage of the cluster core by adiabatic contraction and puffing via collisional relaxation. As a result of this balance, collisions will start to affect the mass function in the Clark, Glover & Klessen (2008) cluster only after 10^3 to 10^4 objects have been formed, further along in its evolution than has yet been simulated.

A final concern is that the Clark *et al.* simulations do not account for the effects of feedback from the protostars that have already formed. Since fragmentation seems to occur on a timescale much shorter than the protostellar Kelvin-Helmholtz timescale, feedback from the protostar is probably unimportant. However, this assumption needs to be verified.

To summarize, there is some evidence, primarily from the work by Bromm, Coppi, & Larson (2002) and Clark, Glover & Klessen (2008), that the number of population III stars that form in the earliest minihalos is higher than the single star that is commonly assumed. However, this conclusion remains controversial, and the evidence is not yet convincing.

3.4. Is there a population III.2?

The final issue that we discuss in this review is the question of whether there might be more than one mode of population III star formation. As we have already discussed, the first Pop III stars form in minihaloes in which H_2 dominates the cooling, with HD generally playing only a minor role. However, work by a number of authors has shown that if molecular hydrogen formation can be more efficiently catalyzed by a higher electron abundance, then HD cooling of the gas can become efficient, and can allow the gas temperature to reach values close to the floor set by the CMB (see e.g. Nakamura & Umemura 2002; Nagakura & Omukai 2005; Johnson & Bromm 2006; Yoshida *et al.* 2007). Efficient HD cooling lowers the characteristic gravitational fragmentation mass scale, and may also reduce the accretion rate of gas onto the fragment or fragments that form (Yoshida *et al.* 2007; McGreer & Bryan 2008). It is therefore reasonable to assume that if HD cooling becomes dominant, lower mass stars will be formed than in the case in which H_2 dominates throughout. If this is true, then it suggests that there may be two distinct sub-populations of stars within population III: a first generation of stars forming from undisturbed primordial gas (termed population III.1 by Tan & McKee 2008) and a subsequent generation forming from gas that remains metal-free but that has an elevated fractional ionization, owing to either the infall of the gas into the deeper potential wells of the first galaxies, or to the effects of feedback from the first generation of stars. Tan & McKee (2008) term this second generation 'population III.2'.

There are various different scenarios that may lead to the formation of population III.2 stars. The simplest scenario involves primordial star formation in the first galaxies (Greif & Bromm 2006). If the virial temperature of such an object exceeds 10^4 K, then most infalling gas will be shock-heated to temperatures high enough for it to become ionized. As this gas subsequently cools and recombines, it will form H_2 at an accelerated rate, owing to the enhanced fractional ionization of the gas (Shapiro & Kang 1987). If enough H_2 is formed to cool the gas to roughly 150 K, then HD cooling will take over, driving the temperature down towards the CMB floor. Recent simulations of the formation of the first galaxies by Greif *et al.* (2008) show this mechanism in operation (see Fig. 2). However, this scenario can only produce population III.2 stars if the gas forming these galaxies has remained metal-free. Greif *et al.* (2008) demonstrate that these galaxies typically have of order 10 or more progenitors that have undergone population III.1 star formation, and so it is likely that most of their gas will already have been contaminated with metals. Nevertheless, population III.2 stars may still be able to form in these objects if metal mixing is inefficient, or if some fraction of the gas that they accrete has remained pristine.

The Pop III.2 star formation mode may also be triggered by the explosions of the first supernovae (SN), as the shocks from these explosions can heat and ionize primordial gas (e.g. Mackey, Bromm & Hernquist 2003; Salvaterra, Ferrara & Schneider 2004; Machida *et al.* 2005; Johnson & Bromm 2006). In a three-dimensional simulation of the explosion of a Pop III pair-instability SN, Greif *et al.* (2007) found that the SN shock-compression of the gas in minihalos can speed its collapse. However, in this simulation the neighboring

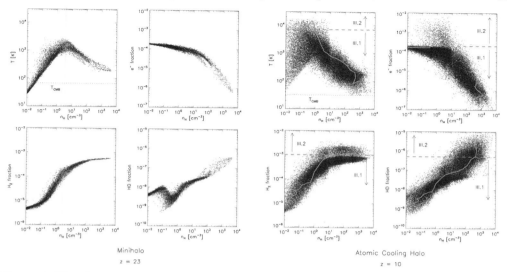

Figure 2. The phase-space distribution of gas inside a minihalo (left-hand panel) and an atomic cooling halo (right-hand panel). We show the temperature, electron fraction, HD fraction and H_2 fraction as a function of hydrogen number density, clockwise from top left to bottom left. *Left-hand panel:* In the minihalo case, adiabatic collapse drives the temperature to $> 10^3$ K and the density to $n_H > 1$ cm^{-3}, where molecule formation sets in and allows the gas to cool to $\simeq 200$ K. At this point, the central clump becomes Jeans-unstable and ultimately forms a Pop III.1 star. *Right-hand panel:* In the galaxy, a second cooling channel has emerged due to an elevated electron fraction at the virial shock, which in turn enhances molecule formation and allows the gas to cool to the temperature of the CMB. The dashed red lines and arrows approximately delineate the resulting Pop III.1 and Pop III.2 channels, while the solid green lines denote the path of a representative fluid element that follows the Pop III.2 channel.

minihalo was not strongly shocked and so did not yield any evidence for the formation of Pop III.2 stars, although it is noted that minihalos within perhaps < 500 pc of such a SN could become strongly ionized. Such minihalos would be the likely sites of the formation of Pop III.2 stars which are formed in the wake of the first SN. Simulations of SN feedback on close-by minihalos which account for the mixing of the metals injected by the supernova with the primordial gas are an important next step (see Cen & Riquelme 2007).

A further location in which population III.2 stars may be formed is in the relic H II regions left behind after the death of the first massive Pop. III stars. Again, the gas starts in a hot and highly ionized state, and forms significant quantities of H_2 as it cools and recombines (Ricotti, Gnedin & Shull 2002; Johnson, Greif, & Bromm 2007; Yoshida, Omukai & Hernquist 2007). However, the H II regions produced by the first massive stars are generally capable of expelling most of the gas from the galaxies containing the stars, and expanding to fill large volumes in the low-density intergalactic medium. Therefore, most of the H_2 that forms within the relic H II regions resides at very low densities and is initially unavailable for star formation.

Finally, Pop. III.2 stars may also be formed in primordial gas irradiated with a sufficiently strong flux of cosmic rays. Stacy & Bromm (2007) studied the effects of cosmic rays on population III star formation and showed that the enhanced fractional ionization that they create would lead to enhanced H_2 and HD production and the cooling of the gas to $T \sim T_{\rm CMB}$ for cosmic ray ionization rates greater than $\zeta \sim 10^{-19}$ s^{-1}. Jasche, Ciardi & Ensslin (2007) reached similar conclusions in a separate study.

As all three scenarios rely on the gas forming more H_2 than in the standard Pop. III.1 formation scenario, additional physical effects that reduce the amount of H_2 formed will lessen the likelihood that any Pop. III.2 stars form. One such effect is radiative cooling by H_2-H^+ and H_2-e^- collisions. Glover & Abel (2008) include these processes in their models of primordial gas cooling and show that, somewhat counter-intuitively, the increased cooling that they provide at early times (while the fractional ionization is high) leads to less H_2 being formed, and hence to less cooling at late times. This surprising result is a consequence of the different temperature dependences of the rate coefficients for H^- formation and for H^+ recombination. Decreasing the temperature decreases the rate at which H^- forms, and hence decreases the H_2 formation rate. However, it also increases the H^+ recombination rate, and so reduces the time available for H_2 formation before the necessary electrons are lost from the gas. Therefore, the faster the gas cools at early times, while the fractional ionization remains large, the less H_2 it will ultimately form. Glover & Abel (2008) show that in their one-zone calculations, this effect does not prevent the gas from cooling below 100 K; however, its effects in more realistic situations have not yet been investigated.

Another obvious candidate for suppressing Pop III.2 star formation is the far ultraviolet background built up by the first stars (Haiman, Abel & Rees 2000; Machacek, Bryan & Abel 2001; Ricotti, Gnedin & Shull 2002; Ahn *et al.* 2008). This dissociates H_2 and HD, and so acts to suppress cooling. Yoshida, Omukai & Hernquist (2007) estimate that a far-UV flux of only 3×10^{-22} erg s^{-1} cm^{-2} Hz^{-1} sr^{-1} is required in order to prevent the gas from cooling below 200 K, thereby completely suppressing the formation of Pop. III.2 stars. However, this estimate is based on simple one-zone calculations that are unlikely to properly capture the dynamics of the gas, and so may be misleading. Furthermore, in galaxies in which the virial temperature exceeds $\sim 10^4$ K the column density of H_2 may become high enough to shield the central regions from the far ultraviolet background radiation, thereby allowing HD molecules to survive and to be an important coolant (see Johnson *et al.* 2008). An investigation of the effects of the ultraviolet background on HD cooling and the formation of Pop. III.2 using a fully three-dimensional approach would be very valuable.

4. Outlook

Of the four open questions discussed in this review, those involving uncertainties in the chemistry and in the cooling of metal-free gas seem the easiest to address. Existing work has gone a long way towards establishing the effects of the chemical uncertainties, and scientists from the atomic and molecular physics communities are rising to the challenge that these uncertainties present. As far as the cooling is concerned, we now have a basic understanding of which processes are important and which are unimportant over a very wide range of scales during protostellar collapse. While there may still be occasional surprises, such as the importance of H_2-H^+ and H_2-e^- collisions in gas with only a slightly elevated fractional ionization, we think it unlikely that any of these will fundamentally change our picture of population III star formation.

The question of how many population III stars form in each minihalo is much further from being settled. There is now reason to believe that the conventional wisdom that only one massive star forms per minihalo may be incorrect. On the other hand, it may be that it is the simulations that are incorrect; they may be giving us a misleading view of what happens owing to the approximations that they make. To settle this question, further numerical study is required, using methods that are capable of following the hydrodynamical evolution of the gas beyond the point at which the first protostar forms, but that do not make as many approximations as in the Clark, Glover & Klessen (2008) study.

Finally, the question of whether Pop. III.2 exists as a distinct sub-population within population III also presents continuing difficulties. Some uncertainties, such as the impact of the revised treatment of H_2 cooling presented by Glover & Abel (2008), will be easy to address with the next generation of numerical simulations, and so should be resolved within the next year or so. However, other issues, such as the impact of the extragalactic far ultraviolet background, involve physics that is difficult to simulate accurately, and it will take far longer before we fully understand its effects. (As an example, consider that after more than ten years of study, there is still not complete agreement regarding the degree to which the ultraviolet background regulates H_2 cooling in minihalos; c.f. Haiman, Rees & Loeb 1997; Haiman, Abel & Rees 2000; Ricotti, Gnedin & Shull 2002; Wise & Abel 2007; O'Shea & Norman 2008). Furthermore, even after these issues are addressed, we will still not be able to claim with confidence that Pop. III.1 and Pop. III.2 differ until we have a better understanding of the processes regulating accretion onto population III stars. Ultimately, this may be a question that is answered as much by stellar archaeology as by theoretical study.

Acknowledgements

The authors would like to thank the organisers of IAU Symposium 255 for organising a very stimulating and enjoyable meeting. RSK acknowledges partial support from the Emmy Noether grant KL 1358/1. RSK, TG, and PCC also acknowledge support from the DFG SFB 439 'Galaxies in the Early Universe'. TG would also like to thank the Heidelberg Graduate School of Fundamental Physics (HGSFP) for financial support. The HGSFP is funded by the excellence initiative of the German government (grant number GSC 129/1). VB acknowledges support from NSF grant AST-0708795.

References

Abel, T., Anninos, P., Zhang, Y., & Norman, M. L. 1997, *New Astron.*, 2, 181
Abel, T., Bryan, G. L., & Norman, M. L. 2002, *Science*, 295, 93
Ahn, K., Shapiro, P. R., Iliev, I. T., Mellema, G., Pen, U.-L. 2008, *ApJ*, submitted; arXiv:0807.2254
Bennett, O. J., Dickinson, A. S., Leininger, T., & Gadéa, F. X. 2003, *MNRAS*, 341, 361
Bate, M. R., Bonnell, I. A., & Price, N. P. 1995, *MNRAS*, 277, 362
Bate, M. R. & Burkert, A. 1997, *MNRAS*, 288, 1060
Bodenheimer, P. 1995, *ARA&A*, 33, 199
Bromm, V., Coppi, P. S., & Larson, R. B. 1999, *ApJ*, 527, L5
Bromm, V., Coppi, P. S., & Larson, R. B. 2002, *ApJ*, 564, 23
Bromm, V. & Larson, R. S. 2004, *ARA&A*, 42, 79
Bromm, V. & Loeb, A. 2003, *ApJ*, 596, 34
Bromm, V. & Loeb, A. 2004, *New Astron.*, 9, 353
Bruhns, H., *et al.* 2008, *AAS Meeting Abstracts*, 212, 03.21
Cen, R. & Riquelme, M. A. 2007, *ApJ*, 674, 644
Clark, P. C., Glover, S. C. O., & Klessen, R. S. 2008, *ApJ*, 672, 757
Clarke, C. J. & Bonnell, I. A. 2008, *MNRAS*, 388, 1171
Cyburt, R. H. 2004, *Phys. Rev. D*, 70, 023505
Dalgarno, A., Kirby, K., & Stancil, P. C. 1996, *ApJ*, 458, 397
Defazio, P., Petrongolo, C., Gamallo, P., & González, M. 2005, *J. Chem. Phys.*, 122, 214303
Diemand, J., Moore, B., & Stadel, J. 2005, *Nature*, 433, 389
Ekström, S., Meynet, G., Chiappini, C., Hirschi, R., & Maeder, A. 2008, *A&A*, accepted; arXiv:0807.0573

Elmegreen, B. G. 2000, *ApJ*, 530, 277

Engel, E. A., Doss, N., Harris, G. J., & Tennyson, J. 2005, *MNRAS*, 357, 471

Flower, D. R., Le Bourlot, J., Pineau des Forêts, G., & Roueff, E. 2000, *MNRAS*, 314, 753

Flower, D. R. & Harris, G. J. 2007, *MNRAS*, 377, 705

Galli, D. & Palla, F. 1998, *A&A*, 335, 403

Gianturco, F. A. & Gori Giorgi, P. 1996, *ApJ*, 479, 560

Glover, S. C. O. 2005, *Space Sci. Rev.*, 117, 445

Glover, S. C. O., 2008, in *First Stars III*, eds. B. O'Shea, A. Heger, & T. Abel, (New York:AIP), 25

Glover, S. C. O. & Abel, T. 2008, *MNRAS*, in press; arXiv:0803.1768

Glover, S., Savin, D., Jappsen, A.-K. 2006, *ApJ*, 640, 553

Glover, S. C. O. & Savin, D. W. 2006, *Phil. Trans. Roy. Soc. Lond. A*, 364, 3107

Glover, S. C. O. & Savin, D. W. 2008, *MNRAS*, submitted

Goodwin, S. P., Whitworth, A. P., & Ward-Thompson, D. 2004, *A&A*, 414, 633

Goodwin, S. P., Whitworth, A. P., & Ward-Thompson, D. 2004, *A&A*, 423, 169

Greif, T. H. & Bromm, V. 2006, *MNRAS*, 373, 128

Greif, T. H., Johnson, J. L., Bromm, V., & Klessen, R. S. 2007, *ApJ*, 670, 1

Greif, T. H., Johnson, J. L., Klessen, R. S., & Bromm, V. 2008, *MNRAS*, 387, 1021

Haiman, Z., Abel, T., & Rees, M. J. 2000, *ApJ*, 534, 11

Haiman, Z., Rees, M. J., & Loeb, A. 1997, *ApJ*, 476, 458

Haiman, Z., Thoul, A., & Loeb, A. 1996, *ApJ*, 464, 523

Heger, A., Woosley, S. E. 2002, *ApJ*, 567, 532

Jasche, J., Ciardi, B., & Ensslin, T. A. 2007, *MNRAS*, 380, 417

Jacobs, T. A., Giedt, R. R., Cohen, N. 1967, *J. Chem. Phys.*, 47, 54

Jappsen, A.-K., Klessen, R. S., Glover, S. C. O., & Mac Low, M.-M. 2007, *ApJ*, submitted; arXiv:0709.3530

Johnson, J. L. & Bromm, V. 2006, *MNRAS*, 366, 247

Johnson, J. L., Greif, T. H., & Bromm, V. 2007, *ApJ*, 665, 85

Johnson, J. L., Greif, T. H., & Bromm, V. 2008, *MNRAS*, 388, 26

Kitsionas, S. & Whitworth, A. P. 2002, *MNRAS*, 330, 129

Kudritzki, R. P. 2002, *ApJ*, 2002, 577, 389

Lepp, S. & Shull, J. M. 1984, *ApJ*, 280, 465

Machacek, M. E., Bryan, G. L., & Abel, T. 2001, *ApJ*, 548, 509

Machida, M. N., Kohji, T., Nakamura, F., & Fujimoto, M. Y. 2005, *ApJ*, 622, 39

Mackey, J., Bromm, V., & Hernquist, L. 2003, *ApJ*, 586, 1

Mac Low, M.-M., & Klessen, R. S. 2004, *Rev. Mod. Phys.*, 76, 125

McGreer, I. D. & Bryan, G. L. 2008, *ApJ*, accepted; arXiv:0802.3918

Miller, S., Achilleos, N., Ballester, G. E., Geballe, T. R., Joseph, R. D., Prangé, R., Rego, D., Stallard, T., Tennyson, J., Trafton, L. M., & Waite, J. H., Jr. 2000, *Phil. Trans. R. Soc.*, 358, 2485

Mizusawa, H., Omukai, K., & Nishi, R. 2005, *PASJ*, 57, 951

Nagakura, T. & Omukai, K. 2005, *MNRAS*, 364, 1378

Nakamura, F. & Umemura, M. 2002, *ApJ*, 569, 549

Naoz, S., Noter, S., & Barkana, R. 2006, *MNRAS*, 373, L98

Neale, L., Miller, S., & Tennyson, J. 1996, *ApJ*, 464, 516

Norman, M. L. 2008, in *First Stars III*, eds. B. O'Shea, A. Heger, & T. Abel, (New York:AIP), 3

Omukai, K. & Nishi, R. 1998, *ApJ*, 508, 141

Omukai, K., Tsuribe, T., Schneider, R., & Ferrara, A. 2005, *ApJ*, 626, 627

O'Shea, B. W. & Norman, M. L. 2007, *ApJ*, 654, 66

O'Shea, B. W. & Norman, M. L. 2008, *ApJ*, 673, 14

Palla, F., Salpeter, E. E., & Stahler, S. W. 1983, *ApJ*, 271, 632

Peebles, P. J. E. & Dicke, R. H. 1968, *ApJ*, 154, 891

Reed, D. S., Bower, R., Frenk, C. S., Gao, L., Jenkins, A., Theuns, T., & White, S. D. M. 2005, *MNRAS*, 363, 393

Ricotti, M., Gnedin, N., & Shull, J. M. 2002, *ApJ*, 575, 49

Ripamonti, E., Haardt, F., Ferrara, A., & Colpi, M. 2002, *MNRAS*, 334, 401

Ripamonti, E. & Abel, T. 2004, *MNRAS*, 348, 1019

Ripamonti, E. 2007, *MNRAS*, 376, 709

Salvaterra, R., Ferrara, A., & Schneider, R. 2004, *New Astron.*, 10, 113

Saslaw, W. C. & Zipoy, D. 1967, *Nature*, 216, 976

Savin, D. W., Krstic, P. S., Haiman, Z., & Stancil, P. C. 2004, *ApJ*, 606, L167; erratum *ApJ*, 607, L147

Shapiro, P. R. & Kang, H. 1987, *ApJ*, 318, 32

Shchekinov, Y. A. & Vasiliev, E. O. 2006, *MNRAS*, 368, 454

Stacy, A. & Bromm, V. 2007, *MNRAS*, 382, 229

Suchkov, A. A. & Shchekinov, Y. A. 1977, *Sov. Astr. Lett.*, 3, 297

Suchkov, A. A. & Shchekinov, Y. A. 1978, *Sov. Astr. Lett.*, 4, 164

Tan, J. C. & McKee, C. F. 2008, in *First Stars III*, eds. B. O'Shea, A. Heger, & T. Abel, (New York:AIP), 47

Tegmark, M., Silk, J., Rees, M., Blanchard, A., Abel, T., & Palla, F. 1997, *ApJ*, 474, 1

Truelove, J. K., Klein, R. I., McKee, C. F., Holliman, J. H., Howell, L. H., & Greenough, J. A. 1997, *ApJ*, 489, L179

Turk, M. J., Abel, T., & O'Shea, B. W. 2008, in *First Stars III*, eds. B. O'Shea, A. Heger, & T. Abel, (New York:AIP), 16

Umeda, H. & Nomoto, K. 2002, *ApJ*, 565, 385

Wise, J. H. & Abel, T. 2007, *ApJ*, 671, 1559

Yoshida, N., Abel, T., Hernquist, L., & Sugiyama, N. 2003, *ApJ*, 592, 645

Yoshida, N., Omukai, K., Hernquist, L., & Abel, T. 2006, *ApJ*, 652, 6

Yoshida, N., Omukai, K., & Hernquist, L. 2007, *ApJ*, 667, L117

Yoshida, N., Oh, S. P., Kitayama, T., & Hernquist, L. 2007, *ApJ*, 663, 687

Zemke, W. T. & Stwalley, W. C. 1980, *J. Chem. Phys.*, 73, 5584

Low-Metallicity Star Formation:
From the First Stars to Dwarf Galaxies
Proceedings IAU Symposium No. 255, 2008
L.K. Hunt, S. Madden & R. Schneider, eds.

Protostar formation in the early universe

Naoki Yoshida[1]

[1] Department of Physics, Nagoya University
Furocho, Chikusa Nagoya 464-8602, Japan
email: nyoshida@a.phys.nagoya-u.ac.jp

Abstract. We study the formation of primordial proto-stars in a ΛCDM universe using ultra high-resolution cosmological simulations. Our approach includes all the relevant atomic and molecular physics to follow the thermal evolution of a prestellar gas cloud to "stellar" densities. We describe the numerical implementation of the physics. We also show the result of a simulation of the formation of primordial stars in a reionized gas.

Keywords. cosmology: theory, molecular processes, radiative transfer

1. Introduction

The study of primordial star formation has a long history. The formation of the first cosmological objects via gas condensation by molecular hydrogen cooling has been studied for many years since the late 1960's. One-dimensional hydrodynamic simulations of spherical gas collapse were performed by a number of researchers (Omukai & Nishi 1998), including a detailed treatment of all the relevant chemistry and radiative processes and thus were able to provide accurate results on the thermal evolution of a collapsing primordial gas cloud up to stellar densities. These authors found that, while the evolution of a spherical primordial gas cloud proceeds in a roughly self-similar manner, there are a number of differences in the thermal evolution from that of present-day, metal- and dust-enriched gas clouds.

Recently, three-dimensional hydrodynamic calculations were performed by several groups. Statistical properties of primordial star-forming clouds and the overall effect of cosmological bias have been studied in detail by Yoshida *et al.* (2003) and Gao *et al.* (2007). Simulations of the formation of primordial proto-stars have been hampered by complexity of physics such as radiative transfer in a very high-density prestellar gas. A critical technique we describe in this contribution is computation of molecular line opacities and continuum opacities. With the implementation of optically thick cooling, the gas evolution can be accurately followed to much higher densities than was possible in the previous studies (Yoshida *et al.* 2006). We show that the method works well in problems of collapsing gas clouds, in terms of accurate computation of radiative cooling rates and of resulting density and temperature structure. We apply this technique to a cosmological simulation of the formation of primordial star.

2. Cosmological simulation

We follow the standard procedure to set up cosmological initial conditions for a flat Λ Cold Dark Matter cosmology with $\Omega_\Lambda = 0.7$, $\Omega_{\mathrm{darkmatter}} = 0.26$, $\Omega_{\mathrm{baryon}} = 0.04$, and $H_0 = 70$ km/sec/Mpc. We use a multi-scale re-simulation technique to achieve a high mass resolution in the region where non-linear objects are formed early. The simulation volume has a side-length of 200 comoving kilo-parsecs. By employing a three-level refinement to this volume, we achieve an *initial* mass resolution of $M_{\mathrm{dm}} = 0.135 M_\odot$

and $M_{\text{gas}} = 0.021 M_\odot$ in the high-density region. We evolve the system until a first star-forming gas cloud is formed in one of the dark matter halos in the high-resolution region.

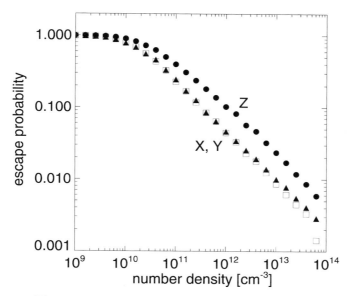

Figure 1. The escape probability for a line photon.

3. Radiative transfer

3.1. *Molecular hydrogen line cooling*

A primordial gas can cool radiatively by molecular hydrogen line cooling at low densities. When the gas density and the molecular fraction are high, the cloud becomes opaque to molecular lines and then H_2 line cooling becomes inefficient. The net cooling rate can be expressed as

$$\Lambda_{H_2,\text{thick}} = \sum_{u,l} h\nu_{ul}\, \beta_{\text{esc},ul}\, A_{ul}\, n_u \,, \qquad (3.1)$$

where n_u is the population density of hydrogen molecules in the upper energy level u, A_{ul} is the Einstein coefficient for spontaneous transition, $\beta_{\text{esc},ul}$ is the probability for an emitted line photon to escape without absorption, and $h\nu_{ul} = \Delta E_{ul}$ is the energy difference between the two levels.

In order to calculate the escape probability, we first evaluate the opacity for each molecular line as

$$\tau_{lu} = \alpha_{lu} L, \qquad (3.2)$$

where L is the characteristic length scale. Since the absorption coefficients α_{lu} are computed in a straightforward manner, although somewhat costly, the remaining key task is the evaluation of the length scale L. By noting that the important quantity we need is the effective gas cooling rate, we can formulate a reasonable and well-motivated approximation. To this end, we decided to use the Sobolev method that is widely used in the study of stellar winds and planetary nebulae.

We calculate the Sobolev length along a line-of-sight as

$$L_r = \frac{v_{\text{thermal}}}{|dV_r/dr|},\tag{3.3}$$

where $v_{\text{thermal}} = \sqrt{kT/m_{\text{H}}}$ is the thermal velocity of H_2 molecules, and V_r is the fluid velocity in the direction. A suitable angle-average must be computed in order to obtain the net escape probability. Details are found in Yoshida *et al.* (2006).

In Fig. 1, we show the direction-dependence of the escape probabilities for one of the strongest H_2 lines. We use an output of our high-resolution cosmological simulation at the time when the central density is $n_{\text{c}} = 10^{14}\,\text{cm}^{-3}$. To make this plot, we configure the coordinate such that the z-direction is aligned to the angular momentum vector of the central $1M_\odot$ portion. The central part cools rapidly by H_2 lines and flattens slightly, to have a prolate shape. The velocity gradient in z-direction is larger than those in the other two orthogonal directions, and hence the molecular lines preferentially escape in the z-direction.

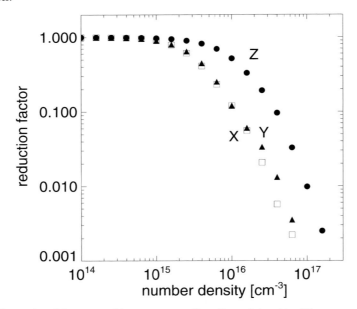

Figure 2. The ratio of $\Lambda_{\text{CIE,thick}}/\Lambda_{\text{CIE,thin}}$ as a function of density. We use an output when the central density reaches $10^{20}\,\text{cm}^{-3}$.

3.2. *Cooling by collision-induced emission*

At densities greater than $n \sim 10^{14}\,\text{cm}^{-3}$, hydrogen molecules collide with each other very frequently, making it possible to interact with photons by induced electric dipoles. This process is known as collision-induced emission (CIE), the opposite process to collision-induced absorption:

$$\text{H}_2(v, J) + \text{H}_2 \rightarrow \text{H}_2(v', J') + \text{H}_2 + h\nu,\tag{3.4}$$

$$\text{H}_2(v, J) + \text{He} \rightarrow \text{H}_2(v', J') + \text{He} + h\nu.\tag{3.5}$$

$$\text{H}_2(v, J) + \text{H} \rightarrow \text{H}_2(v', J') + \text{H} + h\nu.\tag{3.6}$$

$$\text{H} + \text{He} \rightarrow \text{H} + \text{He} + h\nu.\tag{3.7}$$

These processes yield very complex spectra, and have an essentially continuum appearance. We calculate the total emissivity by integrating the contribution from each transition:

$$\eta_{\rm CIE} = \frac{2h\nu^3}{c^2} \, \sigma_{\rm CIE} \, n({\rm H_2}) \exp\left(-\frac{h\nu}{kT}\right). \tag{3.8}$$

We use the updated cross-sections $\sigma_{\rm CIE}$ for $\rm H_2$-$\rm H_2$, $\rm H_2$-H, $\rm H_2$-He, and H-He collisions.

Next, we implement computation of local optical depth to continuum radiation using the Planck opacity table of Lenzuni *et al.* (1991). The continuum opacity is used to evaluate the net CIE cooling rate. We calculate local optical depths in six orthogonal directions from a target point in a smoothed particle hydrodynamics manner (see, e.g., Yoshida *et al.* 2007), by projecting the cubic-spline kernels of surrounding gas particles which have their own densities and temperatures. The net energy transfer rate in each direction scales as $\Lambda \propto 1/(1+\tau)$ when the optical depth is small, whereas, for large optical depths, it should scale as $\Lambda \propto 1/\tau^2$. In practice we adopt a simple double power law:

$$f = \frac{1}{(1+\tau)(1+(\tau/10))} \tag{3.9}$$

which approximates the above two conditions at small and large optical depths. In order to perform a suitable angle averaging, we take the mean of six directions,

$$f_{\rm reduce} = \frac{1}{6}\sum f_i. \tag{3.10}$$

Fig. 2 shows the reduction factor for the CIE cooling rate against local density. We note that, although our method is still approximate, it takes into account local density, temperature, and their structures (geometry) in a self-consistent manner. Our method can be applied to general problems, whereas simple functional fits for cooling rates that are based only on local density will fail in estimating the true cooling rate when the cloud core is prolate and/or has a complex structure such as in a (weak) turbulent velocity field.

4. Primordial protostar

We define a constant density, atomic gas core as a protostar that is pressure-supported. At the final output time, a protostar formed with a mass of just 0.01 solar masses. It had an initial radius of $\sim 5 \times 10^{11}$ cm, similar to that of present-day protostars in theoretical calculations. The central particle number density of the protostar was $\sim 10^{21}$ cm^{-3} and the temperature was well above 10,000 K. Details of the formation process of this protostar is found in Yoshida, Omukai, & Hernquist (2008).

The protostar has an atomic core with mass 0.01 $M_\odot$ within a fully molecular part of $\sim 1 M_\odot$. At the time of protostar formation, the central temperature was so high that almost all the molecules are now collisionally dissociated within an enclosed mass of 0.01 $M_\odot$. The structure is similar to those found in earlier one-dimensional spherically symmetric calculations (e.g. Ripamonti *et al.* 2002).

5. Primordial star-formation in a reionized gas

We employ the technique described in the previous sections in another cosmological simulation. Earlier in Yoshida *et al.* (2007), we used a large cosmological simulation to study the evolution of early relic HII regions until second-generation gas clouds are formed. We further explore the evolution of these prestellar gas clouds. The highest density achieved by the simulation is $\sim 10^{18}$ cm^{-3}, at which point the central core is

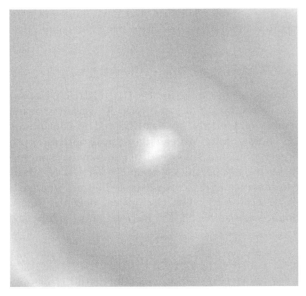

Figure 3. The newly born protostar in our simulation. We show the projected mean density-weighted temperature. The shown region has a side length of 25 astronomical units. The central part has a temperature of ~12,000 K.

optically thick even to continuum. Full-scale dissociation of hydrogen molecules is taking place in the core, which works as an effective *cooling* mechanism. We investigate in detail the structure of such gas clouds. We then compute the gas mass accretion rate and use it as an input to a proto-stellar calculation.

The temperature structure around the second-generation star can be understood by appealing to various atomic and molecular processes. HD line cooling brings the gas temperature below 100 Kelvin. Note that the minimum temperature is set by that of the cosmic microwave background.

The central proto-stellar 'seed' is accreting the surrounding gas at a rate $> 10^{-3} M_\odot/\mathrm{yr}$ and thus a star with mass $\sim 10 M_\odot$ will form within 10^4 years. However, the final stellar mass is determined by processes such as radiative feedback from the protostar. We treat the evolution of a protostar as a sequence of a growing hydrostatic core with an accreting envelope. The ordinary stellar structure equations are applied to the hydrostatic core. The structure of the accreting envelope is calculated under the assumption that the flow is steady for a given mass accretion rate.

Fig. 4 shows the resulting evolution of the protostar. After a transient phase and an adiabatic growth phase at $M_* < 10 M_\odot$, the protostar enters the Kelvin-Helmholtz phase and contracts by radiating its thermal energy. When the central temperature reaches 10^8 K, hydrogen burning by the CNO cycle begins with a slight amount of carbon synthesized by helium burning. This phase is marked by a solid circle in the figure. The energy generation by hydrogen burning halts contraction when the mass is $35 M_\odot$ and its radius is ~ 2.8 solar radii. Soon after, the star reaches the zero-age main sequence (ZAMS). The protostar relaxes to a ZAMS star within about 10^5 years from the birth of the protostellar seed. Accretion is not halted by radiation from the protostar to the end of our calculation.

It is important to point out that the mass of the parent cloud from which the star formed is $M_\mathrm{cloud} \sim 40 M_\odot$. The final stellar mass is likely limited by the mass of the

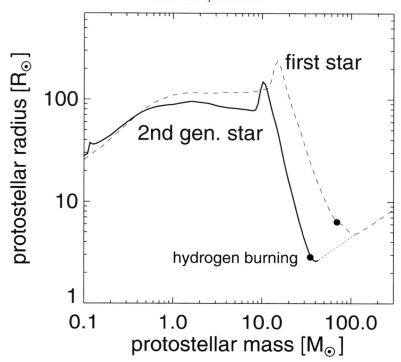

Figure 4. Evolution of the proto-stellar radius and the mass (solid line). The solid circle marks the time when efficient hydrogen burning begins. The dotted line shows the mass (and radius) growth which is calculated under the assumption that a larger amount of gas than the parent cloud can be accreted. For reference we also show the result from Y06 for a first generation star that forms in an initially neutral gas cloud.

gravitationally unstable parent cloud. We thus argue that primordial stars formed from an ionized gas are massive, with a characteristic mass of several tens of solar masses, allowing overall uncertainties in the accretion physics and also the dependence of the minimum gas temperature on redshift. They are smaller than the first stars formed from a neutral gas, but are not low-mass objects as suggested by earlier studies.

The elemental abundance patterns of recently discovered hyper metal-poor stars suggest that they might have been born from the interstellar medium that was metal-enriched by supernovae of these massive primordial stars.

References

Omukai, K. & Nishi, R. 1998, *ApJ*, 508, 141

Yoshida, N., Abel, T, Hernquist, L., & Sugiyama, N., 2003, *ApJ*, 592, 645

Gao, L., Yoshida, N., Abel, T., Frenk, C. S., Jenkins, A., & Springel, V., 2007, *MNRAS*, 378, 449

Yoshida, N., Omukai, K., Hernquist, L., & Abel, T., 2006, *ApJ*, 652, 6 (Y06)

Yoshida, N., Oh, S.-P., Kitayama, T., & Hernquist, L., 2007, *ApJ*, 663, 687

Lenzuni, P., D., Chernoff, D. F., & Salpeter, E. E., 1991, *ApJS*, 76, 759

Yoshida, N., Omukai, K., & Hernquist, L., 2007, ApJL, 667, 117

Yoshida, N., Omukai, K., & Hernquist, L., 2008, *Science*, 321, 669

Ripamonti, E., Haardt, F., Ferrara, A., & Colpi, M., 2002. *MNRAS*, 334, 401

Low-Metallicity Star Formation:
From the First Stars to Dwarf Galaxies
Proceedings IAU Symposium No. 255, 2008
L.K. Hunt, S. Madden & R. Schneider, eds.

© 2008 International Astronomical Union
doi:10.1017/S174392130802454X

Population III.1 stars: formation, feedback and evolution of the IMF

Jonathan C. Tan[1]

[1] Dept. of Astronomy, University of Florida, Gainesville, FL 32611, USA
email: jt@astro.ufl.edu

Abstract. I discuss current theoretical expectations of how primordial, Pop III.1 stars form. Lack of direct observational constraints makes this a challenging task. In particular predicting the mass of these stars requires solving a series of problems, which all affect, perhaps drastically, the final outcome. While there is general agreement on the initial conditions, H_2-cooled gas at the center of dark matter minihalos, the subsequent evolution is more uncertain. In particular, I describe the potential effects of dark matter annihilation heating, fragmentation within the minihalo, magnetic field amplification, and protostellar ionizing feedback. After these considerations, one expects that the first stars are massive $\gtrsim 100 \, M_\odot$, with dark matter annihilation heating having the potential to raise this scale by large factors. Higher accretion rates in later-forming minihalos may cause the Pop III.1 initial mass function to evolve to higher masses.

Keywords. stars: formation, galaxies: formation, dark matter, cosmology: theory

1. Introduction: The Importance of Pop III.1 Stars and their IMF

The first, essentially metal-free (i.e. Population III), stars are expected to have played a crucial role in bringing the universe out of the dark ages: initiating the reionization process, including the local effects of their H II regions in generating shocks and promoting formation of molecular coolants in the relic phase; photodissociating molecules; amplifying magnetic fields to possibly dynamically important strengths; and generating the mechanical feedback, heavy elements and possible neutron star or black hole remnants associated with supernovae. In these ways Pop III stars laid the foundations for galaxy formation, including supermassive black holes and globular clusters. Many of these processes are theorized to depend sensitively on the initial mass function (IMF) of Pop III stars, thus motivating its study. The formation of the first Pop III stars in a given region of the universe is expected to have been unaffected by other astrophysical sources and these have been termed Pop III.1, in contrast to Pop III.2 (McKee & Tan 2008, hereafter MT08). Pop III.1 are important for influencing the initial conditions for future structure formation and for having their properties determined solely by cosmology. There is also the possibility, described in Sect. 2.1, that Pop III.1 star formation may be sensitive to the properties of weakly interacting massive particle (WIMP) dark matter.

Unfortunately, at the present time and in the near future we expect only indirect observational constraints on the Pop III IMF. The epoch of reionization can be constrained by CMB polarization (Page *et al.* 2007) and future high redshift 21 cm HI observations (e.g. Morales & Hewitt 2004). Metals from individual Pop III supernovae may have imprinted their abundance patterns in very low metallicity Galactic halo stars (Beers & Christlieb 2005) or in the Ly-α forest (Schaye *et al.* 2003; Norman, O'Shea, & Pascos 2004). Light from the first stars may contribute to the observed NIR background intensity, (e.g. Santos, Bromm, & Kamionkowski 2002), and its fluctuations (Kashlinsky *et al.* 2004; c.f. Thompson *et al.* 2007). If massive, supernovae marking the deaths of the first stars may be observable by JWST (Weinmann & Lilly 2005). If these supernovae produce gamma-ray bursts then these may already be making a contribution to the population observed by SWIFT (Bromm & Loeb 2002).

The lack of direct observations of Pop III star formation means theoretical models lack constraints, which is a major problem for treating such a complicated, nonlinear process. Numerical simulations have been able to start with cosmological initial conditions and advance to the point of protostar formation (see Yoshida *et al.*, these proceedings), but progressing further through the protostellar accretion phase requires additional modeling of complicated processes, including a possible need to include extra physics such as WIMP annihilation and magnetic fields. Building up a prediction of the final mass achieved by the protostar, i.e. the initial mass (function) of the star (population), is akin to building a house of cards: the reliability of the structure becomes more and more precarious.

In this article we summarize theoretical attempts to understand the formation process and resulting IMF of Pop III.1 stars. We have reviewed much of these topics previously (Tan & McKee 2008), so here we concentrate on a discussion of some of the more uncertain aspects in these models, including the potential effects of WIMP annihilation on Pop III.1 star formation, fragmentation during Pop III.1 star formation, the generation of magnetic fields, the uncertainties in predicting the IMF from feedback models, and the evolution of the Pop III.1 IMF. Note, when discussing possible fragmentation during the formation of a Pop III.1 star, we will consider all stars that result from the same minihalo to be Pop III.1, i.e. they are unaffected by astrophysical sources external to their own minihalo.

2. Initial Conditions and Possible Effects of WIMP Annihilation

The initial conditions for the formation of the first stars are thought to be relatively well understood: they are determined by the growth of small-scale gravitational instabilities from cosmological fluctuations in a cold dark matter universe. The first stars are expected to form at redshifts $z \sim 10-50$ in dark matter "minihalos" of mass $\sim 10^6 \, M_\odot$ (Tegmark *et al.* 1997). In the absence of any elements heavier than helium (other than trace amounts of lithium) the chemistry and thermodynamics of the gas are very simple. Once gas collects in the relatively shallow potential wells of the minihalos, cooling is quite weak and is dominated by the ro-vibrational transitions of trace amounts of H_2 molecules that cool the gas to ~ 200 K at densities $n_H \sim 10^4 \, cm^{-3}$ (Abel, Bryan, & Norman 2002; Bromm, Coppi, & Larson 2002). Glover *et al.* (these proceedings) review the effects of other potential coolants, finding they are small for Pop III.1 star formation.

As the gas core contracts to greater densities, the H_2 cooling becomes relatively inefficient and the temperature rises to ~ 1000 K. At densities $\sim 10^{10} \, cm^{-3}$ rapid 3-body formation of H_2 occurs, creating a fully molecular region that can cool much more efficiently. This region starts to collapse supersonically until conditions become optically thick to the line and continuum cooling radiation, which occurs at densities $\sim 10^{17} \, cm^{-3}$. Recent 3D numerical simulations have advanced to densities of order $10^{21} \, cm^{-3}$ (see contribution by Yoshida *et al.*, these proceedings), but have trouble proceeding further given the short timesteps required to resolve the dynamics of the high density gas of the protostar. Further numerical progress can be achieved by introducing sink particles (Bromm & Loeb 2004) or with 1D simulations (Omukai & Nishi 1998; Ripamonti *et al.* 2002).

Alternatively, given the above initial conditions, the subsequent accretion rate to the protostar can be calculated analytically (Tan & McKee 2004, hereafter TM04). The accretion rate depends on the density structure and infall velocity of the gas core at the point when the star starts to form. Omukai & Nishi (1998) and Ripamonti *et al.* (2002) showed that the accreting gas is isentropic with an adiabatic index $\gamma \simeq 1.1$ due to H_2

cooling; i.e., each mass element satisfies the relation $P = K\rho^\gamma$ with the "entropy parameter" $K = $ const. In hydrostatic equilibrium—and therefore in a subsonic contraction—such a gas has a density profile $\rho \propto r^{-k_\rho}$ with $k_\rho \simeq 2.2$, as is seen in simulations. TM04 describe the normalization of the core density structure via the "entropy parameter"

$$K' \equiv (P/\rho^\gamma)/1.88 \times 10^{12} \text{ cgs} = (T'_{\text{eff}}/300 \text{ K})(n_{\text{H}}/10^4 \text{cm}^{-3})^{-0.1}, \qquad (2.1)$$

where $T'_{\text{eff}} \equiv T + \mu\sigma^2_{\text{turb}}/k$ is an effective temperature that includes the modest effect of subsonic turbulent motions that are seen in numerical simulations (Abel *et al.* 2002).

For the infall velocity at the time of protostar formation, simulations show the gas is inflowing subsonically at about a third of the sound speed (Abel *et al.* 2002). Hunter's (1977) solution for mildly subsonic inflow (Mach number = 0.295) is the most relevant for this case. It has a density that is 1.189 times greater than a singular isothermal sphere (Shu 1977) at $t = 0$, and an accretion rate that is 2.6 times greater.

Feedback from the star, whether due to winds, photoionization, or radiation pressure, can reduce the accretion rate of the star. TM04 and MT08 define a hypothetical star+disk mass, $m_{*d,0}$, and accretion rate, $\dot{m}_{*d,0}$, in the absence of feedback. In this case, the star+disk mass equals the mass of the part of the core (out to some radius, r, that has undergone inside-out collapse) from which it was formed, $m_{*d,0} = M(r)$. The instantaneous and mean star formation efficiencies are $\epsilon_{*d} \equiv \dot{m}_{*d}/\dot{m}_{*d,0}$ and $\bar{\epsilon}_{*d} \equiv m_{*d}/m_{*d,0} = m_{*d}/M$, respectively.

Assuming the Hunter solution applies for a singular polytropic sphere with $\gamma = 1.1$, the accretion rate is then (TM04)

$$\dot{m}_{*d} = 0.026\epsilon_{*d}K'^{15/7}(M/M_\odot)^{-3/7} \ M_\odot \text{ yr}^{-1}, \qquad (2.2)$$

with the stellar mass smaller than the initial enclosed core mass via $m_* \equiv m_{*d}/(1+f_d) = \bar{\epsilon}_{*d}M/(1+f_d)$. We choose a fiducial value of $f_d = 1/3$ appropriate for disk masses limited by enhanced viscosity due to self-gravity.

2.1. *Possible Effects of Dark Matter Annihilation*

Pop III.1 stars form at the centers of dark matter (DM) minihalos. While the mass density is dominated by baryons inside ~ 1 pc, adiabatic contraction ensures that there will still be a peak of DM density co-located with the baryonic protostar. As discussed by Spolyar *et al.* (2008), if the dark matter consists of a weakly interacting massive particle (WIMP) that self annihilates, then this could lead to extra heating that can help support the protostar against collapse. Spolyar *et al.* calculated that, depending on the dark matter density profile, WIMP mass, and annihilation cross section, the local heating rate due to dark matter could exceed the baryonic cooling rate for densities $n_{\text{H}} \gtrsim 10^{14}$ cm^{-3}, corresponding to scales of about 20 AU from the center of the halo/protostar.

Natarajan, Tan, & O'Shea (2008) revisited this question by considering several minihalos formed in numerical simulations. While there was some evidence for adiabatic contraction leading to a steepening of the dark matter density profiles in the centers of the minihalos, this was not well resolved on the scales where heating may become important. Thus various power law ($\rho_\chi \propto r^{-\alpha_\chi}$) extrapolations were considered for the DM density. A value of $\alpha_\chi \simeq 1.5$ was derived based on the numerically well-resolved regions at $r \sim 1$ pc. A steeper value of $\alpha_\chi \simeq 2.0$ was derived based on the inner regions of the simulations. In the limit of very efficient adiabatic contraction, one expects the dark matter density profile to approach that of the baryons, which would yield $\alpha_\chi \simeq 2.2$. For the density profiles with $\alpha_\chi \simeq 2.0$, Natarajan *et al.* (2008) found that dark matter heating inevitably becomes dominant. Natarajan *et al.* also considered the global quasi-equilibrium structures for which the total luminosity generated by WIMP annihilation that is trapped in the protostar, $L_{\chi,0}$, equals that radiated away by the baryons, assuming both density distributions are power laws truncated at some radius, r_c, with a constant density core. This core radius was varied to obtain the equilibrium luminosity. Typical results were $L_{\chi,0} \sim 10^3 \, L_\odot$ and $r_c \simeq$ to a few to a few tens of AU.

These scales at which equilibrium is established are important for determining the subsequent evolution of the protostar, which will continue to gain baryons and probably additional dark matter via adiabatic contraction. Even in the limit where no further dark matter becomes concentrated in the protostar, that which is initially present can be enough to have a major influence on the subsequent protostellar evolution. As the protostar gains baryonic mass it requires a greater luminosity for its support. If there was no dark matter heating, the protostar would begin to contract once it becomes older than its local Kelvin-Helmholz time, i.e. on timescales much longer than the stellar dynamical time. If dark matter is present, it will become concentrated as the protostar contracts, and the resulting annihilation luminosity will grow as $L_\chi \simeq L_{\chi,0}(r_*/r_{*,0})^{-3}$, assuming a homologous density profile. For a starting luminosity of $L_{\chi,0} = 1000 \, L_\odot$ and radius of $r_{*,0} \simeq r_c = 10$ AU, this can mean luminosities that are easily large enough to support $\sim 100 \, M_\odot$ stars, i.e. $\sim 10^6 \, L_\odot$, at sizes of ~ 1 AU, i.e. much greater than their main sequence radii, which would be $\simeq 5 R_\odot = 0.02$ AU. These estimates are of course very sensitive to the initial size of the protostar.

Full treatment of the protostellar evolution (see Freese *et al.* 2008 for an initial model) requires a model for the evolution of the stellar DM content, which grows by accumulation of surrounding WIMPs, but also suffers depletion due to the annihilation process. The mean depletion time in the star is $t_{\rm dep} = (\rho_\chi/\dot{\rho}_\chi) \simeq m_\chi/(\rho_\chi < \sigma_a v >) \rightarrow 105(m_\chi/100 \, {\rm GeV})(\rho_\chi/10^{12} {\rm GeV \, cm}^{-3})^{-1}$ Myr, where we have normalized to typical values of ρ_χ in the initial DM core (Natarajan *et al.* 2008). If the protostar contracts from an initial radius of 10 AU to 1 AU then $t_{\rm dep} \simeq 10^5$ yr. This becomes comparable to the growth time of the protostar (i.e. the time since its formation, its age), $t_* = 2.92 \times 10^4 K'^{-15/7}(m_*/100 \, M_\odot)^{10/7}$ yr (TM04). We see that, if replenishment of WIMPs in the protostar is negligible, then depletion can become important for AU scale protostars of $\sim 100 \, M_\odot$.

Protostars swollen by DM heating would have much cooler photospheres and thus smaller ionizing feedback than if they had followed standard protostellar evolution leading to contraction to the main sequence by about $100 \, M_\odot$. Ionizing feedback is thought to be important in terminating accretion and thus setting the Pop III.1 IMF (MT08; see §5 below). The reduced ionizing feedback of DM-powered protostars may allow them to continue to accrete to much higher masses than would otherwise have been achieved.

3. Protostellar Accretion and Disk Fragmentation

Another process that may affect the IMF of the first stars is fragmentation of the infalling gas after the first protostar has formed. TM04 and MT08 considered the growth and evolution of the protostar in the case of no fragmentation (and no DM heating): the final mass achieved by the protostar is expected to be ~ 100–$200 \, M_\odot$ and set by a balance between its ionizing feedback and its accretion rate through its disk (§5).

The accretion disk of the protostar does present an environment in which density fluctuations can grow, since there will typically be many local dynamical timescales before the gas is accreted to the star. TM04 calculated the expected disk size, $r_d(m_*)$, assuming conservation of angular momentum inside the sonic point, $r_{\rm sp}$, of the inflow, finding

$$r_d = 1280 \left(\frac{f_{\mathrm{Kep}}}{0.5}\right)^2 \left(\frac{m_{*d,2}}{\bar{\epsilon}_{*d}}\right)^{9/7} K'^{-10/7} \mathrm{AU} \to 1850 \left(\frac{f_{\mathrm{Kep}}}{0.5}\right)^2 \frac{m_{*,2}^{9/7}}{K'^{10/7}} \mathrm{AU} \qquad (3.1)$$

where $m_{*d,2} = m_{*d}/100\,M_\odot$, $m_{*,2} = m_*/100\,M_\odot$, the $\to$ is for the case with $f_d = 1/3$ and $f_{\mathrm{Kep}} \equiv v_{\mathrm{rot}}(r_{\mathrm{sp}})/v_{\mathrm{Kep}}(r_{\mathrm{sp}})$, with a typical value of 0.5 seen in numerical simulations.

The high accretion rates of primordial protostars make it likely that the disk will build itself up to a mass that is significant compared to the stellar mass. At this point the disk becomes susceptible to global ($m = 1$ mode) gravitational instabilities (Adams, Ruden, & Shu 1989; Shu *et al.* 1990), which are expected to be efficient at driving inflow to the star, thus regulating the disk mass. Thus TM04 assumed a fixed ratio of disk to stellar mass, $f_d = 1/3$.

Accretion through the disk may also be driven by local instabilities, the effects of which can be approximated by simple Shakura-Sunyaev α_{ss}-disk models. Two dimensional simulations of clumpy, self-gravitating disks show self-regulation with $\alpha_{\mathrm{ss}} \simeq (\Omega t_{\mathrm{th}})^{-1}$ up to a maximum value $\alpha_{\mathrm{ss}} \simeq 0.3$ (Gammie 2001), where Ω is the orbital angular velocity, $t_{\mathrm{th}} \equiv \Sigma k T_{\mathrm{c,d}}/(\sigma T_{\mathrm{eff,d}}^4)$ is the thermal timescale, Σ is the surface density, $T_{\mathrm{c,d}}$ is the disk's central (midplane) temperature, and $T_{\mathrm{eff,d}}$ the effective photospheric temperature at the disk's surface.

Gammie (2001) found that fragmentation occurs when $\Omega t_{\mathrm{th}} \lesssim 3$. This condition has the best chance of being satisfied in the outermost parts of the disk that are still optically thick. However, Tan & Blackman (2004, hereafter TB04) considered the gravitational stability of constant $\alpha_{\mathrm{ss}} = 0.3$ disks fed at accretion rates given by eq. 2.2 and found that the optically thick parts of the disk remained Toomre stable ($Q > 1$) during all stages of the growth of the protostar. Note that the cooling due to dissociation of H_2 and ionization of H was included in these disk models.

We therefore expect that during the early stages of typical Pop III.1 star formation, the accretion disk will grow in mass and mass surface density to a point at which gravitational instabilities, both global and local, act to mediate accretion to the star. The accretion rates that can be maintained by these mechanisms are larger than the infall rates of eq. 2.2, and so the disk does not fragment.

We note that if fragmentation does occur and leads to formation of relatively low-mass secondary protostars in the disk, then one possible outcome is the migration of these objects in the disk until they eventually merge with the primary protostar. The end result of such a scenario would not be significantly different from the case of no fragmentation. Another possibility is that a secondary fragment grows preferentially from the circumbinary disk leading to the formation of a massive twin binary system (Krumholz & Thompson 2007). If both stars are massive, this star formation scenario would be qualitatively similar to the single star case in terms of the effect of radiative feedback limiting accretion. A massive binary system would mean that the accreting gas needs to lose less angular momentum and binary-excited spiral density waves provide an additional, efficient means to transfer angular momentum, compared to the single star case. For close binaries, new stellar evolution channels would be available involving mass transfer and merger, with possible implications for the production of rapidly rotating pre-supernova progenitors and thus perhaps gamma-ray bursts.

Fragmentation will only be significant for the IMF if it occurs vigorously and leads to a cluster of lower mass stars instead of a massive single or binary system. Clark, Glover, & Klessen (2008) claimed such an outcome from the results of their smooth particle hydrodynamical simulation of the collapse of a primordial minihalo. They allowed dense, gravitationally unstable gas to be replaced by sink particles. They found a cluster of 20 or so protostars formed. As discussed by Clark *et al.* (2008) (see also Glover *et al.*, these proceedings), there are a number of caveats associated with this result. The initial conditions (a sphere of radius 0.17 pc with an uniform particle density of $5 \times 10^5 \text{ cm}^{-3}$, and ratios of rotational and turbulent energy to gravitational of 2% and 10%, respectively) were not derived from ab initio simulations of cosmological structure formation. In particular, cosmologically-formed minihalos evolve towards structures that have very steep density gradients, centered about a single density peak. This is likely to allow the first, central protostar to initiate its formation long before other fluctuations have a chance to develop. The development of a massive central object will create tidal forces in the surrounding gas that will make it more difficult for gravitational instabilities to develop. Furthermore, the surrounding gas is infalling on about a local free fall time, so density perturbations have few local dynamical timescales in which to grow. Another caveat with the Clark *et al.* fragmentation results is the use of a simple tabulated equation of state, in which gas can respond instantaneously to impulses that induce cooling. This, and the form of the equation of state used, lead to near isothermal conditions in the fragmenting region.

4. Magnetic Fields and Hydromagnetic Outflows

TB04 considered the growth of magnetic fields in the accretion disk of Pop III.1 protostars. They estimated minimum seed field strengths $\sim 10^{-16} G$. Xu *et al.* (2008) have recently reported field strengths of up to $10^{-9} G$ generated by the Biermann battery mechanism in their simulations of minihalo formation. Such seed fields are expected to be amplified by turbulence in the disk, attaining equipartition strengths by the time the protostar has a mass of a few solar masses or so. If the turbulence generates large scale helicity, as in the model of Blackman & Field (2002), then this can lead to the creation of dynamically-strong fields that are ordered on scales large compared to the disk. Such fields, coupled to the rotating accretion disk, are expected to drive hydromagnetic outflows, such as disk winds (Blandford & Payne 1982).

TB04 then considered the effect of such outflows on the accretion of gas from the minihalo, following the analysis of Matzner & McKee (2000). The force distribution of centrifugally-launched hydromagnetic outflows is collimated along the rotation axes, but includes a significant wider-angled component. Using the sector approximation, TB04 found the angle from the rotation/outflow axis at which the outflow had enough force to eject the infalling minihalo gas. This angle increased as the protostellar evolution progressed, especially as the star contracted to the main sequence, leading to a deeper potential near the stellar surface and thus larger wind velocities. The star formation efficiency due to protostellar outflow winds remains near unity until $m_* \simeq 100 \, M_\odot$, and then gradually decreases to values of 0.3 to 0.7 by the time $m_* \simeq 300 \, M_\odot$, depending on the equatorial flattening of surrounding gas distribution. Comparing these efficiencies to those from ionizing feedback (§5), we conclude that the latter is more important at determining the Pop III.1 IMF (see also Tan & McKee 2008).

5. How Accretion and Feedback Set the IMF

MT08 modeled the interaction of ionizing feedback on the accretion flow to a Pop III.1 protostar. In the absence of WIMP annihilation heating, the protostar contracts to the main sequence by the time $m_* \simeq 100 \, M_\odot$, and from there continues to accrete to higher masses. At the same time, the ionizing luminosity increases, leading to ionization of the infalling envelope above and below the plane of the accretion disk. Once the H II region has expanded beyond the gravitational escape radius for ionized gas from the protostar, pressure forces begin to act to reverse the infall. In the fiducial case, by the stage when $m_* \simeq 100\,M_\odot$ we expect infall to have been stopped from most directions in the minihalo. Only those regions shadowed from direct ionizing flux from the protostar by the accretion disk are expected to remain neutral and be able to accrete.

In these circumstances the protostar starts to drive an ionized wind from its disk (Hollenbach *et al.* 1994). Ionization from the protostar creates an ionized atmosphere above the neutral accretion disk, which then scatters some ionizing photons down on to the shielded region of the outer disk, beyond r_g. An ionized outflow is driven from these regions at a rate

$$\dot{m}_{\rm evap} \simeq 4.1 \times 10^{-5} S_{49}^{1/2} T_{i,4}^{0.4} m_{*d,2}^{1/2} \quad M_\odot\ {\rm yr}^{-1}, \tag{5.1}$$

where S_{49} is the H-ionizing photon luminosity in units of 10^{49} photons s^{-1} and $Ti,4$ is the ionized gas temperature in units of 10^4 K.

MT08 used the condition $\dot{m}_{\rm evap} > \dot{m}_*$ for determining the final mass of the protostar. From numerical models they found it is about $140\,M_\odot$ in the fiducial case they considered, and Table 1 summarizes other cases. MT08 also made an analytic estimate, assuming the H-ionizing photon luminosity is mostly due to the main sequence luminosity of the star:

$$S \simeq 7.9 \times 10^{49}\ \phi_S m_{*,2}^{1.5} \quad {\rm ph\ s}^{-1}, \tag{5.2}$$

which for $\phi_S = 1$ is a fit to Schaerer's (2002) results, accurate to within about 5% for $60\,M_\odot \lesssim m_* \lesssim 300\,M_\odot$. Then the photoevaporation rate becomes

$$\dot{m}_{\rm evap} = 1.70 \times 10^{-4} \phi_S^{1/2} (1 + f_d)^{1/2} \left(\frac{T_{i,4}}{2.5}\right)^{0.4} m_{*,2}^{5/4}\ M_\odot\ {\rm yr}^{-1}. \tag{5.3}$$

The accretion rate onto the star-disk system is given by equation (2.2). Equating this with equation (5.3), we find that the resulting maximum stellar mass is

$$\text{Max } m_{*f,2} = 6.3\ \frac{\epsilon_{*d}^{28/47} \bar{\epsilon}_{*d}^{-12/47} K'^{60/47}}{\phi_S^{14/47} (1 + f_d)^{26/47}} \left(\frac{2.5}{T_{i,4}}\right)^{0.24} \rightarrow 1.45, \tag{5.4}$$

where the $\rightarrow$ assumes fiducial values $\epsilon_{*d} = 0.2$, $\bar{\epsilon}_{*d} = 0.25$, $K' = 1$, $\phi_S = 1$, $f_d = 1/3$, and $T_{i,4} = 2.5$ (see MT08 for details; note also here in eq. 5.4 we have corrected a sign error in the index for ϕ_S). This analytic estimate therefore also suggests that for the fiducial case ($K' = 1$) the mass of a Pop III.1 star should be $\simeq 140\,M_\odot$.

The uncertainties in these mass estimates include: (1) the assumption that the gas distribution far from the star is approximately spherical — in reality it is likely to be flattened towards the equatorial plane, thus increasing the fraction of gas that is shadowed by the disk and raising the final protostellar mass; (2) uncertainties in the disk photoevaporation mass loss rate due to corrections to the Hollenbach *et al.* (1994) rate from the flow starting inside r_g and from radiation pressure corrections; (3) uncertainties in the H II region breakout mass due to hydrodynamic instabilities and 3D geometry effects; (4) uncertainties in the accretion rate at late times, where self-similarity may break down (Bromm & Loeb 2004); (5) the simplified condition, $\dot{m}_{\rm evap} > \dot{m}_{*d}$, used to mark the end of accretion; (6) the possible effect of protostellar outflows (discussed above); (7) the neglect of WIMP annihilation heating (discussed above) and (8) the effect of rotation on protostellar models, which will lead to cooler equatorial surface temperatures and thus a reduced ionizing flux in the direction of the disk.

Table 1. Mass Scales of Population III.1 Protostellar Feedback

K'	f_{Kep}	$T_{i,4}$	$m_{*,pb}$ $(M_\odot)$[1]	$m_{*,eb}$ $(M_\odot)$[2]	$m_{*,evap}$ $(M_\odot)$[3]
1	0.5	2.5	45.3	50.4	137[4]
1	0.75	2.5	37	41	137
1	0.25	2.5	68	81	143
1	0.125	2.5	106	170	173
1	0.0626	2.5	182	330[5]	256
1	0.5	5.0	35	38	120
1	0.25	5.0	53.0	61	125
0.5	0.5	2.5	23.0	24.5	57
2.0	0.5	2.5	85	87	321

Notes:
[1] Mass scale of HII region polar breakout.
[2] Mass scale of HII region near-equatorial breakout.
[3] Mass scale of disk photoevaporation limited accretion.
[4] Fiducial model.
[5] This mass is greater than $m_{*,evap}$ in this case because it is calculated without allowing for a reduction in $\dot{m}_*$ during the evolution due to polar HII region breakout (see MT08).

Here we discuss briefly the last of these effects. Using the results of Ekström *et al.* (2008) and Georgi *et al.* (these proceedings), we estimate that for a zero age main sequence protostar with $\Omega/\Omega_{crit} = 0.99$ (i.e. rotating very close to break-up), at an angle 80° from the pole (i.e. the direction relevant for the accretion disks modeled by MT08) the surface temperature is reduced by a factor of 0.7. For $m_* = 140\,M_\odot$ this would cause $T_{eff,*}$ to be reduced from 1.0×10^5 K to 7×10^4 K causing a reduction in the ionizing flux (and thus also ϕ_S) by a factor of about 3. From eq. 5.4 we see that the mass of Pop III.1 star formation would be increased by about a factor of 1.4, to $200\,M_\odot$ in the fiducial case.

6. Evolution of the Pop III.1 IMF

As the universe evolves and forms more and more structure, regions of Pop III.1 star formation will become ever rarer. Indeed, because the effects of radiation from previous stellar generations can propagate relatively freely compared to the spreading and mixing of their metals in supernovae, most metal-free star formation may be via Pop III.2 (Greif & Bromm 2006). Nevertheless, understanding Pop III.1 star formation is necessary as it establishes the initial conditions of what follows.

O'Shea & Norman (2007) studied the properties of Pop III.1 pre-stellar cores as a function of redshift. They found that cores at higher redshift are hotter in their outer regions, have higher free electron fractions and so form larger amounts of H_2 (via H^-), although these are always small fractions of the total mass. As the centers of the cores contract above the critical density of 10^4 cm^{-3}, those with higher H_2 fractions are able to cool more effectively and thus maintain lower temperatures to the point of protostar formation. The protostar thus accretes from lower-temperature gas and the accretion rates, proportional to $c_s^3 \propto T^{3/2}$, are smaller. Measuring infall rates at the time of protostar formation at the scale of $M = 100\,M_\odot$, O'Shea & Norman find accretion rates of $\sim 10^{-4}\,M_\odot$ yr^{-1} at $z = 30$, rising to $\sim 2 \times 10^{-2}\,M_\odot$ yr^{-1} at $z = 20$. If Hunter's (1977) solution applies, the mass accretion rates to the protostar will be higher by a factor of 3.7 by the time $m_{*d} = 100\,M_\odot$. These accretion rates then correspond to K'=0.37 ($z = 30$) to 4.3 ($z = 20$). A naive application of eq. 5.4 would imply a range of masses of $40\,M_\odot$ to $900\,M_\odot$. This suggests that the very first Pop III.1 stars were relatively low-mass massive stars, e.g. below the mass required for pair instability supernovae (140–260 $M_\odot$ in the models of Heger & Woosley 2002). Such stars would have had relatively little influence on their cosmological surroundings, thus allowing Pop III.1 star formation to continue to lower redshifts. It is not yet clear from simulations when Pop III.1 star formation was finally replaced by other types, since this depends on the early IMFs of Pop III.1, III.2 and II stars. This transition presumably occurred before reionization was complete.

Acknowledgements

We thank the organizers of IAU255 for a very stimulating meeting. The research of JCT is supported by NSF CAREER grant AST-0645412.

References

Abel, T., Bryan, G. L., & Norman, M. L. 2002, *Science*, 295, 93
Adams, F. C., Ruden, S. P, & Shu, F. H. 1989, *ApJ*, 347, 959
Beers, T. C. & Christlieb, N. 2005, *ARA&A*, 43, 531
Blackman, E. G. & Field, G. B. 2002, Phys. Rev. Lett., 89, 265007
Blandford R. D. & Payne D. G. 1982, MNRAS, 199, 883
Bromm, V., Coppi, P. S., & Larson, R. B. 2002, *ApJ*, 564, 23
Bromm, V. & Loeb, A. 2002, *ApJ*, 575, 111–116
Bromm, V. & Loeb, A. 2004, *New Astron.*, 9, 353
Clark, P. C., Glover, S. C. O., & Klessen, R. S. 2008, *ApJ*, 672, 757
Ekström, S., Meynet, G., Chiappini, C., Hirschi, R., & Maeder, A. 2008, A&A, in press (arXiv:0807.0573)
Freese, K., Bodelheimer, P., Spolyar, D., & Gondolo, P. 2008, arXiv: 0806.0617
Gammie, C. F. 2001, *ApJ*, 553, 174
Greif T. H. & Bromm, V. 2006, *MNRAS*, 373, 128
Heger, A. & Woosley, S. E. 2002, *ApJ*, 567, 532
Hollenbach, D., Johnstone, D., Lizano, S., & Shu, F. 1994, *ApJ*, 428, 654
Hunter, C. 1977, *ApJ*, 218, 834
Kashlinsky, A. Arendt, R., Gardner, J. P. *et al.* 2004, *ApJ*, 608, 1
Krumholz, M. R. & Thompson, T. A. 2007, *ApJ*, 661, 1034
Matzner, C. D. & McKee, C. F. 2000, *ApJ*, 545, 364
McKee, C. F. & Tan, J. C. 2008, *ApJ*, 681, 771 (MT08)
Morales, M. F. & Hewitt, J. 2004, *ApJ*, 615, 7
Natarajan, A., Tan, J. C., & O'Shea, B. W. 2008, *ApJ*, submitted (arXiv:0807.3769)
Norman, M. L., O'Shea, B. W., & Paschos, P. 2004, *ApJ*, 601, L115
Omukai, K. & Nishi, R. 1998, *ApJ*, 508, 141
O'Shea, B. W. & Norman, M. L. 2007, *ApJ*, 654, 66
Page, L., Hinshaw, G., Komatsu, E. *et al.* 2007, *ApJS*, 170, 335
Ripamonti, E., Haardt, F., Ferrara, A., & Colpi, M. *MNRAS*, 334, 401
Santos, M. R., Bromm, V., & Kamionkowski, M. 2002, *MNRAS*, 336, 1082
Schaerer, D. 2002, *A&A*, 382, 28
Schaye, J., Aguirre, A., Kim, T-S. *et al.* 2003, *ApJ*, 596, 768
Shu, F. H. 1977, *ApJ*, 214, 488
Shu, F. H., Tremaine, S., Adams, F. C., & Ruden, S. P. 1990, *ApJ*, 358, 495
Spolyar, D., Freese, K., & Gondolo, P., 2008, Physical Review Letters, 100, 051101
Tan, J. C. & Blackman, E. G. 2004, *ApJ*, 603, 401 (TB04)
Tan, J. C. & McKee, C. F. 2004, *ApJ*, 603, 383 (TM04)
Tan, J. C. & McKee, C. F. 2008, First Stars III, eds. O'Shea *et al.*, AIP Conf. Proc., 990, p47
Tegmark, M., Silk, J., Rees, M.J., Blanchard, A., Abel, T., & Palla, F. 1997, *ApJ*, 474, 1
Thompson, R. I., Eisenstein, D., Fan, X. *et al.* 2007, *ApJ*, 657, 669
Xu, H., O'Shea, B. W., Collins, D. C., *et al.* 2008, *ApJ*, submitted (arXiv:0807.2647)
Weinmann, S. M. & Lilly, S. J. 2005, *ApJ*, 624, 526

Low-Metallicity Star Formation:
From the First Stars to Dwarf Galaxies
Proceedings IAU Symposium No. 255, 2008
L.K. Hunt, S. Madden & R. Schneider, eds.

The formation of the first galaxies and the transition to low-mass star formation

T. H. Greif[1,2]**, D. R. G. Schleicher**[1]**, J. L. Johnson**[2]**, A.-K. Jappsen**[3]**,
R. S. Klessen**[1]**, P. C. Clark**[1]**, S. C. O. Glover**[1,4]**,
A. Stacy**[2] **and V. Bromm**[2]

[1]Institut für theoretische Astrophysik, Albert-Ueberle Strasse 2, 69120 Heidelberg, Germany

[2]Department of Astronomy, University of Texas, Austin, TX 78712, USA

[3]School of Physics and Astronomy, Cardiff University, Queens Buildings, The Parade, Cardiff CF24 3AA, UK

[4]Astrophysikalisches Institut Potsdam, An der Sternwarte 16, 14482 Potsdam, Germany

Abstract. The formation of the first galaxies at redshifts $z \sim 10 - 15$ signaled the transition from the simple initial state of the universe to one of ever increasing complexity. We here review recent progress in understanding their assembly process with numerical simulations, starting with cosmological initial conditions and modelling the detailed physics of star formation. In this context we emphasize the importance and influence of selecting appropriate initial conditions for the star formation process. We revisit the notion of a critical metallicity resulting in the transition from primordial to present-day initial mass functions and highlight its dependence on additional cooling mechanisms and the exact initial conditions. We also review recent work on the ability of dust cooling to provide the transition to present-day low-mass star formation. In particular, we highlight the extreme conditions under which this transition mechanism occurs, with violent fragmentation in dense gas resulting in tightly packed clusters.

Keywords. cosmology: theory, galaxies: formation, galaxies: high-redshift, ISM: HII regions, galaxies: intergalactic medium, stars: formation, ISM: supernova remnants

1. Introduction

One of the key goals in modern cosmology is to study the assembly process of the first galaxies, and understand how the first stars and stellar systems formed at the end of the cosmic dark ages, a few hundred million years after the Big Bang. With the formation of the first stars, the so-called Population III (Pop III), the universe was rapidly transformed into an increasingly complex, hierarchical system, due to the energy and heavy elements they released into the intergalactic medium (IGM; for recent reviews, see Barkana & Loeb 2001; Miralda-Escudé 2003; Bromm & Larson 2004; Ciardi & Ferrara 2005; Glover 2005). Currently, we can directly probe the state of the universe roughly a million years after the Big Bang by detecting the anisotropies in the cosmic microwave background (CMB), thus providing us with the initial conditions for subsequent structure formation. Complementary to the CMB observations, we can probe cosmic history all the way from the present-day universe to roughly a billion years after the Big Bang, using the best available ground- and space-based telescopes. In between lies the remaining frontier, and the first galaxies are the sign-posts of this early, formative epoch.

There are a number of reasons why addressing the formation of the first galaxies and understanding second-generation star formation is important. First, a rigorous connection between well-established structure formation models at high redshift and the properties of present-day galaxies is still missing. An understanding of how the first galaxies formed could be a crucial step towards undertanding the formation of more massive systems. Second, the initial burst of Pop III star formation may have been rather brief due to the strong negative feedback effects that likely acted to self-limit this formation mode (Madau *et al.* 2001; Ricotti & Ostriker 2004; Yoshida *et al.* 2004; Greif & Bromm 2006). Second-generation star formation, therefore, might well have been cosmologically dominant compared to Pop III stars. Despite their importance for cosmic evolution, e.g., by possibly constituting the majority of sources for the initial stages of reionization at $z > 10$, we currently do not know the properties, and most importantly the typical mass scale, of the second-generation stars that formed in the wake of the very first stars. Finally, a subset of second-generation stars, those with masses below $\simeq 1\ M_\odot$, would have survived to the present day. Surveys of extremely metal-poor Galactic halo stars therefore provide an indirect window into the Pop III era by scrutinizing their chemical abundance patterns, which reflect the enrichment from a single, or at most a small multiple of, Pop III supernovae (SNe; Christlieb *et al.* 2002; Beers & Christlieb 2005; Frebel *et al.* 2005). Stellar archaeology thus provides unique empirical constraints for numerical simulations, from which one can derive theoretical abundance patterns to be compared with the data.

Focusing on numerical simulations as the key driver of structure formation theory, the best strategy is to start with cosmological initial conditions, follow the evolution up to the formation of a small number ($N < 10$) of Pop III stars, and trace the ensuing expansion of SN blast waves together with the dispersal and mixing of the first heavy elements, towards the formation of second-generation stars out of enriched material (Greif *et al.* 2007; Wise & Abel 2007a). In this sense some of the most pressing questions are: How does radiative and mechanical feedback by the very first stars in minihalos affect the formation of the first galaxies? How and when does metal enrichment govern the transition to low-mass star formation? Is there a critical metallicity at which this transition occurs? How does turbulence affect the chemical mixing and fragmentation of the gas? These questions have been addressed with detailed numerical simulations as well as analytic arguments over the last few years, and we here review some of the most recent work. For consistency, all quoted distances are physical, unless noted otherwise.

2. Feedback by Population III.1 Stars in Minihalos

Feedback by the very first stars in minihalos plays an important role for the subsequent build-up of the first galaxies. Among the most prominent mechanisms are ionizing and molecule-dissociating radiation emitted by massive Pop III.1 stars, as well as the mechanical and chemical feedback exerted by the first SNe. In the next few sections, we briefly discuss these mechanisms in turn.

2.1. *Radiative Feedback*

Star formation in primordial gas is believed to produce very massive stars. During their brief lifetimes of $\simeq$2–3 Myr, they produce $\sim 4 \times 10^4$ ionizing photons per stellar baryon and thus have a significant impact on their environment (Bromm *et al.* 2001b; Schaerer 2002). Based on the large optical depth measured by the *Wilkinson Microwave Anisotropy Probe* (*WMAP*) after one year of operation, Wyithe & Loeb (2003) suggested that the universe was reionized by massive metal-free stars. Even though the reionization depth according to *WMAP* 5 decreased significantly (Komatsu *et al.* 2008; Nolta *et al.* 2008), recent reionization studies indicate that massive stars are still required (Schleicher *et al.* 2008a). Considering different reionization scenarios with and without additional physics like primordial magnetic fields, they showed that stellar populations according to a Scalo-type initial mass function (IMF; Scalo 1998) are ruled out within 3σ, unless very high star formation efficiencies of order 10% are adopted. On the contrary, populations of very massive stars or mixed populations can easily provide the required optical depth.

Apart from their ionizing flux, Pop III.1 stars also emit a strong flux of H_2-dissociating Lyman-Werner (LW) radiation (Bromm *et al.* 2001b; Schaerer 2002). Thus, the radiation from the first stars dramatically influences their surroundings, heating and ionizing the gas within a few kiloparsec around the progenitor, and destroying the H_2 and HD molecules locally within somewhat larger regions (Ferrara 1998; Kitayama *et al.* 2004; Whalen *et al.* 2004; Alvarez *et al.* 2006; Abel *et al.* 2007; Johnson *et al.* 2007). Additionally, the LW radiation emitted by the first stars could propagate across cosmological distances, allowing the build-up of a pervasive LW background radiation field (Haiman *et al.* 2000). The impact of radiation from the first stars on their local surroundings has important implications for the numbers and types of Pop III stars that form. The photoheating of gas in the minihalos hosting Pop III.1 stars drives strong outflows, lowering the density of the primordial gas and delaying subsequent star formation by up to 100 Myr (Whalen *et al.* 2004; Johnson *et al.* 2007; Yoshida *et al.* 2007). Furthermore, neighboring minihalos may be photoevaporated, delaying star formation in such systems as well (Shapiro *et al.* 2004; Susa & Umemura 2006; Ahn & Shapiro 2007; Greif *et al.* 2007; Whalen *et al.* 2008a). The photodissociation of molecules by LW photons emitted from local star-forming regions will, in general, act to delay star formation by destroying the main coolants that allow the gas to collapse and form stars.

The photoionization of primordial gas, however, can ultimately lead to the production of copious amounts of molecules within the relic H II regions surrounding the remnants of Pop III.1 stars (Figure 1; see also Ricotti *et al.* 2001; Oh & Haiman 2002; Nagakura & Omukai 2005; Johnson & Bromm 2007). Recent simulations tracking the formation of, and radiative feedback from, individual Pop III.1 stars in the early stages of the assembly of the first galaxies have demonstrated that the accumulation of relic H II regions has two important effects. First, the HD abundance that develops in relic H II regions allows the primordial gas to re-collapse and cool to the temperature of the CMB, possibly leading to the formation of Pop III.2 stars in these regions (Johnson *et al.* 2007; Yoshida *et al.* 2007; Greif *et al.* 2008b). Second, the molecule abundance in relic H II regions, along with their increasing volume-filling fraction, leads to a large optical depth to LW photons over physical distances of the order of several kiloparsecs. The development of a high optical depth to LW photons over such short length-scales suggests that the optical depth to LW photons over cosmological scales may be very high, acting to suppress the build-up of a background LW radiation field, and mitigating negative feedback on star formation.

Even absent a large optical depth to LW photons, Pop III.1 stars in minihalos may readily form at $z > 15$. While star formation in more massive systems may proceed relatively unimpeded, through atomic line cooling, during the earliest epochs of star formation these atomic-cooling halos are rare compared to the minihalos which host individual Pop III stars. Although the process of star formation in atomic cooling halos is not well understood, for a broad range of models the dominant contribution to the LW background is from Pop III.1 stars formed in minihalos at $z \geqslant 15-20$. Therefore, at these redshifts the LW background radiation may be largely self-regulated, with Pop III.1 stars producing the very radiation which, in turn, suppresses their formation. Johnson *et al.* (2008) argue that there is a critical value for the LW flux, $J_{\mathrm{LW,crit}} \sim 0.04$, at which Pop III.1 star formation occurs self-consistently, with the implication that the Pop III.1 star formation rate in minihalos at $z > 15$ is decreased by only a factor of a few. Simulations of the formation of the first galaxies at $z \geqslant 10$ which take into account the effect of a

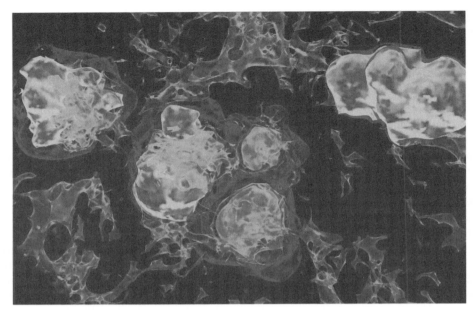

Figure 1. The chemical interplay in relic H II regions. While all molecules are destroyed in and around active H II regions, the high residual electron fraction in relic H II regions catalyzes the formation of an abundance of H_2 and HD molecules. The light and dark shades of blue denote regions with a free electron fraction of 5×10^{-3} and 5×10^{-4}, respectively, while the shades of green denote regions with an H_2 fraction of 10^{-4}, 10^{-5}, and 3×10^{-6}, in order of decreasing brightness. The regions with the highest molecule abundances lie within relic H II regions, which thus play an important role for subsequent star formation, allowing molecules to become shielded from photodissociating radiation and altering the cooling properties of the primordial gas (see Johnson *et al.* 2007).

LW background at $J_{\rm LW,crit}$ show that Pop III.1 star formation takes place before the galaxy is fully assembled, suggesting that the formation of the first galaxies does indeed take place after chemical enrichment by the first SN explosions (Greif *et al.* 2007; Wise & Abel 2007a; Johnson *et al.* 2008; Whalen *et al.* 2008b).

2.2. *Mechanical Feedback*

Numerical simulations have indicated that Pop III.1 stars might become as massive as 500 $M_\odot$ (Omukai & Palla 2003; Bromm & Loeb 2004; Yoshida *et al.* 2006; O'Shea & Norman 2007). After their main-sequence lifetimes of typically 2–3 Myr, stars with masses below $\simeq 100$ $M_\odot$ are thought to collapse directly to black holes without significant metal ejection, while in the range $\simeq 140 - 260$ $M_\odot$ a pair-instability supernova (PISN) disrupts the entire progenitor, with explosion energies ranging from $10^{51} - 10^{53}$ ergs, and yields of order 50% (Heger & Woosley 2002; Heger *et al.* 2003). Less massive primordial stars with a high degree of angular momentum might explode with similar energies, but as jet-like hypernovae (Umeda & Nomoto 2002; Tominaga *et al.* 2007). The significant mechanical and chemical feedback effects exerted by such explosions have been investigated with a number of detailed calculations, but these were either performed in one dimension (Salvaterra 2004; Kitayama & Yoshida 2005; Machida *et al.* 2005; Whalen *et al.* 2008b), or did not start from realistic initial conditions (Bromm *et al.* 2003; Norman *et al.* 2004). Recent work treated the full three-dimensional problem in a cosmological context at the cost of limited resolution, finding that the SN remnant propagated for a Hubble time at $z \simeq 20$ to a final mass-weighted mean shock radius of 2.5 kpc, roughly half the size of the

H II region (Greif *et al.* 2007). Due to the high explosion energy, the host halo was entirely evacuated. Additional simulations in the absence of a SN explosion were performed to investigate the effect of photoheating and the impact of the SN shock on neighboring minihalos. For the case discussed in Greif *et al.* (2007), the SN remnant exerted positive mechanical feedback on neighboring minihalos by shock-compressing their cores, while photoheating marginally delayed star formation. Although a viable theoretical possibility, secondary star formation in the dense shell via gravitational fragmentation (e.g. Machida *et al.* 2005; Mackey *et al.* 2003; Salvaterra *et al.* 2004) was not observed, primarily due to the previous photoheating by the progenitor and the rapid adiabatic expansion of the shocked gas.

2.3. *Chemical Feedback*

The dispersal of metals by the first SN explosions transformed the IGM from a simple, pure H/He gas to one with ubiquituous metal enrichment. The resulting cooling ultimately enabled the formation of the first low-mass stars – the key question is then when and where this transition occurred. As indicated in the previous section, such a transition could only occur well after the explosion, as cooling by metal lines or dust requires the gas to re-collapse to high densities. Furthermore, the distribution of metals becomes highly anisotropic, since the shocked gas expands preferentially into the voids around the host halo. Due to the high temperature and low density of the shocked gas, dark matter (DM) halos with $M_{\rm vir} \sim 10^8$ M$_\odot$ must be assembled to efficiently mix the gas. For this reason, the first galaxies likely mark the formation environments of the first low-mass stars and stellar clusters (see Section 5).

3. The First Galaxies and the Onset of Turbulence

How massive were the first galaxies, and when did they emerge? Theory predicts that DM halos containing a mass of $\sim 10^8$ M$_\odot$ and collapsing at $z \sim 10$ were the hosts for the first bona fide galaxies. These systems are special in that their associated virial temperature exceeds the threshold, $\simeq 10^4$ K, for cooling due to atomic hydrogen (Oh & Haiman 2002). These so-called 'atomic-cooling halos' did not rely on the presence of molecular hydrogen to enable cooling of the primordial gas. In addition, their potential wells were sufficiently deep to retain photoheated gas, in contrast to the shallow potential wells of minihalos (Madau *et al.* 2001; Mori *et al.* 2002; Dijkstra *et al.* 2004). These are arguably minimum requirements to set up a self-regulated process of star formation that comprises more than one generation of stars, and is embedded in a multi-phase interstellar medium. In this sense, we will term all objects with a viral temperature exceeding 10^4 K as a 'first galaxy' (see Figure 2).

An important consequences of atomic cooling is the softening of the equation of state below the virial radius, allowing a fraction of the potential energy to be converted into kinetic energy (Wise & Abel 2007b). Perturbations in the gravitational potential can then generate turbulent motions on galactic scales, which are transported to the center of the galaxy. In this context the distinction between two fundamentally different modes of accretion becomes important. Gas accreted directly from the IGM is heated to the virial temperature and comprises the sole channel of inflow until cooling in filaments becomes important. This mode is termed hot accretion, and dominates in low-mass halos at high redshift. The formation of the virial shock and the concomitant heating are visible in Figure 3, where we show the hydrogen number density and temperature of the central $\simeq 40$ kpc (comoving) around the center of a first galaxy (Greif *et al.* 2008b). This case also reveals a second mode, termed cold accretion. It becomes important as soon as filaments are massive enough to enable molecule reformation, which allows the gas to cool and flow into the nascent galaxy with high velocities. These streams create a multitude of unorganized shocks near the center of the galaxy and could trigger the gravitational collapse of individual clumps (Figure 4). In concert with metal enrichment by previous star formation in minihalos, chemical mixing might be highly efficient and could lead to the formation of the first low-mass star clusters (Clark *et al.* 2008), in extreme cases possibly even to metal-poor globular clusters (Bromm & Clarke 2002). Some of the extremely iron-deficient, but carbon and oxygen-enhanced stars observed in the halo of the Milky Way may thus have formed as early as redshift $z \simeq 10$.

4. Importance of Initial Conditions and Metal Enrichment

Related to the issue of metal enrichment, an important question is what controls the transition from a population of very massive stars to a distribution biased towards low-mass stars. In a seminal paper, Bromm *et al.* (2001a) performed simulations of the collapse of cold gas in a top-hat potential that included the metallicity-dependent effects of atomic fine-structure cooling. In the absence of molecular cooling, they found that fragmentation suggestive of a present-day IMF only set in at metallicities above a threshold value of $Z \simeq 10^{-3.5}$ $Z_\odot$. However, they noted that the neglect of molecular cooling could be significant. Omukai *et al.* (2005) argued, based on the results of their detailed one-zone models, that molecular cooling would indeed dominate the cooling over many orders of magnitude in density.

The effects of molecular cooling at densities up to $n \simeq 500$ cm^{-3} have been discussed by Jappsen *et al.* (2007a) in three-dimensional collapse simulations of warm ionized gas in minihalos for a wide range of environmental conditions. This study used a time-dependent chemical network running alongside the hydrodynamic evolution as described in Glover & Jappsen (2007). The physical motivation was to investigate whether minihalos that formed within the relic H II regions left by neighboring Pop III stars could form subsequent generations of stars themselves, or whether the elevated temperatures and fractional ionizations found in these regions suppressed star formation until larger halos formed. In this study, it was found that molecular hydrogen dominated the cooling of the gas for abundances up to at least 10^{-2} $Z_\odot$. In addition, there was no evidence for fragmentation at densities below 500 cm^{-3}. Jappsen *et al.* (2007b) showed that gas in simulations with low initial temperature, moderate initial rotation, and a top-hat DM overdensity, will readily fragment into multiple objects, regardless of metallicity, provided that enough H_2 is present to cool the gas. Rotation leads to the build-up of massive disk-like structures in these simulations, which allow smaller-scale fluctuations to grow and become gravitationally unstable. The resulting mass spectrum of fragments peaks at a few hundred solar masses, roughly corresponding to the thermal Jeans mass in the disk-like structure (see Figure 5). These results suggest that the initial conditions adopted by Bromm *et al.* (2001a) may have determined the result much more than might have been appreciated at the time. To make further progress in understanding the role that metal-line cooling plays in promoting fragmentation, it is paramount to develop a better understanding of how metals mix with pristine gas in the wake of the first galaxies.

5. Transition from Population III to Population II

The discovery of extremely metal-poor subgiant stars in the Galactic halo with masses below one solar mass (Christlieb *et al.* 2002; Beers & Christlieb 2005; Frebel *et al.* 2005) indicates that the transition from primordial, high-mass star formation to the 'normal'

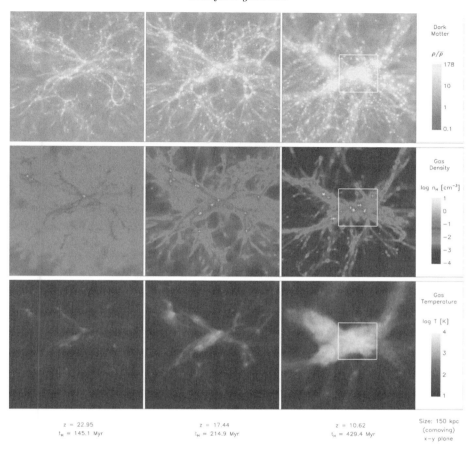

Figure 2. The DM overdensity, hydrogen number density and temperature averaged along the line of sight within the central $\simeq 150$ kpc (comoving) of a simulation depicting the assembly of a first galaxy, shown at three different output times. White crosses denote Pop III.1 star formation sites in minihalos, and the insets approximately delineate the boundary of the galaxy, further enlarged in Figures 3 and 4. *Top row:* The hierarchical merging of DM halos leads to the collapse of increasingly massive structures, with the least massive progenitors forming at the resolution limit of $\simeq 10^4$ $M_\odot$ and ultimately merging into the first galaxy with $\simeq 5 \times 10^7$ $M_\odot$. The brightest regions mark halos in virial equilibrium according to the commonly used criterion $\rho/\bar{\rho} > 178$. Although the resulting galaxy is not yet fully virialized and is still broken up into a number of sub-components, it shares a common potential well and the infalling gas is attracted towards its center of mass. *Middle row:* The gas generally follows the potential set by the DM, but pressure forces prevent collapse in halos below $\simeq 2 \times 10^4$ $M_\odot$ (cosmological Jeans criterion). Moreover, star formation only occurs in halos with virial masses above $\simeq 10^5$ $M_\odot$, as densities must become high enough for molecule formation and cooling. *Bottom row:* The virial temperature of the first star-forming minihalo gradually increases from $\simeq 10^3$ K to $\simeq 10^4$ K, at which point atomic cooling sets in (see Greif *et al.* 2008b).

mode of star formation that dominates today occurs at abundances considerably smaller than the solar value. At the extreme end, these stars have iron abundances less than 10^{-5} $Z_\odot$, and carbon or oxygen abundances that are still $\leqslant 10^{-3}$ the solar value. These stars are thus strongly iron deficient, which could be due to unusual abundance patterns produced by enrichment from Pop III stars (Umeda & Nomoto 2002), or due to mass transfer from a close binary companion (Ryan *et al.* 2005; Komiya *et al.* 2007). Recent

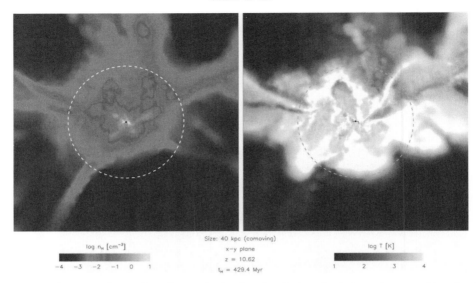

Figure 3. The central $\simeq 40$ kpc (comoving) of a simulation depicting the assembly of a first galaxy. Shown is the hydrogen number density (*left-hand side*) and temperature (*right-hand side*) in a slice centered on the galaxy. The dashed lines denote the virial radius at a distance of $\simeq 1$ kpc. Hot accretion dominates where gas is accreted directly from the IGM and shock-heated to $\simeq 10^4$ K. In contrast, cold accretion becomes important as soon as gas cools in filaments and flows towards the center of the galaxy, such as the streams coming from the left- and right-hand side. They drive a prodigious amount of turbulence and create transitory density perturbations that could in principle become Jeans-unstable. In contrast to minihalos, the initial conditions for second-generation star formation are highly complex (see Greif *et al.* 2008b).

work has shown that there are hints for an increasing binary fraction with decreasing metallicity (Lucatello *et al.* 2005). However, if metal enrichment is the key to the formation of low-mass stars, then logically there must be some critical metallicity Z_{crit} at which the formation of low-mass stars first becomes possible. However, the value of Z_{crit} is a matter of ongoing debate. As discussed in the previous sections, some models suggest that low-mass star formation becomes possible only once atomic fine-structure line cooling from carbon and oxygen becomes effective (Bromm *et al.* 2001a; Bromm & Loeb 2003; Santoro *et al.* 2006; Frebel *et al.* 2007), setting a value for Z_{crit} at around $10^{-3.5}$ $Z_\odot$. Another possibility is that low-mass star formation is a result of dust-induced fragmentation occurring at high densities, and thus at a very late stage in the protostellar collapse (Schneider *et al.* 2002; Omukai *et al.* 2005; Schneider *et al.* 2006; Tsuribe & Omukai 2006). In this model, $10^{-6} \leqslant Z_{\mathrm{crit}} \leqslant 10^{-4}$ $Z_\odot$, where much of the uncertainty in the predicted value results from uncertainties in the dust composition and the degree of gas-phase depletion (Schneider *et al.* 2002, 2006).

In recent work, Clark *et al.* (2008) focused on dust-induced fragmentation in the high-density regime, with 10^5 cm$^{-3} \leqslant \mathrm{n} \leqslant 10^{17}$ cm^{-3}. They modeled star formation in the central regions of low-mass halos at high redshift adopting an equation of state (EOS) similar to Omukai *et al.* (2005), finding that enrichment of the gas to a metallicity of only $Z = 10^{-5}$ $Z_\odot$ dramatically enhances fragmentation. A typical time evolution is illustrated in Figure 6. It shows several stages in the collapse process, spanning a time interval from shortly before the formation of the first protostar (as identified by the formation of a sink particle in the simulation) to 420 years afterwards. During the initial contraction, the cloud builds up a central core with a density of about $n = 10^{10}$ cm^{-3}. This core is supported by a combination of thermal pressure and rotation. Eventually, the core

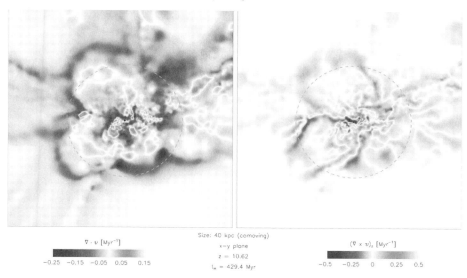

Figure 4. The central $\simeq 40$ kpc (comoving) of a simulation depicting the assembly of a first galaxy. Shown is the divergence (*left-hand side*) and z-component of the vorticity (*right-hand side*) in a slice centered on the galaxy. The dashed lines denote the virial radius at a distance of $\simeq 1$ kpc. The most pronounced feature in the left-hand panel is the virial shock, where the ratio of infall speed to local sound speed approaches unity and the gas decelerates over a comparatively small distance. In contrast, the vorticity at the virial shock is almost negligible. The high velocity gradients at the center of the galaxy indicate the formation of a multitude of shocks where the bulk radial flows of filaments are converted into turbulent motions on small scales (see Greif *et al.* 2008b).

reaches high enough densities to go into free-fall collapse, and forms a single protostar. As more high angular momentum material falls to the center, the core evolves into a disk-like structure with density inhomogeneities caused by low levels of turbulence. As it grows in mass, its density increases. When dust-induced cooling sets in, it fragments heavily into a tightly packed protostellar cluster within only a few hundred years. One can see this behavior in particle density-position plots in Figure 7. The simulation is stopped 420 years after the formation of the first stellar object (sink particle). At this point, the core has formed 177 stars. The time between the formation of the first and second protostar is roughly 23 years, which is two orders of magnitude higher than the free-fall time at the density where the sinks are formed. Note that without the inclusion of sink particles, one would only have been able to capture the formation of the first collapsing object which forms the first protostar: the formation of the accompanying cluster would have been missed entirely.

The fragmentation of low-metallicity gas in this model is the result of two key features in its thermal evolution. First, the rise in the EOS curve between densities 10^9 cm^{-3} and 10^{11} cm^{-3} causes material to loiter at this point in the gravitational contraction. A similar behavior at densities around $n = 10^3$ cm^{-3} is discussed by Bromm *et al.* (2001a). The rotationally stabilized disk-like structure, as seen in the plateau at $n \simeq 10^{10}$ cm^{-3} in Figure 7, is able to accumulate a significant amount of mass in this phase and only slowly increases in density. Second, once the density exceeds $n \simeq 10^{12}$ cm^{-3}, the sudden drop in the EOS curve lowers the critical mass for gravitational collapse by two orders of magnitude. The Jeans mass in the gas at this stage is only $M_{\rm J} = 0.01$ M$_\odot$. The disk-like structure suddenly becomes highly unstable against gravitational collapse and fragments vigorously on timescales of several hundred years. A very dense cluster of embedded

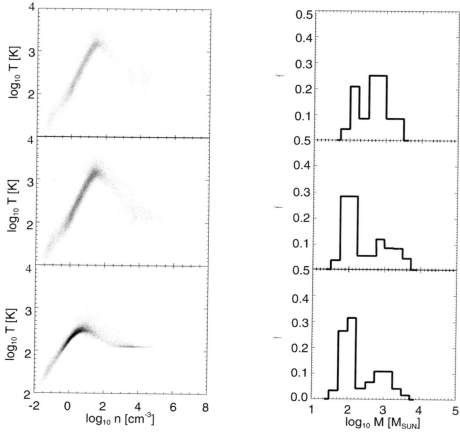

Figure 5. Gas temperature versus hydrogen density (*left-hand side*) and mass distribution of clumps (*right-hand side*) for gas collapsing in a typical minihalo, shown for the primordial case and pre-enriched to $Z = 10^{-3}$ and 10^{-1} $Z_\odot$, from top to bottom (see also Jappsen 2007b). The temperature evolution of primordial gas is very similar to the $Z = 10^{-3}$ $Z_\odot$ case, showing that metal-line cooling becomes important only for very high metallicities. The fraction of low-mass fragments increases with higher metallicity, since more gas can cool to the temperature of the CMB before becoming Jeans-unstable. However, the fragments are still very massive, suggesting that metal-line cooling might not be responsible for the transition to low-mass star formation. Instead, this transition might be governed by dust cooling occurring at higher densities. For more definitive conslusions, one must perform detailed numerical simulations that follow the collapse to higher densities in a realistic cosmological environment.

low-mass protostars builds up, and the protostars grow in mass by accretion from the available gas reservoir. The number of protostars formed by the end of the simulation is nearly two orders of magnitude larger than the initial number of Jeans masses in the cloud setup.

Because the evolutionary timescale of the system is extremely short – the free-fall time at a density of $n = 10^{13}$ cm^{-3} is of the order of 10 years – none of the protostars that have formed by the time that the simulation is stopped have yet commenced hydrogen burning. This justifies neglecting the effects of protostellar feedback in this study. Heating of the dust due to the significant accretion luminosities of the newly-formed protostars will occur (Krumholz 2006), but is unlikely to be important, as the temperature of the dust at the onset of dust-induced cooling is much higher than in a typical Galactic protostellar core ($T_{\text{dust}} \sim 100$ K or more, compared to ~ 10 K in the Galactic case). The

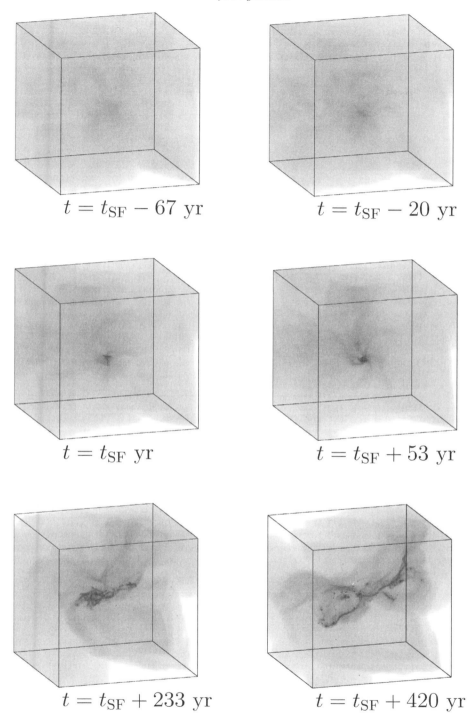

$t = t_{\mathrm{SF}} - 67$ yr

$t = t_{\mathrm{SF}} - 20$ yr

$t = t_{\mathrm{SF}}$ yr

$t = t_{\mathrm{SF}} + 53$ yr

$t = t_{\mathrm{SF}} + 233$ yr

$t = t_{\mathrm{SF}} + 420$ yr

Figure 6. Time evolution of the density distribution in the innermost 400 AU of the gas cloud shortly before and shortly after the formation of the first protostar at t_{SF}. Only gas at densities above 10^{10} cm^{-3} is plotted. The dynamical timescale at a density of $n = 10^{13}$ cm^{-3} is of the order of 10 years. Dark dots indicate the location of protostars as identified by sink particles forming at $n \geqslant 10^{17}$ cm^{-3}. Note that without usage of sink particles to identify collapsed protostellar cores one would not have been able to follow the build-up of the protostellar cluster beyond the formation of the first object. There are 177 protostars when we stop the calculation at $t = t_{\mathrm{SF}} + 420$ yr. They occupy a region roughly a hundredth of the size of the initial cloud. (see Clark *et al.* 2008).

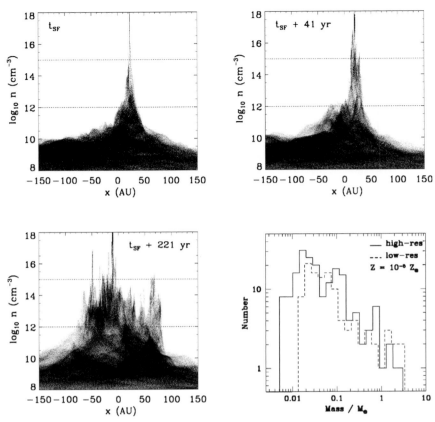

Figure 7. To illustrate the onset of the fragmentation process in the $Z = 10^{-5}$ Z$_\odot$ simulation, the graphs show the densities of the particles, plotted as a function of their position. Note that for each plot, the particle data has been centered on the region of interest. Results are plotted for three different output times, ranging from the time that the first star forms (t_{SF}) to 221 years afterwards. The densities lying between the two horizontal dashed lines denote the range over which dust cooling lowers the gas temperature. We also plot the mass function for a metallicity of $Z = 10^{-5}$ Z$_\odot$ and mass resolution 0.002 M$_\odot$ and 0.025 M$_\odot$, respectively. Note the similarity between the results of the low-resolution and high-resolution simulations. The onset of dust cooling in the $Z = 10^{-5}$ Z$_\odot$ cloud results in a stellar cluster which has a mass function similar to that for present-day stars, in that the majority of the mass resides in the low-mass objects. This contrasts with the $Z = 10^{-6}$ Z$_\odot$ and primordial clouds, in which the bulk of the cluster mass is in high-mass stars (see Clark *et al.* 2008).

rapid collapse and fragmentation of the gas also leaves no time for dynamo amplification of magnetic fields (Tan & Blackman 2004), which in any case are expected to be weak and dynamically unimportant in primordial and very low metallicity gas (Widrow 2002). However, other authors suggest that the Biermann battery effect may amplify weak initial fields such that the magneto-rotational instability can influence the further collapse of the star (Silk & Langer 2006). Simulations by Xu *et al.* (2008) show that this effect yields peak magnetic fields of 1 nG in the center of star-forming minihalos. Jets and outflows may reduce the final stellar mass by $3 - 10\%$ (Machida *et al.* 2006). In the presence of primordial fields, the magnetic pressure may even prevent star formation in minihalos and thus increase the mass scale of star-forming objects (Schleicher *et al.* 2008a,b).

The mass functions of the protostars at the end of the $Z = 10^{-5}$ Z$_\odot$ simulations (both high and low resolution cases) are shown in Figure 7. When the simulation is terminated, collapsed cores hold ~19 M$_\odot$ of gas in total. The mass function peaks somewhere below 0.1 M$_\odot$ and ranges from below 0.01 M$_\odot$ to about 5 M$_\odot$. This is not the final protostellar mass function. The continuing accretion of gas by the cluster will alter the mass function, as will mergers between the newly-formed protostars (which cannot be followed using our current sink particle implementation). Protostellar feedback in the form of winds, jets and H II regions may also play a role in determining the shape of the final stellar mass function. However, a key point to note is that the chaotic evolution of a bound system such as this cluster ensures that a wide spread of stellar masses will persist. Some stars will enjoy favorable accretion at the expense of others that will be thrown out of the system (as can be seen in Figure 6), thus having their accretion effectively terminated (see the discussions in Bonnell & Bate 2006; Bonnell *et al.* 2007). The survival of some of the low-mass stars formed in the cluster is therefore inevitable.

The forming cluster represents a very extreme analogue of the clustered star formation that we know dominates in the present-day universe (Lada & Lada 2003). A mere 420 years after the formation of the first object, the cluster has formed 177 stars (see Figure 6). These occupy a region of only around 400 AU, or 2×10^{-3} pc, in size, roughly a hundredth of the size of the initial cloud. With ~19 M$_\odot$ accreted at this stage, the stellar density is 2.25×10^9 M$_\odot$ pc^{-3}. This is about five orders of magnitude greater than the stellar density in the Trapezium cluster in Orion (Hillenbrand & Hartmann 1998) and about a thousand times greater than that in the core of 30 Doradus in the Large Magellanic Cloud (Massey & Hunter 1998). This means that dynamical encounters will be extremely important during the formation of the first star cluster. The violent environment causes stars to be thrown out of the denser regions of the cluster, slowing down their accretion. The stellar mass spectrum thus depends on both the details of the initial fragmentation process (e.g. as discussed by Clark & Bonnell 2005; Jappsen *et al.* 2005) as well as dynamical effects in the growing cluster (Bonnell *et al.* 2001, 2004). This is different to present-day star formation, where the situation is less clear-cut and the relative importance of these two processes may vary strongly from region to region (Krumholz *et al.* 2005; Bonnell & Bate 2006; Bonnell *et al.* 2007). In future work, it will be important to assess the validity of the initial conditions adopted for the present study, ideally by performing cosmological simulations that simultaneously follow the formation of the first galaxies and the metal enrichment by primordial SNe in minihalos.

6. Summary

Understanding the formation of the first galaxies marks the frontier of high-redshift structure formation. It is crucial to predict their properties in order to develop the optimal search and survey strategies for the *JWST*. Whereas *ab-initio* simulations of the very first stars can be carried out from first principles, and with virtually no free parameters, one faces a much more daunting challenge with the first galaxies. Now, the previous history of star formation has to be considered, leading to enhanced complexity in the assembly of the first galaxies. One by one, all the complex astrophysical processes that play a role in more recent galaxy formation appear back on the scene. Among them are external radiation fields, comprising UV and X-ray photons, as well as local radiative feedback that may alter the star formation process on small scales. Perhaps the most important issue, though, is metal enrichment in the wake of the first SN explosions, which fundamentally alters the cooling and fragmentation properties of the gas. Together with the onset of turbulence (Wise & Abel 2007b; Greif *et al.* 2008b), chemical mixing might be highly efficient and could lead to the formation of the first low-mass stars and stellar clusters (Clark *et al.* 2008).

In this sense a crucial question is whether the transition from Pop III to Pop II stars is governed by atomic fine-structure or dust cooling. Theoretical work has indicated that molecular hydrogen dominates over metal-line cooling at low densities (Jappsen *et al.* 2007a,b), and that fragment masses below ~ 1 $M_\odot$ can only be attained via dust cooling at high densities (Omukai *et al.* 2005; Clark *et al.* 2008). On the other hand, observational studies seem to be in favor of the fine-structure based model (Frebel *et al.* 2007), even though existing samples of extremely metal-poor stars in the Milky Way are statistically questionable (Christlieb *et al.* 2002; Beers & Christlieb 2005; Frebel *et al.* 2005). Moreover, these studies assume that their abundances are related to primordial star formation – a connection that is still debated (Lucatello *et al.* 2005; Ryan *et al.* 2005; Komiya *et al.* 2007). In light of these uncertainties, it is essential to push numerical simulation to ever lower redshifts and include additional physics in the form of radiative feedback, metal dispersal (Greif *et al.* 2008a), chemisty and cooling (e.g. Glover & Jappsen 2007) and the effects of magnetic fields (e.g. Xu *et al.* 2008). We are confident that a great deal of interesting discoveries, both theoretical and observational, await us in the rapidly growing field of early galaxy formation.

Acknowledgements

The authors would like to thank the organisers of the IAU Symposium 255 for a very enjoyable and stimulating conference. DRGS thanks the LGFG for financial support. PCC acknowledges support by the Deutsche Forschungsgemeinschaft (DFG) under grant KL 1358/5. RSK thanks for partial support from the Emmy Noether grant KL 1358/1. VB acknowledges support from NSF grant AST-0708795. DRGS, PCC, RSK and TG acknowledge subsidies from the DFG SFB 439, Galaxien im frühen Universum. DRGS and TG would like to thank the Heidelberg Graduate School of Fundamental Physics (HGSFP) for financial support. The HGSFP is funded by the excellence initiative of the German government (grant number GSC 129/1).

References

Abel, T., Wise, J. H., & Bryan, G. L., 2007, ApJ, 659, L87
Ahn, K. & Shapiro, P. R., 2007, MNRAS, 375, 881
Alvarez, M. A., Bromm, V., & Shapiro, P. R., 2006, ApJ, 639, 621
Barkana, R. & Loeb, A., 2001, Phys. Rept., 349, 125
Beers, T. C. & Christlieb, N., 2005, ARA&A, 43, 531
Bonnell, I. A. & Bate, M. R., 2006, MNRAS, 370, 488
Bonnell, I. A., Clarke, C. J., Bate, M. R., & Pringle, J. E., 2001, MNRAS, 324, 573
Bonnell, I. A., Larson, R. B., & Zinnecker, H., 2007, in Reipurth, B., Jewitt, D., Keil, K., eds,
    Protostars and Planets V The Origin of the Initial Mass Function. pp 149–164
Bonnell, I. A., Vine, S. G., & Bate, M. R., 2004, MNRAS, 349, 735
Bromm, V. & Clarke, C. J., 2002, ApJ, 566, L1
Bromm, V., Ferrara, A., Coppi, P. S., & Larson, R. B., 2001a, MNRAS, 328, 969
Bromm, V., Kudritzki, R. P., & Loeb, A., 2001b, ApJ, 552, 464
Bromm, V. & Larson, R. B., 2004, ARA&A, 42, 79
Bromm, V. & Loeb, A., 2003, Nature, 425, 812
Bromm, V. & Loeb, A., 2004, New Astronomy, 9, 353
Bromm, V., Yoshida, N., & Hernquist, L., 2003, ApJ, 596, L135
Christlieb N., *et al.*, 2002, Nature, 419, 904
Ciardi B. & Ferrara, A., 2005, Space Science Reviews, 116, 625
Clark P. C. & Bonnell, I. A., 2005, MNRAS, 361, 2
Clark, P. C., Glover, S. C. O., & Klessen, R. S., 2008, ApJ, 672, 757

Dijkstra, M., Haiman, Z., Rees, M. J., & Weinberg, D. H., 2004, ApJ, 601, 666

Ferrara, A., 1998, ApJ, 499, L17+

Frebel, A., *et al.*, 2005, Nature, 434, 871

Frebel, A., Johnson, J. L. & Bromm, V., 2007, MNRAS, 380, L40

Glover, S., 2005, Space Science Reviews, 117, 445

Glover, S. C. O. & Jappsen, A.-K., 2007, ApJ, 666, 1

Greif, T. H. & Bromm, V., 2006, MNRAS, 373, 128

Greif, T. H., Glover, S. C. O., Bromm, V., & Klessen, R. S., 2008a, MNRAS, submitted (arXiv:0808.0843)

Greif, T. H., Johnson, J. L., Bromm, V., & Klessen, R. S., 2007, ApJ, 670, 1

Greif, T. H., Johnson, J. L., Klessen, R. S., & Bromm, V., 2008b, MNRAS, 387, 1021

Haiman, Z., Abel, T., & Rees, M. J., 2000, ApJ, 534, 11

Heger, A., Fryer, C. L., Woosley, S. E., Langer, N., & Hartmann, D. H., 2003, ApJ, 591, 288

Heger, A. & Woosley, S. E., 2002, ApJ, 567, 532

Hillenbrand, L. A. & Hartmann, L. W., 1998, ApJ, 492, 540

Jappsen, A.-K., Glover, S. C. O., Klessen, R. S., & Mac Low, M.-M., 2007a, ApJ, 660, 1332

Jappsen, A.-K., Klessen, R. S., Glover, S. C. O., & Mac Low, M.-M., 2007b, ApJ, submitted (arXiv:0709.3530)

Jappsen, A.-K., Klessen, R. S., Larson, R. B., Li, Y., & Mac Low, M.-M., 2005, A&A, 435, 611

Johnson, J. L. & Bromm, V., 2007, MNRAS, 374, 1557

Johnson, J. L., Greif, T. H., & Bromm, V., 2007, ApJ, 665, 85

Johnson, J. L., Greif, T. H., & Bromm, V., 2008, MNRAS, 388, 26

Kitayama, T. & Yoshida, N., 2005, ApJ, 630, 675

Kitayama, T., Yoshida, N., Susa, H., & Umemura, M., 2004, ApJ, 613, 631

Komatsu, E., *et al.*, 2008, ApJS, submitted (arXiv:0803.0547)

Komiya, Y., Suda, T., Minaguchi, H., Shigeyama, T., Aoki, W., & Fujimoto, M. Y., 2007, ApJ, 658, 367

Krumholz M. R., 2006, ApJ, 641, L45

Krumholz, M. R., McKee, C. F., & Klein, R. I., 2005, Nature, 438, 332

Lada, C. J. & Lada, E. A., 2003, ARA&A, 41, 57

Lucatello, S., Tsangarides, S., Beers, T. C., Carretta, E., Gratton, R. G., & Ryan S. G., 2005, ApJ, 625, 825

Machida, M. N., Omukai, K., Matsumoto, T., & Inutsuka, S.-i., 2006, ApJ, 647, L1

Machida, M. N., Tomisaka, K., Nakamura, F., & Fujimoto, M. Y., 2005, ApJ, 622, 39

Mackey, J., Bromm, V., & Hernquist, L., 2003, ApJ, 586, 1

Madau, P., Ferrara, A., & Rees, M. J., 2001, ApJ, 555, 92

Massey, P. & Hunter, D. A., 1998, ApJ, 493, 180

Miralda-Escudé, J., 2003, Science, 300, 1904

Mori, M., Ferrara, A., & Madau, P., 2002, ApJ, 571, 40

Nagakura, T. & Omukai, K., 2005, MNRAS, 364, 1378

Nolta, M. R., *et al.*, 2008, ApJS, submitted (arXiv:0803.0593)

Norman, M. L., O'Shea, B. W., & Paschos, P., 2004, ApJ, 601, L115

Oh, S. P. & Haiman, Z., 2002, ApJ, 569, 558

Omukai, K. & Palla, F., 2003, ApJ, 589, 677

Omukai, K., Tsuribe, T., Schneider, R., & Ferrara, A., 2005, ApJ, 626, 627

O'Shea, B. W. & Norman, M. L., 2007, ApJ, 654, 66

Ricotti, M., Gnedin, N. Y. & Shull, J. M., 2001, ApJ, 560, 580

Ricotti, M. & Ostriker, J. P., 2004, MNRAS, 350, 539

Ryan, S. G., Aoki, W., Norris, J. E. & Beers, T. C., 2005, ApJ, 635, 349

Salvaterra, R., Ferrara, A., & Schneider, R., 2004, New Astronomy, 10, 113

Santoro, F. & Shull, J. M., 2006, ApJ, 643, 26

Scalo, J., 1998, ASP Conference Series (arXiv:astro-ph/9712317)

Schaerer, D., 2002, A&A, 382, 28

Schleicher, D. R. G., Banerjee, R., & Klessen, R. S., 2008a, Phys. Rev. D, submitted (arXiv:0807.3802)

Schleicher, D. R. G., Banerjee, R., & Klessen, R. S., 2008b, ApJ, submitted (arXiv:0808.1461)

Schneider, R., Ferrara, A., Natarajan, P., & Omukai, K., 2002, ApJ, 571, 30

Schneider, R., Omukai, K., Inoue, A. K., & Ferrara, A., 2006, MNRAS, 369, 1437

Shapiro, P. R., Iliev, I. T., & Raga, A. C., 2004, MNRAS, 348, 753

Silk, J. & Langer, M., 2006, MNRAS, 371, 444

Susa, H. & Umemura, M., 2006, ApJ, 645, L93

Tan, J. C. & Blackman, E. G., 2004, ApJ, 603, 401

Tominaga, N., Umeda, H., & Nomoto, K., 2007, ApJ, 660, 516

Tsuribe, T. & Omukai, K., 2006, ApJ, 642, L61

Umeda, H. & Nomoto, K., 2002, ApJ, 565, 385

Whalen, D., Abel, T., & Norman, M. L., 2004, ApJ, 610, 14

Whalen, D., O'Shea, B. W., Smidt, J., & Norman, M. L., 2008a, ApJ, 679, 925

Whalen, D., van Veelen, B., O'Shea, B. W., & Norman, M. L., 2008b, ApJ, 682, 49

Widrow, L. M., 2002, Reviews of Modern Physics, 74, 775

Wise, J. H. & Abel, T., 2007a, ApJ, accepted (arXiv:0710.3160)

Wise, J. H. & Abel, T., 2007b, ApJ, 665, 899

Wyithe, J. S. B. & Loeb, A., 2003, ApJ, 588, L69

Xu, H., O'Shea, B. W., Collins, D. C., Norman, M. L., Li, H., & Li, S., 2008, ApJ, submitted (arXiv:0807.2647)

Yoshida, N., Bromm, V., & Hernquist, L., 2004, ApJ, 605, 579

Yoshida, N., Oh, S. P., Kitayama, T., & Hernquist, L., 2007, ApJ, 663, 687

Yoshida, N., Omukai, K., Hernquist, L., & Abel, T., 2006, ApJ, 652, 6

Low-Metallicity Star Formation:
From the First Stars to Dwarf Galaxies
Proceedings IAU Symposium No. 255, 2008
L.K. Hunt, S. Madden & R. Schneider, eds.

Low-metallicity Star Formation: the characteristic mass and upper mass limit

Kazuyuki Omukai

National Astronomical Observatory of Japan, 2-21-1 Osawa, Mitaka, Tokyo 181-8588, Japan
email: omukai@th.nao.ac.jp

Abstract. We consider the fragmentation mass scale of low-metallicity clouds based on their thermal evolution. Then we present the protostellar evolution in the main accretion phase, and discuss the upper limit on the stellar mass by the stellar feedback.

Keywords. stars: formation, stars: Population II

1. Characterictic Stellar Mass by Fragmentation

The first stars in the universe were very massive, with a characteristic mass of $\gtrsim 100 M_\odot$. This statement has been confirmed by many numerical simulations (e.g., Bromm *et al.* 2002; Abel *et al.* 2002; Yoshida *et al.* 2008) and by analytical works (Omukai & Palla 2002; McKee & Tan 2008), and now has almost become a part of the standard theory. In the solar neighborhood, on the other hand, the initial mass function exhibits a peak around $0.1 - 1 M_\odot$. The origin of these characteristic masses can be attributed to the thermal evolution of pre-stellar clouds (e.g., Larson 2005).

In Figure 1, we show the temperature evolution of star-forming cores with different metallicities (Omukai *et al.* 2005). This is calculated by using a one-zone model assuming that the clouds collapse almost at a free-fall rate, the size of the cores is about the Jeans length, and the dust to metal ratio is the same as the local ISM value. The dotted lines indicate the constant Jeans mass.

The temperature of metal-free gas has a minimum around 10^4 cm^{-3}, where the rotational levels of H_2 reach LTE and its cooling rate saturates. Up to this point, the rapid cooling allows the temperature to decrease with increasing density, i.e., the effective ratio of specific heat $\gamma = d\log p / d\log \rho < 1$. In such a condition, clouds in shapes other than the sphere, e.g., filaments, which eventually fragment into smaller pieces, can also collapse gravitationally. However, once the temperature begins to increase with density, filamentary clouds no longer collapses due to increasing pressure support, and only spherical clouds can collapse thereafter. As a consequence, spherical clouds are formed at this moment. Since spherical clouds are hard to fragment, the typical epoch of fragmention is around this temperature minimum, and the Jeans mass there is imprinted as the characteristic fragmentation mass scale. In the case of zero-metallicity gas, this characteristic fragmentation mass is given by the Jeans mass around 10^4 cm^{-3} and is about $1000 M_\odot$.

On the other hand, the solar-metallicity gas has a temperature minimum at $\sim 10^5$ cm^{-3}. This is due to the cooling of the dust by infrared emission, which makes the temperature decrease until the thermal coupling of the gas and dust is reached. The Jeans mass at this temperature minimum is $\sim M_\odot$, which agrees with the observed characteristic mass in the solar neighborhood.

What are the processes in the slightly metal-enriched clouds? Below the metallicity $[M/H] = -6$, the temperature evolution is the same as that of zero-metallicity. With

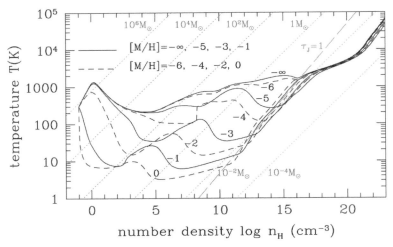

Figure 1. Temperature evolution of pre-stellar clouds with different metallicities. Those with metallicities [M/H] $= -\infty$ (Z=0), $-5, -3$, and $-1(-6, -4, -2,$ and 0) are shown by solid (dashed) lines. Only the present-day (3K) CMB is included as an external radiation field. The lines for constant Jeans mass are indicated by thin dotted lines. The positions where the central part of the clouds becomes optically thick to continuum self-absorption is indicated by the thin solid line. To the right of this line, the clouds are optically thick. (After Omukai *et al.* 2005)

metallicity of [M/H] $= -5$, a temperature dip appears in high density $\sim 10^{14}$ cm^{-3} owing to the dust cooling. With more metallicity of [M/H] $= -4$, temperature in the density range 10^{5-8} cm^{-3} begins to deviate as a result of the enhanced H$_2$ fraction due to its formation on dust grains. Finally, with metallicity of [M/H] $= -3$, this lower-density dip becomes deeper by the cooling by the fine-structure lines of C and O. Although addition of further metallicity results in even deeper temperature dips, the overall nature of the thermal evolution remains the same. The evolutionary track of the temperature has two minima. One at lower density is by line-emission cooling, i.e., H$_2$ and HD cooling at low ([M/H] $\lesssim -3.5$) metallicity, while C, O fine-structure lines at higher metallicity. The other at higher density is by the dust cooling. Existence of the two temperature minima corresponds to two fragmentation epochs. Note that the fragmentation mass scales, which are given by the Jeans mass at the temperature minimum, depend on metallicity only moderately. The line-induced fragmentation is very high, 100–$1000 M_\odot$, while the dust-induced fragmentation is low, 0.1–$1 M_\odot$. Only the dust cooling enables low-mass fragmentation. Therefore, *the dust is indispensable for low-mass star formation in a low-metallicity gas.*

How much metallicity, or correctly speaking, amount of dust, is needed to cause fragmentation and thus produce low-mass cores? This threshold value is often called *the critical metallicity* in the literature (e.g., Schneider *et al.* 2002, Bromm & Loeb 2003). To pin down this value, we have studied evolution of pre-stellar cores in the dust-cooling phase by SPH simulations (Tsuribe & Omukai 2006; 2008). The results for [M/H] $= -4.5$ and -5.5 are shown in Figure 2. Here, for the dust model, we have assumed that produced in the supernovae of the first stars as in Schneider *et al.* (2006). In this model, the dust effects tend to be more emphasized than in the standard Galactic dust model used in Figure 1, owing to the smaller size (larger area for the same amount of mass) and more refractory composition. We assumed that the cores are initially moderately elongated and have some random velocity perturbations (a-1, b-1 in Fig. 2).

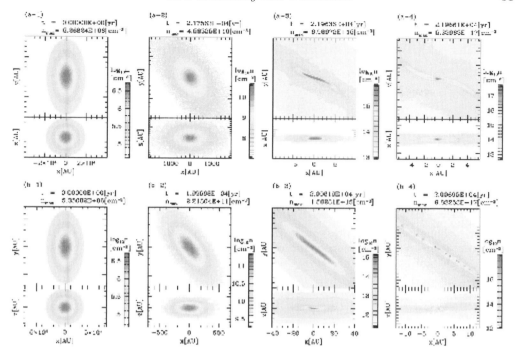

Figure 2. Evolution of cores with (a) [M/H] = −4.5 and (b) −5.5. The initial angular velocity is $0.1\sqrt{4\pi G\rho_c}$, where ρ_c is the central density. Four different stages (panels 1–4) are shown from left to right. Density distributions in the $z = 0$ plane (top), and the $y = 0$ plane (bottom) are shown in each panel. The panels 1 show the initial states. The gray scale denotes the density in the logarithmic scale. The maximum number density and elapsed time are indicated on the top of each panel. (After Tsuribe & Omukai 2008)

In the case of [M/H] = −5.5, which is shown in the bottom row (b-1..4), the core becomes more elongated (b-2,3) during the dust-cooling phase and finally fragments into many pieces (b-4). Similar fragmentation events are observed also at metallicity [M/H] = −6, but not at lower metallicity. Thus, for the first dust model, the critical metallicity is $[M/H]_{cr} \simeq −6$. If we use the standard dust model, more dust is needed to cause fragmentation. Taking this into account, we conclude that the critical metallcity is in the range $[M/H]_{cr} = −6$ to −5. *Only with slight metal enrichment, low-mass fragments can be produced as long as a sizable fraction of metals have condensed in the dust.*

However, the cases between metallicity [M/H] = −5 and -4 are exceptions. As an example, the case of [M/H] =-4.5 is shown in the top row of Figure 2 (a-1..4). In this model, no fragmentation is observed in the dust-cooling phase, and thus only massive stars are expected to form. This absence of fragmentation is caused by a sudden temperature jump by the three-body H_2 formation at 10^8 cm^{-3}, which can be seen also for [M/H] = −4 in Figure 1. This heating and resultant pressure increase makes the core spherical. Without the initial elongation, the core does not fragment subsequently.

2. Accretion Evolution and Upper Limit on the Stellar Mass

So far, we have seen the evolution during the pre-stellar collapse. Dense molecular cores are formed as a result of fragmentation, and inside of them, protostars are eventually

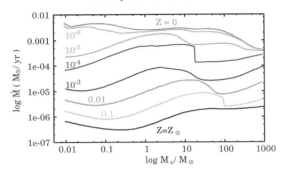

Figure 3. The fiducial protostellar accretion rates for different metallicities. (After Hosokawa & Omukai 2009)

formed. Although the initial mass of the protostars is very small $\sim 10^{-2} M_\odot$, subsequently they grow in mass by accretion of the remnant of the dense core, or the envelope, and finally reach the main sequence stars.

One of the most important quantities in the accretion phase is the accretion rate, which is given by $\dot{M} \simeq c_s^3/G \propto T^{3/2}$, where c_s is the sound speed in the pre-stellar cores. Using this relation, we can convert the temperature evolution of Figure 1 into the accretion rate as a function of the protostellar mass. We also assume that the cores are gravitationally unstable marginally, and then the instantaneous mass of the protostars M_* is given by the Jeans mass in the cores. The "fiducial" accretion rate calculated in this way is shown in Figure 3. In low metallicity gas, the accretion rates are higher reflecting the higher temperature in pre-stellar cores. We calculated the protostellar evolution under the fiducial accretion rates by using the method of Palla & Stahler (1991). In this method, the structure of a protostar is solved by solving ordinary stellar structure equations, while that of an accreting envelope is found by assuming the stationary flow. These two solutions are matched at the stellar surface by using the radiative shock condition. By increasing the protostellar mass in each time step, we can obtain an evolutionary sequence of the protostellar structure.

Figure 4 presents the evolution of the protostellar radius as a function of protostellar mass, which increases with time by accretion. The cases of metallicity [M/H] = $-\infty, -6, -4, -2$, and 0 are shown from the top to bottom. The protostars have larger radii for lower metallicity, i.e., higher accretion rate. For higher accretion rate, the accreting envelope becomes denser, and the radiative cooling is less effective. Consequently, higher entropy gas is accreted and the star has a larger entropy and thus a larger radius. Despite this difference, the evolutionary features are rather similar in all cases, and can be characterized by four phases of (i) the adiabatic accretion, (ii) swelling, (iii) Kelvin-Helmholtz contraction, and (iv) main-sequence accretion phases. Initially, the Kelvin-Helmholtz (KH) time is longer than the accretion time and the stars remain adiabatic. In this adiabatic accretion phase, the radius gradually increases. In the $Z_\odot$ case, the deuterium burning somewhat enhances the radial expansion in this phase. At some point, the luminosity increases, and then the KH time becomes comparable to the accretion time. The stars swell rapidly in the process of releasing the heat contained inside and subsequently begin the KH contraction. The central temperature continues to increse with the protostellar mass. Finally, the protostars begin the hydrogen burning and become the ordinary zero-age main-sequence stars. Note that, at lower metallicity, the protostars become more massive before the onset of hydrogen burning.

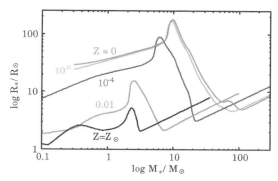

Figure 4. Evolution of the protostellar radii for different metallicities $Z = 0, 10^{-6}, 10^{-4}, 10^{-2}$ and $1Z_{\odot}$. The fiducial accretion rates are adopted. (After Hosokawa & Omukai 2009)

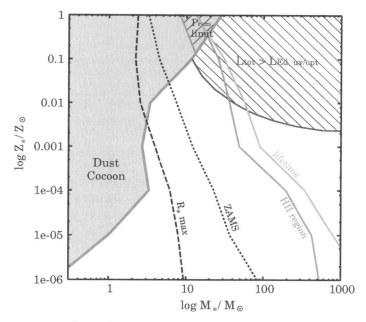

Figure 5. The upper mass limit of forming stars set by stellar feedback for different metallicities. The left gray region (with "dust cocoon") shows where the dust is optically thick. The right of it is optically thin region. The range where the accretion is halted by the radiation pressure onto the dusty envelope is indicated by the upper right hatched region ("$P_{\rm ram}$ limit" for the optically thick case, and "$L_{\rm tot} > L_{\rm Ed, uv/opt}$" for the optically thin case). The curve with "HII region" shows the limit by the expansion of the HII regions. The curve with "lifetime" (rightmost thin curve) is for the lifetime limit, where the MS lifetime exceeds the formation time. Also shown are the epochs of maximum protostellar radii in the swelling phase (dashed, with "R_* max") and the MS arrival (dotted, with "ZAMS"). (After Hosokawa & Omukai 2009)

Finally, we discuss the upper mass limit by the stellar feedback. Although the characteristic mass of dense cores is given by the fragmentation process, the cores are considred to be formed in a spectrum of mass. After the dust cooling has already caused the transition to the low-mass star formation mode, some fraction of massive cores can be formed. Even if the mass reservoir in the dense core is sufficient, however, the accretion cannot continue unlimitedly. When the central protostar becomes massive enough, its feedback onto the accreting envelope becomes very strong and stops the accretion eventually.

Here we consider the two kinds of stellar feedback processes to halt the accretion: the radiation pressure on the dusty envelope and the expansion of the HII regions. The stellar mass limit by these processes is shown in Figure 5. In low-metallicity gas, the higher accretion rate and lower opacity result in weaker feedback and then higher upper mass limit. This upper limit decreases with increasing metallicity gradually without remarkable sudden drops. For example, for metallicity above $0.01Z_\odot$, the upper limit is set by radiation pressure, which is about 10 to $40M_\odot$, assuming spherical accretion. For lower metallicity, the upper limit is set by the HII region expansion, which is as high as 40 to $500M_\odot$.

3. Summary

We have studied the thermal evolution of low-metallicity gas and discussed its fragmentation properties. The line-emission cooling affects the thermal evolution only at low densities where the Jeans mass is still high (> 10 to $100M_\odot$). The dust grains cause a sudden temperature drop at high density where $M_{\rm Jeans} \lesssim 1M_\odot$, which induces low-mass fragmentation. The critical metallicity for the dust-induced fragmentation is $[{\rm M/H}]_{\rm cr} \sim -6$ to -5. However, H_2-formation heating prohibits low-mass fragmentation in the range $[{\rm M/H}] \sim -5$ to -4.

Also, we have studied the evolution of low-metallicity protostars and discussed the upper limit on the mass by stellar feedback. In low-metallicity gas, the high temperatures in star-forming clouds result in high accretion rates. Protostellar evolution is characterized by four evolutionary stages for all metallicities: 1. the adiabatic accretion, 2. swelling, 3. KH contraction, and 4. MS accretion phases. Lower-metallicity protostars become more massive before the arrival to the main-sequence due to higher accretion. The upper limit on the stellar mass is 10 to $40M_\odot$ set by radiation pressure feedback for $Z > 0.01Z_\odot$, while it is 40 to $500M_\odot$ set by expansion of HII regions for $Z < 0.01Z_\odot$.

Acknowledgements

I thank the organizers for invitation to the beautiful town of Rapallo and a very enjoyable meeting.

References

Abel, T., Bryan, G. L., & Norman, M. L. 2002, Science, 295, 93
Bromm, V., Coppi, P. S., & Larson, R. B. 2002, ApJ, 564, 23
Bromm, V. & Loeb, A. 2003, Nature, 425, 812
Hosokawa, T. & Omukai, K., 2009, in preparation
Larson, R. B. 2005, MNRAS, 359, 211
McKee, C. & Tan, J. 2008, ApJ, 681, 771
Omukai, K. & Palla, F. 2003, ApJ, 589, 677
Omukai, K., Tsuribe, T., Schneider, R., & Ferrara, A. 2005, ApJ, 626, 627
Palla, F. & Stahler, S. W. 1991, ApJ, 375, 288

Schneider, R., Ferrara, A., Natarajan, P., & Omukai, K., 2002, ApJ, 571, 30
Schneider, R., Omukai, K., Inoue, A.-K., & Ferrara, A. 2006, MNRAS, 369, 1437
Tsuribe, T. & Omukai, K. 2006, ApJ, 642, L61
Tsuribe, T. & Omukai, K. 2008, ApJ, 676, L45
Yoshida, N., Omukai, K., & Hernquist, L. 2008, Science, 321, 669

Low-Metallicity Star Formation:
From the First Stars to Dwarf Galaxies
Proceedings IAU Symposium No. 255, 2008
L.K. Hunt, S. Madden & R. Schneider, eds.

© 2008 International Astronomical Union
doi:10.1017/S1743921308024575

Dark Stars: Dark matter in the first stars leads to a new phase of stellar evolution

Katherine Freese[1], Douglas Spolyar[2], Anthony Aguirre[2], Peter Bodenheimer[3], Paolo Gondolo[4], J. A. Sellwood[5], and Naoki Yoshida[6]

[1] Michigan Center for Theoretical Physics, University of Michigan, Ann Arbor, MI 48109, USA
email: ktfreese@umich.edu

[2] Dept. of Physics, University of California, Santa Cruz, CA 95064, USA
email: dspolyar@physics.ucsc.edu, aguirre@scipp.ucsc.edu

[3] Dept. of Astronomy, University of California, Santa Cruz, CA 95064, USA
email: peter@ucolick.org

[4] Physics Dept., University of Utah, Salt Lake City, UT 84112, USA
email: paolo@physics.utah.edu

[5] Dept. of Physics and Astronomy, Rutgers Univ., Piscataway, NJ 08854, USA
email: sellwood@physics.rutgers.edu

[6] Inst. for the Physics and Math. of the Universe, Univ. of Tokyo, Kashiwa, Chiba, Japan
email: nyoshida@a.phys.nagoya-u-ac.jp

Abstract. The first phase of stellar evolution in the history of the universe may be Dark Stars, powered by dark matter heating rather than by fusion. Weakly interacting massive particles, which are their own antiparticles, can annihilate and provide an important heat source for the first stars in the the universe. This talk presents the story of these Dark Stars. We make predictions that the first stars are very massive ($\sim 800 M_\odot$), cool (6000 K), bright ($\sim 10^6 L_\odot$), long-lived ($\sim 10^6$ years), and probable precursors to (otherwise unexplained) supermassive black holes. Later, once the initial DM fuel runs out and fusion sets in, DM annihilation can predominate again if the scattering cross section is strong enough, so that a Dark Star is born again.

Keywords. cosmology: dark matter, stars: evolution, stars: fundamental parameters

---

1. Introduction

In October 2006, as guests of the Galileo Galilei Institute in Florence, two of us began a new line of research: the effect of Dark Matter particles on the very first stars to form in the universe. We found a new phase of stellar evolution: the first stars to form in the universe may be "Dark Stars:" dark matter powered rather than fusion powered. We first reported on this work in a paper (Spolyar, Freese, & Gondolo 2008) submitted to the arxiv in April 2007 (hereafter Paper I). When we presented this work at The First Stars conference in Santa Fe soon after (Freese, Gondolo & Spolyar 2008), many questions were raised, which we have addressed in the subsequent year. In this talk I review the basic ideas as well as report on the followup work we performed over the past year.

The Dark Matter particles considered here are Weakly Interacting Massive Particles (WIMPs) (such as the Lightest Supersymmetric Particle), which are one of the major motivations for building the Large Hadron Collider at CERN that will begin taking data very soon. These particles are their own antiparticles; they annihilate among themselves in the early universe, leaving the correct relic density today to explain the dark matter in the universe. These particles will similarly annihilate wherever the DM density is high. The first stars are particularly good sites for annihilation because they form at high

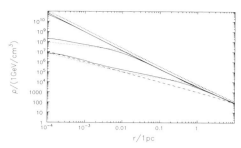

Figure 1. Adiabatically contracted DM profiles in the first protostars for an initial NFW profile (dashed line) using (a) the Blumenthal method (dotted lines) and (b) an exact calculation using Young's method (solid lines), for $M_{\rm vir} = 5 \times 10^7 M_\odot$, $c = 2$, and $z = 19$. The four sets of curves correspond to a baryonic core density of $10^4, 10^8, 10^{13}$, and 10^{16} cm^{-3}. The two different approaches to obtaining the DM densities find values that differ by less than a factor of two.

redshifts (density scales as $(1 + z)^3$) and in the high density centers of DM haloes. The first stars form at redshifts $z \sim 10 - 50$ in dark matter (DM) haloes of $10^6 M_\odot$ (for reviews see e.g. Ripamonti & Abel 2005, Barkana & Loeb 2001, Bromm & Larson 2003; see also Yoshida *et al.* 2006). One star is thought to form inside one such DM halo.

As our canonical values, we will use the standard annihilation cross section, $\langle \sigma v \rangle = 3 \times 10^{-26}$ cm^3/sec, and $m_\chi = 100$ GeV for the particle mass; but we will also consider a broader range of masses and cross-sections. Paper I found that DM annihilation provides a powerful heat source in the first stars, a source so intense that its heating overwhelms all cooling mechanisms; subsequent work has found that the heating dominates over fusion as well once it becomes important at later stages (see below). Paper I (Spolyar, Freese, & Gondolo 2008) suggested that the very first stellar objects might be "Dark Stars," a new phase of stellar evolution in which the DM – while only a negligible fraction of the star's mass – provides the power source for the star through DM annihilation.

2. Three Criteria

Paper I (Spolyar, Freese, & Gondolo 2008) outlined the three key ingredients for Dark Stars: 1) high dark matter densities, 2) the annihilation products get stuck inside the star, and 3) DM heating wins over other cooling or heating mechanisms. These same ingredients are required throughout the evolution of the dark stars, whether during the protostellar phase or during the main sequence phase.

First criterion: High Dark Matter density inside the star. To find the DM density profile, we start with an overdense region of $\sim 10^6 M_\odot$ with an NFW (Navarro, Frenk & White 1996) profile for both DM and gas, where the gas contribution is 15% of that of the DM. Originally we used adiabatic contraction $(M(r)r = {\rm constant})$ (Blumenthal *et al.* 1985) and matched onto the baryon density profiles given by Abel, Bryan & Norman (2002) and Gao *et al.* (2007) to obtain DM profiles. This method is overly simplified: it considers only circular orbits of the DM particles. Our original DM profile matched that obtained numerically in Abel, Bryan, & Norman (2002) with $\rho_\chi \propto r^{-1.9}$, for both their earliest and latest profiles; see also Natarajan, Tan, & O'Shea (2008) for a recent discussion. Subsequent to our original work, we have done an exact calculation (which includes radial orbits, Freese, Gondolo, Sellwood & Spolyar 2008) and found that our original results were remarkably accurate, to within a factor of two. Our resultant DM profiles are shown in Fig. 1. At later stages, we also consider possible further enhancements due to capture of DM into the star (discussed below).

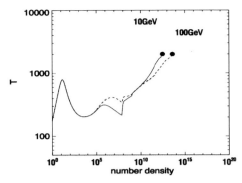

Figure 2. Temperature (in degrees K) as a function of hydrogen density (in cm^{-3}) for the first protostars, with DM annihilation included, for two different DM particle masses (10 GeV and 100 GeV). Moving to the right in the figure corresponds to moving forward in time. Once the "dots" are reached, DM annihilation wins over H2 cooling, and a Dark Star is created.

Second Criterion: Dark Matter Annihilation Products get stuck inside the star. WIMP annihilation produces energy at a rate per unit volume $Q_{\mathrm{ann}} = \langle \sigma v \rangle \rho_\chi^2 / m_\chi \simeq 1.2 \times 10^{-29}$ erg/cm^3/s $(\langle \sigma v \rangle/(3 \times 10^{-26}$ cm^3/s))(n/cm^{-3})^{1.6}(m_\chi/(100\,\mathrm{GeV}))^{-1}$. In the early stages of Pop III star formation, when the gas density is low, most of this energy is radiated away (Ripamonti, Mapelli & Ferrara 2006). However, as the gas collapses and its density increases, a substantial fraction f_Q of the annihilation energy is deposited into the gas, heating it up at a rate $f_Q Q_{\mathrm{ann}}$ per unit volume. While neutrinos escape from the cloud without depositing an appreciable amount of energy, electrons and photons can transmit energy to the core. We have computed estimates of this fraction f_Q as the core becomes more dense. Once $n \sim 10^{11}$ cm^{-3} (for 100 GeV WIMPs), e$^-$ and photons are trapped and we can take $f_Q \sim 2/3$.

Third Criterion: DM Heating is the dominant heating/cooling mechanism in the star. We find that, for WIMP mass $m_\chi = 100\,\mathrm{GeV}$ (1 GeV), a crucial transition takes place when the gas density reaches $n > 10^{13}$ cm^{-3} ($n > 10^9$ cm^{-3}). Above this density, DM heating dominates over all relevant cooling mechanisms, the most important being H$_2$ cooling (Hollenbach & McKee 1979).

Figure 2 shows evolutionary tracks of the protostar in the temperature-density phase plane with DM heating included (Yoshida *et al.* 2008), for two DM particle masses (10 GeV and 100 GeV). Moving to the right on this plot is equivalent to moving forward in time. Once the black dots are reached, DM heating dominates over cooling inside the star, and the Dark Star phase begins. The protostellar core is prevented from cooling and collapsing further. The size of the core at this point is $\sim$17 A.U. and its mass is $\sim 0.6 M_\odot$ for 100 GeV mass WIMPs. A new type of object is created, a Dark Star supported by DM annihilation rather than fusion.

3. Building up the Mass

Recently, we have found the stellar structure of the dark stars (hereafter DS) (Freese, Bodenheimer, Spolyar, & Gondolo 2008). Though they form with the properties just mentioned, they continue to accrete mass from the surrounding medium. In our paper we build up the DS mass as it grows from $\sim 1 M_\odot$ to $\sim 1000 M_\odot$. As the mass increases, the DS contracts and the DM density increases until the DM heating matches its radiated luminosity. We find polytropic solutions for dark stars in hydrostatic and thermal equilibrium. We start with a few $M_\odot$ DS and find an equilibrium solution. Then we build

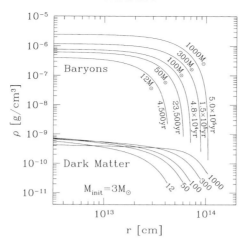

Figure 3. Evolution of a dark star (n=1.5) as mass is accreted onto the initial protostellar core of 3 $M_\odot$. The set of upper (lower) curves correspond to the baryonic (DM) density profile at different masses and times. Note that DM constitutes $< 10^{-3}$ of the mass of the DS.

up the DS by accreting $1M_\odot$ at a time with an accretion rate of $2 \times 10^{-3} M_\odot/\mathrm{yr}$, always finding equilibrium solutions. We find that initially the DS are in convective equilibrium; from $(100\text{--}400)M_\odot$ there is a transition to radiative; and heavier DS are radiative. As the DS grows, it pulls in more DM, which then annihilates. We continue this process until the DM fuel runs out at $M_{DS} \sim 800 M_\odot$ (for 100 GeV WIMPs). Figure 3 shows the stellar structure. One can see "the power of darkness:" although the DM constitutes a tiny fraction ($<10^{-3}$) of the mass of the DS, it can power the star. The reason is that WIMP annihilation is a very efficient power source: 2/3 of the initial energy of the WIMPs is converted into useful energy for the star, whereas only 1% of baryonic rest mass energy is useful to a star via fusion.

4. Results and Predictions

Our final result (Freese, Bodenheimer, Spolyar, & Gondolo 2008), is **very large first stars!** E.g., for 100 GeV WIMPs, the first stars have $M_{DS} = 800 M_\odot$. Once the DM fuel runs out inside the DS, the star contracts until it reaches 10^8 K and fusion sets in. A possible end result of stellar evolution will be **large black holes**. The Pair Instability SN (Heger & Woosley 2002) that would be produced from 140–260 $M_\odot$ stars (and whose chemical imprint is not seen) would not be as abundant. Indeed this process may help to explain the supermassive black holes that have been found at high redshift ($10^9 M_\odot$ BH at z = 6) and are, as yet, unexplained (Li *et al.* 2006; Pelupessy *et al.* 2007).

The lifetime of this new DM powered phase of stellar evolution, prior to the onset of fusion, is $\sim 10^6$ years. The stars are very bright, $\sim 10^6 L_\odot$, and relatively cool, (6000-10,000)K (as opposed to standard Pop III stars whose surface temperatures exceed $30,000K$). Reionization during this period is likely to be slowed down, as these stars can heat the surroundings but not ionize them. One can thus hope to find DS and differentiate them from standard Pop III stars; perhaps some even still exist to low redshifts.

5. Later stages: Capture

There is another possible source of DM in the first stars: capture of DM particles from the ambient medium. Whereas capture is negligible during the pre-mainsequence phase, once fusion sets in it can be important, depending on the value of the scattering cross section of DM with the gas. Two simultaneous papers (Freese, Spolyar, & Aguirre 2008, Iocco 2008) found the same basic idea: the DM luminosity from captured WIMPs can be larger than fusion for the DS. Two uncertainties exist here: the scattering cross section, and the amount of DM in the ambient medium to capture from†. DS studies including capture have assumed the maximal scattering cross sections allowed by experimental bounds and ambient DM densities that are never depleted. With these assumptions, DS evolution models with DM heating after the onset of fusion have now been studied in several papers (Iocco *et al.* 2008, Taoso *et al.* 2008, Yoon *et al.* 2008) which were posted during the conference and will be discussed in the talk by Fabio Iocco. We have been pursuing similar research with Alex Heger on DS evolution after the onset of fusion.

6. Conclusion

The line of research we began in Florence almost two years ago is reaching a very fruitful stage of development. Dark matter can play a crucial role in the first stars. The first stars to form in the universe may be Dark Stars: powered by DM heating rather than by fusion. Our work indicates that they may be very large ($850 M_\odot$ for 100 GeV mass WIMPs). The connections between particle physics and astrophysics are ever growing!

References

Spolyar, D., Freese, K., & Gondolo, P. astro-ph/0705.0521, 2008, *Phys. Rev. Lett.*, 100, 051101

K. Freese, P. Gondolo, & Spolyar, D. 2008, AIP Conf.Proc., 990, 42

Ripamonti, E. & Abel, T. astro-ph/0507130.

Barkana, R. & Loeb, A. 2001, *Phys. Rep.*, 349, 125

Bromm, V. & Larson, R. B. 2004, *ARAA*, 42, 79

Yoshida, N., Omukai, K., Hernquist, L., & Abel, T. 2006, *ApJ*, 652, 6

Navarro, J. F., Frenk, C. S., & White, S. D. M. 1996, *ApJ*, 462, 563

Blumenthal, G. R., Faber, S. M., Flores, R., & Primack, J. R. 1986, *ApJ*, 301, 27

Abel, T., Bryan, G. L., & Norman, M. L. 2002, *Science*, 295, 93

Gao, L., Abel, T., Frenk, C. S., Jenkins, A., Springel, V., & Yoshida, N. 2007, *MNRAS*. 378, 449

Natarajan, A., Tan, J., & O'Shea, B. 2008, arXiv:0807.3769 [astro-ph]

Freese, K., Gondolo, P., Sellwood, J., & Spolyar, D. 2008, arxiv:0805.3540 [astro-ph]

Ripamonti, E., Mapelli, M., & Ferrara, A. 2007, *MNRAS*, 375, 1399

Hollenbach, D. & McKee, C. F. 1979, *ApJ* Suppl., 41, 555

Yoshida, N., Freese, K., Gondolo, P., & Spolyar, D. 2008, in preparation

Freese, K., Bodenheimer, P., Spolyar, D., & Gondolo, P. 2008, arxiv:0806.0617 [astro-ph]

Heger, A. & Woosley, A. 2002, *ApJ*, 567, 532

Li, Y. X. *et al.* 2007, *ApJ*, 665, 187

Pelupessy, F. I., Matteo, T. Di, & Ciardi, B. 2007, arXiv:astro-ph/0703773.

Freese, K., Spolyar, D., & Aguirre, A. 2008, arXiv:0802.1724 [astro-ph]

Iocco, F. 2008, *ApJ* Letters, 677, 1

Iocco, F., Bressan, A., Ripamonti, E., Schneider, R., Ferrara, A., & Marigo, P. 2008, arXiv:0805.4016

Taoso, M., Bertone, G., Meynet, G., & Ekstrom, S. 2008, arXiv:0806.2681 [astro-ph]

Yoon, S., Iocco, F., & Akiyama, S. 2008, arxiv:0806.2662 [astro-ph]

---

† Unlike the annihilation cross section, which is set by the relic density, scattering is to some extent a free parameter set only by bounds from direct detection experiments.

Low-Metallicity Star Formation:
From the First Stars to Dwarf Galaxies
Proceedings IAU Symposium No. 255, 2008
L.K. Hunt, S. Madden & R. Schneider, eds.

Effects of dark matter annihilation on the first stars

F. Iocco[1], A. Bressan[2,3], E. Ripamonti[4], R. Schneider[1], A. Ferrara[3], and P. Marigo[5]

[1]INAF–Oss.Astr.di Arcetri; Largo E. Fermi 5, 50125 Firenze, Italy

[2]INAF/Osservatorio Astronomico di Padova; Vicolo dell'Osservatorio 5, Padova, Italy

[3]SISSA; Via Beirut 4, Trieste, Italy

[4]Università degli Studi dell'Insubria, Dip. di Scienze Chimiche, Fisiche e Naturali;
Via Valleggio 12, Como, Italy

[5]Università degli Studi di Padova, Dip. di Astronomia; Vicolo dell'Osservatorio 3, Padova, Italy

Abstract. We study the evolution of the first stars in the universe (Population III) from the early pre–Main Sequence (MS) until the end of helium burning in the presence of WIMP dark matter annihilation inside the stellar structure. The two different mechanisms that can provide this energy source are the contemporary contraction of baryons and dark matter, and the capture of WIMPs by scattering off the gas with subsequent accumulation inside the star. We find that the first mechanism can generate an equilibrium phase, previously known as a *dark star*, which is transient and present in the very early stages of pre–MS evolution. The mechanism of scattering and capture acts later, and can support the star virtually forever, depending on environmental characteristics of the dark matter halo and on the specific WIMP model.

Keywords. cosmology: early universe, cosmology: dark matter, stars: formation

1. Introduction

Within the scenario of a ΛCDM cosmology, it has been recently recognized that if the dark matter (DM) dominant component is a weakly interacting particle (WIMP), its annihilation could play a relevant role on the formation and evolution of the first baryonic objects in our universe. Spolyar, Freese, and Gondolo (2008) noticed that during the proto–stellar phase, the cooling of the baryonic gas could be overcome by the energy deposition following the annihilation of DM concentrated in the star formation site. This is a consequence of the peculiar formation characteristics of the first stars, at the center of a minihalo whose gas cooling is dominated by the little efficient primordial chemistry; the authors suggested that this phase (called a *dark star*) could prevent the formation of the first stars, and be a new phase of stellar evolution. Iocco (2008) and Freese, Spolyar, and Aguirre (2008) noticed that, if a star does eventually form, the process of WIMP capture by scattering could be so efficient that their subsequent accumulation and annihilation inside the celestial object may provide an energy source comparable or even exceeding its nuclear luminosity. Motivated by these works, Iocco *et al.* (2008) and Freese *et al.* (2008b,c) have studied the early pre–MS phase, in which the baryonic structure is sustained by the annihilation of the DM accreted inside it by gravitational contraction, with the help of numerical codes. Both the groups find this phase is transient, although the techniques adopted are different and the details of the treatment lead to different duration estimates; they both conclude, however, that the collapse must continue at the end of this process, which for the sake of simplicity we call the Adiabatic Contraction (*AC*) phase. In Iocco *et al.* (2008) (hereafter, I08), we have also studied the pre–MS phase of these stars in presence of annihilating scattered and captured DM (*SC* phase) and followed the evolution of stellar models of different mass until the end of the helium burning, in different DM environments. We find the duration of the MS is dramatically prolonged, up to a potentially everlasting phase (depending on the choice of parameters, see later): the energy released by the annihilating DM can support the core at temperatures low enough that nuclear reactions are never ignited. Yoon, Iocco, and Akiyama (2008) and Taoso *et al.* (2008) also studied the *SC* phase, confirming our results with different codes and carrying their analysis further. These proceedings are based on the results obtained by I08, to which we address the reader for detailed referencing and more quantitative details: here we aim to a more qualitative description of the physical processes at the basis of this class of objects.

2. Adiabatic Contraction phase

Early stars are thought to form in halos of $M_h \sim 10^6$ $M_\odot$ and virial temperature T $\lesssim$ 10^4 K at redshift z $\sim$ 20. The primordial, metal free composition of the gas, the absence of strong magnetic fields and turbulence make early star formation very different from the one in the older universe. Simulations tell us that the result of such peculiar environment are massive (30–300 $M_\odot$) stars that form in the very center of the halo (accurate review and referencing in the First Stars III proceedings, 2008). The collapse of the baryonic material "pulls" also the collisional DM towards the center, thus contributing to the build–up of a central "spike" of DM; if that is indeed made of self-annihilating particles (as WIMP neutralinos are), its higher concentration in the center of the halo causes a huge enhancement of the annihilation term:

$$\frac{dL_{\rm DM}}{dV} = f c^2 \frac{\rho^2 \langle \sigma v \rangle}{m_\chi}; \qquad (2.1)$$

with ρ the local DM density, m_χ the neutralino mass –which we take to be 100GeV–, $\langle \sigma v \rangle$ the thermally–averaged annihilation rate. It is worth noting that this energy term depends on the self–annihilation rate, whose value can be quite safely established through cosmological arguments (see e.g. the latest review on particle DM physics, Bertone *et al.* 2005) as $\langle \sigma v \rangle = 3 \times 10^{-26}$ cm^3 s^{-1}. A fraction $(1 - f)$ of the whole energy is emitted in the form of neutrinos, and therefore lost by the system, whereas all the other products of annihilation can quickly thermalize inside the gas already at densities $n \sim 10^{13} \#/{\rm cm}^3$, as shown by Spolyar *et al.* (2008); we take $f \sim 2/3$, a typical value for a neutralino annihilation. In order to study this peculiar object we have modified the *Padova* Stellar Evolution code to take into account the energy released by DM annihilation. The initial models have been prepared as follows: (*i*) the "stellar" profile (namely the baryonic structure) has been obtained by "pumping" a Zero Age Main Sequence star with an artificial energy source (as much as permitted by the stability of our numerical code); for a 100$M_\odot$ star, the resulting structure is an object of radius $R_* = 1.2 \times 10^{14}$cm and effective temperature $T \sim 5 \times 10^3$ K; (*ii*) the DM profile has been obtained in the approximation of an adiabatically contracted DM profile (from an original NFW) which has been matched to the baryonic structure in the center, which dictates the gravitational potential. For details on the characteristics of the halo and on the adiabatic contraction approximation used we address the reader to I08. This adiabatically contracted DM profile represents the initial model, at the time we start our analysis: we have implemented in our code a routine which allows to follow the adiabatic contraction of the DM inside our stellar object, account for the energy released by its annihilation, and so self–consistently describe the evolution of the DM coupled to the baryons. In this way we can follow the

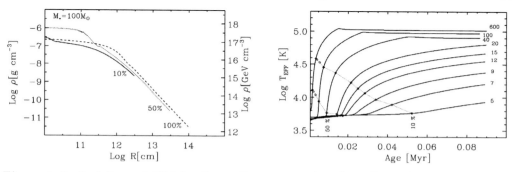

Figure 1. In the left panel: DM density profiles, truncated at the stellar radius, for a 100M$_\odot$ at different times during the *AC phase*. In the right panel: evolution of the effective temperature for all stellar models, benchmarked by the relative contribution of DM annihilation to the total luminosity of the star. See text for details.

contraction of the protostar toward the Main Sequence including all relevant energy sources (gravitational, DM annihilation, and possibly nuclear) and adopt a complete equation of state for the gas.

For the adopted AC parameters, DM annihilation is able to support the baryonic structure in the very early phases of gravitational contraction, in all our selected stellar models (5–600 M$_\odot$) The protostars contract "pulling in" more and more DM until an equilibrium is reached in very short times, $\mathcal{O}(10^2\mathrm{yr})$.

However, this phase is doomed to be transient. As the annihilation proceeds, the star contracts because of the shrink of DM luminosity caused by its consumption. But this further contraction is not able to restore the previous DM density profile, and the star finds its equilibrium at a lower luminosity. In this phase the star descends slowly along the Hayashi line. Besides consumption due to annihilation, the star continuously *loses* the external shells of DM, because of the different response of baryons and DM to the growing gravitational potential. Eventually the contraction of DM becomes not fast enough to restore an "efficient" cusp and the DM luminosity is no more able to balance the stellar energy losses. At this point the star keeps on contracting on a typical Kelvin-Helmholtz time-scale, terminating the AC phase and moving toward the main sequence. By defining the duration of this phase τ_{AC} as the the time needed from the DM annihilation to scale from 100% to 50% of the total luminosity of the object, we find typical values of order $\tau_{AC} \sim 10^3\mathrm{yr}$, ranging from $\tau_{AC} = 2.1 \times 10^3\mathrm{yr}$ for a 600$M_\odot$ star up to $\tau_{AC} = 1.8 \times 10^4$ yr for a 9$M_\odot$ star. The rate of gravitational energy release is larger in more massive stars, so it comes as no surprise that the stalling phase is shorter for higher mass stars. In Figure 1 we show the effective temperature evolution for different stellar models and the DM profile *inside* the baryonic structure for a 100M$_\odot$ star at different times during the stellar collapse. The benchmark values correspond to the fractional contribution of DM annihilation to the total luminosity.

3. Scattering and Capture phase

The other physical mechanism able to concentrate DM inside a star is the capture of WIMPs by means of elastic scattering with the gas particles that constitute the object. The captured WIMPs thermalize with the gas and eventually reach a thermally relaxed state inside the star: the density profile n_χ is

$$n_\chi(R) = n_\chi^c \exp(-R^2/r_\chi^2), \quad n_\chi^c = \frac{C_* \tau_\chi}{\pi^{3/2} r_\chi^3}, \quad r_\chi = c\left(\frac{3kT_c}{2\pi G \rho_c m_\chi}\right)^{1/2}; \quad (3.1)$$

where T_c and ρ_c are the stellar core temperature and density, respectively, and C_* the WIMP capture rate, discussed in the following. It can be seen that r_χ is in general much smaller than the star core, $r_\chi \sim 10^9$ cm for a $100M_\odot$ star. For each volume element, the DM annihilation energy released is given by Equation 2.1; however, capture and annihilation reach the equilibrium on timescales much shorter than the stellar lifetime (see Iocco 2008 and I08 for a more detailed discussion of this issue), and the amount of energy released inside the star, the "dark luminosity" L_{DM} reads $L_{DM} = f m_\chi C_*$, being dictated by the capture rate, C_*. The latter has the characteristic of a scattering term, and can therefore be cast as:

$$C_* \propto M_* v_{esc}^2 \frac{\sigma_0 \rho}{m_\chi}. \quad (3.2)$$

with σ_0 the elastic scattering cross section between WIMP and baryons, ρ the DM density *outside* the star, v_{esc} the escape velocity at the surface of the star, and the subscript $*$ referring to stellar quantities throughout this paper. We refer to I08 and references therein for a more quantitative discussion of the capture rate, but we wish here to stress on few peculiarities of the *SC* process. The energy is indeed provided by means of DM *annihilation* in the bosom of the star; however, the bottleneck of the capture/scattering mechanism is given by capture: the star cannot burn more WIMPs than it accretes. Therefore, at the equilibrium, the annihilation is dictated by the capture and shows the parameter dependence of a scattering process. Also, this process is sensitive to the WIMPs that stream through the star, and thus the halo plays the role of a "reservoir", virtually unexhaustable†. It depends on σ_0 rather than $\langle \sigma v \rangle$, and *linearly* on the *environmental* DM density ρ. However, when the object is in the *AC* phase, the *SC* mechanism is playing little role due to the very low density of the object (which makes it an inefficient "net" for capturing DM) and of the very long equilibrium timescale at these stages, much longer than the Kelvin-Helmholtz time. By contracting, the star enhances the efficiency of the *SC* process and DM annihilation contributes to the stellar luminosity.

For $\sigma_0 = 10^{-38}$ (at the level of the current upper limit for the spin–dependent elastic scattering cross section (Desai *et al.* 2004), and an environment density $\rho = 10^{12}$ GeV/cm^3 (a likely value achieved around the star, see Figure 1), the energy provided by the DM annihilation inside the star is able to support all our models before they get to the ZAMS, therefore not igniting nuclear reactions. Their locus on the HR diagram can be observed in Figure 2, compared with their ZAMS one. Lower values of ρ or σ_0 make the DM contribution smaller, and the star can contract increasingly, thus progressively enabling the ignition of nuclear reactions. Stars in the white region of the right panel of Figure 1 do never ignite nuclear reactions; stars below the solid curve, in the shaded area do. The closer to the curve, the longer their lifetimes will be, as energy contribution required by nuclear burning is lower, and their chemical evolution is slower, see I08.

---

† The DM mass-energy content of the halo is much bigger than the energy used by the star: only approximately $10M_\odot = 10^{-5}M_h$ are entirely converted into energy in 10 Gyr, at a rate 10^{39} erg/s. Only $0.1M_\odot$ of DM can support a 100 $M_\odot$ star for ten times longer than its natural lifetime, 10^6 yr.

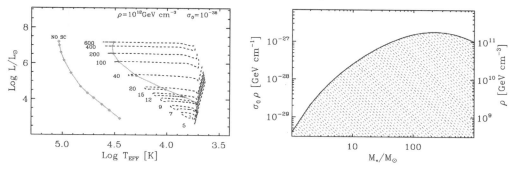

Figure 2. In the left panel: HR diagram for metal–free stars in presence of a DM environmental density $\rho = 10^{12}$ GeV, compared to their normal position in the HR; sold curve labeled *NO SC*. For all masses, the ZAMS is never reached, as stars are entirely supported by DM annihilation. In the left panel: stars in the white region, above the solid line never ignite nuclear reactions, being "frozen" in their evolution. In the shaded area they do evolve at different rates, being only partially supported by DM burning. See text for details.

4. Conclusions

We have studied the effects of WIMP dark matter annihilation from the late stages of stellar formation through the end of helium burning. Addressing the reader to I08 for quantitative details, here we wish to stress the fundamental physical difference between the *SC* mechanism, which needs a weak process to mediate the capture and thermalization of DM particles (and therefore the DM density profile inside the star), and the *AC* one, where the DM density profile is dictated by gravitational contraction only. The difference in the two physical processes dictates diverse characteristics for the two phases: the *AC* is transient, and takes place early in the pre–Main Sequence evolution of the star, when it is almost at the proto–stellar stage. The *SC* phase becomes active when the star is at the bottom of the Hayashi track or later in the pre–MS evolution, and it can dramatically extend the stellar life, up to orders of magnitude more than its standard lifespan, depending on parameters.

References

Bertone, G., Hooper, D., & Silk, J., 2005, Phys. Rep. 405, 279

Desai, S. *et al.* [Super-Kamiokande Collaboration], 2004, Phys. Rev. D, 70, 109901

"First Stars III", 2008, AIP Conf. Proc., 990, T. Abel, A. Heger, and B. W. 'O. Shea eds

Freese, K., Spolyar, D., & Aguirre, A., arXiv:0802.1724 [astro-ph]

Freese, K., Gondolo, P., Sellwood, J. A., & Spolyar, D., [Freese *et al.* (2008b)] arXiv:0805.3540 [astro-ph]

Iocco, F., 2008, ApJ, 677, L1

Iocco, F., Bressan, A., Ripamonti, E., Schneider, R., Ferrara, A., & Marigo, P., MNRAS in press, arXiv:0805.4016 [astro-ph]

Yoon, S. C., Iocco, F., & Akiyama, S., arXiv:0806.2662 [astro-ph]

Spolyar, D., Freese, K., & Gondolo, P., 2008, Phys. Rev. Lett., 100, 051101

Taoso, M., Bertone, G., Meynet, G., & Ekstrom, S., arXiv:0806.2681 [astro-ph].

Low-Metallicity Star Formation:
From the First Stars to Dwarf Galaxies
Proceedings IAU Symposium No. 255, 2008
L.K. Hunt, S. Madden & R. Schneider, eds.

© 2008 International Astronomical Union
doi:10.1017/S1743921308024599

Searching for Pop III stars and galaxies at high redshift

Daniel Schaerer[1,2]

[1] Geneva Observatory, University of Geneva, 51, Ch. des Maillettes,
CH-1290 Versoix, Switzerland

[2] Laboratoire d'Astrophysique de Toulouse-Tarbes, Université de Toulouse, CNRS, 14 Avenue
E. Belin, F-31400 Toulouse, France

Abstract. We review the expected properties of Pop III and very metal-poor starbursts and the behaviour the Lyα and He II λ1640 emission lines, which are most likely the best/easiest signatures to single out such objects. Existing claims of Pop III signatures in distant galaxies are critically examined, and the searches for He II λ1640 emission at high redshift are summarised. Finally, we briefly summarise ongoing and future deep observations at $z > 6$ aiming in particular at detecting the sources of cosmic reionisation as well as primeval/Pop III galaxies.

Keywords. galaxies: high-redshift, galaxies: stellar content, (cosmology:) early universe

1. Introduction

Finding the first stars and galaxies remains a major observational challenge in astrophysics. Searches for metal-free (Pop III) or extremely metal-poor individual stars or for ensembles of stars (clusters, proto-galaxies, populations of galaxies, etc.) are ongoing both nearby (Pop III stars in the halo of our Galaxy), and in galaxies out to the highest redshifts currently known.

Astronomers have been quite inventive in searching for Pop III or metal-poor stars, exploiting many possible direct and indirect signatures; some of these methods will be discussed below. A more detailed account on primeval galaxies, some of the physics related to these objects, an overview on searches etc. is presented in the lectures of Schaerer (2007).

Although some studies may have found signatures of Pop III stars, none of those claims is very strong (see discussion below), and my personal opinion – taking a somewhat conservative attitude – is that Pop III stars still remain to be found. However, there is good hope that this goal should be reached in the near future with current or forthcoming instrumentation, as I will sketch below.

2. Expected properties of Pop III and very metal-poor populations

Many authors have modeled the evolution of Pop III stars in the 1980s and before. Since then the interest in these stars has been revived, and new generations of models computed. Subsequently I will use predictions from the stellar evolution and non-LTE atmosphere models as well as the evolutionary synthesis models computed by Schaerer (2002, 2003). For a detailed description of the input physics and references to earlier work the reader is referred to these papers.

One of the main distinctive features of Pop III stars is their extreme compactness close to and on the zero age main sequence (ZAMS), implying much higher effective temperatures (up to ~ 100 kK) than usual. For a given mass this leads to a considerable

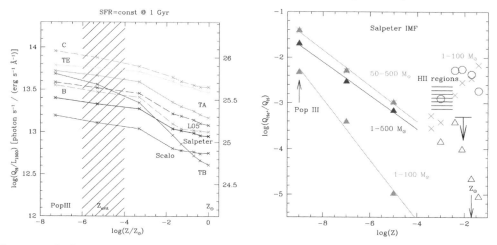

Figure 1. Left: Predicted ionising photon flux as a function of metallicity for different IMFs. The ionising output is normalised here to the UV luminosity. The shaded area labeled Z_{crit} indicates the domain where the typical stellar mass / IMF may change (cf. Schneider *et al.* 2004). **Right:** Dependence of the hardness of the ionising flux, expressed by $Q_{\mathrm{He^+}}/Q_H$, on metallicity and IMF. Figure from Schaerer (2003).

increase of the ionising flux and to a much harder ionising spectrum (cf. Tumlinson *et al.* 2001, Schaerer 2002). In addition, if the conditions in metal-poor environments favour the formation of massive stars (cf. Bromm & Larson 2004, and contributions by Abel, Bromm, Tan, and others in these proceedings), the integrated spectrum of young, zero and very low metallicity stellar populations can be quite different than that of populations at "normal" metallicities, as shown e.g. by Schaerer (2002, 2003). In particular, the strong Lyα line emission and the presence of nebular He II emission lines are probably the best/easiest features to search for very metal-poor and Pop III objects.

For example in Figure 1 (left) we show the increase of the ionising (Lyman continuum) photon flux Q_H normalised to the UV continuum luminosity with decreasing metallicity for different IMFs (Salpeter, Scalo; B, C as in Schaerer 2003, and other IMFs). For a fixed IMF the increase from solar to zero metallicity is typically a factor of 2; assuming that more extreme IMFs may be valid for Pop III, this increase can be up to a factor ~ 10 (compared to solar and Salpeter). Note that this prediction also depends on the exact star-formation (SF) history, which affects, in particular, the UV continuum output. The right panel of Fig. 1 shows the increase of the hardness of the ionising flux, expressed by the ratio of He$^+$ to H ionising flux, $Q_{\mathrm{He^+}}/Q_H$, with decreasing metallicity.

As already mentioned, the strong Lyman continuum flux at low metallicities implies a strong intrinsic Lyα emission (cf. Schaerer 2002, 2003). For example, the *maximum Lyα equivalent width* predicted for an integrated stellar population increases from $\sim$ 200-300 Å to 500-1000 Å or even larger (depending on the IMF) from solar to zero metallicity, as shown in Fig. 2. Of course it must be reminded that after its emission inside the ionised region surrounding the starburst, Lyα photons are scattered and may be absorbed, thereby altering the intrinsic value of W (see e.g. Schaerer 2007).

The increase of the hardness of the ionising radiation shown in Fig. 1 (right) translates into the increase of He II recombination lines, such as He II $\lambda1640$, shown in Fig. 2 (right). For obvious reasons their strength depends on the exact SF history, age, and in particular

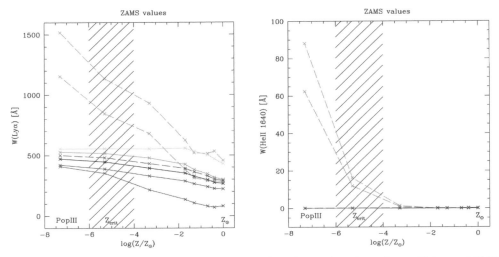

Figure 2. Left: Predicted maximum value of $W(\text{Ly}\alpha)$ as a function of metallicity and IMF (same colors as Fig. 1). **Right:** Same as left for $W(\text{He II }\lambda 1640)$. See discussion in text.

on the IMF, as shown by these figures (see also Schaerer 2003). While clearly a promising signature of very metal-poor populations as pointed out earlier†, this strong sensitivity to the presence of the most massive stars – i.e. to the IMF and age – may complicate the interpretation of He II lines, especially when dealing with non-detections (cf. below).

Once the properties of individual Pop III clusters or galaxies are known, important questions are of course how many such objects are expected and what is their distribution, e.g. in mass and luminosity? Answering these questions requires in particular a description of the transport and mixing of metals to determine the duration (locally) of Pop III SF events. Several authors have addressed this using different approaches (e.g. Scannapieco *et al.* 2003, Yoshida *et al.* 2004, Tornatore *et al.* 2007). The predicted Pop III SF rate, defined by SF in regions with $Z < Z_{\text{crit}}$, as a function of redshift from the recent computations of Tornatore *et al.* (2007) is shown in Fig. 3. Predicted number counts from Choudhury & Ferrara (2007) are shown in Fig. 5 for illustration.

3. Have we found Pop III?

It is possible that galaxies containing Pop III or very metal-poor stars have already been found in different samples. Here I will critically examine some of these results.

3.1. *Have we found Pop III in Lyman Break Galaxies?*

For example, Jimenez & Haiman (2006) have recently proposed an explanation to several apparent puzzles concerning Lyman Break Galaxies (LBGs) and related objects at $z \sim$ 3–4. They propose that their stellar populations contain $\sim$ 10–30 % of primordial stars, which would explain the observed He II $\lambda 1640$ emission in LBGs, the existence of some

---

† Caveat: He II $\lambda 1640$ emission can also of different origin (stellar cf. Sect. 3.1, AGN or shocks) and nebular He II is observed in some normal H II regions (see e.g. Thuan & Izotov 2005) or in other peculiar objects (cf. Fosbury *et al.* 2003 and discussion in Schaerer 2003b).

galaxies with very high W(Lyα), the excess of Lyman continuum flux seen in some LBGs, and the nature of Lyα blobs.

Although the idea to tackle simultaneously four problems is in principle attractive, the proposed solution is probably not unique and not the most likely one, and some of the 4 problems may not be robust observationally. First, the strength of the He II λ1640 line in the composite spectrum of LBGs is quite modest ($W(1640) = 1.3 \pm 0.3$ Å), and small enough to be explained by Wolf-Rayet stars in normal stellar populations (cf. models of Schaerer & Vacca 1998, and Brinchmann *et al.* 2008). Furthermore the observed line is broad, as observed in Wolf-Rayet stars. Second, the existence of a population of galaxies at $z \sim 4$–5 with high Lyα equivalent widths found by the LALA survey, does not seem to be confirmed by other groups or is at least very uncertain (see Sect. 3.2). Third, other explanations are proposed for extended Lyα blobs (e.g. Matsuda *et al.* 2007). Finally, the proposed explanation may also have difficulties with mixing time scales (Pan & Scalo 2007). The excess of Lyman continuum flux in some LBGs remains still puzzling (see also Shapley *et al.* 2006, Iwata *et al.* 2008)

3.2. *Have we found Pop III in Lyα emitters?*

A few years ago Malhotra & Rhoads (2002) found Lyα emitters (LAE) at $z \sim 4.5$ with an unusual Lyα equivalent width distribution from their LALA survey. They suggested that their large fraction of objects with a high W(Lyα) (> 200-300 Å) could be due to very metal-poor or Pop III objects, a very unusual IMF, or to AGN. Follow-up observations of these objects have been undertaken, also at X-rays, and several papers have already adressed these results (see discussion in Schaerer 2007). The AGN hypothesis has been rejected (Wang *et al.* 2004). Deep spectroscopy aimed at detecting other emission lines, including the He II λ1640 line indicative of a Pop III contribution (cf. Sect. 4), have been unsuccessful (Dawson *et al.* 2004), although the achieved depth may not be sufficient. If taken at face value, the origin of these high W(Lyα) remains thus unclear. However, there is some doubt on the reality of these equivalent widths measured from NB and broad-band imaging, or at least on them being so numerous even at $z = 4.5$. First of all the objects with the highest $W(\text{Lyα})$ have very large uncertainties since the continuum is faint or non-detected. Second, the determination of $W(\text{Lyα})$ from a NB and a centered broad-band filter (R-band in the case of Malhotra & Rhoads 2002) may be quite uncertain, e.g. due to unknowns in the continuum shape, the presence of a strong spectral break within the broad-band filter etc. (see Hayes & Oestlin 2006 for a quantification, and Shimasaku *et al.* 2006). Furthermore other groups have not found such high W objects (e.g. Hu *et al.* 2004, Ajiki *et al.* 2003 and compilation in Verhamme *et al.* 2008) suggesting also that this may be related to insufficient depth of the LALA photometry.

Recently Shimasaku *et al.* (2006) and Dijkstra & Wyithe (2007) have again, used $W(\text{Lyα})$ distributions to argue for a non-negligible contribution of Pop III stars in LAEs, this time in objects at $z \geqslant 5.7$. In fact the *observed* restframe values of $W(\text{Lyα})$ LAE samples at $z = 5.7$ and 6.5 obtained with SUBARU are considerably lower than those of Malhotra & Rhoads at $z = 4.5$, and only few objects show $W^{\text{rest}}_{\text{obs}}(\text{Lyα}) \gtrsim 200$ Å, the approximate limit expected for "normal" stellar populations (cf. Fig. 2). However, if the IGM transmission T_α affects half of the Lyα line, as often assumed (see also Hu *et al.* 2004), this would imply true intrinsic equivalent widths $2/T_\alpha$ times higher than the observed value, i.e. typically 4 times higher at $z = 5.7$! For this reason Shimasaku *et al.* suggest that these may be young galaxies or objects with Pop III contribution. Based on this reasoning and on modeling the Lyα luminosity function and IGM transmission, Dijkstra & Wyithe (2007) argue for the presence of ~ 4–10 % of Pop III SF in $\sim$ half of Lyα selected galaxies and in few percent of i-drop galaxies at $z \gtrsim 5.7$.

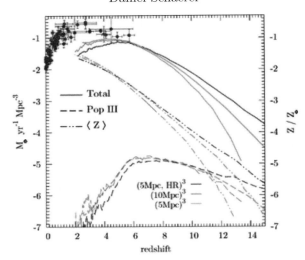

Figure 3. Predicted SFR density and average metallicity as a function of redshift showing the contribution of Pop III (defined as metallicity $Z < Z_{\mathrm{crit}}$, cf. above) and other stars. Recently Nagao *et al.* (2008) have derived $SFRD_{\mathrm{PopIII}} \lesssim 5.\,10^{-6}$ $\mathrm{M_\odot\ yr^{-1}\ Mpc^{-3}}$ at $4.0 \lesssim z \lesssim 4.6$ close to the predicted value at this redshift. Figure from Tornatore *et al.* (2007).

Although possible, this conclusion rests strongly on the assumption that the IGM transmission (due to individual or overlapping Lyα forest clouds) really cuts out a significant fraction of the Lyα line emerging from the galaxy. This is by no means clear, e.g. since it implies the need for cold gas close (and/or infalling) to the galaxies, which has not been observed yet. E.g. in the LBG cB58 studied in depth, Savaglio *et al.* (2002) find no Lyα absorption close to the galaxy (at $\Delta v \lesssim 2000$ km s^{-1}). Also Verhamme *et al.* (2008) find no need for IGM absorption in their Lyα profile modeling of LBG/LAE at $z \sim 3$–5. Furthermore, absorbing a fraction of Lyα by the IGM is even more difficult if the intrinsic Lyα profile emerging from the galaxy is already redshifted (and asymmetric), as expected if outflows are as ubiquitous as in LBGs at lower redshift. Assessing these important issues remains to be done. In the meantime conclusions depending strongly on corrections to the observed Lyα equivalent widths (or LF) should probably be taken with caution.

4. Searches for He II emission

Several groups have tried to use the expected He II λ1640 emission (cf. Sect. 2) to search for Pop III stars, so far with no positive detection. However, the limits obtained from most recent survey begin to provide interesting constraints.

4.1. *Upper limits from individual objects or composite spectra*

Follow-up spectroscopy of LALA sources by Dawson *et al.* (2004) (cf. Sect. 3.2) have yielded an upper limit of $W(\mathrm{Ly}\alpha) < 25$ Å (3 σ) and He II λ1640/Lyα< 0.13–0.20 at 2–3 σ from their composite spectrum (11 objects). Nagao *et al.* (2005) searched for He II λ1640 with deep spectroscopy of one strong Lyα emitter at $z = 6.33$. Currently the lowest limit on He II emission is that measured by Ouchi *et al.* (2008) in composite spectra of 36 and 25 $z = 3.1$ and 3.7 LAEs. They reach $I(\mathrm{He\ II}\ \lambda1640)/I(\mathrm{Ly}\alpha) < 0.02$ and 0.06 (3σ) at these two redshifts, which translates to $\log(Q_{\mathrm{He^+}}/Q_H) < -1.9$ and < -1.4 respectively†. Although no definite conclusions can be drawn from that, comparison with Fig. 1 (right) show that these limits are already close to the maximum or slightly below the predicted values for metal-free populations with IMFs including very massive stars.

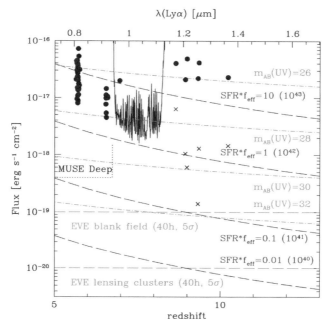

Figure 4. Observed Lyα flux as a function of redshift (observed wavelength on top) achieved with 10m class telescopes. Observations from Shimasaku *et al.* (2006), Kashikawa *et al.* (2006), and Ota *et al.* (2008) at $z = 5.7, 6.5$, and 7, as well as the $z \sim 8.5$ to 10 lensed candidates of Stark *et al.* (2007) are shown by the filled circles (observed fluxes, and crosses: intrinsic fluxes for lensed objects). The black line shows the flux limit from deep NIRSPEC/Keck spectroscopy of Richard *et al.* (2008) from 0.95 to 1.1 μm. Blue dashed lines show the fluxes corresponding to values of SFR$\times f_{\mathrm{eff}}$ from 0.01 to 10 M$_\odot$ yr^{-1}, where f_{eff} denotes the effective Lyα transmission, including the IGM transmission and other possible losses, e.g. inside the galaxy. Green dash-dotted lines show the expected Lyα flux for star-forming galaxies with rest-frame UV continuum magnitudes from 26 to 32 in the AB system, assuming $f_{\mathrm{eff}} = 1$. In both cases the conversion to Lyα flux was computed assuming a standard SFR calibration based on Kennicutt (1998) and case B. Deep observations with MUSE, a 2nd generation instrument for the VLT, and observations with EVE, a proposed multi-object spectrograph for the E-ELT, will allow to push the current limits down by ~ 1–2 magnitudes!

4.2. *The first constraint on the He II $\lambda 1640$ flux density at high redshift*

Using an original approach combining narrow/intermediate-band filters to select dual Lyα + He II $\lambda 1640$ emitters at $z \sim 4.0$ and 4.6, Nagao *et al.* (2008) have carried out a survey of 875 arcmin^2 using SUBARU in the Subaru Deep Field (see also Nagao, these proceedings, for more details). No such dual emitters were found, down to a flux limit of $(6 - 7) \times 10^{-18}$ erg s^{-1} cm^{-2} for He II $\lambda 1640$. With some assumptions on the IMF and using the models of S03, the observed flux limits translate to an upper limit for the Pop III SFR of $\gtrsim 2$ M$_\odot$ yr^{-1}, and a SFR density of $SFRD_{\mathrm{PopIII}} \lesssim 5.\,10^{-6}$ M$_\odot$ yr^{-1}

---

† The approximate translation between line ratios and ionising flux ratios is $Q_{\mathrm{He^+}}/Q_H = 0.6\,I(\mathrm{He\ II}\ \lambda 1640)/I(\mathrm{Ly}\alpha)$ (cf. S03).

Mpc^{-3} at $4.0 \lesssim z \lesssim 4.6$ (Nagao *et al.* 2008). Interestingly this first Pop III SFR density determination turns out to be very close to the theoretical prediction of Tornatore *et al.* (2007) shown in Fig. 3. Clearly further work and deeper observations are required to track the elusive Population III at high-redshift and future theoretical work may provide a more detailed/accurate picture of the expectations.

5. Ongoing and future deep observations at $z > 6$

The impressive depth – down to few times 10^{-18} erg s^{-1} cm^{-2} – reached already in emission line searches with SUBARU, VLT, and Keck is illustrated in Fig. 4, where Lyα measurements and upper limits obtained in the visible and near-IR domain are compiled. In the case of lensed galaxies the intrinsic depths can be considerably larger, as e.g. shown for the candidate high-z galaxies of Stark *et al.* (2007). In several cases of $z > 6$ galaxies or candidates, spectroscopic follow-up observations have already been undertaken to search for various emission lines, including the potential Pop III indicator He II λ1640 discussed above (e.g. Pelló *et al.* 2005, Stark *et al.* 2007, Richard *et al.* 2008, in preparation).

Since the average metallicity of galaxies must decrease with increasing redshift, searches for primeval galaxies and/or Pop III galaxies are also naturally focussed towards the highest redshift. At the same time galaxy searches at $z > 6$ are also of great interest to identify the sources of cosmic reionisation.

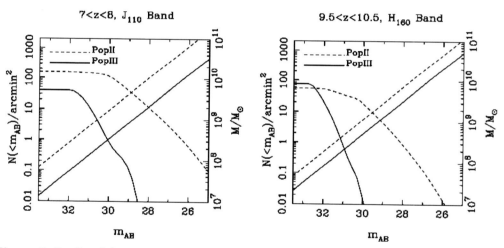

Figure 5. Predicted (cumulative) number density of "normal" (Pop II) and Pop III dominated sources for $z \sim 7.5$ (left) and $z \sim 10$ (right). Figure from Choudhury & Ferrara (2007).

Reviewing the status of galaxy searches/observations at $z > 6$ is beyond the scope of the present contribution. For an overview see e.g. Schaerer (2007 and references therein). The most recent results from searches for $z > 7$ galaxies are presented by Bouwens *et al.* (2008) and Richard *et al.* (2008), who use surveys in blank fields or fields with massive galaxy clusters benefiting from strong gravitational lensing. At the bright end of the LF the number density of galaxies at $z \gtrsim 7$, and hence also the total SFR density, remains controversial (cf. Richard *et al.* 2006, Bouwens *et al.* 2008). At fainter magnitudes ($m_{AB} \gtrsim 27$) the results from these two approaches (blank fields and lensing clusters) give consistent number densities (see Richard *et al.* 2008). The lensing cluster technique allows one to extend the observations to the faintest levels, currently down to an effective magnitude of $m_{AB} \sim 29$–30. Clearly the two techniques, one with an additional gain in depth by lensing over a small area and the other over wide fields, are complementary and both approaches will likely play an important role in finding distant, primeval, maybe even Pop III dominated objects.

The lensing studies show a source density of 3–100 $\text{arcmin}^{-2}(\Delta z = 1)^{-1}$ for $7 < z < 8$ from $m_{AB} = 28$ to 30 (Richard *et al.* 2008). This source density is also in agreement with the theoretical predictions from Stiavelli *et al.* (2004) and Choudhury & Ferrara (2007), shown in Fig. 5 . If spectroscopy of objects down to magnitudes of 30 or even 32 become feasible with extremely large telescopes (cf. Fig. 4) this means that we should expect very high source densities, e.g. several hundred objects(!) in a field of view of several arcmin^2 as foreseen for ELT instruments. High multiplex multi-object spectrographs reaching with an optical and near-IR coverage up to $\sim 1.7\mu\text{m}$, such as EVE proposed for the E-ELT, should therefore be very efficient in detecting Lyα and He II $\lambda1640$ lines up to redshift 13 and 9.4 respectively. Great progress can be expected from searches for the sources of cosmic reionisation and primeval/Pop III galaxies in the fairly near future!

References

Ajiki, M., *et al.*, 2003, *AJ*, 126, 2091

Bouwens, R. J., Illingworth, G. D., Franx, M., & Ford, H. 2008, arXiv:0803.0548

Brinchmann, J., Pettini, M., & Charlot, S. 2008, *MNRAS*, 385, 769

Bromm, V., & Larson, R. B. 2004, *ARAA*, 42, 79

Choudhury, T. R., & Ferrara, A. 2007, *MNRAS*, 380, L6

Dawson, S., *et al.* 2004, *ApJ*, 617, 707

Dijkstra, M., & Wyithe, J. S. B. 2007, *MNRAS*, 379, 1589

Fosbury, R. A. E., *et al.* 2003, *ApJ*, 596, 797

Hayes, M., Östlin, G. 2006, *A&A*, 460, 681

Hu, E. M., *et al.* 2004, *AJ*, 127, 563

Iwata, I., *et al.*, 2008, ApJL, submitted, arXiv:0805.4012

Jimenez, R., & Haiman, Z. 2006, Nature, 440, 501

Kashikawa, N., *et al.* 2006, ApJ, 648, 7

Kennicutt, R.C., 1998, *ARAA*, 36, 189

Malhotra, S., Rhoads, J.E., 2002, *ApJ*, 565, L71

Matsuda, Y., *et al.*, & Petitpas, G. R. 2007, *ApJ*, 667, 667

Nagao, T., *et al.* 2008, *ApJ*, 680, 100

Ouchi, M., *et al.* 2008, *ApJs*, 176, 301

Pan, L., & Scalo, J. 2007, *ApJ*, 654, L29

Pelló, R., Schaerer, D., Richard, J., Le Borgne, J.-F., & Kneib, J.-P. 2005, in "Gravitational Lensing Impact on Cosmology", IAU Symp. 225, 373

Richard, J., Stark, D. P., Ellis, R. S., *et al.* & Smith, G. P. 2008, *ApJ*, in press, arXiv:0803.4391

Savaglio, S., Panagia, N., & Padovani, P. 2002, *ApJ*, 567, 702

Scannapieco, E., Schneider, R., & Ferrara, A. 2003, *ApJ*, 589, 35

Schaerer, D. 2002, *A&A*, 382, 28

Schaerer, D. 2003, *A&A*, 397, 527

Schaerer, D. 2003b, in "Multi-wavelength cosmology", astro-ph/0309528

Schaerer, D., 2007, in "The emission line Universe", XVIII Canary Islands Winter School of Astrophysics, Ed. J. Cepa, Cambridge Univ. Press, arXiv.0706.0139

Schaerer, D., & Vacca, W. D. 1998, *ApJ*, 497, 618

Schneider, R., Ferrara, A., & Salvaterra, R. 2004, *MNRAS*, 351, 1379

Shapley, A. E., Steidel, C. C., Pettini, M., Adelberger, K. L., & Erb, D. K. 2006, *ApJ*, 651, 688

Shimasaku, K., *et al.* 2006, *PASJ*, 58, 313

Stark, D.P., Ellis, R.S., Richard, J., Kneib, J.-P., Smith, G.P., Santos, M.R., 2007, *ApJ*, 663, 10

Stiavelli, M., Fall, S. M., & Panagia, N. 2004, *ApJ*, 600, 508

Thuan, T. X., & Izotov, Y. I. 2005, *ApJS*, 161, 240

Tornatore, L., Ferrara, A., & Schneider, R. 2007, *MNRAS*, 382, 945

Tumlinson, J., Giroux, M.L., Shull, J.M., 2001, *ApJ*, 550, L1

Verhamme, A., Schaerer, D., Atek, H., & Tapken, C. 2008, *A&A*, in press, arXiv:0805.3601

Wang, J. X., *et al.* 2004, *ApJ*, 608, L21

Yoshida, N., Bromm, V., & Hernquist, L. 2004, *ApJ*, 605, 579

Low-Metallicity Star Formation:
From the First Stars to Dwarf Galaxies
Proceedings IAU Symposium No. 255, 2008
L.K. Hunt, S. Madden & R. Schneider, eds.

The search for Population III stars

Sperello di Serego Alighieri[1], Jaron Kurk[2], Benedetta Ciardi[3],
Andrea Cimatti[4], Emanuele Daddi[5] and Andrea Ferrara[6]

[1]INAF – Osservatorio Astrofisico di Arcetri, Largo E. Fermi 5, Firenze, Italy
email: sperello@arcetri.astro.it

[2]Max Planck Institut für Astronomie. Königstuhl 17, Heidelberg, Germany

[3]Max Planck Institut für Astrophysik, Karl Schwarzschild Str. 1, Garching, Germany

[4]Università di Bologna, Via Ranzani 1, Bologna, Italy

[5]CEA/Saclay, Gif sur Yvette, France

[6]SISSA, Via Beirut 4, Trieste, Italy

Abstract. Population III stars, the first generation of stars formed from primordial Big Bang material with a top–heavy IMF, should contribute substantially to the Universe reionization and they are crucial for understanding the early metal enrichment of galaxies. Therefore it is very important that these objects, foreseen by theories, are detected by observations. However PopIII stars, searched through the HeII 1640Å line signature, have remained elusive. We report about the search for the HeII line in a galaxy at z = 6.5, which is a very promising candidate. Unfortunately we are not yet able to show the results of this search. However we call attention to the possible detection of PopIII stars in a lensed HII dwarf galaxy at z = 3.4, which appeared in the literature some years ago, but has been overlooked.

Keywords. cosmology: observation, galaxies: formation, stars: chemically peculiar

1. Introduction

Theoretical models foresee that the first generation of stars, forming from primordial Big Bang material, should have unusually massive stars (a *top–heavy* IMF), with masses up to about 500 $M_\odot$ (Schneider *et al.* 2002). These stars are called Population III (PopIII) stars, and at least some of them are expected to quickly release metals in the interstellar medium (ISM); therefore later generations would soon be metal enriched. The metallicity threshold in the ISM for producing PopIII stars should be around 10^{-5}–$10^{-4} Z_\odot$. The very massive PopIII stars would then produce a short phase of unusually hard and strong UV radiation ($T_{eff} \sim 100000K$), resulting in a specific line emission signature (Schaerer 2002). The most prominent and unique emission line is expected to be HeII 1640Å, which can become as strong as 1/3 of Lyman α, making it observable with current 8–10m telescopes in the highest redshift galaxies known (Scannapieco *et al.* 2003).

The detection of PopIII stars would be extremely important for understanding the early chemical evolution of galaxies, and because they should be crucial contributors to the reionization of the Universe (Ciardi *et al.* 2003). PopIII stars have however remained elusive: searches for HeII 1640Å through stacking of spectra of Lyman α emitting galaxies (Dawson *et al.* 2004 and Ouchi *et al.* 2008), through deep spectroscopy of an individual galaxy (Nagao *et al.* 2005), or through dual (Lyα and HeII) narrow–band imaging (Nagao *et al.* 2008) have failed. Jimenez & Haiman (2006) have ascribed to PopIII stars the HeII 1640Å line detected in the composite spectrum of $\sim$1000 Lyman break galaxies at $z \sim 3$, but the line is only about 1/10 of Lyman α and is considerably broader: it could therefore be attributed to a stellar wind feature associated with massive WR stars (Shapley *et al.*

2003). The upper limits to the star formation rate (SFR) for PopIII stars, which can be set from these negative results are now quite close to the rates expected from the models of Tornatore *et al.* (2007) (see the contribution by T. Nagao to these Proceedings).

We report about the search for the HeII 1640Å line in a galaxy at z = 6.5 and about the possible detection of PopIII stars in a lensed dwarf HII galaxy at z = 3.4, which has appeared in the literarure (Fosbury *et al.* 2003), but has so far been overlooked.

2. The search for PopIII stars in KCS 1166

KCS 1166 is a Lyman α emitting galaxy at z = 6.518, discovered with slitless spectroscopy in the atmospheric window at $\lambda \sim 9100$Å (Kurk *et al.* 2004). In this object the Lyman α line is clearly asymmetric, being steeper on the blue side, has a very large equivalent width, at least 80Å in the rest frame, and a luminosity of $1.1 \times 10^{43} erg\ s^{-1}$. No continuum is detected at shorter wavelengths, while it is present redward of the line. These characteristics and the fact that the HeII 1640Å line is expected in a region of the J–band relatively free of sky emissions make KCS 1166 a very good candidate for the search of PopIII stars.

We have therefore carried out near IR spectroscopy in the J–band of KCS 1166, using SINFONI, an Integral Field Spectrograph at the VLT (Bonnet *et al.* 2004), with a total on–source exposure time of 9 hours, equally spread over 3 consecutive nights (Fig. 1). Thanks to the excellent efficiency in the J–band and the lack of OH Suppressor and of slit losses of SINFONI, we estimate that our observations of KCS 1166 should be a factor of about 1.5 more efficient in detecting the eventual presence of the HeII 1640Å line than those of SDF J132440.6+273607 at z = 6.3 by Nagao *et al.* (2005), although we had a slightly shorter exposure time.

However at the time of this Conference we are yet unable to report about the results of our observations, since the data reduction, which is rather complicated for an Integral Field Spectrograph like SINFONI, is not yet finished.

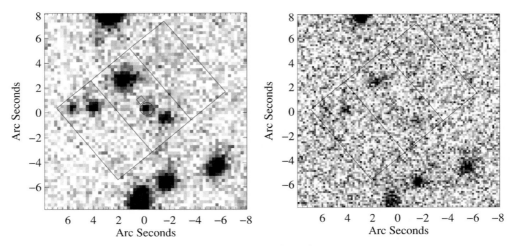

Figure 1. FORS z–band (left) and ISAAC J–band (right) images of the field of KCS 1166. The large boxes mark the two squared areas covered by the Integral Field of view of SINFONI in the two observing positions, shifted by half of the field size to improve the sky subtraction and to leave KCS 1166 in the overlapping area observed for the whole observing time. The galaxy is marked by a small circle and is clearly visible only in the z–band, which contains the Lyman α line.

3. The possible detection of PopIII stars in the Lynx arc

We take this opportunity to draw the attention of the Conference participants and of those interested in PopIII stars to the results obtained by Fosbury *et al.* (2003) on the Lynx arc, since these could well mark the first detection of PopIII stars, and are completely overlooked by the relevant literature. Although we recommend the interested reader to read the paper by Fosbury *et al.* directly, we give here a summary of their work.

The Lynx arc is a dwarf HII galaxy at z = 3.357, lensed by the z = 0.570 cluster RX J0848+4456, and it has been discovered serendipitously by Holden *et al.* (2001) during spectroscopic follow–up of the cluster. The HII galaxy is magnified by a factor of about 10, as evaluated by a detailed analysis of the complex cluster environment from HST WFPC2 images. The unusual emission line spectrum has been thoroughly studied with the LRIS, ESI and NIRSPEC at the Keck telescopes, covering the rest–frame ranges 900–2500 and 3300–5700Å. It shows strong and self–absorbed Lyman α line and CIV 1548,1551Å doublet, strong intercombination lines of NIV] 1483,1487Å, CIII] 1907,1909Å, and OIII] 1661,1666Å, moderate HeII 1640Å, absent [OII] 3727,3729Å. The doublet SiIII] 1883,1892Å is clearly detected in the ESI spectrum, and is 40 times brighter, relative to Hβ, than foreseen by models with scaled solar abundances. The intercombination lines have a very narrow width ($\sigma \sim 30 - 35 km s^{-1}$), indicating a small gravitating mass of about $10^9 M_\odot$, typical of a dwarf galaxy. The absence of NV 1240Å and the weakness of NIII] 1750Å indicate that the ionizing source is a blackbody, rather than a power law, as it would be expected in case of ionization by an AGN.

The intensity of the continuum observed longward of Lyman α is completely explained by the nebular continuum, as accurately predicted from the observed strength of Hβ. The continuum produced by an instantaneous burst of $10^7 M_\odot$, Salpeter IMF, Z = 0.05×$Z_\odot$, age of 1 to 5 Myr would be seen in the data, if it were present, but would produce an ionizing flux 20 times lower than the one necessary to produce the observed emission lines: this type of continuum, therefore, cannot be the source of ionization.

The photoionization model reproducing all the observed features has a black–body ionizing source with: a) a black–body temperature of 80000K, much higher than the effective temperature of Galactic compact HII regions, which does not exceed 40000K; b) an ionization parameter U = 0.1, also higher than in local HII regions; c) a low *nebular* metallicity Z = 0.05×$Z_\odot$. The lack of stellar continuum longward of Lyman α, the strong necessary ionizing flux, and the small stellar mass all indicate that the ionizing flux should be produced by fewer than 10^6 hot stars, formed with a top–heavy IMF, and most likely with a metallicity still lower than the nebular one. The substantial overabundance of silicon in the nebula indicate enrichment by past pair instability supernovae, as those resulting from the total disruption of stars with 140–260 $M_\odot$ (Hegel & Woosley 2002).

All these characteristics point strongly to the presence of PopIII stars, some of which might have already exploded to partially enrich the ISM. However in a dwarf galaxy like the Lynx arc, the formation of PopII stars could have been delayed long enough to make the uncontaminated UV signature of PopIII stars detectable at intermediate redshift. In fact this uncontaminated signature of PopIII stars might have a much shorter duration in the massive galaxies, which are the only observable galaxies at high redshift. This might be the reason why PopIII stars have not yet been detected at high redshift, where they have been searched so far, following the obvious paradigm that primordial material is more abundant in the very young Universe.

In a contribution to a conference Schaerer (2004) finds it unlikely that the Lynx arc contains an extremely metal poor cluster, because the ISM metallicity is 1/20 solar and there are no known cases of the stellar metallicity lower than the nebular one, and the alternative explanation of an obscured AGN (Binette *et al.* 2003) seems to be capable of reproducing the observed spectrum. However, it should not be surprising that new phenomena are observed when dealing with a new class of objects, like PopIII stars, and Binette *et al.* (2003) find it plausible that the stellar metallicity might be lower than the nebular one. Furthermore, of the 5 models presented by Binette *et al.* (2003) the one that best fits the observed line ratios of the Lynx arc is the hot star (80000 K) model, which is relevant for PopIII and which has only one inconsistent line, Si III] 1883,1892Å, whose discrepancy could in any case be due to the nucleosynthetic signature of pair instability supernovae, as explained above. The other 4 models also have discrepant Si III] 1883,1892Å, but in addition have at least two more lines inconsistent with the observations. Therefore the most likely explanations for the observed spectrum of the Lynx arc remains the very hot star model expected for PopIII stars.

4. Final remarks

Although we are yet unable to report on the results of our observations of the HeII 1640Å line in KCS 1166, which is a good candidate for PopIII search at z = 6.518, it is well possible that PopIII stars have already been detected by Fosbury *et al.* (2003) in a dwarf star–forming galaxy at z = 3.357, thanks to the fortunate combination of the longer uncontaminated PopIII phase in dwarf galaxies and of the opportunity given by the lensing magnification to identify PopIII stars, even if not at their peak activity.

Clearly further detections of PopIII stars, or stringent upper limits, would be extremely important to understand the connection between reionization and metal production, since PopIII stars should be major players on both scenes, and could spoil the simple proportionality between ionizing photons and metals, claimed by some (see e.g. the contribution by A. Ferrara to these Proceedings).

5. Acknowledgements

We would like to thank Bob Fosbury and Raffaella Schneider for very useful comments.

References

Binette, L., Groves, B., Villar-Martin, M., Fosbury, R.A.E. & Axon, D.J. 2003, *A&A*, 405, 975
Bonnet, H., Abuter, R., Baker, A. *et al.* 2004, *ESO Messenger*, 117, 17
Ciardi, B., Ferrara. A. & White, S.D.M. 2003, *MNRAS*, 344, L7
Dawson, S., Rhoads, J.E., Malhotra, S. *et al.* 2004, *ApJ*, 617,707
Fosbury, R.A.E., Villar-Martin, M., Humphrey, A. *et al.* 2003, *ApJ*, 596, 797
Heger, A. & Woosley, S.E. 2002, *ApJ*, 567, 532
Holden, B.P., Stanford, S.A., Rosati, P. *et al. AJ*, 122, 629
Jimenez, R. & Haiman, Z. 2006, *Nature*, 441, 120
Kurk, J.D., Cimatti, A., di Serego Alighieri, S. *et al.* 2004, *A&A*, 422, L13
Nagao, T., Motohara, K., Maiolino, R. *et al.* 2005, *ApJ*, 631, L5
Nagao, T., Sasaki, S.S., Maiolino, R. *et al.* 2008, *ApJ* 680, 100
Ouchi, M., Shimasaku, K., Akiyama, M. *et al.* 2008, *ApJS*, 176, 301
Scannapieco, E., Schneider, R. & Ferrara, A. 2003, *ApJ*, 589, 35
Schaerer, D. 2002, *A&A*, 382, 28
Schaerer, D. 2003, in: M. Plionis (ed.), *Multiwavelength Cosmology* (Kluwer), p. 219
Schneider, R., Ferrara, A., Natarajan, P., Omukai, K. 2002, *ApJ*, 571, 30
Shapley, A.E., Steidel, C.C., Pettini, M. & Adelberger, K.L. 2003, *ApJ*, 588, 65
Tornatore, L., Ferrara, A. & Schneider, R. 2007, *MNRAS*, 382, 945

Low-Metallicity Star Formation:
From the First Stars to Dwarf Galaxies
Proceedings IAU Symposium No. 255, 2008
L.K. Hunt, S. Madden & R. Schneider, eds.

Observational Search for Population III Stars in High-Redshift Galaxies

Tohru Nagao[1]

[1]Research Center for Space and Cosmic Evolution, Ehime University
2-5 Bunkyo-cho, Matsuyama 790-8577, Japan
email: tohru@cosmos.ehime-u.ac.jp

Abstract. In this contribution we present our new photometric search for high-z galaxies hosting Population III (PopIII) stars based on deep intermediate-band imaging observations, by using Supreme-Cam on the Subaru Telescope. By combining our new data with the existing broad-band and narrow-band data in the target field, we searched for galaxies which emit strongly both in Lyα and in HeIIλ1640 ("dual emitters") that are promising candidates for PopIII-hosting galaxies, at $4 \lesssim z \lesssim 5$. Although we found 10 "dual emitters", most of them turn out to be [OII]-[OIII] dual emitters or Hβ-Hα dual emitters at $z < 1$, as inferred from their broad-band colors and from the ratio of the equivalent widths. No convincing candidate of Lyα-HeII dual emitter with $SFR_{\rm PopIII} \gtrsim 2M_\odot$ yr^{-1} was found. This result disfavors low feedback models for PopIII star clusters, and implies an upper limit of the PopIII SFR density of $SFRD_{\rm PopIII} < 5\times10^{-6}M_\odot$ yr^{-1} Mpc^{-3}. This new selection method to search for PopIII-hosting galaxies should be useful in future surveys for the first observational detection of PopIII-hosting galaxies at high redshift.

Keywords. early universe, galaxies: evolution, galaxies: formation, stars: early-type

1. Introduction

Population III (PopIII) stars are those formed out of primordial gas, enriched only through Big-Bang nucleosynthesis. Since massive PopIII stars are promising candidates as sources for cosmic reionization and an important population for early phases of the cosmic chemical evolution, their properties have been extensively investigated from the theoretical point of view. PopIII stars have not been discovered yet; obviously, their direct detection and the observational studies of their properties would provide a completely new and important step toward understanding the evolution of galaxies. The expected observables of high-z galaxies hosting PopIII stars have been theoretically investigated in recent years. Such galaxies are expected to show strong Lyα emission, with an extremely large equivalent width (EW), and moderately strong HeIIλ1640 emission (e.g., Tumlinson & Shull 2000; Tumlinson *et al.* 2001; Oh *et al.* 2001; Schaerer 2002, 2003; Tumlinson *et al.* 2003), due to the high effective temperature up to $\sim 10^5$ K of PopIII stars.

Most models predict that PopIII stars dominated the re-ionization of the universe at $7 \lesssim z \lesssim 15$. However, they also predict that PopIII stars may still exist at redshifts currently accessible with 8–10m-class telescopes, i.e., $z < 7$ (e.g., Scannapieco *et al.* 2003, Tornatore *et al.* 2007). Some observations have found Lyα emitters (LAEs) at $z > 4$ with a very large EW, which is hard to explain through star-formation without PopIII (e.g., Malhotra & Rhoads 2002; Nagao *et al.* 2004, 2005a, 2007; Shimasaku *et al.* 2006; Dijkstra & Wyithe 2006). However, the search for HeIIλ1640 emission as direct evidence for PopIII in such galaxies is far more controversial. Jimenez & Haiman (2006) pointed out the possible HeIIλ1640 signature in the composite spectrum of LBGs at $z \sim 3$ made by Shapley *et al.* (2003), although it may be attributed to a stellar wind feature associated with massive stars as mentioned by Shapley *et al.* (2003). On the other hand, other searches for HeIIλ1640 in higher-z galaxies have failed, through stacking analysis of LAEs (Dawson *et al.* 2004; Ouchi *et al.* 2008) or through ultra-deep near-infrared spectroscopy of an individual LAE (Nagao *et al.* 2005b).

Nevertheless, the HeIIλ1640 emission from PopIII-hosting galaxies may already be detected in current deep narrow-band (NB) surveys (mostly aiming for LAE searches) as NB-excess objects, but not identified as HeII emitters (Tumlinson *et al.* 2001) since NB surveys are more sensitive to faint emission lines than spectroscopic observations. If a NB-excess object is due to HeIIλ1640 emission, then the same object should show stronger Lyα emission at a shorter wavelength, since the PopIII-hosting galaxies should emit Lyα with $EW_{\mathrm{rest}} > 500$Å (e.g., Schaerer 2003). Therefore, by performing additional NB (or intermediate-band) imaging observations whose wavelength is matched to the redshifted Lyα, we may be able to find "Lyα-HeII dual emitters" that are promising candidates for PopIII-hosting galaxies. Motivated by these considerations, we performed new intermediate-band imaging observations (see Nagao *et al.* 2008 for details).

2. Observations

The field investigated in this project is the Subaru Deep Field (SDF: Kashikawa *et al.* 2004; Taniguchi *et al.* 2005). Among some existing NB images in the SDF, we focus on NB816 and NB921 that can be used to search for HeII emitters at $3.93 \lesssim z \lesssim 4.01$ or $4.57 \lesssim z \lesssim 4.65$, respectively. If there are HeII emitters in these redshift ranges, they should show very strong Lyα emission at 5992Å $\lesssim \lambda_{\mathrm{obs}} \lesssim 6089$Å or 6769Å $\lesssim \lambda_{\mathrm{obs}} \lesssim 6867$Å. To detect possible Lyα emission in these wavelengths, we observed the SDF on 22 April 2007 (UT) with Suprime-Cam on the Subaru Telescope, using two intermediate-passband filters, IA598 and IA679. By combining these new data with the existing imaging data, we can search for "Lyα-HeII dual emitters" in photometric way, as schematically shown in Fig 1.

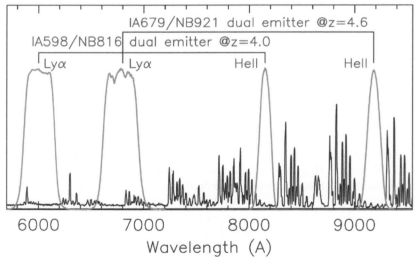

Figure 1. Schematic view of our selection method for Lyα-HeII dual emitters. The black solid spectrum denotes a typical sky spectrum. Red solid curves denote the filter transmission curves of IA598, IA679, NB816, and NB921. The dual excess of the combination of IA598 and NB816, and that of IA679 and NB921 corresponds to $z \sim 4.0$ and $z \sim 4.6$, respectively.

3. Results

To search for Lyα-HeII dual emitters, we first identified the IA-excess objects (i.e., emission-line galaxies) by adopting an IA-excess criterion of 0.3 mag, that corresponds to the emission-line EW of $EW_{obs} \sim 100$Å. Note that this limiting EWs are lower than intrinsic EWs of Lyα theoretically expected for PopIII-hosting galaxies. The numbers of the identified IA-excess objects are 133 and 234 for IA598 and IA679, respectively. We then investigated possible NB816 excesses for IA598-excess objects, and also possible NB921 excesses for IA679-excess objects. By adopting the NB816-excess criterion of 0.3 mag (i.e., $EW_{obs} \gtrsim 45$Å) for IA598-excess objects, we found 4 IA598-NB816 dual excess objects. In addition, by adopting the NB921-excess criterion of 0.15 mag (i.e., $EW_{obs} \gtrsim 20$Å) for IA679-excess objects, we also found 6 IA679-NB921 dual excess objects.

Fig 2 shows the SEDs of 6 IA679-NB921 dual-excess objects. These SEDs are apparently inconsistent with the interpretation that they are galaxies at $z > 4$. This is because the objects shown in Fig 2 show relatively blue $B - V$ colors ($B - V < 1$), unlike star-forming galaxies at $z > 4$ that should show B-band dropout due to the Lyman-limit absorption (i.e., $B - V \gtrsim 2$). Instead they are more consistent to star-forming galaxies at $z < 1$. This is also true for IA598-NB816 dual-excess objects. The possible low-z contamination in IA-NB dual emitter sample is from [OII]-[OIII] dual emitters and Hβ-Hα dual emitters, because the wavelength ratios of Lyα/HeII, [OII]/[OIII], and Hβ/Hα are so similar (~ 0.741, ~ 0.744, and ~ 0.741, respectively). Note that several IA-NB dual

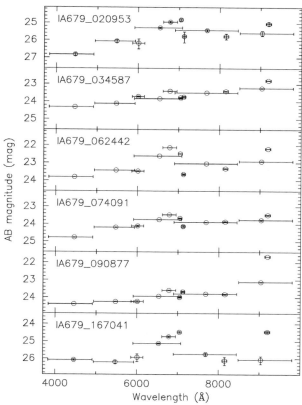

Figure 2. SEDs of the IA679-NB921 dual emitters. Error bars in the y-axis direction denote the 1σ photometric errors. The ID of each objects is shown at the upper-left corner of each panel.

emitters have a large ratio of the NB-excess flux to the IA-excess flux. Lyα-He II dual emitters cannot have such a large flux ratios of He II/Lyα (which should instead be $\lesssim$0.1, depending on the adopted PopIII models; e.g., Schaerer 2003). In contrast, star-forming galaxies have the flux ratio of [O III]/[O II] $\sim$ 0.1–10 (depending on the gas metallicity and/or the ionization parameter; e.g., Kewley & Dopita 2002; Nagao et $al.$ 2006) and those of Hα/Hβ $\gtrsim$ 3. Therefore, the ratios of the IA excess to the NB excess observed in IA-NB dual-excess objects are again consistent with star-forming galaxies at $z < 1$, rather than PopIII-hosting galaxies.

4. Discussion

Schaerer (2003) investigated the temporal evolution of EW(He II) for PopIII stellar clusters by assuming IMFs with a Salpeter slope and some combinations of lower and upper mass cut-offs ($M_{\rm low}$, $M_{\rm up}$). Here we focus on the predictions of EW(He II) in the case of ($M_{\rm low}$, $M_{\rm up}$) = ($50M_\odot$, $500M_\odot$). Then the He II luminosity can be written as L(He II) = 6.01×10^{41} ($SFR_{\rm PopIII}/M_\odot{\rm yr}^{-1}$). Taking the 3σ limiting fluxes for the NB excesses into account, our survey can detect PopIII-hosting galaxies if their SFR is higher than $\sim 2M_\odot$ yr^{-1}. Therefore, the non-detection of Lyα-He II dual emitters suggests that there are no PopIII-hosting galaxies with $SFR_{\rm PopIII} \gtrsim 2M_\odot$ yr^{-1} at $4.0 \lesssim z \lesssim 4.6$ toward the SDF, in a volume of 4.03×10^5 Mpc^3. This result implies an upper-limit of the PopIII SFR density of $SFRD_{\rm PopIII} < 5 \times 10^{-6} M_\odot$ yr^{-1} Mpc^{-3}, if taking only galaxies with $SFR_{\rm PopIII} > 2M_\odot$ yr^{-1} into account. Note that the inferred upper limit on $SFR_{\rm PopIII}$ is uncertain, since the predicted flux of He II for a given $SFR_{\rm PopIII}$ strongly depends on the assumed IMF (e.g., Schaerer 2003). It also depends on the evolutionary processes of PopIII stars, especially the mass loss during their evolution (e.g., Tumlinson et $al.$ 2001; Schaerer 2002).

Some theoretical studies suggest that the volume-averaged IGM metallicity quickly reached $Z_{\rm crit} = 10^{-4}Z_\odot$ at $z > 10$ (e.g., Tornatore et $al.$ 2007), where $Z_{\rm crit}$ is the critical metallicity, below which very massive stars could be formed. However, this does not necessarily suggest that the formation of PopIII stars was terminated at such a high redshift, because of the inhomogeneous metal distribution in the early universe (e.g., Scannapieco et $al.$ 2003; Tornatore et $al.$ 2007). As demonstrated by Scannapieco et $al.$ (2003), the redshift evolution of the $SFR_{\rm PopIII}$ density in the universe depends sensitively on some PopIII model parameters, especially the feedback efficiency that is closely related to the PopIII IMF. Low-feedback models of Scannapieco et $al.$ (2003) predict a large fraction ($\sim$30%) of PopIII-hosting galaxies among LAEs at $4.0 \lesssim z \lesssim 4.6$ with log L(Lyα) $\sim 10^{43}$ ergs s^{-1}. A similarly large fraction of PopIII-hosting galaxies among high-z LAEs is also inferred by Dijkstra & Wyithe (2007). Since the number density of LAEs with this luminosity at similar redshifts is $\sim 10^{-5} - 10^{-4}$ Mpc^{-3} (e.g., Ouchi et $al.$ 2008), the number of PopIII-hosting galaxies in our survey, expected by such low-feedback models, is roughly 1 to 10. Therefore, the non-detection in our Lyα-He II dual emitters survey may suggest that low-feedback models are not appropriate, and that PopIII stars may instead by characterized by a relatively large feedback efficiency.

This photometric survey for Lyα-He II dual emitters demonstrated that wide and deep imaging observations, combining narrow-band and/or intermediate-band filters, are potentially a powerful tool to search or constrain the properties of PopIII-hosting galaxies at high redshifts. The data recently obtained by sensitive narrow-band near-infrared surveys and similar wide and deep surveys planned in future may be useful to search for Lyα-He II dual emitters at $z > 6$, by adding data of narrow- or intermediate-band observations at corresponding wavelengths to check strong Lyα emission. Such a survey is promising, since $SFRD_{\mathrm{PopIII}}$ increases at higher redshifts (see, e.g., Dijkstra & Wyithe 2007). In future observational searches for Lyα-He$\mathsc{ii}$ dual emitters, serious sources of contamination would be [O$\mathsc{ii}$]-[O$\mathsc{iii}$] and Hβ-Hα dual emitters, as demonstrated in this paper. In addition to broad-band color criteria, the flux (or EW) ratio of the dual excesses is also a powerful diagnostic to discriminate the populations and to identify Lyα-He$\mathsc{ii}$ dual emitters among the photometric candidates.

References

Dawson, S., Rhoads, J. E., Malhotra, S., *et al.* 2004, *ApJ*, 617, 707

Dijkstra, M., & Wyithe, J. S. B. 2007, *MNRAS*, 379, 1589

Jimenez, R., & Haiman, Z. 2006, *Nature*, 441, 120

Kashikawa, N., Shimasaku, K., Yasuda, N., *et al.* 2004, *PASJ*, 56, 1011

Kewley, L. J., & Dopita, M. A. 2002, *ApJS*, 142, 35

Malhotra, S., & Rhoads, J. E. 2002, *ApJ*, 565, L71

Nagao, T., Kashikawa, N., Malkan, M. A., *et al.* 2005a, *ApJ*, 634, 142

Nagao, T., Maiolino, R., & Marconi, A. 2006, *A&A*, 459, 85

Nagao, T., Motohara, K., Maiolino, R., *et al.* 2005b, *ApJ*, 631, L5

Nagao, T., Murayama, T., Maiolino, R., *et al.* 2007, *A&A*, 468, 877

Nagao, T., Sasaki, S. S., Maiolino, R., *et al.* 2008, *ApJ*, 680, 100

Nagao, T., Taniguchi, Y., Kashikawa, N., *et al.* 2004, *ApJ*, 613, L9

Oh, S. P., Haiman, Z., & Rees, M. J. 2001, *ApJ*, 553, 73

Ouchi, M., Shimasaku, K., Akiyama, M., *et al.* 2008, ApJS, 176, 301

Scannapieco, E., Schneider, R., & Ferrara, A. 2003, *ApJ*, 589, 35

Schaerer, D. 2002, *A&A*, 382, 28

Schaerer, D. 2003, *A&A*, 397, 527

Shapley, A. E., Steidel, C. C., Pettini, M., *et al.* 2003, *ApJ*, 588, 65

Shimasaku, K., Kashikawa, N., Doi, M., *et al.* 2006, *PASJ*, 58, 313

Taniguchi, Y., Ajiki, M., Nagao, T., *et al.* 2005, *PASJ*, 57, 165

Tornatore, L., Ferrara, A., & Schneider, R. 2007, *MNRAS*, 382, 945

Tumlinson, J., Giroux, M. L., & Shull, J. M. 2001, *ApJ*, 550, L1

Tumlinson, J., & Shull, J. M. 2000, *ApJ*, 528, L65

Tumlinson, J., Shull, J. M., & Venkatesan, A. 2003, *ApJ*, 584, 608

Session II

Metal enrichment, chemical evolution, and feedback

Low-Metallicity Star Formation:
From the First Stars to Dwarf Galaxies
Proceedings IAU Symposium No. 255, 2008
L.K. Hunt, S. Madden & R. Schneider, eds.

Cosmic metal enrichment

Andrea Ferrara[1]

[1]SISSA/ISAS, Via Beirut 2-4, 34014 Trieste, Italy
email: ferrara@sissa.it

Abstract. I review the present understanding of the process by which the universe has been enriched in the course of its history with heavy elements produced by stars and transported into the surrounding intergalactic medium. This process goes under the name of "cosmic metal enrichment" and presents some of the most challenging puzzles in present day physical cosmology. These are reviewed along with some proposed explanations that all together form a coherent working scenario.

Keywords. (galaxies:) intergalactic medium, cosmology: theory, galaxies: high-redshift

1. Preliminaries

After the Big Bang the cosmic gas had a composition which was (virtually) free of heavy elements. Yet, every single parcel of gas that we can probe today with our most powerful telescopes shows the sign of considerable amounts of metals, up to the highest redshift. When these metals where first produced, by what sources, and how these atomic species traveled away from their production sites are among the most challenging puzzles in current cosmological scenarios.

A very simple, and yet robust, estimate of the amount of metals present at the end of reionization can be made. This is based on the fact that the sources of heavy elements and photons with energy > 1 Ryd responsible for hydrogen reionization are massive stars. Hence, as we know that reionization was complete by redshift $z = 6$, we can compute the metallicity of the cosmic gas associated with the ionizing photons required to reionize the universe, in the hypothesis that such process has been powered by stellar radiation. Obviously, as different type of sources, as quasars and/or decaying/annihilating dark matter particles may have contributed, the result of this simple exercise is an upper limit to the amount of metals. However, many arguments suggest that stars are by far the most viable reionization source candidates. In this case, we find that the mean metallicity of the cosmic gas (in practice, the intergalactic medium [IGM] which contains most of the baryons) at $z = 6$ is

$$Z = \frac{\langle y \rangle \nu C}{\eta} = 3.5 \times 10^{-4} C \; Z_\odot, \tag{1.1}$$

where $\langle y \rangle$ is the mean supernova metal yield, ν is the number of supernovae per unit stellar mass formed, $C \approx 1 - 10$ is the IGM clumping factor and η is the number of ionizing photons per baryon into stars. The above result has been derived for a Salpeter stellar IMF but the value of Z is only mildly dependent on such quantity. The main point is that metal production associated with cosmic reionization is already substantial and comparable with the value derived from QSO absorption line experiments at lower redshifts.

The next question concerns the sources that predominantly produced the metals and photons. Current data from a number of different experiments including CMB polarization anisotropies, Lyα and Lyβ Gunn-Peterson tests, cosmic star formation history and galaxy number counts constrain reionization history in a very stringent way (Choudhury & Ferrara 2007). Thus it is possible from those models to determine which sources were contributing most of the ionizing photons (and hence metals) at a given redshift by considering the fractional instantaneous contribution of halos above a certain mass,

$$f_\gamma(> M, z) \equiv \frac{\dot{n}_\gamma(> M, z)}{\dot{n}_\gamma(z)}, \qquad (1.2)$$

The main conclusion from that study is that $> 80\%$ of the ionizing power at $z \geqslant 7$ is provided by halos with masses $< 10^9 M_\odot$ which are predominantly harboring metal-free (PopIII) stars. A turnover to a PopII-dominated reionization phase occurs shortly after, with this population, residing in $M > 10^9 M_\odot$ haloes, producing $\approx 60\%$ of the ionizing photons at $z = 6$. In conclusion, PopIII stars and *small* galaxies initiate reionization at high redshift and remain important until they are overcome by PopII stars and QSOs below $z = 7$.

It then appears that cosmic metal enrichment is a gradual process that proceeds along with the increase of the mean ionized fraction in the universe and that the heavy elements were predominantly expelled at high redshifts ($z > 6$) by dwarf-like galaxies. These galaxies had probably larger ejection efficiencies with respect to "normal" ones; in addition such early enrichment has interesting properties and implications that we discuss in the following.

2. Early enrichment by dwarf galaxies

Mechanical feedback results from the energy deposited in the surrounding medium by winds from massive stars and supernova explosions. The powerful shocks originating from such events heat and accelerate the gas, driving outflows of heavily metal-enriched gas into the IGM. The central question of the problem concerns the fraction of produced metals that escape from the galaxy, i.e. the metal escape fraction, δ_B. A solution has been proposed by Ferrara, Pettini & Shchekinov (2000) who noticed that SNe in "normal" galaxies (as for example the Milky Way) are distributed in the disk and clustered in OB associations; thus, these explosion sites act incoherently. In low-mass galaxies, on the contrary, the size of the galaxy is comparable to the size of individual SN-driven bubbles and therefore the energy deposited can work to drive coherently the same outflow. This makes the metal escape fraction from the latter objects much larger than from large galaxies. To see this point quantitatively, we recall that in nearby galaxies it is found that the luminosity function of OB associations is well approximated by a power-law,

$$\phi(N) = \frac{d\mathcal{N}_{OB}}{dN} = AN^{-\beta}, \qquad (2.1)$$

with $\beta \approx 2$. Here $\mathcal{N}_{OB}$ is the number of associations containing N OB stars; normalization of $\phi(N)$ to unity requires $A = 1$. Thus the probability for a cluster of OB stars to host N SNe is $\propto N^{-2}$, where $N = L_{OB} t_{OB}/\epsilon_0$, and $t_{OB} = 40$ Myr is the time at which the lowest mass ($\approx 8M_\odot$) SN progenitors expire. The total mechanical luminosity, is then found to be

$$L_t(z) = \int_{N_m}^{N_M} L_{OB}(N) \phi \, dN, \qquad (2.2)$$

where $N_m = 1$ (N_M) is the minimum (maximum) possible number of SNe in an association. This gives

$$L_t(z) = \text{const.} \; \frac{\epsilon_0}{t_{OB}} \ln \frac{N_M}{N_m}. \tag{2.3}$$

The contribution to the total luminosity from clusters powerful enough to lead to blowout
is

$$L_B(z, > L_c) = \text{const.} \; \frac{\epsilon_0}{t_{OB}} \ln \frac{N_M}{N_c}, \tag{2.4}$$

where N_c is the number of SNe in a cluster with mechanical luminosity equal to L_c, i.e.

$$N_c = \frac{L_c t_{OB}}{\epsilon_0}. \tag{2.5}$$

Thus, the fraction of mechanical energy (and metals) that can be blown out is

$$\delta_B = \frac{\ln(N_M/N_c)}{\ln(N_M/N_m)} < 1. \tag{2.6}$$

Clearly, N_M (and therefore δ_B) is an intrinsically stochastic number. To determine its
dependence on the total number of supernovae $N_t = L_t(z) t_{OB}/\epsilon_0$ produced by a galaxy
during the lifetime of an OB association, we have used a Monte Carlo procedure applied
to the distribution function in eq. 2.1. For low values of N_t the quantity N_M is larger
than N_c, implying that in every galaxy at least some superbubbles are able to blow
out. However, near $N_t = 10^4$ N_M flattens and eventually becomes equal to N_c at $N_t \simeq$
$45\,000$. Above this limit (corresponding to a galaxy with $\dot{M}_\star \approx 0.35$ $M_\odot$ yr^{-1} or $M_h \approx$
$10^{12}(1+z)^{-3/2} M_\odot$) blowout is inhibited. The fraction δ_B is a decreasing function of N_t;
an approximate analytical form is

$$\delta_B(N_t) = 1 \quad \text{for } N_t < 100 \tag{2.7}$$
$$\delta_B(N_t) = a + b \ln(N_t^{-1}) \quad \text{for } N_t > 100,$$

with $a = 1.76$, $b = 0.165$. Clearly, in small galaxies even the smallest associations are
capable of producing blowout, and therefore $\delta_B \to 1$, whereas for large galaxies there
are less and less OB associations able to vent their metals and hot gas into the IGM,
i.e. $\delta_B \to 0$. The transition occurs roughly when the mass of the galaxy is $\approx 1/10$ of the
Milky Way mass; to be precise, we use the term "dwarfs" here to identify galaxy with
mass below that value.

   If most of the IGM metals by $z = 6$ have been ejected from dwarf galaxies at high
redshifts, such early enrichment scenario has two important implications, that we discuss
in turn.

2.1. *The filling factor argument*

First, the volume filling factor of enriched material, quantified by the porosity factor,
Q, becomes large if pollutants are dwarf galaxies. This can be seen as follows. In a
ΛCDM universe, structure formation is a hierarchical process in which nonlinear, massive
structures grow via the merger of smaller initial units. Large numbers of low–mass galaxy
halos are expected to form at early times in these popular cosmogonies, perhaps leading
to an era of widespread pre–enrichment and preheating. The Press–Schechter (hereafter
PS) theory for the evolving mass function of dark matter halos predicts a power–law
dependence, $dN/d\ln m \propto m^{(n_{\text{eff}}-3)/6}$, where n_{eff} is the effective slope of the CDM power
spectrum, $n_{\text{eff}} \approx -2.5$ on subgalactic scales. As hot, metal–enriched gas from SN–driven

winds escapes its host halo, shocks the IGM, and eventually forms a blast wave, it sweeps a region of intergalactic space which increases with the 3/5 power of the energy E injected into the IGM (in the adiabatic Sedov–Taylor phase). The total fractional volume or porosity, Q, filled by these 'metal bubbles' per unit explosive energy density $E\,dN/d\ln m$ is then

$$Q \propto E^{3/5}\,dN/d\ln m \propto (dN/d\ln m)^{2/5} \propto m^{-11/30}. \tag{2.8}$$

Within this simple scenario it is the star–forming objects with the smallest masses which will arguably be the most efficient pollutant of the IGM on large scales.

2.2. *The cooling argument*

The second point concerns the ability of the shocked gas to cool. This is necessary as at $z = 3 - 3.5$, Lyα clouds show a spread of at most an order of magnitude in their metallicity, and their narrow line widths require that they be photoionized and cold rather than collisionally ionized and hot. At these redshifts, hot rarefied gas, exposed to a metagalactic ionizing flux, will not be able to radiatively cool within a Hubble time. The simple formula below, gives the redshift span required for a gas of primordial composition (at the metallicities present in the Lyα forest contribution from metal cooling is virtually negligible) at the mean cosmic density heated at some redshift z_i to cool down:

$$\Delta z = 231(\Omega_m h^2)^{1/2}(1 + z_i)^{-3/2} \approx 3 \left(\frac{1 + z_i}{10}\right)^{-3/2}. \tag{2.9}$$

Hence, a gas that has been shock-heated at $z_i = 9$ will be already cooled by $z = 6$, but if heating occurs at $z \leqslant 5$ the cooling time will exceed the Hubble time.

In conclusion early pre-enrichment by dwarfs offers the double advantage of a large metal filling factor and of efficient cooling of the metal enriched gas ejected by galactic outflows. While it is possible that some metals were dispersed in intergalactic space at late times, as hot pressurized bubbles of shocked wind and SN ejecta escaped the grasp of massive galaxy halos and expanded, cooling adiabatically, into the surrounding medium, such a delayed epoch of galactic super–winds would have severely perturbed the IGM (since the kinetic energy of the ejecta is absorbed by intergalactic gas), raising it to a higher adiabat and producing variations of the baryons relative to the dark matter: Lyα forest clouds would not then be expected to closely reflect gravitationally induced density fluctuations in the dark matter distribution, and the success of hydrodynamical simulations in matching the overall observed properties of Lyα absorption systems would have to be largely coincidental.† In contrast, the observed narrow Doppler widths could be explained if the ejection of heavy elements at velocities exceeding the small escape speed of subgalactic systems were to take place at very high redshifts.

3. Metal ejection from dwarfs: observational support

Several theoretical arguments and numerical simulations in the literature have shown that metals and energy are easily expelled from dwarf galaxies. This does not necessarily

† Assume, for example, that the chemical enrichment of intergalactic gas was due to the numerous population of Lyman–break galaxies (LBGs) observed at $z = 3$. With a comoving space density above $m_* + 1 = 25.5$ of $0.013\,h^3$ Mpc^{-3} (Steidel *et al.* 1999), a 1% filling factor would be obtained if each LBG produced a metal–enriched bubble of proper radius equal to about $140\,h^{-1}$kpc. To fill such a bubble in 5×10^8 yr, the ejecta would have to travel at an average speed close to 600 km s^{-1}, with characteristic post-shock temperatures in excess of 2 million degrees.

imply a correspondingly high gas mass loss, i.e. the galaxy is typically able to retain some fraction of its original gas which might be as high as 50%. In addition, quenching of star formation associated with the SN energy deposition (mechanical feedback) is strongly advocated in order to prevent too many stars to form at high redshift (overcooling problem). What are the observational evidences in support of such metal ejection scenario?

The first one is provided by the so-called Mass-Metallicity relation. The huge local galaxy catalog put together by the Sloan Digital Sky Survey (Panter *et al.* 2008) has allowed to construct the relation between the metallicity, Z, obtained from suitable indicators and the stellar mass, M_*, of each galaxy in the sample. Such curve shows a metallicity increase with M_* up to a break point, located at $M_* = 10^{10} M_\odot$, beyond which the curve flattens considerably. A reasonable fit to the median line can be obtained using a tanh function over the mass range that contains more than 2000 galaxies $(10^{8.8} M_\odot < M_* < 10^{11.8} M_\odot)$ of the form:

$$\log(Z/Z_\odot) = A + B \tanh \left(\frac{\log M_* - \log M_c}{\Delta} \right), \tag{3.1}$$

where $\log M_c = 9.66, \Delta = 1.04, A = -0.45, B = 0.57$ and masses are expressed in solar mass units. The behavior of the relation can be readily interpreted in the framework of the basic early enrichment picture. Low-mass galaxies are prone to metal loss and therefore their metallicity is suppressed by a factor that is inversely proportional to their mass. As the gravitational potential increases – along with the loss of coherent action of individual superbubbles in driving the outflow – moving towards more massive and larger objects, Z approaches the effective yield value expected from a simple closed box model which describes well the flat trend deduced for galaxies with mass $M > 10^{10} M_\odot$. It is noticeable that a very similar mass-metallicity relation starts to emerge from observations at higher redshifts, up to $z = 3.3$ in agreement with numerical simulations which predict it to hold even at $z \approx 6$.

The second strong argument in favor of the fact that dwarf galaxies are the most likely source of the metals detected in the Lyα forest comes from the M_* dependence on the galaxy host halo mass, M_h. The general trend of this relation is one in which the dark-to-visible mass ratio steadily decreases as a function of M_* from the smallest galaxies in the sample up to $M_* \approx 10^{10.5} M_\odot$ and then remains essentially constant above the scale. A handy fit to the $M_* - M_h$ curve can be cast in the following functional form:

$$M_* = 2.3 \times 10^{10} M_\odot \frac{(M_h/3 \times 10^{11} M_\odot)^{3.1}}{1 + (M_h/3 \times 10^{11} M_\odot)^{2.2}}. \tag{3.2}$$

The interpretation of this behavior is quite straightforward. Dwarf galaxies have typical dark-to-visible mass ratios $(M_h/M_*) > 30$, a value largely exceeding the cosmological value $\Omega_M/\Omega_b \approx 6$, as a substantial amount of their gas has been ejected (along with their metals), whereas Milky Way-size galaxies have been able to retain most of their baryons and evolve in an almost closed-box mode.

These two evidences hence represent very strong arguments in favor of the dominant role played by dwarf galaxies in the metal enrichment process, which occurred along with reionization during the first billion year. The next question concerns when metals were mostly injected in the IGM and their specific abundance evolution. To answer these questions we need to resort a combination of data and theory, as explained in the next Section.

4. Metal enrichment evolution

A growing evidence has been accumulating in the last decade from the analysis of metal absorption line spectra of high redshift ($2 < z < 6$) quasars that the evolution of the most readily observed species as C IV and Si IV is close to flat. This result is particularly surprising since C IV abundance depends on both the overall carbon abundance and its ionization fraction into C IV, which in turn depends on the evolution of the UV background intensity. Fig. 1 shows a compilation of the data points showing the evolution of Ω_{CIV} together with the predicted behavior of the metallicity with redshift adopted from Scannapieco, Ferrara & Madau (2002). Apart from the uncertain conversion between C IV and C abundance, and the assumption of solar metallicity abundances in order to obtain the value of IGM metallicity Z, the striking flatness of the data is very well reproduced by the model. The ability of the model to reproduce such feature is due to the implementation of the prescription eq. 2.8 into the hierarchical build up of galaxies. The result is that in the initial phases of structure formation the most common objects (say 2σ fluctuations of the density field) are dwarfs that are able to spread their metals; hence the metallicity grows rapidly. However, when the characteristic mass scales shift into the regime of "normal" galaxies for which $\delta_B \to 0$, metals are retained within the potential well of the galaxy and the metallicity levels off to an almost constant level. One critical point concerning this explanation has to do with the fact that the UV background ionizing radiation changes by a factor ≈ 10 between $z = 2$ and $z = 6$. Yet, even assuming that Z remains constant, one would expect that the abundances of ionized species like

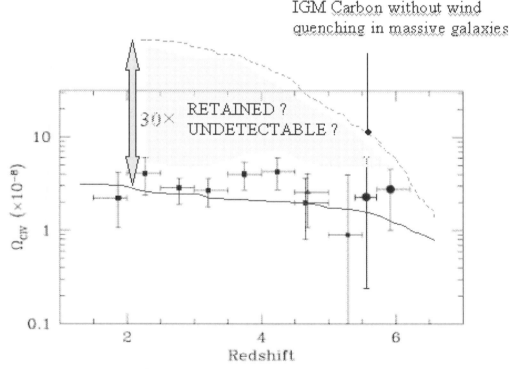

Figure 1. Data compilation for the redshift evolution of Ω_{CIV} (points) superposed to the theoretical prediction by Scannapieco, Ferrara & Madau (2002) shown by the continuous line. The dotted line shows the evolution of the same quantity without accounting for the wind quenching in massive galaxies (see text).

C IV and Si IV would change as a consequence of the different photoionization rate to which they are exposed. Nothing similar is observed. This might advocate a changing ionization correction which conspires to balance the UVB evolution to yield the flat redshift dependence observed. Such huge correction (as high as a factor of 30) can be achieved by collisional ionizations if the gas is hot, with temperatures of several 10^5 K. Although this is generally possible, it might cause difficulties related to an overproduction of species like O VI ; in addition, and more worrying, it might broaden the absorption lines to widths that exceed by far the observed ones. An alternative possibility is that the metal ionization is not dominated by the UVB but rather by a proximity of nearby star forming galaxies. There is clear evidence for clustering of metals (Pichon 2003) in regions that are typically 2–3 comoving Mpc away from large galaxies ($M \approx 10^{12} M_\odot$) and therefore it is likely that within this region the flux from the galaxy is dominating over the UVB one. Note that the such clustering *does not* imply that metals have been ejected from the large galaxy itself at the redshift at which they are observed ($z = 3 - 4$), but they could have been produced earlier by similarly biased (but low-mass) galaxies. Equally good fits to the clustering distribution of all observed ionized species are found in a similarly biased high-redshift enrichment model in which metals are placed within 2.4 comoving Mpc of $3 \times 10^9 M_\odot$ sources at $z = 7.5$.

5. A viable picture

Based on the previous considerations we propose a global scenario, illustrated schematically in Fig. 2, for the metal enrichment which is consistent with all the observations and physical processes than need to be accounted for and included in the model. Such scenario predicts that the IGM is a two-phase medium, with a hot phase at a temperature of about a million degree K and a cooler one with T of the order of several 10^4 K. The hot phase is constituted by gas that is shock-heated by supernova-driven wind from the most active galaxies at redshift $z = 3 - 5$; these galaxies are likely to be identified with the LBGs. This gas contains freshly synthesized metals, and it is located within galactic halos, thus in bubbles with comoving sizes < 1 Mpc; due to their high temperatures, low filling factor, and collisional ionization corrections, these metals hardly appear in UV/optical QSO absorption studies using O VI (let alone C IV and Si IV). The colder, photoionized component instead contains metals that were produced at higher redshifts by the dwarf galaxy population. As a result, their filling factor is about 10 times larger than the hotter component and it is much easier to detect. This simple scenario emerges from all the data discussed above, resulting in a viable metal enrichment picture. However, it can make additional predictions that concern the so-called "missing metals" problem that we are going to discuss next.

6. The missing metals

In its original formulation (Pettini 1999), the "missing metals" problem was stated as follows. Studies of the comoving luminosity density of distant galaxies allow us to trace the cosmic star formation density (or history, SFH), $\dot{\rho}_\star(z)$, up to redshift $z_{max} \approx 7$. Assuming an initial mass function of such stars (IMF), one can compute the specific fraction of heavy elements ('metals') they produce, y, and derive the metal production rate $\dot{\rho}_Z(z) = y\dot{\rho}_\star(z)$, whose integral from z_{max} gives the density of cosmic metals in units of the critical density, Ω_Z^{sfh}, at any given z. Early searches in cosmic structures for

rest being found in the hot phase; (ii) 1%-6% (3%-30%) of the observed C IV (O VI) is in the hot phase. We conclude that more than 90% of the metals produced during the star forming history can be placed in a hot phase of the IGM, without violating any observational constraint. To further constrain the hot phase parameter range, we have searched in the LP C IV line list for components with large Doppler parameters. We find no lines with $b_{CIV} \geqslant 26.5$ km s^{-1}, corresponding to $\log T_h > 5.7$; this result seems to exclude the high density and temperature region of the allowed parameter space in the middle panel of Fig. 3. We checked that the above findings are insensitive to variations of Γ_{12} of $\pm 50\%$; however, O VI /C IV ratios in the cold phase might depend on the UVB shape around 4 Ryd.

The derived values of T_h and Δ_h are suggestive of regions likely to be found around galaxies; moreover, 10^6 K gas temperature would have a scale height of > 10 kpc, hence it cannot be confined in the disk. To test this hypothesis we resort to cosmological simulations. As an illustration, Fig. 4 shows the temperature and velocity structure in a 2D cut through the center of a simulated galaxy (we used the multiphase version [Marri & White 2003] of the GADGET2 code to simulate a comoving $10h^{-1}$ Mpc^3 cosmic volume) at redshift $z = 3.3$; its total (dark + baryonic) mass is $2 \times 10^{11} M_\odot$, the star formation rate $\approx 20 M_\odot$ yr^{-1}. This galaxy has been selected to match LBG properties, but it is not unusual in the simulation volume. As often observed in LBGs, a strong galactic wind is visible, whose expansion is counteracted by energy losses due to cooling and gravity, and ram pressure exerted by the infalling gas. Infall is particularly effective at confining the wind into a broadly spherical region of physical radius ≈ 300 kpc, into which cold infalling streams of gas penetrate. Inside such wind-driven bubble the temperature (Fig. 2) is roughly constant $T \approx 10^6$ K, whereas the density spans values of $0 < \log \Delta < 5$ [$\Delta(z = 3.3) = 1$ corresponds to $\approx 2 \times 10^{-5}$ cm^{-3}]. The cool phase is evident in the outer boundary of the bubble, where cooling interfaces arise from the interaction with infalling streams. Hence halos of LBGs seem to meet the requirements as repositories of missing metals.

Additional support for this conclusion comes from studies of the correlation properties of C IV and O VI absorbers (Pichon *et al.* 2003, Aracil *et al.* 2004), which conclude that: (i) O VI absorption in the lowest density gas is usually (about 2/3 of the times) located within $\approx 300 - 400$ km s^{-1} of strong H I absorption lines; (ii) the C IV correlation function is consistent with metals confined within bubbles of typical (comoving) radius $\approx 1.4h^{-1}$ Mpc in halos of mass $M \geqslant 5 \times 10^{11} M_\odot$ at $z = 3$. If each of such objects hosts one bubble, the cosmic volume filling factor of metals is $f_Z = 11\%$; it follows that halo metallicity is $\Omega_Z^{sfh}/f_Z \Omega_b = 0.165 Z_\odot$. A temperature of $\log T_h = 5.8$ corresponds to H I (O VI) Doppler parameters $b_{HI} = 102$ ($b_{OVI} = 25.5$) km s^{-1} and to $N_{OVI}/N_{HI} = 3$; absorbers with $\log N_{OVI} = 13$ are detectable for $b_{OVI} = 25.5$ km s^{-1} but the corresponding $\log N_{HI} = 12.4$ ones for $b_{HI} = 102$ km s^{-1} are not. This raises the possibility of finding O VI absorbers without associated H I .

7. Implications

The scenario proposed leads to several interesting consequences. First, metals produced by LBGs do not seem to be able to escape from their halos, due to the confining mechanisms mentioned above. This is consistent with the prediction (Ferrara, Pettini & Shchekinov 2000) that galaxies of total mass $\mathcal{M} > 10^{12}(1 + z)^{-3/2} M_\odot$ do not eject their metals into the IGM. Interestingly, the metallicity-mass relation recently derived from the SDSS (Tremonti *et al.* 2004) shows that galaxies with *stellar* masses above $3 \times 10^{10} M_\odot$ (their total mass corresponds to $\mathcal{M}$ for a star formation efficiency $f_\star = 0.2$)

chemically evolve as "closed boxes," *i.e.* they retain their heavy elements. Second, the nearly constant $(2 \leqslant z \leqslant 5, Z \approx 3.5 \times 10^{-4} Z_\odot)$ metallicity of the low column density IGM (Songaila 2001) is naturally explained by the decreasing efficiency of metal loss from larger galaxies. Early pollution from low-mass galaxies allows a sufficient time for metals to cool after ejection; however, the majority of metals in LBG halos at lower redshifts are still too hot to be detected. Hence their contribution to the metallicity evolution of the IGM cannot be identified by absorption line experiments, which mostly sample the cool phase of the forest. Third, the rapid deceleration of the wind results either in a quasi-hydrostatic halo or in a 'galactic fountain' if radiative losses can cool the halo gas. In both cases this material is very poorly bound and likely to be stripped by ram pressure if, as it seems reasonable, the galaxy will be incorporated in the potential well of a larger object (galaxy group or cluster) corresponding to the next level of the hierarchical structure growth. Turbulence and hydrodynamic instabilities associated with this process are then likely to efficiently mix the metals into the surrounding gas within approximately a sound crossing time of ~ 1 Gyr, or $\Delta z \approx 0.5$. If metals produced and stored in LBG halos by $z = 2.3$ end up in clusters, than the average metallicity of the intracluster medium is $Z_{ICM} = \Omega_Z^{sfh}/\Omega_{ICM} = 0.31 Z_\odot$, having assumed (Fukugita, Hogan & Peebles 1998) $\Omega_{ICM} = 0.0026 h_{70}^{-1.5}$. Not only is this number tantalizingly close to the observed value at $z = 1.2$ (Tozzi *et al.* 2003), but we also predict that little evolution will be found in the ICM metallicity up to $z \approx 2$ as essentially all the metals that could have escaped galaxies during cosmic evolution had already done so by this epoch.

References

Aracil, B., Petitjean, P., Pichon, C. & Bergeron, J. 2004, A&A, 419, 811
Bahcall, J. & Peebles, J. 1969, ApJ, 156, L7
Bergeron, J., Aracil, B., Petitjean, P. & Pichon, C. 2002, A&A, 396, L11
Bergeron, J. & Herbert-Fort, S. 2005, astro-ph/0506700
Bolton, J. S., Haehnelt, M. G., Viel, M., Springel, V. 2005, MNRAS, 257, 1178
Bouwens, R. J. *et al.* 2004, astro-ph/0409488 (2004)
Carswell, B., Schaye, J. & Kim, T.-S. 2002, ApJ, 578, 43
Davé *et al.* 2001, ApJ, 552, 473
Ferrara, A., Pettini, M. & Shchekinov, Y. 2000, MNRAS, 319, 539
Fukugita, M., Hogan, C. J. & Peebles, P. J. E. 1998, ApJ, 503, 518
Haardt, F. & Madau, P. 1996, ApJ, 461, 20
Marri, S. & White, S. D. M. 2003, MNRAS, 345, 561
Panter, B. *et al.* 2008, preprint, astro-ph/0804.3091
Petitjean, P., Webb, J. K., Rauch, M., Carswell, R. F., & Lanzetta, K. 1993, MNRAS, 262, 499
Pettini, M. *ESO Workshop, Chemical Evolution from Zero to High Redshift*, 233-247 (1999)
Pichon, C., Scannapieco, E., Aracil, B., Petitjean, P., Aubert, D., Bergeron, J. & Colombi, S. 2003, ApJ, 597, L97
Prochaska, J. X. & Wolfe, A. M. 2000, ApJ, 533, L5
Prochaska, J. X., Gawiser, E., Wolfe, A. M., Castro, S. & Djorgovski, S. G. 2003, ApJL, 595, L9
Rao, S. M. & Turnshek, D. A. 2000, ApJS, 130, 1
Reddy, N. A. & Steidel, C. C. 2004, ApJ, 603, L13
Savage, B. D. & Semback, K. R. 1996, ARA&A, 34, 279
Scannapieco, E., Ferrara, A. & Madau, P. 2002, 574, 590
Scannapieco, E. *et al.*, MNRAS, 365, 615
Schaye, J., Aguirre, A., Kim, T.-S., Theuns, T., Rauch, M. & Sargent, W. L. W. 2003, ApJ, 596, 768

Shapley, A. E., Steidel, C. C., Erb, D. K., Reddy, N. A., Adelberger, K. L., Pettini, M., Barmby, P., Huang, J. 2005, ApJ, 626, 698

Simcoe, R. A., Sargent, W. L. W. & Rauch, M. 2004, ApJ, 606, 92

Songaila, A. 2001, ApJ, 561, L153

Spergel, D. N. *et al.* 2003, ApJ, 148, 175

Telfer, R. C., Kris, G. A., Zheng, W., Davidsen, D. A. & Tytler, D. 2002, ApJ, 579, 500

Tozzi, P., Rosati, P., Ettori, S., Borgani, S., Mainieri, V. & Norman, C. 2003, ApJ, 593, 705

Tremonti, C. A. *et al.* 2004, ApJ, 613, 898

Wolfe, C. A., Gawiser, E. & Prochaska, J. X. 2003, ApJ, 593, 235

Low-Metallicity Star Formation:
From the First Stars to Dwarf Galaxies
Proceedings IAU Symposium No. 255, 2008
L.K. Hunt, S. Madden & R. Schneider, eds.

Insights into the Origin of the Galaxy Mass-Metallicity Relation

Henry Lee[1], Eric F. Bell[2], and Rachel S. Somerville[2,3]

[1] Gemini Observatory, AURA Chile
Colina El Pino s/n, La Serena, Chile
email: hlee@gemini.edu

[2] Max-Planck-Institut für Astronomie
Königstuhl 17, D-69117 Heidelberg, Germany
email: bell@mpia.de

[3] Present address: Space Telescope Science Institute
3700 San Martin Drive, Baltimore, MD 21218 USA
email: somer@stsci.edu

Abstract. We examine mass–metallicity relations for nearby ($D < 2$ Mpc) gas-rich and gas-poor dwarf galaxies. We derived stellar and baryonic masses using photometric data and used average stellar iron abundances as the metallicity indicator. With the inclusion of available data for massive galaxies, we find a continuous mass–metallicity relation for galaxies spanning nine orders of magnitude in mass, and that the mass–metallicity relations are the same for both gas-rich and gas-poor dwarf galaxies. We derive stellar effective yields from the stellar abundances, finding that gas-poor dwarf galaxies form a single sequence with mass, whereas gas-rich dwarf galaxies have higher yields at comparable mass. Simple chemical evolution models show that a mass-dependent star-formation efficiency can simultaneously account for the correlations between metallicity, gas fraction, and stellar effective yield with mass. In agreement with recent and independent results, we conclude that a key driver of the mass-metallicity relation is the variation of star-formation efficiency with galaxy mass, modulated by galaxy mass-dependent outflows and/or stellar IMF variations, and coupled with environmental gas-removal processes.

Keywords. galaxies: abundances, galaxies: dwarf, galaxies: evolution, galaxies: stellar content, Local Group

1. Introduction

It has long been recognized that the relationship between the chemical abundances and the properties of galaxies is a key diagnostic of the physical processes that shape their evolution. The best-known of these relations is the luminosity-metallicity (L–Z) relation, observed in the nearby universe across a large dynamic range in luminosity from dwarf to giant galaxies for both star-forming gas-rich types (e.g., Tremonti et $al.$ 2004, Lee et $al.$ 2006) and quiescent gas-poor types (e.g., Grebel et $al.$ 2003, Gallazzi et $al.$ 2005). Although the methods most frequently used to measure metallicity depend on whether galaxies have gas (i.e., nebular oxygen abundances in gas-rich systems, and stellar iron abundances in gas-poor systems), the sense of the L–Z relations is the same: more luminous galaxies are more metal-rich.

Many different physical processes may contribute to the chemical evolution of galaxies; e.g., variations in star formation efficiency, infall of pristine gas diluting the more metal-rich interstellar medium of the galaxy, outflows driven by supernovae, ram-pressure stripping, metallicity-dependent supernova yields, systematic changes in the galaxy-scale stellar initial mass function. Many of these physical processes are demonstrated to account for the observed *L–Z* relations without the influence of other physical processes, all of which makes convergence towards a unique model of chemical evolution challenging.

2. Mass–Metallicity (*M–Z*) Relations

We took a sample of 65 nearby ($D \leqslant 2$ Mpc) dwarf galaxies, based upon the compilation in Grebel *et al.* (2003; GGH03) and the recent discovery of dwarfs in the Local Group; see Lee *et al.* (2008; LBS08) for a complete list of references. We included a sample of massive galaxies taken from the Sloan Digital Sky Survey Data Release 2 (SDSS DR2), using stellar metallicities from Gallazzi *et al.* (2005) and estimates of stellar and gas masses from Bell (2003). For the dwarf galaxies, the methods by which masses are derived and how stellar metallicities have been assigned (especially for gas-rich dwarfs) are discussed in much greater detail in LBS08.

The fits for mean stellar metallicity against stellar and baryonic mass are expressed as

$$[\text{Fe/H}] = (-3.74 \pm 0.16) + (0.354 \pm 0.025) \log(M_*/M_\odot), \tag{2.1}$$

$$[\text{Fe/H}] = (-3.68 \pm 0.16) + (0.338 \pm 0.024) \log(M_b/M_\odot). \tag{2.2}$$

There is a clear correlation between stellar metallicity and stellar (or baryonic) mass. We

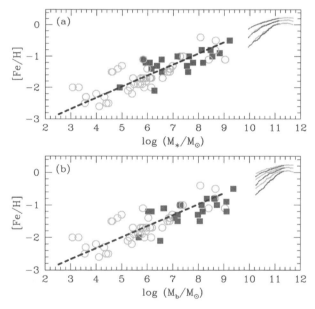

Figure 1. Mass-metallicity relations spanning ~ 9 dex in galaxy mass. Mean stellar iron abundance versus stellar mass (a) and baryonic mass (b). Open circles in orange denote nearby ($D \leqslant 2$ Mpc) gas-poor dwarf galaxies. Filled squares in blue denote nearby gas-rich dwarf galaxies. The median metallicity and upper and lower quartiles as a function of stellar mass are also plotted for massive galaxies from the SDSS DR2. The high-mass sample is coded into gas-rich (blue) and gas-poor (orange) populations. The fits in Equation (2.1) and (2.2) are shown as dashed lines in each panel.

place the *L–Z* relation on a more physical basis when this relation is expressed in terms of stellar or baryonic mass instead of galaxy luminosity. The mass-metallicity correlation has some scatter, but the scatter appears relatively constant at all masses $\leqslant 10^9 \, M_\odot$. The offset found previously in *L–Z* relations no longer appears, because we have accounted for various methods by which stellar iron abundances are obtained in dwarf galaxies with a larger mix of recent or present-day star formation (see LBS08).

3. Stellar Effective Yield

To place gas-rich galaxies and gas-poor galaxies on the same scale, we derive the *stellar effective yield*. By analogy to the gas effective yield (Pagel 1997), stellar effective yield is defined by

$$y_{\mathrm{eff}}^* = \frac{Z_{\mathrm{Fe}}/Z_{\mathrm{Fe},\odot}}{<z>} = \frac{10^{[\mathrm{Fe/H}]}}{<z>} , \qquad (3.1)$$

$$<z> = 1 + \frac{\mu \ln \mu}{1 - \mu} , \qquad (3.2)$$

where $Z_{\mathrm{Fe}}/Z_{\mathrm{Fe},\odot}$ is the iron mass fraction relative to solar, $[\mathrm{Fe/H}] = \log(\mathrm{Fe/H}) - \log(\mathrm{Fe/H})_\odot$, and $<z>$ is the mean metallicity of a stellar population with remaining cold-gas fraction μ if the galaxy evolves as a closed box with yield y_{eff}^* (i.e., as $\mu \to 0$, $<z> \to 1$ and $\log y_{\mathrm{eff}}^* \to [\mathrm{Fe/H}]$; see Smith 1985 and Pagel 1997), under the assumption that the true yield is constant as a function of time and metallicity. The stellar effective yield is the mass of metals produced per unit stellar mass formed, normalized so that solar metallicity is unity. In essence, y_{eff}^* is the metals yield required to place a given galaxy on a "closed box" chemical evolution track, and attempts to encapsulate the "potential" of the galaxy to form metals in the future.

We have plotted stellar effective yields derived from Equation (3.1) against stellar mass and baryonic mass in Fig. 2. For galaxies with no measured gas masses, we assume

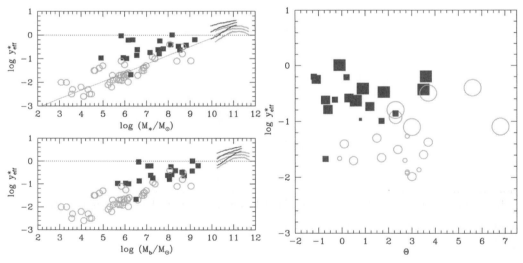

Figure 2. LEFT PANEL: Stellar effective yield versus (a) stellar mass and (b) baryonic mass. The dotted horizontal line in each panel represents $y_{\mathrm{eff}}^* = 1$. Plot symbols and colors are the same as in Fig. 1. Uncertainties have not been plotted to improve clarity. Stellar effective yields correlate well with mass for gas-poor dwarfs; the fit to these dwarfs expressed in Equation (3.3) is shown as a thin solid line in panel (a). RIGHT PANEL: Stellar effective yield versus tidal index. Tidal indices are taken from Karachentsev *et al.* (2004). Plot symbols and colors are the same as in Fig. 1. The size of the plot symbols increases with stellar mass. That gas-rich dwarfs with higher stellar effective yields are found in relative isolation is very similar to the result that gas-rich dwarfs are found at large projected galactocentric distances, e.g., GGH03.

their effective yields are equivalent to their stellar iron abundances. For the gas-poor dwarfs, there is a strong correlation between mass and effective yield, because it is almost equivalent to the M–Z relation in the case for gas-poor galaxies. The resulting ordinary least-squares bisector fit to the gas-poor dwarfs is

$$\log y_{\mathrm{eff}}^* = (-3.80 \pm 0.21) + (0.367 \pm 0.034) \log(M_*/M_\odot). \tag{3.3}$$

where M is either stellar or baryonic mass. Stellar effective yields for gas-rich dwarfs are *higher* than yields for gas-poor dwarfs at comparable mass, and are less sensitive to the dependence on stellar mass than gas-poor dwarfs.

We plot the stellar effective yield as a function of tidal index in Fig. 2, using available tidal indices from Karachentsev *et al.* (2004), where the tidal index of galaxy is a measure of proximity to a massive neighbor. Galaxies with negative tidal indices are isolated in the field, and those with positive tidal indices are found within higher-density (e.g., group) environments, where tidal interactions are more likely. We note that the tidal index is simply a measure of the *present-day* environment of a galaxy. The stellar effective yield is a function of environment: isolated (low tidal-index) gas-rich galaxies have higher stellar effective yields at a given stellar mass than group (high tidal-index) galaxies.

4. Implications

Taken together with the well-documented tendency for gas-poor dwarf galaxies to be found only in the vicinity of giant galaxies (as noted by GGH03), the present result lends support to the notion that most gas-poor dwarf galaxies have had their gas contents removed by (primarily hydrodynamical) stripping. This result helps to remove a key objection to a possible stripping origin for present-day gas-poor dwarf galaxies having evolved from gas-rich dwarf galaxies.

We explored the predictions of a number of simple models of chemical evolution, all of which reproduced the M_*–Z relation, but for different physical reasons; see LBS08 for discussion. We found that while outflows alone can successfully imprint a correct M–Z relation, such models are not capable of reproducing trends in stellar effective yields, cold-gas fractions, and star formation efficiencies. However, models in which the primary driver of the M–Z correlation is a strong variation in star formation efficiency with mass are successful at simultaneously accounting for all of these observed trends. In Fig. 3, we show how the M–Z relation is driven in equal measure by SFE variations and metal-enriched winds. There is freedom to choose model parameters in such a hybrid model, but from the scatter in the relations, it is not clear that it is a meaningful exercise to identify a "favorite" model. However, a combination of mass-dependent SFE and outflows (or potentially galaxy-scale IMF variations) can provide a framework from which one can meaningfully begin to interpret the behavior of nearby gas-rich galaxies.

We summarize with these three points. First, gas-rich and gas-poor dwarf galaxies have the same M–Z relation, which when combined with a sample of more massive galaxies spans nine orders of magnitude in galaxy mass. Second, the M–Z relation is defined primarily by the physics of gas-rich galaxies. Third, the primary driver of the M–Z relation is the variation of star-formation efficiency with galaxy mass, modulated by galaxy mass-dependent outflows and/or variations in stellar IMF, and coupled with environmental gas-removal processes (e.g., stripping) at relatively late times. These results are

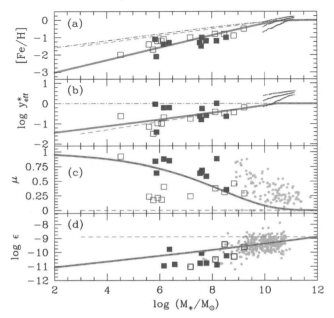

Figure 3. Trends in stellar metallicity (a), stellar effective yield (b), cold-gas fraction (c), and star-formation efficiency (d; SFE, in units of yr^{-1}) with stellar mass for gas-rich galaxies. Blue squares represent the present sample of gas-rich dwarf galaxies; open (filled) symbols indicate less (more) isolated galaxies with tidal indices larger (smaller) than +0.7. Small filled circles in grey in panel c show gas fractions for a sample of galaxies from Bell & de Jong (2000), whereas the small filled circles in grey in panel d denote SFEs for a combined sample taken from Kennicutt (1998), van Zee (2001), and Bell (2003). Toy models are shown, where both the SFE and metal-enriched winds are a strong function of mass. The solid line denotes the "combined" model, the dash-dotted line shows the effect of mass-dependent SFE only, and the dashed line shows the contribution of the mass-dependent metal-enriched winds.

consistent with the conclusions independently drawn, for example, by Dalcanton (2007) and Ellison *et al.* (2008).

Acknowledgements

HL thanks the members of the conference LOC & SOC for an excellent conference which was held at such a beautiful seaside venue. HL also acknowledges support from the Max-Planck Institute for Astronomy, the University of Minnesota, and Gemini Observatory.

References

Bell, E. F. 2003, *ApJ*, 586, 794

Bell, E. F. & de Jong, R. S. 2000, *MNRAS*, 312, 497

Dalcanton, J. J. 2007, *ApJ*, 658, 941

Ellison, S. L., Patton, D. R., Simard, L., & McConnachie, A. W. 2008, *ApJ*, 672, L107

Gallazzi, A., Charlot, S., Brinchmann, J., White, S. D. M., & Tremonti, C. A. 2005, *MNRAS*, 362, 41

Grebel, E. K., Gallagher, J. S., & Harbeck, D. 2003, *AJ*, 125, 1926 (GGH03)

Karachentsev, I. D., Karachentseva, V. E., Huchtmeier, W. K., & Kakarov, D. I. 2004, *AJ*, 127, 2031

Kennicutt, R. C. 1998, *ApJ*, 498, 541

Lee, H., Skillman, E. D., Cannon, J. M., Jackson, D. C., Gehrz, R. D., Polomski, E. F., & Woodward, C. E. 2006, *ApJ*, 647, 970

Lee, H., Bell, E. F., & Somerville, R. S. 2008, *ApJ*, submitted (LBS08)

Pagel, B. E. J. 1997, *Nucleosynthesis and the Chemical Evolution of Galaxies* (Cambridge: CUP)

Tremonti, C. A., *et al.* 2004, *ApJ*, 613, 898

van Zee, L. 2001, *AJ*, 121, 2003

Low-Metallicity Star Formation:
From the First Stars to Dwarf Galaxies
Proceedings IAU Symposium No. 255, 2008
L.K. Hunt, S. Madden & R. Schneider, eds.

© 2008 International Astronomical Union
doi:10.1017/S1743921308024654

LSD and AMAZE: the mass-metallicity relation at z > 3

F. Mannucci[1] and R. Maiolino[2]†

[1]INAF - IRA, Largo E. Fermi 5, I-50125, Firenze, Italy
email: filippo@arcetri.astro.it

[2]INAF - OAR, Roma

Abstract. We present the first results on galaxy metallicity evolution at z > 3 from two projects, LSD (Lyman-break galaxies Stellar populations and Dynamics) and AMAZE (Assessing the Mass Abundance redshift Evolution). These projects use deep near-infrared spectroscopic observations of a sample of ~40 LBGs to estimate the gas-phase metallicity from the emission lines. We derive the mass-metallicity relation at z > 3 and compare it with the same relation at lower redshift. Strong evolution from z = 0 and z = 2 to z = 3 is observed, and this finding puts strong constraints on the models of galaxy evolution. These preliminary results show that the effective oxygen yields do not increase with stellar mass, implying that the simple outflow model does not apply at z > 3.

Keywords. galaxies: abundances, galaxies: formation, galaxies: high-redshift, galaxies: starburst

1. Introduction

Metallicity is one of the most important property of galaxies, and its study is able to shed light on the detailed properties of galaxy formation. It is an integrated property, not related to the present level of star formation the galaxy, but rather to the whole past history of the galaxy. In particular, metallicity is sensitive to the fraction of baryonic mass already converted into stars, i.e., to the evolutive stage of the galaxy. Also, metallicity is affected by the presence of inflows and outflows, i.e., by feedback processes and the interplay between the forming galaxy and the intergalactic medium.

It is well known that local galaxies follow a well-defined mass-metallicity relation, where galaxies with larger stellar mass have higher metallicities (Tremonti *et al.* 2004, Lee *et al.* 2006). The origin of the relation is uncertain because several effects can be, and probably are, active. It is well known that in the local universe starburst galaxies eject a significant fraction of metal-enriched gas into the intergalactic medium because of the energetic feedback from exploding SNe, both core-collapse and, possibly, type Ia (Mannucci *et al.* 2006). Outflows are expected to be more important in low-mass galaxies, where the gravitational potential is lower and a smaller fraction of gas is retained. As a consequence, higher mass galaxies are expected to be more metal rich (see, for example, Edmunds 1990, Garnett 2002). A second possibility is related to the well known effect of "downsizing" (e.g., Cowie *et al.* 1996), i.e., lower-mass galaxies form their stars later and on longer time scales. At a given time, lower mass galaxies have formed a smaller fraction of their stars, therefore are expected to show lower metallicities. Other possibilities exist, for example some properties of star formation, as the initial mass function (IMF), could change systematically with galaxy mass (Köppen *et al.* 2007).

† on behalf of the LSD and AMAZE collaborations

All these effects have a deep impact on galaxy formation, and the knowledge of their relative contributions is of crucial importance. Different models have been built to reproduce the shape of the mass-metallicity relation in the local universe, and different assumptions produce divergent predictions at high redshifts (z > 2). To explore this issue several groups have observed the mass-metallicity relation in the distant universe, around z = 0.7 (Savaglio *et al.* 2005) and z = 2.2 (Erb *et al.* 2006). They have found a clear evolution with cosmic time, with metallicity for a given stellar mass decreasing with increasing redshift.

For several reasons, it is very interesting to explore even higher redshifts. The redshift range at z~3–4 is particularly interesting: it is before the peak of the cosmic star formation density (see, for example, Mannucci *et al.* 2007), only a small fraction (15%, Pozzetti *et al.* 2007) of the total stellar mass has already been created, the number of mergers among the galaxies is much larger than at later times (Conselice *et al.* 2007). As a consequence, the prediction of the different models tend to diverge above z = 3, and it is important to sample this redshift range observationally. The observations are really challenging because of the faintness of the targets and the precision required to obtain a reliable metallicity. Nevertheless, the new integral-field unit (IFU) instruments on 8-m class telescopes are sensitive enough to allow for the project.

2. LSD and AMAZE

Metallicity at z~3 can be obtained by measuring the fluxes of the main optical emission lines ([OII], Hβ, [OIII], Hα), whose ratios have been calibrated against metallicity in the local universe (Nagao *et al.* 2006, Kewley & Ellison 2008). Of course, this method can be applied only to line-emitting galaxies, i. e., to low-extinction, star-forming galaxies, whose line can be seen even at high-redshifts. In contrast, the gas metallicity of more quiescent and/or dust extinced galaxies, like EROs (Mannucci *et al.* 2002), DRGs (Franx *et al.* 2003), and SMGs (Chapman *et al.* 2005), cannot be easily measured at high redshifts †.

For the **AMAZE** project we observed 30 galaxies at z~3.3 from various sources (see Maiolino *et al.* 2008 for details). We only selected galaxies having a good SPITZER/IRAC photometry (3.6–8 μm), an important piece of information to derive a reliable stellar mass. These galaxies were observed, in seeing-limited mode, with the integral-field unit (IFU) spectrometer SINFONI on ESO/VLT, with integration times between 3 and 6 hours. When computing line ratios, it is important that all the lines are extracted from exactly the same aperture, to avoid differential slit losses that could spoil the line ratio. In this, the use of an IFU is of great help. We observed the H and K bands simultaneously with spectral resolution R~1500, providing a full coverage of all the most important lines. This paper is based on about 1/3 of the full data sample, while the remaining fraction is still under analysis.

For **LSD**, we extracted 10 galaxies from the UV-selected Lyman Break Galaxies (LBG) sample by Steidel *et al.* 2004. The aim of this project is not only to measure metallicities, but also to obtain spatially-resolved spectra to measure dynamics and spectral gradients. For this reason we used adaptive optics to obtain diffraction-limited images and spectra in the near-IR. The target galaxies were chosen to be within 30" of bright foreground stars, needed to drive the adaptive-optics system. As the presence of a nearby bright star is the only request, this sample, albeit small, is expected to be representative of the full population of the LBGs. For LSD we used SINFONI with the same resolution of

---

† Stellar metallicities, measured by absorption lines, can also be measured if enough observing time is provided, and will be the subject of a future work.

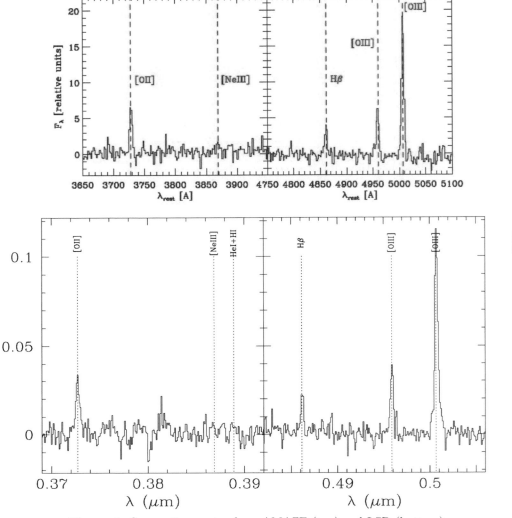

Figure 1. Composite spectra from AMAZE (top) and LSD (bottom).

AMAZE and similar integration times. About half of the LSD galaxies have been already analyzed. The typical spatial resolution is about 0.2".

Fig. 1 shows the composite spectra of AMAZE (top) and LSD (bottom). For both AMAZE and LSD, the main optical lines are detected in most of the galaxies. A few targets also show the [NeIII]3869 line, an important piece of information to derive a robust measurement of metallicity (Nagao *et al.* 2006).

3. The Mass-metallicity relation and the effective yields

Stellar masses are derived by fitting the spectral-energy distributions (SEDs) with spectrophotometric models of galaxy evolutions, as detailed in Pozzetti *et al.* (2007) and Grazian *et al.* (2007). These fits also provide estimates of the age and dust extinction of the dominant stellar population. The presence, for most of the objects, of good IRAC photometry, allows the determination of reliable stellar masses as the SED is sampled up to the rest-frame J band.

Metallicities are derived by a simultaneous fit of all the available line ratios, as explained in Maiolino *et al.* (2008). In practice, the derived value is determined by the R23 indicator or, similarly, by the [OIII]5007/Hβ ratio, while the [OIII]5007/[OII]3727 ratio is used to discriminate between the two possible branches of these ratios. The [NeIII]3869/[OII]3727 line ratio is also very important, when both lines are detected. The uncertainties on metallicity are due to both the spread of the calibration and to the observational error on the line ratio, and on average amount to 0.2–0.3 dex.

Fig. 2 shows the resulting mass-metallicity relation, compared to the same relation as measured at lower redshift. A quantitative interpretation of this result will be given when the whole data sample will be analyzed. Nevertheless, a strong metallicity evolution can be seen, i.e., galaxies at z~3.1 have metallicities ~6 times lower than galaxies of similar stellar mass in the local universe.

The presence, at z > 3, of galaxies with relatively high stellar masses (log(M/M$\odot$) = 9-11) and low metallicity put strong constraints on the process dominating galaxy formation. Several published models (e.g., de Rossi *et al.* 2007, Kobayashi *et al.* 2007) cannot account for such a strong evolution, and the physical reason for this can be traced to be due to some inappropriate assumption, for example about feedback processes or

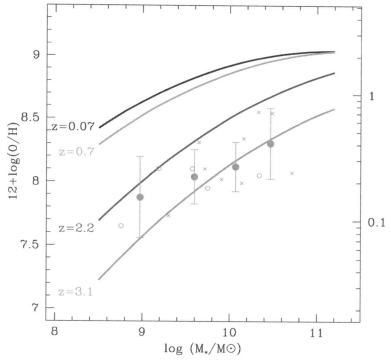

Figure 2. Evolution of the mass-metallicity relation from z = 0.07 (Kewley & Ellison 2008), to z = 0.7 (Savaglio *et al.* 2005), z = 2.2 (Erb *et al.* 2006) and z = 3.1 (AMAZE+LSD). All data have been calibrated to the same scale in order to make all the different results directly comparable. Crosses show the AMAZE galaxies, empty dots the LSD galaxies. Solid dots with error bars show the average metallicity of the galaxies in each stellar-mass bin, with the associated dispersion. The lines show quadratic fits to the data. The line corresponding to z = 3.1 shows the fit in Maiolino *et al.* (2008).

merging history. When taken at face value, some other models (e.g., Brooks *et al.* 2007, Tornatore *et al.* 2007) provide a better match with the observations, but a meaningful comparison can only be obtained by taking into account all the selection effects and observational biases.

Important hints on the physical processes shaping the mass-metallicity relation can be understood by considering the effective yields, i.e., the amount of metals produced and retained in the galaxy per unit mass of formed stars. If outflows are the main effect in shaping the mass-metallicity relation, then the effective yields are expected to increase with stellar mass because all the galaxies have converted, on average, the same fraction of baryonic mass into stars, but lower mass galaxies have lost a larger fraction of metals into the intergalactic medium. This is what is observed in the local universe by Tremonti *et al.* (2004) (but not by Lee *et al.* 2006). For LSD and AMAZE, preliminary results show that the effective yields tend to decrease, rather than increase, with stellar mass, as in Erb *et al.* (2006). Such a decrease is partly due to selection effects (see Mannucci *et al.* in preparation, for a full explanation), but the observed trend seems to be larger than what can be attributed to the effects of biases. The presence of higher yields at lower stellar mass imply that outflows decreasing with galaxy mass are not the main driver of the mass-metallicity relation at z > 3, and different possibilities, related to the efficiency of star formation (e.g., Brooks *et al.* 2007) or different outflow schemes (Erb *et al.* 2006) must be considered.

References

Brooks, A. M., *et al.*, 2007, *ApJ*, 655, L17

Chapman, S. C., *et al.*, 2005, *ApJ*, 662, 772

Conselice, S. C., *et al.*, 2007, *MNRAS*, 386, 909

Cowie, L. L., Songaila, A., Hu, E. M., & Cohen, J. G., 1996, *AJ*, 112, 839

de Rossi, M. E., Tissera, P. B., & Scannapieco, C., 2007, *MNRAS*, 374, 323

Edmunds D., 1990, *MNRAS*, 246, 678

Erb D., 2006, *ApJ*, 644, 813

Franx, M., *et al.*, 2003, *ApJ*, 587, 79

Garnett D., 2002, *ApJ*, 581, 1019

Grazian, A., *et al.*, 2007, *A&A*, 465, 393

Kewley, L. & Ellison, S. L. 2008, *ApJ*, 681, 1183

Kobayashi, C., Springel, V., & White, S. D. M. 2007, *MNRAS*, 376, 1465

Köppen, J., Weidner, C., & Kroupa, P. 2007, MNRAS, 375, 673

Lee, H., *et al.*, 2006, *ApJ*, 647, 970

Maiolino, R., *et al.*, 2008, *A&A*, 329, 57

Mannucci, F., *et al.* 2002, *MNRAS*, 329, 57

Mannucci, F., *et al.*, 2006, *MNRAS*, 360, 773

Mannucci, F., *et al.*, 2007, *A&A*, 461, 423

Nagao, F., *et al.*, 2006, *A&A*, 459, 85

Pozzetti, F., *et al.*, 2007, *A&A*, 474, 443

Savaglio, S., *et al.*, 2005, *ApJ*, 635, 260

Steidel, C., *et al.*, 2004, *ApJ*, 604, 534

Tornatore, L. *et al.*, 2007 *MNRAS*, 382, 945

Tremonti, C. A., *et al.*, 2004, *ApJ*, 613, 898

Low-Metallicity Star Formation:
From the First Stars to Dwarf Galaxies
Proceedings IAU Symposium No. 255, 2008
L.K. Hunt, S. Madden & R. Schneider, eds.

© 2008 International Astronomical Union
doi:10.1017/S1743921308024666

Three Modes of Metal-Enriched Star Formation at High Redshift

Britton D. Smith[1], Matthew J. Turk[2], Steinn Sigurdsson[3], Brian W. O'Shea[4], and Michael L. Norman[5]

[1] Center for Astrophysics & Space Astronomy, Department of Astrophysical & Planetary Sciences, University of Colorado, Boulder, CO, 80309
email: brittons@origins.colorado.edu

[2] Kavli Institute for Particle Astrophysics and Cosmology, 2575 Sand Hill Rd., Mail Stop 29, Menlo Park, CA 94025
email: mturk@slac.stanford.edu,

[3] Department of Astronomy & Astrophysics, 525 Davey Laboratory, The Pennsylvania State University, University Park, PA 16802
email: steinn@astro.psu.edu,

[4] Department of Physics & Astronomy, Michigan State University, East Lansing, MI 48824
email: bwoshea@lanl.gov

[5] Center for Astrophysics and Space Sciences, University of California at San Diego, La Jolla, CA 92093
email: mlnorman@ucsd.edu

Abstract. It is generally accepted that the very first stars in the universe were significantly more massive and formed much more in isolation than stars observed today. This suggests that there was a transition in star formation modes that was most likely related to the metallicity of the star-forming environment. We study how the addition of heavy elements alters the dynamics of collapsing gas by performing a series of numerical simulations of primordial star formation with various levels of pre-enrichment, using the adaptive mesh refinement, hydrodynamic + N-body code, Enzo. At high redshifts, the process of star formation is heavily influenced by the cosmic microwave background (CMB), which creates a temperature floor for the gas. Our results show that cloud-collapse can follow three distinct paths, depending on the metallicity. For very low metallicities ($\log_{10}(Z/Z_\odot) < -3.5$), star formation proceeds in the primordial mode, producing only massive, singular objects. For high metallicities ($\log_{10}(Z/Z_\odot) > -3$), efficient cooling from the metals cools the gas to the CMB temperature when the core density is still very low. When the gas temperature reaches the CMB temperature, the core becomes very thermally stable, and further fragmentation is heavily suppressed. In our simulations with $\log_{10}(Z/Z_\odot) > -3$, only a single object forms with a mass-scale of a few hundred $M_\odot$. We refer to this as the CMB-regulated star formation mode. For metallicities between these two limits ($-3.5 < \log_{10}(Z/Z_\odot) < -3$), the gas cools efficiently, but never reaches the CMB temperature. In this mode, termed the metallicity-regulated star formation mode, the minimum gas temperature is reached at much higher densities, allowing the core to fragment and form multiple objects with mass-scales of only a few $M_\odot$. Our results imply that the stellar initial mass function was top-heavy at very high redshift due to stars forming in the CMB-regulated mode. As the CMB temperature lowers with time, the metallicity-regulated star formation mode (producing multiple low-mass stars) operates at higher metallicities and eventually becomes the sole mode of star formation.

Keywords. stars: formation, (cosmology:) early universe

1. Simulations

We perform a series of 24 primordial star formation simulations using the Eulerian adaptive mesh refinement hydrodynamics + N-body code, Enzo (Bryan & Norman 1997,

Table 1. Summary of simulations performed. All simulations in each set have identical initial conditions. z_{col} is the redshift of collapse for the metal-free run in that set. The final column lists the metallicities of all runs performed in that set. In this work, each simulation is referred to by the letter 'r', then the number of the initial conditions, followed by the letter 'Z' and the log of the metallicity. The metal-free runs are referred to with the letters 'mf'. For example, run r2_Z−3.5 refers to the simulation with $Z = 10^{-3.5} Z_\odot$ and using the second set of initial conditions.

Set	z_{col}	$\log_{10}(Z/Z_\odot)$
1	14.8	Metal-free, -6, -5, -4.25, -4, -3.75, -3.5, -3.25, -3, -2.5, -2
2	17.4	Metal-free, -4, -3.5, -3, -2.5, -2
3	23.9	Metal-free, -4, -3.5, -3, -2.5, -2, -2*

Notes:
*This run was performed with $Z = 10^{-2} Z_\odot$, but without the CMB.

O'Shea *et al.* 2004). Excluding our three metal-free control runs, the gas in each simulation is homogeneously pre-enriched to some non-zero metallicity. The nature of the initial conditions for our simulations are identical to those used in Smith & Sigurdsson (2007). All of the simulations have the following parameters: box size of 300 h^{-1} kpc, initial redshift, $z_i = 99$, $\Omega_M = 0.3$, $\Omega_\Lambda = 0.7$, $\Omega_B = 0.04$, Hubble constant, $h = 0.7$, in units of 100 km s^{-1} Mpc^{-1}, and $\sigma_8 = 0.9$. We use three different sets of initial conditions, each with the parameters described above, but created with unique random seeds. The first set is the one used by Smith & Sigurdsson (2007). The second and third sets were initially used by O'Shea & Norman (2007) and correspond to the runs named L0_30A and L0_30D in that work. A short list of the simulations performed is given in Table 1.

   We use the second implementation of the optically-thin metal cooling method of Smith *et al.* (2008). This method uses tabulated cooling functions created with the photoionization software, Cloudy (Ferland *et al.* 1998), for all elements heavier than He, up to atomic number 30 (Zn). A solar abundance pattern is used for the metals in all of the simulations.

2. Results

   The gas phase critical metallicity, Z_{cr}, has been estimated analytically by calculating the chemical abundance required for the cooling time to equal the dynamical time at the minimum temperature reached in the collapse of metal-free gas, $n \sim 10^{3-4}$ cm^{-3} and $T \sim 200$ K (Bromm & Loeb 2003, Santoro & Shull 2006). Since this temperature minimum occurs at slightly different temperatures and densities for each of the three metal-free runs (see the solid, black curves in Figure 3), we calculate individual values of Z_{cr} for each run. The values of Z_{cr} for each run are: Set 1, $10^{-4.08} Z_\odot$; Set 2, $10^{-3.90} Z_\odot$; and Set 3, $10^{-3.85} Z_\odot$.

   In Figure 1, we show projections of baryon density of the central 0.5 pc surrounding the point of maximum baryon density for all runs in Set 1. For the runs with metallicities near or below Z_{cr} ($10^{-4.08} Z_\odot$ for this set of simulations), the central cores appear quite round and show no clear signs of forming more than one object. In the metallicity range from $10^{-3.75} Z_\odot$ to $10^{-3.25} Z_\odot$, the cores appear increasing asymmetric, with at least one additional density maximum present. However, at metallicities at or above $10^{-3} Z_\odot$, the cores return to a more spherical shape.

   In Figure 2, we plot a histogram of all the number of clumps found in each run, as a function of the metallicity of the run. Figure 2 confirms what is seen in Figure 1. In all

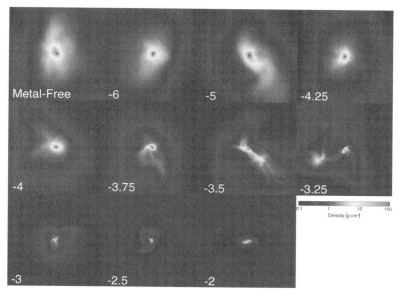

Figure 1. Density projections for the final output of all runs in Set 1. Each projection is centered on the location of maximum density in the simulation box and has a width of 0.5 pc proper. The labels in each panel indicate the log of the metallicity with respect to solar for that run.

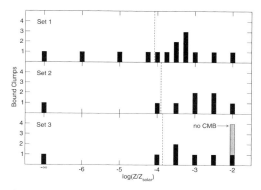

Figure 2. The number of bound clumps found within a sphere of radius 1 pc, centered on the point of maximum density, as a function of metallicity for all of the simulations performed. The grey bar in the bottom panel denotes the run where the temperature floor created by the CMB was not included. In each panel, the vertical dashed line represents the estimated value of Z_{cr} for each set.

runs with metallicities below Z_{cr}, only a single bound clump is found. As the metallicity increases, the number of clumps increases, then decreases back to only a single clump for the highest metallicities. The range of metallicities where fragmentation occurs is consistent between Sets 1 and 3, but offset by 0.5 dex toward higher metallicities for Set 2. It is not clear what causes this offset, but the qualitative trend of increasing and then decreasing number of clumps exists in all 3 sets. It is also worthwhile to note that runs r1_Z-4, r1_Z-3.75, and r2_Z-3.5, while slightly above Z_{cr}, do not show fragmentation.

In Figure 3, we plot the number density vs. gas temperature for the final output of each simulation in Sets 1 and 3. In all the runs with $Z < Z_{cr}$, the cooling is too low to prevent the temperature from rising at the H_2 thermalization density, $n \sim 10^4$ cm^{-3}. Therefore, the minimum fragmentation mass for these runs, set by the Jeans mass as the temperature

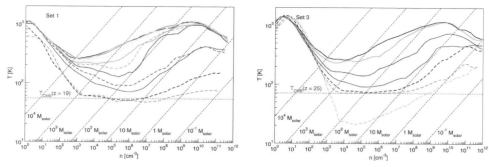

Figure 3. Mass-weighted, average gas temperature as a function of number density for all runs in Set 1 (left-panel) and all runs in Set 3 (right-panel). The metallicities are $Z = 0$ (solid-black), $10^{-6} Z_\odot$ (solid-red), $10^{-5} Z_\odot$ (yellow), $10^{-4.25} Z_\odot$ (dashed-yellow), $10^{-4} Z_\odot$ (green), $10^{-3.75} Z_\odot$ (dashed-green), $10^{-3.5} Z_\odot$ (blue), $10^{-3.25} Z_\odot$ (dashed-blue), $10^{-3} Z_\odot$ (purple), $10^{-2.5} Z_\odot$ (dashed-black), $10^{-2} Z_\odot$ (dashed-red), and $10^{-2} Z_\odot$ with the CMB removed (orange-dashed). The thin, black, dashed lines indicate lines of constant Jeans mass. The horizontal, blue, dashed lines denote the CMB temperature at the approximate redshifts of collapse.

minimum, is well over 1,000 $M_\odot$, which is nearly equivalent to the total enclosed mass. Even though runs r1_Z-4 and r1_Z-3.75 are above Z_{cr}, the additional cooling provided by the metals is not sufficient to significantly lower the minimum fragmentation mass. For runs r1_Z-3.5 and r1_-3.25, the more efficient cooling lowers the minimum fragmentation mass to just over 100 $M_\odot$, which is approximately a factor of 10 lower than the total mass within 1 pc.

For the runs with the highest metallicities, as in runs r1_Z-2.5 and r1_Z-2, the gas cools all the way to the temperature of the CMB. The cooling proceeds so efficiently that the gas has not had sufficient time to reach high densities before hitting the temperature floor of the CMB. Fragmentation can only continue as long as the temperature decreases with increasing density (Larson 1985, Larson 2005). Although the temperature decreases slightly in runs r1_Z-2.5 and r1_Z-2 for densities greater than 10^3 cm^{-3}, the temperature minimum is effectively at $n = 10^3$ cm^{-3}, where the gas reaches the CMB temperature. Near the CMB temperature, the value of the cooling rate, Λ, effectively becomes $(\Lambda(T) - \Lambda(T_{CMB}))$. Therefore, when the gas reaches the CMB temperature, the cooling rate drops to zero, and the cooling time becomes infinite. The gas cloud becomes extremely thermally stable, preventing further fragmentation. To verify that the CMB is indeed suppressing fragmentation, we run one simulation, r3_Z-2_noCMB, with the CMB temperature floor removed. With the CMB temperature floor in place, r3_Z-2 formed a single bound clump. When the CMB was removed, the gas was able to cool to roughly 20 K (dashed, orange curve in right panel of Figure 3) and form 4 bound clumps (Figure 2).

3. Discussion

The mass scale of collapsing clumps can be estimated from the Jeans mass at the end of the cooling phase. This implies the existence of three distinct metallicity regimes for star formation. In the first regime, $Z < Z_{cr}$, which we refer to as the 'primordial' mode, metals do not provide enough additional cooling to allow the gas temperature to continue to decrease monotonically with increasing density when the core reaches the H_2 thermalization density. In this case, the collapse proceeds in a similar way to the metal-free scenario, resulting in the formation of a single, massive object. At the other extreme, we define Z_{CMB} as the metallicity at which the gas can cool to the CMB temperature.

When $Z \gg Z_{CMB}$, the cloud-core will efficiently cool to the temperature of the CMB when the central density is still relatively low. In this scenario, fragmentation is limited by cooling rapidly to the CMB temperature, as the mass scale is determined by the Jeans mass at the density when the core first reaches the CMB temperature. We refer to this as the CMB-regulated star formation mode, similar to Tumlinson (2007). As fragmentation is severely limited in this mode, these stars will most likely be more massive on average than the characteristic mass of stars forming today. Finally, our simulations have shown that there exists a special range in metallicity, $Z_{cr} \leqslant Z < Z_{CMB}$, where the core does not reheat at the metal-free stalling point, but also cannot cool all the way to the CMB temperature. The minimum temperature is set only by the balance of radiative cooling and adiabatic heating. The mass scale is not regulated externally by the CMB, but rather internally by the metallicity-dependent gas-cooling. Hence, we term this the metallicity-regulated star formation mode. This mode produces the lowest mass stars of the three modes mentioned.

The CMB-regulated star formation mode creates a means by which a higher number of massive stars are formed in the very early universe, when the CMB temperature was much higher. As the universe evolves, the CMB temperature will slowly decrease, which will increase the metallicity required to reach the CMB temperature, referred to here as Z_{CMB}. The decrease in the CMB temperature also means that the fragmentation mass scale will be lower at the point where the gas reaches the cooling floor. Thus, the characteristic mass of stars produced by the CMB-regulated mode will slowly decrease with time. Such a phenomenon was predicted by Larson(1998). As the metallicity threshhold for the CMB-regulated mode advances to higher metallicity, the range of operation of the metallicity-regulated mode will extend to take its place. Observations of nearby star-forming clouds show that the minimum achievable temperature in the local universe is roughly 10 K (e.g., Evans(1999)). This implied that the CMB-regulated star formation mode is in operation up to $z \sim 2.7$, at the absolute latest.

References

Bromm, V. & Loeb, A. 2003, *Nature*, 425, 812

Bryan, G. & Norman, M. L. 1997, in Workshop on Structured Adaptive Mech Refinement Grid Methods, ed. N. Chrisochoides, IMA Volumes in Mathematics No. 117 (Springer-Verlag)

Evans, II, N. J. 1999, *ARAA*, 37, 311

Ferland, G. J., Korista, K. T., Verner, D. A., Ferguson, J. W., Kingdon, J. B., & Verner, E. M. 1998, *PASP*, 110, 761

Larson, R. B. 1985, *MNRAS*, 214, 379

—. 1998, *MNRAS*, 301, 569

—. 2005, *MNRAS*, 359, 211

O'Shea, B. W., G., B., Bordner, J., Norman, M. L., Abel, T., Harknes, R., & Kritsuk, A. 2004, in Lecture Notes in Computational Science and Engineering, Vol. 41, Adaptive Mesh Refinement - Theory and Applications, ed. T. Plewa, T. Linde, & V. G. Weirs

O'Shea, B. W. & Norman, M. L. 2007, *ApJ*, 654, 66

Santoro, F. & Shull, J. M. 2006, *ApJ*, 643, 26

Smith, B., Sigurdsson, S., & Abel, T. 2008, *MNRAS*, 385, 1443

Smith, B. D. & Sigurdsson, S. 2007, *ApJl*, 661, L5

Tumlinson, J. 2007, *ApJl*, 664, L63

Low-Metallicity Star Formation:
From the First Stars to Dwarf Galaxies
Proceedings IAU Symposium No. 255, 2008
L.K. Hunt, S. Madden & R. Schneider, eds.

Primordial Supernovae and the Assembly of the First Galaxies

Daniel Whalen[1], Bob Van Veelen[2], Brian W. O'Shea[3], and Michael L. Norman[4]

[1] Applied Physics (X-2), Los Alamos National Laboratory, Los Alamos, NM 87545
email: dwhalen@lanl.gov

[2] Astronomical Institute Utrecht, Princetonplein 5, Utrecht, The Netherlands

[3] Theoretical Astrophysics (T-6), Los Alamos National Laboratory, Los Alamos, NM 87545

[4] Center for Astrophysics and Space Sciences, University of California at San Diego,
La Jolla, CA 92093

Abstract.
Current numerical studies suggest that the first protogalaxies formed a few stars at a time and were enriched only gradually by the first heavy elements. However, these models do not resolve primordial supernova (SN) explosions or the mixing of their heavy elements with ambient gas, which could result in intervening, prompt generations of low-mass stars. We present multiscale 1D models of Population III supernovae in cosmological minihalos that evolve the blast from its earliest stages as a free expansion. We find that if the star ionizes the halo, the ejecta strongly interacts with the dense shell swept up by the H II region, potentially cooling and fragmenting it into clumps that are gravitationally unstable to collapse. If the star fails to ionize the halo, the explosion propagates metals out to 20 - 40 pc and then collapses, heavily enriching tens of thousands of solar masses of primordial gas, in contrast to previous models that suggest that such explosions 'fizzle'. Rapid formation of low-mass stars trapped in the gravitational potential well of the halo is unavoidable in these circumstances. Consequently, it is possible that far more stars were swept up into the first galaxies, at earlier times and with distinct chemical signatures, than in present models. Upcoming measurements by the *James Webb Space Telescope* (*JWST*) and *Atacama Large Millimeter Array* (*ALMA*) may discriminate between these two paradigms.

Keywords. cosmology: theory, (cosmology:) early universe, hydrodynamics, stars: early-type, supernovae

1. Introduction

Adaptive mesh refinement (AMR) and smooth particle hydrodynamics methods indicate that primordial stars form in the first cosmological dark matter halos to reach masses of a few 10^5 $M_\odot$ at $z \sim 20$ - 30, and that due to inefficient H_2 cooling they are likely very massive, 30 - 500 $M_\odot$ (Abel *et al.* 2002, Bromm *et al.* 2001, Nakamura & Umemura 2001). With surface temperatures of $\sim 10^5$ K and ionizing emissivity rates of 10^{50} s^{-1}, these stars profoundly alter the halos that give birth to them, creating H II regions 2.5 - 5 kpc in radius and sweeping half of the baryons in the halo into a dense shell that grows to the virial radius of the halo by the end of the life of the star, as first pointed out by Whalen *et al.* (2004) and Kitayama *et al.* (2004). Recent work shows that a second lower-mass star can form in the relic H II region of its predecessor in the absence of a supernova (Yoshida *et al.* 2007).

While single stars form consecutively in halos, gravitational mergers consolidate them into larger structures. Numerical attempts to follow this process find that primordial supernovae expel their heavy elements into low-density voids, where star formation is not possible. The metals return to their halos of origin on timescales of 50 - 100 Myr by accretion infall and mergers and, having been diluted by their expulsion into the IGM, are taken up into later generations of stars relatively slowly. However, the large computational boxes required to follow the formation of halos from cosmological initial conditions prevents them from fully resolving the explosions. Furthermore, the blasts themselves are not properly initialized, being set up as static bubbles of thermal energy rather than the free expansions that actually erupt through the atmosphere of the star.

If the explosion is initialized in an H II region and one is not concerned with the transport of metals or fine structure cooling, thermal 'bombs' yield an acceptable approximation to gas motion on kiloparsec scales. However, this approach cannot capture the mixture of heavy elements with ambient gas or the subsequent formation of dense enriched clumps due to metal line cooling for two reasons. First, the thermal bubble does create large pressure gradients that accelerate gas outward into the surrounding medium, but because the ejecta has no initial momentum and dynamically couples to surrounding gas at low density ($\lesssim 1$ cm^{-1}), not all the metals are launched out into the halo. Our numerical experiments demonstrate that unphysically large amounts of heavy elements remain at the origin of the coordinate grid when the explosion is implemented in this manner. Second, the early evolution of the thermal pulse is different from that of a free expansion, which exhibits the formation of reverse shocks and contact discontinuities that would destabilize in 3D and cause mixing at fairly small radii. Since dynamical instabilities in the blast that are mediated by line cooling are highly nonlinear, it is crucial to follow them from their earliest stages, something that cannot be accomplished with thermal bubbles.

If instead the progenitor fails to ionize the halo, densities at the center remain high, in excess of 10^8 cm^{-3}. When the supernova is initialized with thermal energy in this environment, all the energy of the blast is radiated away before any of the surrounding gas can be driven outward. Since the energy of the explosion is deposited as heat at the center of the grid, the temperature skyrockets to billions of degrees, instantly ionizing the gas collisionally. Bremsstrahlung and inverse Compton cooling time scales at these densities and temperatures are extremely short, leading earlier studies to erroneously conclude that such blasts 'fizzle'. In reality, even though 90% of the energy of the explosion may be lost in the first 2 - 3 years, the momentum of the free expansion cannot be radiated away, guaranteeing some propagation of ejecta out into the halo. We present a new series of Population III supernova models in both neutral and ionized cosmological minihalos to address these difficulties.

2. Code Algorithm/Models

We examined the explosion of 15, 40, and 200 $M_\odot$ stars (Type II supernovae, hypernovae, and pair-instability supernovae–PISN) in 5.9×10^5, 2.1×10^6, and 1.2×10^7 $M_\odot$ dark matter halos, which span the range in mass in which stars are expected to form by H$_2$ cooling (Whalen *et al.* 2008). Each model was carried out in two stages: first, spherically-averaged halo baryon profiles computed from cosmological initial conditions in the Enzo AMR code were imported into the ZEUS-MP code (Whalen & Norman 2006) and photoionized by the star, which is placed at the center of the halo. We then set off the explosion in the H II region of the star, using the Truelove & McKee (1999) free expansion solution for the blast profile. Each free expansion was confined to 0.0012 pc to ensure that the profile enclosed less ambient gas than ejecta mass. The explosion was then evolved with primordial gas chemistry self-consistently coupled to hydrodynamics to follow energy losses from the remnant due to line, bremsstrahlung, and inverse Compton emission in gas swept up by the ejecta. We show H II region and blast profiles in Figure 1.

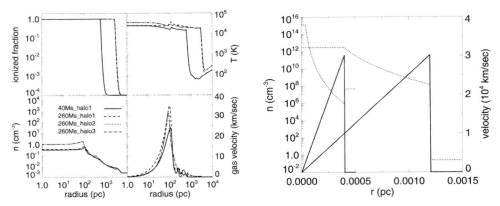

Figure 1. Left panel: H II region profiles of stars that fully ionize their halos. Right panel: Truelove & McKee free-expansion density and velocity (triangular) profiles.

The 200 $M_\odot$ star fully ionizes the first two halos and partially ionizes the most massive; the 15 and 40 $M_\odot$ stars fail to ionize the third halo, but either fully or partially ionize the two less massive halos. The explosions evolve along two distinct pathways according to whether they occur in H II regions or in neutral halos.

3. Explosions in H II Regions

Profiles of density, temperature, ionization fraction, and velocity for a 200 $M_\odot$ PISN in the 5.9×10^5 halo are shown at 31.7, 587, and 2380 yr, respectively, in Figure 2 a-d. At 31.7 yr a homologous free expansion is still visible in the density profile, which retains a flat central core and power-law dropoff. By 2380 yr the remnant has swept up more than its own mass, forming a reverse shock that is separated from the forward shock by a contact discontinuity. In 3D this contact discontinuity will break down into Rayleigh-Taylor instabilities, mixing the surrounding pristine gas with metals at small radii, 15 pc or less. Thus, mixing will occur well before the remnant collides with the shell, which we show in Figure 2 e-h at 19.8 and 420 kyr. The 400 km s^{-1} shock overtakes the 25 km s^{-1} H II region shell at $r = 85$ pc at 61.1 kyr. Its impact is so strong that a second reverse shock forms and separates from the forward shock at 420 kyr. Both shocks are visible in the density and velocity profiles of Figure 2 at 175 and 210 pc. In reality, the interaction of the SN and shell is more gradual: the remnant encounters the tail of the shell at 60 pc at 19.8 kyr, at which time the greatest radiative losses begin, tapering off by 7 Myr with the formation of another reverse shock. More than 80% of the energy of the blast is radiated away upon collision with the shell. Hydrogen Ly-α radiation dominates, followed by inverse Compton scattering, collisional excitation of He$^+$ and bremsstrahlung, but the remnant also collisionally ionizes H and He in the dense shell. It is clear that the impact of the remnant with the shell will result in a second episode of violent mixing with metals, although dynamical instabilities in the blast will set in well before the collision.

4. Explosions in Neutral Halos

In Figure 3 we show hydrodynamical profiles for a 40 $M_\odot$ hypernova in the 1.2×10^7 $M_\odot$ halo. In panels (a) - (d) are shown the formation of the reverse shock between 7.45 and 17.4 yr. The Chevalier phase is again evident, in which a reverse shock backsteps from the forward shock with an intervening contact discontinuity, but it occurs at much earlier times because much more gas resides at the center of the halo. Heavy element mixing sets in at very small radii in neutral halos. Expansion and fallback of the remnant is evident in panels (e) - (f), in which profiles are taken at 2.14 and 6.82 Myr. The hot bubble

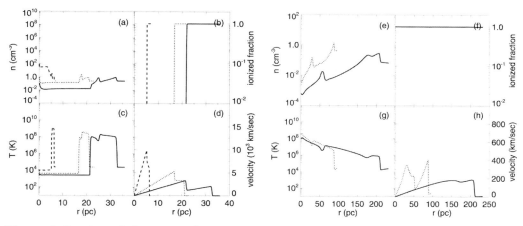

Figure 2. Panels a - d: formation of a reverse shock in a 260 $M_\odot$ PISN. Dashed: 31.7 yr; dotted: 587 yr; solid: 2380 yr. Panels e - h: collision of the remnant with the H II region shell. Dotted: 19.8 kyr; solid: 420 kyr.

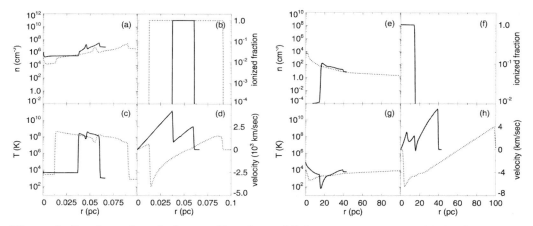

Figure 3. Panels a - d: early flow profiles of a 40 $M_\odot$ hypernova in a neutral halo. Solid: 7.45 yr; dotted: 17.4 yr. Panels e - h: collapse of the remnant. Dotted: 2.14 Myr; solid: 6.82 Myr.

reaches a final radius of $\sim$ 40 pc and then recollapses toward the center of the halo. The remnant undergoes several subsequent cycles of expansion and contraction in the gravitational potential of the dark matter, with episodes of large central accretion rates, which we show in the left panel of Figure 4. A few tens of thousands of solar masses will become enriched with metals above the threshold for low-mass star formation, with the likely result being a swarm of low-mass stars gravitationally bound to the dark matter potential of the halo. We show in the right panel of Figure 4 the final outcome of each explosion model.

5. Conclusion

Our 1D survey of primordial supernova remnant energetics in cosmological halos strongly suggest that when metals and metal line cooling are included in the next generation of 3D models, dynamical instabilities will strongly mix the surrounding gas with heavy elements. This may lead to a second, prompt generation of stars forming either in the enriched dense shell of the relic H II region or deeper within a neutral halo. If so, the first protogalaxies may have had far more stars than in current models. Large infall rates

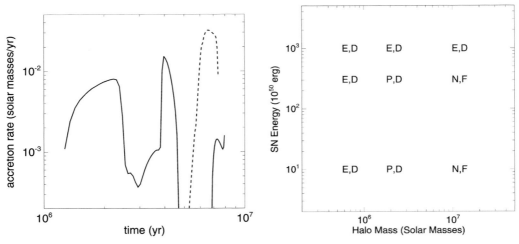

Figure 4. Left panel: Infall rates associated with fallback of the 15 (solid) and 40 (dashed) $M_\odot$ remnants in the most massive of the three halos. Right panel: eventual fate of a halo given the indicated explosion energy. The first letter refers to the final state of the halo prior to the explosion; E: photoevaporated; P: partly ionized, defined as the I-front not reaching the virial radius; N: neutral, or a failed H II region. The second letter indicates outcome of the explosion; D: destroyed, or F: fallback.

in trapped explosions may also be efficient at fueling the growth of the compact remnant of less massive progenitors, possibly providing the seeds of the supermassive black holes found in most large galaxies today.

References

Abel, T., Bryan, G. L., & Norman, M. L. 2002, Science, 295, 93

Abel, T., Wise, J. H., & Bryan, G. L. 2007, *ApJL*, 659, L87

Alvarez, M. A., Bromm, V., & Shapiro, P. R. 2006,

Bromm, V., Ferrara, A., Coppi, P. S., & Larson, R. B. 2001, *MNRAS*, 328, 969

Kitayama, T., Yoshida, N., Susa, H., & Umemura, M. 2004, *ApJ*, 613, 631

Nakamura, F. & Umemura, M. 2001, *ApJ*, 548, 19

Truelove, J. K. & McKee, C. F. 1999, *ApJS*, 120, 299

Yoshida, N., Oh, S. P., Kitayama, T., & Hernquist, L. 2007, *ApJ*, 663, 687

Whalen, D., Abel, T., & Norman, M. L. 2004, *ApJ*, 610, 14

Whalen, D. & Norman, M. L. 2006, *ApJS*, 162, 281

Whalen, D., van Veelen, B., O'Shea, B. W., & Norman, M. L. 2008, ArXiv e-prints, 801, arXiv:0801.3698

Low-Metallicity Star Formation:
From the First Stars to Dwarf Galaxies
Proceedings IAU Symposium No. 255, 2008
L.K. Hunt, S. Madden & R. Schneider, eds.

Damped Lyα systems as probes of chemical evolution over cosmological timescales

Miroslava Dessauges-Zavadsky[1]

[1]Geneva Observatory, University of Geneva, 51, Ch. des Maillettes, 1290 Sauverny, Switzerland
email: miroslava.dessauges@unige.ch

Abstract. We review the current state of knowledge of damped Lyα systems (DLAs) selected in absorption on quasar sightlines. These objects are extremely useful to study the interstellar medium of high-redshift galaxies and the nucleosynthesis in the early Universe. The characteristics of this galaxy population has been investigated for years and slowly we are getting information on their puzzling nature. Imaging at $z < 1$ shows that DLAs are associated with a mixing bag of galaxies with no especially large contribution from dwarf galaxies. Evidence for a mild evolution of the cosmic mean metallicity with time is observed. The star formation histories of these high-redshift galaxies begin to be accessible and indicate that DLAs tend to be young, gas-dominated galaxies with low star formation rates per unit area. Finally, indirect estimation of the DLA stellar masses from the mass-metallicity relations observed for emission-selected star-forming galaxies at $z = 2 - 3$ points to intermediate-mass galaxies with $M_* < 10^9$ M$_\odot$.

Keywords. ISM: abundances, galaxies: evolution, galaxies: high-redshift, quasars: absorption lines

1. Introduction

Prior to the advent of the 8–10 m class telescopes, the knowledge of galaxies at $z > 0.5$ relied almost exclusively on galaxies detected in absorption, and in particular on the so-called damped Lyα systems (DLAs), detected in QSO sightlines with the highest H I column densities ($N(\text{H I}) > 2 \times 10^{20}$ cm^{-2}). The first dedicated large survey for DLAs was done by Wolfe *et al.* (1986), and nowadays more than 1000 DLAs have been identified in the Sloan Digital Sky Survey (SDSS, Prochaska *et al.* 2005†). The study of DLAs is extremely valuable, since DLAs contain most of the neutral hydrogen available for star formation in the Universe, arise in the interstellar medium (ISM) of protogalaxies, and are believed to be the progenitors of present-day normal galaxies. Moreover, the selection of these high-redshift galaxies solely on their H I column density presents several advantages compared to the selection techniques in emission (see Wolfe *et al.* 2005 for a review): (i) it is independent from the galaxy's brightness and leads to a large diversity of luminosities and masses; (ii) it provides a better sampling of the global star formation history of the Universe; (iii) it allows to characterize the physical properties of the ISM of high-redshift galaxies; and (iv) it gives access to comprehensive samples of elemental abundances that give constraints on the galactic chemical evolution. The major drawback is the single line of sight at disposal to characterize the global properties of the high-redshift galaxy.

What have we learnt so far on this galaxy population ? Various DLA studies—imaging, neutral gas mass density census, analysis of chemical properties, kinematics—have been undertaken with the common goal that aimed at understanding the nature of this galaxy population and its link with other high-redshift galaxies identified with other selection techniques. However, despite decades of studies, it is still not clear whether the observed

† http://www.ucolick.org/~xavier/SDSSDLA/index.html

cold neutral gas associated with DLAs traces the ISM of spiral galaxies, of halos of galaxies, or of dwarf galaxies. In what follows, we will try to review the different results obtained so far.

2. DLA imaging

DLA imaging is extremely challenging, because of the strong glare of the background QSO which makes it very difficult to detect a foreground faint galaxy located only a few arcsec away from the QSO sightline. However, imaging provides crucial information in the understanding of the DLA nature by constraining the morphological type, the star formation rate, and the impact parameter which probes the extension of the neutral gas around these high-redshift galaxies.

At $z < 1$, the identification of the DLA counterparts has been successful for 13 systems so far (e.g. Le Brun *et al.* 1997; Chen & Lanzetta 2003; Rao *et al.* 2003; Chen *et al.* 2005). It appears that DLAs are associated with a mixing bag of galaxies, 45% being disk dominated, 22% bulge dominated, 11% irregulars, and 22% being located in galaxy groups. As a consequence, DLA galaxies at $z < 1$ seem to be representative of the field galaxy population, with an especially large contribution from dwarf galaxies being not necessary. Their average luminosity, $L_B \simeq 0.5 L_B^*$, points to a sub-L^* galaxy population and the impact parameters reach $r = 24 - 30 \ h^{-1}$ kpc.

At $z > 1$, the identification of the DLA counterparts is even more challenging and so far only 7 DLA galaxies have been detected, among them 3 with $z_{DLA} \approx z_{QSO}$ (e.g. Warren *et al.* 2001; Møller *et al.* 2002; Møller *et al.* 04; Weatherley *et al.* 2005). The observed Lyα luminosity varies from 1 to 10×10^{42} erg s^{-1}, corresponding to $L_B \simeq 0.6 L_B^*$. This leads to a star formation rate (SFR) of 10 to 30 M$_\odot$ yr^{-1}. Møller *et al.* (2004) suggested the existence of a possible luminosity-metallicity relation with the detected high-redshift DLA galaxies showing higher metallicities than the undetected ones.

3. Census of the neutral gas mass density with DLAs

The neutral gas mass density is defined as:

$$\Omega_g = \frac{\mu m_H H_0}{c \rho_c} \frac{\Sigma N(\text{H I})}{\Delta X}$$

where μ is the mean molecular mass of the gas, H_0 the Hubble constant, ρ_c the critical mass density, and ΔX the total path length. Ω_g has an important cosmological significance. Its evolution with cosmic time constrains the build-up of structure within hierarchical cosmology and represents a key ingredient in models of galaxy formation. It serves as the neutral gas reservoir for star formation at high redshift and describes the competition between gas accretion and star formation.

The first DLA surveys have reported statistical error on Ω_g of about 30% in redshift intervals of $\Delta z \approx 0.5$ at high redshift (e.g. Storrie-Lombardi & Wolfe 2000). With the SDSS, Prochaska *et al.* (2005) managed to derive Ω_g with a precision better than 10% (see Fig. 1). They found the first evidence for significant evolution with a 50% decrease of Ω_g from $z = 3.3$ to $z = 2.3$. It is, however, still not clear how this decrease can be explained—with star formation it seems unlikely, since the SFR peaks below $z = 2$, and with ionization it seems unlikely as well, since the intensity of the extragalactic background radiation field is constant—and how it can be reconciled with a flat gas mass density from $z = 2$ to today as supported by the low-redshift DLA survey made by Rao *et al.* (2006).

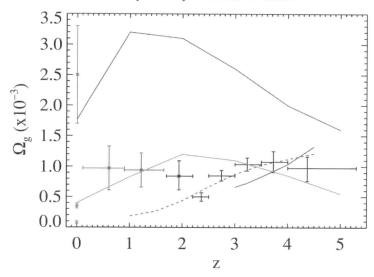

Figure 1. DLA neutral gas mass density obtained at $z > 1.5$ from the SDSS by Prochaska *et al.* (2005) and at $z < 1.5$ by Rao *et al.* (2006). These observations are compared to different theoretical models. The data points at $z = 0$ correspond from top to bottom to the stellar mass density, the neutral gas mass density of local galaxies, and the mass density of dwarf irregular galaxies. Figure reproduced from Prochaska *et al.* (2005).

The comparison with the stellar mass density shows that it is 3 times higher than the DLA gas mass density at $z \sim 3$ and suggests that every galaxy is (or was) a DLA. On the other hand, the local gas mass density matches the one of DLAs at $z \sim 2$ and indicates that the SFR seems to balance the accretion. Finally, the dwarf mass density appears to be 10 times smaller than the DLA gas mass density, showing that the majority of DLAs cannot evolve into dwarf galaxies.

4. DLA chemical properties

4.1. *DLA metallicity evolution*

Evidence for metallicity evolution as traced by DLAs has been investigated by several groups (e.g. Pettini *et al.* 1999; Vladilo *et al.* 2000; Prochaska & Wolfe 2000; Prochaska *et al.* 2003; Kulkarni & Fall 2002; Kulkarni *et al.* 2005). Currently, metallicity measurements of 162 DLAs are available from moderate- to high-resolution spectroscopy. They are assembled in Fig. 2 and show the following characteristics: (i) a large scatter of about 1 dex that is roughly constant with redshift and that can be explained by a combination of gas cross-section selection, metallicity gradients and/or different star formation histories; (ii) very few DLAs with solar metallicities in contrast with emission-selected galaxies like the Lyman-break galaxies (LBGs, e.g. Erb *et al.* 2006a); and (iii) a metallicity floor at $[M/H] \sim -3$, a value that yet significantly exceeds the detection limit of the current instruments. This questions the existence of primordial gas in the ISM of high-redshift galaxies, but on the other hand supports the idea that DLAs are linked to current/recent star formation that takes place within these galaxies.

The data show a mild evolution with redshift in both the unweighted mean metallicities (Fig. 2, large red crosses) and in the weighted mean metallicities by the H I column densities (Fig. 2, large black crosses). A decrease by -0.25 dex per redshift-bin is observed, which corresponds to a metallicity decrease by a factor of 2 per Gyr. A key implication

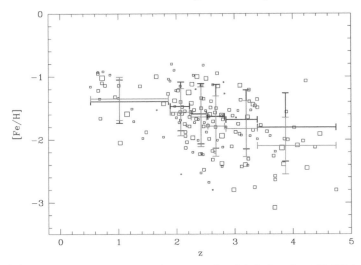

Figure 2. [Fe/H] metallicity plotted as a function of redshift for the 162 DLA measurements currently available (open squares with size proportional to the measurement errors). The large black crosses correspond to the H I weighted mean metallicities and the large red crosses to the unweighted mean metallicities. Both highlight a mild metallicity evolution with redshift.

of this result comes to light when comparing the average DLA metallicity at $z = 2.5$ with the integral under the DLA cosmic star formation history at $z > 2.5$ (Wolfe *et al.* 2003. The integrated star formation rate implies $3 - 10$ times the mass density of metals observed in DLAs, highlighting the so-called "missing metals problem", first pointed out by Pettini (1999). This suggests several possible implications: (i) DLAs may not be representative of the global galaxy population; (ii) DLAs may be dominated by dwarf or low surface brightness galaxies, as supported by the cross-section selection effect which tends to reveal faint galaxies that have larger cross-sections than bright galaxies, but is in contradiction with the H I studies at $z \sim 0$ showing that most H I resides in large spirals and with DLA imaging at $z < 1$; and (iii) the DLA selection may be biased against metal-rich systems.

Pei *et al.* (1991) was the first to suggest that DLAs with the highest metal column densities may be those with the highest dust content and may obscure the background QSO. This led Ellison et al. to build up a survey of DLAs in front of radio-loud QSOs which selection should be free from the dust obscuration effect. They showed that radio-loud QSOs point to a mild reddening and to small differences with optically-selected QSOs, both in the DLA neutral gas mass density, the number density of strong Mg II systems and the DLA average metallicity (Ellison *et al.* 2001; Ellison *et al.* 2004; Akerman *et al.* 2005). These results were further confirmed by Murphy & Liske (2004) who reviewed all the SDSS QSOs with and without DLAs and found that the dust-reddening caused by foreground DLAs is small at $z \sim 3$ with $E(B - V) < 0.02$ mag.

4.2. *DLA abundance measurement uncertainties*

There are two main effects that harm elemental gas-phase abundance measurements: dust and ionization. Indeed, a fraction of elements may not be detected in gas-phase, if locked in dust grains and/or if ionized. Dust depletion effects have been highlighted in DLAs when looking at the abundance ratios of two elements with the same nucleosynthetic origins—Fe-peak elements or α-elements. They tend to resemble the dust depletion effects observed in the Galactic warm halo gas (e.g. Hou *et al.* 2001). There are three possible ways to tackle the problem of dust depletion: (i) the study of dust-free DLAs only with [Zn/Fe] < 0.3 dex; (ii) the analysis of mildly refractory elements solely (N, O, S, Ar, and Zn); and (iii) the computation of dust depletion corrections (prescriptions of e.g. Vladilo 2002).

The high H I column densities observed in DLAs mostly imply that the bulk of the neutral gas is self-shielded. The dominant ionization states hence are the neutral one for elements with the first ionization potential > 13.6 eV (O^0, N^0) and the singly ionized one for elements with the first ionization potential < 13.6 eV (Si^+, Fe^+). The ionization effects are generally estimated as unimportant with an ionization fraction $x = H^+/(H^0 + H^+) < 10$ and ionization corrections lower than abundance measurement errors (e.g. Viegas 1995; Howk & Sembach 1999; Vladilo *et al.* 2001). However, $> 10\%$ of DLAs are certainly significantly ionized (Prochaska *et al.* 2002). Consequently, to derive intrinsic abundance measurements of a DLA, dust depletion and ionization effects have to be carefully evaluated.

4.3. *DLA abundance patterns*

In high-resolution and good signal-to-noise ratio spectra (S/N > 20 per pixel) obtained on a 8–10 m class telescope, column densities of more than 50 ions and abundances of 16 elements are accessible in DLAs: 5 α-elements O, Mg, Si, S, Ti; 6 Fe-peak elements Cr, Mn, Fe, Co, Ni, Zn; and 5 other elements C, N, Al, P, Ar (see Dessauges-Zavadsky *et al.* 2006). Is this information sufficient to constrain the star formation histories (SFHs) of these high-redshift galaxies?

On the theoretical point of view, the information at disposal is sufficient, since the SFH is entirely determined by the relative abundances versus absolute abundances and the age is constrained by the relative abundances versus the redshift. Indeed, the relative abundances depend on the stellar lifetimes, the yields and the initial mass function and hence play the role of cosmic clocks when considering two elements produced on different timescales ([α/Fe-peak], [N/α]), while the absolute abundances directly depend on the SFH type. This implies that a weak star formation or a strong+fast star formation will yield to different abundance patterns (see the contribution by Francesca Matteucci).

In practice, things are not so straightforward, mainly because of dust and ionization effects that may mimic nucleosynthetic effects. There is, for instance, an old controversy on the real α-enhancement in DLAs, usually traced by the [Si/Fe] ratio known to be contaminated by dust depletion effects (Prochaska & Wolfe 2002; Dessauges-Zavadsky *et al.* 2006). Systems with low dust contents selected with [Zn/Fe] < 0.3 dex show [Si/Fe] ratios between 0.15 to 0.35 dex, while systems with [Zn/Fe] > 0.3 dex show [Si/Fe] ratios reaching up to $0.8 - 0.9$ dex. However, the large number of elements and ions accessible in the available QSO spectra now allow to accurately apprehend the dust depletion and ionization effects, when DLAs are studied individually.

4.4. *DLA star formation histories*

To determine the SFHs and ages of the DLA galaxies, a grid of chemical evolution models has first to be built up. The models are then fitted to reproduce the observed DLA abundance patterns. Two types of chemical evolution models identified as the "spiral models" (Chiappini *et al.* 1997) and the "dwarf irregular models" (Bradamante *et al.* 1998), according to the type of galaxies they do match best, are usually considered for DLAs. The spiral models are constructed on the basis of the inside-out scenario for the disk formation with the star formation proceeding faster in the inner regions than in the outer regions. This implies that different abundance patterns are expected at different galactocentric distances. In dwarf irregular models, star formation may proceed

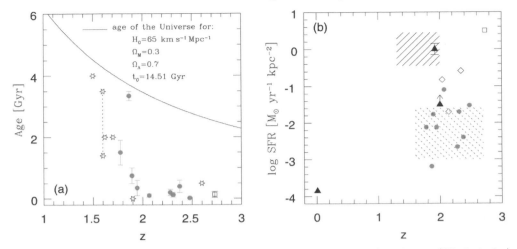

Figure 3. *Left.* Age distribution as a function of redshift for DLA galaxies (filled circles), emission-selected galaxies (stars), and the LBG MS 1512-cB58 (square). *Right.* Star formation rate per unit area distribution as a function of redshift. The DLAs are shown by filled circles. The dotted-shaded area corresponds to the SFRs obtained for DLAs by Wolfe *et al.* (2003) from the C II fine-structure line and the dashed-shaded area corresponds to the SFRs obtained for the emission-selected galaxies. The triangles represent the three DLAs for which we have an estimation of their SFR from emission lines, the square corresponds to the LBG MS 1512-cB58, and the diamonds to the gamma-ray burst host galaxies.

either in bursts separated by quiescent periods or at a low regime but continuously. This implies the need to constrain three free parameters: the star formation efficiency, the burst duration, and the time of occurrence of the burst.

This exercise has been successfully applied to 9 DLAs so far (Dessauges-Zavadsky *et al.* 2004; Dessauges-Zavadsky *et al.* 2007). No single SFH explains the diverse sets of DLA abundance patterns: 6/9 DLAs have SFHs typical of dwarf irregular bursting star formation, 1/9 DLA has a SFH typical of dwarf irregular continuous star formation (continuous SFH), and only 2/9 DLAs (each at $z < 2$) have SFHs typical of spiral outer disks (episodic bursting SFH). The common characteristic is that DLAs are weak star-forming galaxies, models with high star formation efficiencies are ruled out. The derived DLA star formation rates per unit area are moderate or low, with values between $-3.2 < \log \mathrm{SFR} < -1.1$ $M_\odot$ yr^{-1} kpc^{-2} (see Fig. 3, right panel). They are in good agreement with the DLA SFRs obtained from the fine-structure C II line by Wolfe *et al.* (2003), however they are $10 - 1000$ times lower than the SFRs of the emission-selected galaxies at the same redshifts (Savaglio *et al.* 2004; Erb *et al.* 2006b). DLAs are dominated by young objects with ages between $20 - 600$ Myr (see Fig. 3, left panel). One DLA at $z_{\mathrm{abs}} = 1.864$ in our sample has an age estimated to more than 3 Gyr, suggesting that galaxies were already formed at $z > 10$.

5. DLA kinematics

In QSO high-resolution spectra obtained on a 8–10 m class telescope, the metal-line profiles are resolved into clouds that trace the velocity fields of the DLA high-redshift galaxies. The reconstruction of the nature of these velocity fields—rotation, outflows, turbulence—is, however, very challenging with individual one-dimensional sightlines. The velocity widths, ΔV, are measured from unsaturated low-ionization lines and are often defined as the interval that contains 90% of the integrated optical width.

The large data sample currently available allowed recently to confirm the existence of a velocity-metallicity correlation for DLA galaxies (Ledoux *et al.* 2006; Prochaska *et al.* 2008). This correlation is shown in Fig. 2 from Ledoux *et al.* (2006) and the best-fit linear relation yields: $[X/H] = 1.55(\pm0.12)\log\Delta V - 4.33(\pm0.23)$. The observed dispersion, larger than the metallicity measurement uncertainties, can be explained by (i) different sightline impact parameters, (ii) different galaxy inclinations, and (iii) radial gradients in metallicity. The data also support the evidence for a redshift evolution of the velocity-metallicity relation with the median metallicity and median velocity width increasing with decreasing redshift: $[X/H] = -1.59$, $\Delta V = 69$ km s^{-1} at $z > 2.4$ and $[X/H] = -1.15$, $\Delta V = 92$ km s^{-1} at $z < 2.4$.

It is very tempting to interpret this DLA velocity-metallicity correlation, observed over more than a factor of 100 spread in metallicity, as the consequence of an underlying mass-metallicity relation for the galaxies associated with the DLA absorption. Several arguments favor this interpretation. First, peculiar ejection of hot gas should primarily affect the kinematics of high-ionization lines such as C IV and Si IV, the measurements of velocity widths based on low-ionization lines should instead be dominated by motions on galactic scale governed, or induced, by gravity. Second, Prochaska & Wolfe (1997) proposed that the observed asymmetries in the line profiles may be explained by rotating large disks. Third, the observed line profile velocity broadening may also result from the merging of proto-galactic clumps, in this case ΔV is also connected to gravity, being a good indicator of the circular velocity of the underlying dark-matter halo (Haehnelt *et al.* 1998). Fourth, as pointed out in Sect. 2, DLAs successfully imaged at $z > 2$ have high metallicities, suggesting that stellar mass correlates with enrichment. There are two counterarguments to interpreting the measured velocity widths as the tracers of the gravitational potential. First, it may be unrealistic to explain $\Delta V > 200$ km s^{-1} in terms of a single galactic potential, additional velocity fields certainly contribute. These could include galactic-scale outflows, but also peculiar motions of multiple galaxies along the sightline (e.g. Maller *et al.* 2001). Second, Bouché *et al.* (2006) reported an anti-correlation between Mg II equivalent width and the cluster length of these absorbers with large red galaxies that they interpreted as the result of a wind scenario: systems exhibiting larger equivalent widths occur in less massive, star-bursting galaxies.

On balance, the interpretation of the velocity-metallicity relation as the mass-metallicity relation is favored, moreover that a mass-metallicity relation for local galaxies is well established (e.g. Tremonti *et al.* 2004). A mass-metallicity relation is also observed for galaxies at $0.4 < z < 1$ from the Gemini Deep Deep Survey and the Canada-France Redshift Survey (Savaglio *et al.* 2005) and for UV-selected galaxies at $z \sim 2.2$ and at $z \sim 3.5$ (Erb *et al.* 2006a and Maiolino *et al.* 2008, respectively). For discussion on the evolution of the mass-metallicity relation with redshift, we refer the readers to the contribution by Filippo Mannucci. We would like just to point out that the observed redshift evolution of the DLA velocity-metallicity relation is consistent with the evolution of the mass-metallicity relation. Assuming DLAs at $z = 2 - 3$ follow a similar mass-metallicity relation as the high-redshift UV-selected galaxies, we can infer an upper limit to their stellar masses of 10^9 M$_\odot$. This suggests that DLAs are intermediate-mass galaxies.

References

Akerman, C. J., Ellison, S. L., Pettini, M., & Steidel, C. C. 2005, *A&A*, 440, 499

Bouché, N., Murphy, M. T., Péroux, C., Csabai, I., & Wild, V. 2006, *MNRAS*, 371, 495

Bradamante, F., Matteucci, F., & D'Ercole, A. 1998, *A&A*, 337, 338

Chen, H.-W. & Lanzetta, K. M. 2003, *ApJ*, 597, 706

Chen, H.-W., Kennicutt, R. C. Jr., & Rauch, M. 2005, *ApJ*, 620, 703

Chiappini, C., Matteucci, F., & Gratton, R. 1997, *ApJ*, 477, 765

Dessauges-Zavadsky, M., Calura, F., Prochaska, J. X., D'Odorico, S., & Matteucci, F. 2004, *A&A*, 416, 79

Dessauges-Zavadsky, M., Prochaska, J. X., D'Odorico, S., Calura, F., & Matteucci, F. 2006, *A&A*, 445, 93

Dessauges-Zavadsky, M., Calura, F., Prochaska, J. X., D'Odorico, S., & Matteucci, F. 2007, *A&A*, 470, 431

Ellison, S. L., Yan, L., Hook, I. M., Pettini, M., Wall, J. V., & Shaver, P. 2001, *A&A*, 379, 393

Ellison, S. L., Churchill, C. W., Rix, S., & Pettini, M. 2004, *ApJ*, 615, 118

Erb, D. K., Shapley, A. E., Pettini, M., Steidel, C. C., Reddy, N. A., & Adelberger, K.L. 2006, *ApJ*, 644, 813

Erb, D. K., Steidel, C. C., Shapley, A. E., Pettini, M., Reddy, N. A., & Adelberger, K. L. 2006, *ApJ*, 647, 128

Haehnelt, M. G., Steinmetz, M., & Rauch, M. 1998, *ApJ*, 495, 647

Hou, J. L., Boissier, S., & Prantzos, N. 2001, *A&A*, 370, 23

Howk, J. C. & Sembach, K. R. 1999, *ApJ*, 523, L141

Kulkarni, V. P. & Fall, S. M. 2002, *580*, 732

Kulkarni, V. P., Fall, S. M., Lauroesch, J. T., York, D. G., Welty, D. E., Khare, P., & Truran, J. W. 2005, *ApJ*, 618, 68

Le Brun, V., Bergeron, J., Boissé, P., & Deharveng, J. M. 1997, *A&A*, 321, 733

Ledoux, C., Petitjean, P., Fynbo, J. P., Møller, P., & Srianand, R. 2006, *A&A*, 457, 71

Maiolino, R., *et al.*, 2008, *A&A*, submitted [arXiv:0806.2410]

Maller, A. H., Prochaska, J. X., Somerville, R. S., & Primack, J. R. 2001, *MNRAS*, 326, 1475

Møller, P., Warren, S. J., Fall, S. M., Fynbo, J. U., & Jakobsen, P. 2002, *ApJ*, 574, 51

Møller, P., Fynbo, J. P. U., & Fall, S. M. 2004, *A&A*, 422, L33

Murphy, M. T. & Liske, J. 2004, *MNRAS*, 354, L31

Pei, Y. C., Fall, S. M., & Bechtold, J. 1991, *ApJ*, 378, 6

Pettini, M., Ellison, S. L., Steidel, C. C., & Bowen, D. V. 1999, *ApJ*, 510, 576

Pettini, M. 1999, in: J. R. Walsch, M. R. Rosa (eds), *ESO Workshop*, Springer-Verlag, p. 233

Prochaska, J. X. & Wolfe, A. M. 2000, *ApJ*, 533, L5

Prochaska, J. X., Henry, R. B. C., O'Meara, J. M., Tytler, D., Wolfe, A. M., Kirkman, D., Lubin, D., & Suzuki, N. 2002, *PASP*, 114, 933

Prochaska, J. X. & Wolfe, A. M. 2002, *ApJ*, 566, 68

Prochaska, J. X., Gawiser, E., Wolfe, A. M., Castro, S., & Djorgovski, S. G. 2003, *ApJ*, 595, L9

Prochaska, J. X., Herbert-Fort, S., & Wolfe, A. M. 2005, *ApJ*, 635, 123

Prochaska, J. X., Chen, H.-W., Wolfe, A. M., Dessauges-Zavadsky, M., & Bloom, J. S. 2008, *ApJ*, 672, 59

Rao, S. M., Nestor, D. B., Turnshek, D. A., Lane, W. M., Mornier, E. M., & Bergeron, J. 2003, *ApJ*, 595, 94

Rao, S. M., Turnshek, D. A., & Nestor, D. B. 2006, *636*, 610

Savaglio, S., *et al.*, 2004, *ApJ*, 602, 51

Savaglio, S., *et al.*, 2005, *ApJ*, 635, 260

Storrie-Lombardi, L. J. & Wolfe, A. M. 2000, *ApJ*, 543, 552

Tremonti, C. A., *et al.* 2004, *ApJ*, 613, 898

Viegas, S. M. 1995, *MNRAS*, 276, 268

Vladilo, G., Bonifacio, P., Centurión, M., & Molaro, P. 2000, *ApJ*, 543, 24

Vladilo, G., Centurión, M., Bonifacio, P., & Howk, J. C. 2001, *ApJ*, 557, 1007

Vladilo, G. 2002, *A&A*, 391, 407

Warren, S. J., Møller, P., Fall, S. M., & Jakobsen, P. 2001, *MNRAS*, 326, 759

Weatherley, S. J., Warren, S. J., Møller, P., Fall, S. M., Fynbo, J. U., & Croom, S. M. 2005, *MNRAS*, 358, 985

Wolfe, A. M., Turnshek, D. A., Smith, H. E., & Cohen, R. D. 1986, *ApJS*, 61, 249

Wolfe, A. M., Gawiser, E., & Prochaska, J. X. 2003, *ApJ*, 593, 235

Wolfe, A. M., Gawiser, E., & Prochaska, J. X. 2003, *ARA&A*, 43, 861

Low-Metallicity Star Formation:
From the First Stars to Dwarf Galaxies
Proceedings IAU Symposium No. 255, 2008
L.K. Hunt, S. Madden & R. Schneider, eds.

Connecting high-redshift galaxy populations through observations of local Damped Lyman Alpha dwarf galaxies

Regina E. Schulte-Ladbeck

Department of Physics and Astronomy, University of Pittsburgh,
Pittsburgh, PA 15260, USA
email: rsl@pitt.edu

Abstract. I report on observations of the $z = t\, 0.01$ dwarf galaxy SBS1543 + 593 which is projected onto the background QSO HS1543 + 5921. As a star-forming galaxy first noted in emission, this dwarf is playing a pivotal role in our understanding of high-redshift galaxy populations, because it also gives rise to a Damped Lyman Alpha system. This enabled us to analyze, for the first time, the chemical abundance of α elements in a Damped Lyman Alpha galaxy using both, emission and absorption diagnostics. We find that the abundances agree with one another within the observational uncertainties. I discuss the implications of this result for the interpretation of high-redshift galaxy observations. A catalog of dwarf-galaxy–QSO projections culled from the Sloan Digital Sky Survey is provided to stimulate future work.

Keywords. line: formation, methods: data analysis, techniques: spectroscopic, galaxies: ISM, galaxies: abundances, galaxies: dwarf, galaxies: starburst, quasars: absorption lines

1. The Problem

We now know a wealth of galaxy populations at high redshifts. Damped Lyman Alpha (DLA) systems are identified from the strong neutral Hydrogen gas absorption they cause in background QSO spectra, while Lyman Break Galaxies, Lyman Alpha Emitters, and Gamma Ray Burst (GRB) host galaxies are discovered through emission from their stars and ionized gas. What we would yet like to understand is how these populations relate to each other, and to the galaxies we see today. While the study of DLA systems has made great contributions to our empirical knowledge of the chemical evolution of galaxies, unfortunately, despite decades of observational searches it has proven extremely difficult to detect the host galaxies directly in emission (cf. Dessauges-Zavadsky, this volume). Our understanding of high redshift galaxies remains biased by how they were selected, via absorption or emission.

Figure 1 shows α element abundances for star-forming galaxies (SFG) and HII regions derived from emission lines (these are O abundances), as well as for DLA systems determined from absorption lines (O, S, or Si abundances). This figure illustrates several key points which have entered in the debate over the abundances of emission- versus absorption-selected systems. First, emission diagnostics tend to dominate the low-redshift abundance data while absorption diagnostics are still the method of choice for deriving high-redshift abundances. And second, there appears to be very little overlap in the α element abundances derived using emission versus absorption techniques. What does that mean for the galaxy populations probed?

2. Proposed Solutions

Four solutions to this problem have been thought of at this time. They are briefly summarized below.

Diagnostics Hypothesis. The diagnostics hypothesis posits that the absorption lines give lower metallicities than the emission lines because this is an inherent difference in the diagnostics. Verifying or falsifying this hypothesis has been the focus of our work, and will be the main topic of this paper.

Different Radial Biases Hypothesis. The different radial biases hypothesis predicts that lower abundances will be measured in absorption than in emission if the negative radial abundance gradient observed in local spirals (Searle 1971) holds for DLAs in general. Star-formation tends to occur in the centers of galaxies; QSO sightlines intercept foreground systems at any radii where the HI column density is above DLA threshold, thus, on average, they will be probing larger radii.

Distinct Populations Hypothesis. The distinct population hypothesis postulates that low abundances for DLAs result if they are comprised of low-mass or dwarf (proto-)galaxies (Haehnelt *et al.* 1998). Mass-metallicity relations have now been shown to exist out to redshifts of about 3 (cf. Mannucci, Lee, this volume).

Dust Obscuration Bias Hypothesis. A metal-rich interstellar medium tends to be dusty as well (cf. Hunter, Spaans, this volume). The dust obscuration bias hypothesis predicts low abundances for DLAs result if optically selected QSO samples are biased toward QSOs with little foreground extinction (Ostriker & Heisler 1984).

3. Testing the Diagnostics Hypothesis

The emission-line technique is based on the measurement of forbidden-line fluxes. Line fluxes need to be corrected for stellar and foreground dust absorptions. We then determine electron temperature, density, and ionic abundances, which, with appropriate ionization corrections, lead to element abundances. When the temperature cannot be derived, and for high-redshift galaxies, we often use a strong-line method to evaluate abundances rather than the direct or Te method. The emission-line technique for deriving abundances is described more fully in Stasinska (this volume).

The absorption-line technique is based on the measurement of absorption-line profiles. Their optical depths lead to ionic column densities. Dust depletion and ionization corrections must be considered; then we determine element abundances. When high-resolution spectra are not available, such as is the case with FUSE and HST data of DLAs, we use the curve-of-growth method to estimate an ion's column density.

What do we know about whether the physics of the line diagnostics works for astronomical objects? Williams *et al.* (2008) have recently compared the abundances of planetary nebulae measured in emission with the abundances measured in absorption toward their central stars. They find good agreement between forbidden emission-line and absorption-line abundances for a wide range of elements.

What do we know about how the diagnostics work on the scale of galaxies? And wouldn't the abundances depend on whether and how well the different phases of the interstellar medium in which they are measured are mixed? In the Galaxy, HII regions abundances interpolated for the distance of the solar circle (Deharveng *et al.* 2000) agree very well with abundances in the local ISM measured on sightlines toward nearby hot stars (Moos *et al.* 2002). This leads to the expectation that emission and absorption diagnostics should also give concordant abundances when applied to other galaxies.

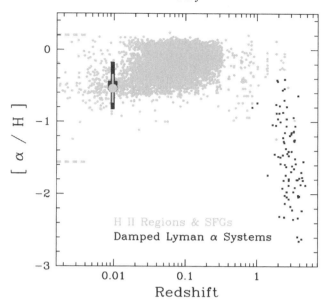

Figure 1. Alpha element abundances for SFGs and DLAs (adapted from König *et al.* 2006). The dashed lines indicate the highest and lowest metallicities measured in the local galaxy population. Our measurements for the $[O/H]_{II}$ and $[S/H]_I$ abundances of SBS1543 + 593 are shown as well.

3.1. *Test Designs*

There are two experimental designs that have been applied to the study of emission- and absorption-based abundances in external galaxies. The first method uses an internal starcluster or starburst within a galaxy (cf. contributions by Lebouteiller, Aloisi, and Thuan in this volume). The advantage of doing so is that the HII and the HI gas are both on the sightline to the starcluster. The disadvantage is that the starcluster may contain thousands of hot stars and is spatially extended; it also lies behind a complex foreground screen. The size of the aperture used for the absorption spectroscopy usually results in an integration over these many, diverse sightlines. This is a potential problem that could be fixed if a single hot star could be picked out from the cluster, but that has not been possible due to the high spatial resolution that would be required.

The second method employs the classical absorption technique by using a QSO as the background source. The advantage is that the background source is clearly a point source. The disadvantage is that the probability of finding a QSO sightline that is aligned with an HII region in a foreground galaxy is extremely small. Recall it has been hard to find DLA galaxies at all. We therefore cannot, at this time, measure the HII abundances on the same sightlines as the HI abundances. This could be fixed if we could find a QSO behind an HII region. A possible workaround is to use an HII region in a dwarf galaxy, because, to the best of our knowledge, dwarf galaxies do not exhibit radial abundance gradients (cf. Stasinska, this volume). For this to work, one has to find suitable dwarf DLA galaxy–QSO projections. Until recently, there was only one known case.

3.2. *SBS1543+593*

The alignment of SBS1543 + 593 (z = 0.0096) with the QSO HS1543 + 5921 was discovered by Reimers & Hagen (1998). Previously, the galaxy had been thought of as having a Seyfert nucleus. The discovery that the galaxy intercepts the QSO at a mere 0.5 kpc finally enabled the test of diagnostics hypothesis. Bowen *et al.* (2001) found that the sightline to the QSO gives rise to a DLA. Schulte-Ladbeck *et al.* (2004) classified the galaxy an Sdm, determined its absolute B magnitude (-16.8) and estimated its star-formation rate (0.006 $M_\odot yr^{-1}$). We also derived $12 + \log(O/H)_{II}$ is 8.2 ± 0.2, or $[O/H]_{II}$ is -0.54 ± 0.2. In Schulte-Ladbeck *et al.* (2005) we also derive a Sulphur abundance in emission, and determine $[S/H]_{II}$ is -0.27 ± 0.3. Our analysis of HST spectra, on the other hand, yielded $[O/H]_I > -2.14$ (the line tends to be saturated) and $[S/H]_I = -0.50 \pm 0.33$ (this may be a lower limit as the lines were unresolved and may therefore contain hidden saturated components). The $[O/H]_{II}$ and $[S/H]_I$ data are plotted in Fig. 1.

3.3. *Result*

Our experiment indicates that the emission- and absorption-derived α element abundances of SBS1543 + 593 agree within the errors. The **diagnostics hypothesis** has been **falsified**. In other words, the hypothesis that QSO absorption lines give lower abundances than HII emission lines for (proto-)galactic systems is not true.

3.4. *Verification*

Analysis of the same HST data by Bowen *et al.* (2005) resulted in $[O/H]_I > -0.9$ and $[S/H]_I = -0.41\pm0.06$, in good agreement with Schulte-Ladbeck *et al.* (2004, 2005). There has been no independent confirmation of the ionized gas abundances in SBS1543 + 593.

3.5. *Implications*

Our result indicates that abundances derived via HII-region emission lines and QSO absorption lines are directly comparable. The difference between the abundances measured in SFGs and DLAs is real. Our result also implies that the neutral gas of SBS1543 + 593, which extends out to 15 times its optical radius (Rosenberg *et al.* 2006), has the same abundance as the ionized gas found within the confines of the galaxy.

Abundances for high-redshift GRB host galaxies are now routinely determined from absorption lines. This is analogous to the QSO experiment since the GRB is a point source. Swift recently discovered a wealth of low-redshift GRBs. With HST lacking UV spectroscopic ability, the abundances of the host galaxies have been detemined using emission lines. Savaglio (this volume) shows an abundance-redshift plot for GRB host galaxies, combining the low-redshift emission abundances with the high-redshift absorption abundances, to discuss trends in metallicity evolution of GRBs. Our result indicates that this is indeed a valid comparison.

4. Future Work: A Catalog of Dwarf Galaxies on QSO Sightlines

It would be helpful to obtain confirmation of our result through observation of more dwarf galaxies on QSO sightlines. Table 1 is a catalog of SDSS dwarf galaxy ($M_B > -18$) & QSO pairs from B. Cherinka's thesis project. It gives the SDSS galaxy, QSO name, redshifts, galaxy luminosity (k- and foreground absorption corrected, in the g band), and ratio of impact parameter to galaxy r-band radius. Based on the strength of their CaII or NaI QSO absorption lines, potential candidates for DLA galaxies are SDSS J032803.11 + 002055.1, J125700.31 + 010143.3 (=UGC 8066), J170330.32 + 240330.8, and J221216.89 + 003243.9. It should be noted that the SDSS QSO spectra are of quite low S/N, therefore, our absorption line measurements are quite uncertain. UGC 8066, an LSB galaxy with $\log M_{HI}[M_\odot] = 9.29$ and $M_{HI}/L_B \approx 2$ (Burkholder *et al.* 2001), is perhaps our best candidate.

Table 1. Dwarf Galaxies on QSO sightlines in SDSS

SDSS Galaxy Name	z_{GAL}	SDSS QSO Name	z_{QSO}	$L_g^{k,f}$ $[L_g^*]$	b/r_{petro}
J001233.41+010014.2	0.08543	J001233.34+010010.3	1.21324	0.16	1.11
J021734.23−002637.2	0.04069	J021734.63−002641.9	1.55744	0.10	1.33
J023818.88−003030.5	0.03724	J023819.26−003029.3	2.60503	0.14	0.85
J024329.07+003833.7	0.02796	J024328.86+003831.2	2.75318	0.07	0.59
J024421.09+004031.3	0.00932	J024420.36+004029.2	2.21791	0.09	0.42
J032758.83+001652.0	0.03702	J032759.51+001713.1	2.06170	0.04	0.84
J032803.11+002055.1	0.02363	J032801.70+002100.1	0.32205	0.02	1.91
J075010.54+304106.3	0.01477	J075010.17+304032.3	1.89210	0.07	1.82
J112023.20+574429.5	0.00697	J112020.12+340555.3	0.76961	0.01	1.30
J113955.50+132802.0	0.01193	J113955.97+132713.3	1.99390	0.09	1.12
J115115.25+485331.0	0.02564	J115118.58+485331.1	1.07180	0.11	1.18
J122754.83+080525.4	0.00207	J122752.60+080526.6	1.62126	0.01	1.94
J123636.73+141333.1	0.00374	J123637.35+141316.2	1.60133	0.002	1.37
J125700.31+010143.3	0.00930	J125703.67+010132.0	0.958967	0.04	1.18
J131529.74+472958.7	0.00086	J131531.57+473054.6	1.72172	0.001	1.69
J170330.32+240330.8	0.03083	J170331.83+240339.8	0.95816	0.16	1.84
J221216.89+003243.9	0.02971	J221217.27+003227.0	2.25410	0.08	0.99
J233724.00+002330.0	0.00932	J233722.01+002238.9	1.37617	0.07	0.87

Acknowledgements. I thank my department chair, David Turnshek, for approving a travel grant that helped offset some of the cost of my conference participation. Brian Cherinka and I acknowledge the use of SDSS data (see sdss.org/collaboration/credits.html).

References

Bowen, D. V., Tripp, T. M., & Jenkins, E. B. 2001, *AJ*, 121, 1456
Bowen, D. V., Jenkins, E. B., Pettini, M., & Tripp, T. M. 2005, *ApJ*, 635, 880
Burkholder, V., Impey, C., & Sprayberry, D. 2001, *AJ*, 122, 2318
Deharveng, L., Peña, M., Caplan, J., & Costero, R. 2000, *MNRAS*, 311, 329
König, B., Schulte-Ladbeck, R. E., & Cherinka, B. 2006, *AJ*, 132, 1844
Haehnelt, M. G., Steinmetz, M., & Rauch, M. 1998, *ApJ*, 495, 647
Moos, H. W., *et al.* 2002, *ApJS*, 140, 3
Ostriker, J. P. & Heisler, J. 1984, *ApJ*, 278, 1
Reimers, D. & Hagen, H.-J. 1998, *A&A*, 329, L25
Rosenberg, J. L., Bowen, D. V., Tripp, T. M., & Brinks, E. 2006, *AJ*, 132, 478
Schulte-Ladbeck, R. E., Rao, S. M., Drozdovsky, I. O., Turnshek, D. A., Nestor, D. B., & Pettini, M. 2004, *ApJ*, 600, 613
Schulte-Ladbeck, R. E., König, B., Miller, C. J., Hopkins, A. M., Drozdovsky, I. O., Turnshek, D. A., & Hopp, U. 2005, *ApJL*, 625, L79
Searle, L. 1971 *ApJ*, 168, 327
Williams, R., Jenkins, E. B., Baldwin, J. A., Zhang, Y., Sharpee, B., Pellegrini, E., & Phillips, M. 2008, *ApJ*, 677, 1100

Low-Metallicity Star Formation:
From the First Stars to Dwarf Galaxies
Proceedings IAU Symposium No. 255, 2008
L.K. Hunt, S. Madden & R. Schneider, eds.

© 2008 International Astronomical Union
doi:10.1017/S1743921308024708

Chemical enrichment and feedback in low metallicity environments: constraints on galaxy formation

Francesca Matteucci[1,2]

[1]Dipartimento di Astronomia, Trieste University
Via G.B. Tiepolo 11, 34131 Trieste, Italy
[2]Italian National Institute for Astrophysics (INAF), Trieste
Via G.B. Tiepolo 11, 34131 Trieste, Italy
email: `matteucc@oats.inaf.it`

Abstract. Chemical evolution models for dwarf metal poor galaxies, including dwarf irregulars and dwarf spheroidals will be presented. The main ingredients necessary to build detailed models of chemical evolution including stellar nucleosynthesis, supernova progenitors, stellar lifetimes and stellar feedback will be discussed. The stellar feedback will be analysed in connection with the development of galactic winds in dwarf galaxies and their effects on the predicted abundances and abundance ratios. Model results concerning α-elements (O, Mg, Si, Ca), Fe and s-and r-process elements will be discussed and compared with the most recent observational data for metal poor galaxies of the Local Group. We will show how the study of abundance ratios versus abundances can represent a very powerful tool to infer constraints on galaxy formation mechanisms. In this framework, we will discuss whether, on the basis of their chemical properties, the dwarf galaxies of the Local Group could have been the building blocks of the Milky Way.

Keywords. ISM: abundances, galaxies: abundances, galaxies: evolution

1. Introduction

Galactic chemical evolution studies how the abundances of the elements in the interstellar medium (ISM) evolve in space and time. In order to do that we need to build models containing important prescriptions relative to the most important physical processes acting in galaxies. In particular the main parameters are:

• The initial conditions, namely whether the gas which forms a galaxy has a primordial chemical composition or has been pre-enriched by an early stellar generation. Then one should decide whether the system is closed (no exchange of gas with the surrounding) or open (infall, outflow).

• The birthrate function is defined as the product of the star formation rate (SFR), $\psi(t)$ and the initial mass function (IMF), $\varphi(m) = IMF$. The SFR can be parametrized in several ways, such as exponentially decreasing in time with a typical timescale which should be fixed by reproducing the present time properties of the studied object. However, the most common parametrization of the SFR is the Schmidt (1959) law where the SFR is proportional to some power ($k = 1$–2) of the surface gas density:

$$SFR = \nu \sigma_{gas}^{k} \tag{1.1}$$

where ν is the efficiency of star formation, namely the inverse of the star formation timescale and it is fixed by reproducing the present time SFR.

The IMF is a probability function and is normally described by a power law with one or more slopes:

$$\varphi(M) = aM^{-(1+x)} \tag{1.2}$$

and is the number of stars with masses in the interval $M, M + dM$.

The IMF has been derived only for the stars in the solar vicinity and we recall the one of Salpeter (1955) with only one slope $x = 1.35$ for stars $> 10M_\odot$ and the more recent ones, extended to low mass stars with more than one slope (Scalo 1986, 1998; Kroupa *et al.* 1993). We do not have any information about the IMF in other galaxies than the Milky Way.

• Stellar evolution and nucleosynthesis, in other words the stellar yields. Stars contribute in a different way to the galactic chemical enrichment according to their mass:

–Low and Intermediate mass stars ($0.8 \leqslant M/M_\odot \leqslant 8.0$) produce ^{4}He, C, N (primary and secondary) and s-process ($A > 90$) elements.

–Massive stars ($M \geqslant 10M_\odot$) which die as core-collapse SNe (SNe II and SNe Ib/c) produce mainly α-elements (O, Ne, Mg, Si, S, Ca), some Fe-peak elements, s-process elements ($A < 90$) and r-process elements during the SN explosion.

–White dwarfs (WD) of C and O (therefore originating in the low and intermediate mass range) in binary systems can give rise to Type Ia SNe which produce mainly Fe and Fe-peak elements. The most common scenarios for the progenitors of these SNe are: i) the Single Degenerate (SD) scenario and ii) the Double Degenerate (DD) scenario. The SD scenario, namely the classical scenario of Whelan and Iben (1973), consists in the C-deflagration in a C-O WD reaching the Chandrasekhar mass, M_{Ch}, after accreting material from a red giant companion. The minimum time scale for the first explosion to occur in this scenario is $t_{SNIa_{min}} = 0.03$ Gyr (Greggio and Renzini 1983; Matteucci & Recchi 2001) and the maximum is as long as a Hubble time. The DD scenario consists in the merging of two C-O WDs, due to gravitational wave radiation, which explode by C-deflagration when M_{Ch} is reached (Iben and Tutukov 1984). The minimum timescale for the first explosion here is only slightly longer ~ 0.031 Gyr (Greggio 2005), while the maximum timescale can be several Hubble times. Recently an empirical bimodal SN Ia rate (Mannucci *et al.* 2005;2006) has been proposed: $\sim 50\%$ of all Type Ia SNe arise from binary systems with lifetimes $< 10^8$ years, whereas the rest comes from smaller progenitors with a much broader distribution of stellar lifetimes (see also Matteucci *et al.* 2006).

–Very Massive Objects (VMO) ($M > 100M_\odot$, Pop III), if they ever existed, could form only up to a metallicity of $(10^{-6} - 10^{-4})Z_\odot$ (Schneider *et al.* 2006) and therefore their effect on the element production was negligible (Ballero *et al.* 2006).

• Gas infall and/or outflow from the system. For the infall rate (IR) normally an exponential function is assumed:

$$IR = Ce^{-t/\tau}, \tag{1.3}$$

where τ is a typical timescale. For the galactic wind the most common parametrization is to adopt that the wind rate (WR) is several times the star formation rate:

$$WR = \lambda SFR. \tag{1.4}$$

The constants C and λ are suitable parameters chosen to reproduce the present time properties of the studied galaxies.

Chemodynamical models of dwarf galaxies have shown that galactic winds are easily produced in these galaxies and that they are metal-enriched (e.g. McLow & Ferrara, 1999; D'Ercole & Brighenti 1999; Recchi *et al.* 2001).

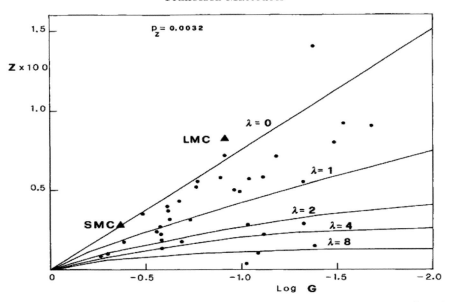

Figure 1. The Z-*logG* diagram. Solution b.)(see text) from Matteucci & Chiosi (1983). Note that the solution of the Simple Model ($\lambda = 0$) corresponds to the effective yield p_Z, which is the highest possible yield per stellar generation.

Here we will describe the chemical evolution of low metallicity dwarf galaxies, in particular of dwarf irregulars and blue compact galaxies where both the infall and outflow are taken into account.

2. Dwarf Irregulars

Dwarf Irregular (DIG) and Blue Compact galaxies (BCG) are important objects where to test galaxy evolution. They have a low metal content (abundances measured from HII regions) and large gas fractions suggesting that they either suffered isolated bursts of star formation and/or a low but continuous star formation (Searle *et al.* 1973). Both DIG and BCG show a distinctive spread in their chemical properties, although this spread is decreasing with the new more accurate data, but also a definite mass-metallicity relation.

From the point of view of chemical evolution, Matteucci and Chiosi (1983) first studied the evolution of DIG and BCG by means of analytical chemical evolution models including either outflow or infall and concluded that: closed-box models cannot account for the Z-log $G(G = M_{gas}/M_{tot})$ distribution even if the number of bursts varies from galaxy to galaxy and suggested possible solutions to explain the observed spread. The possible solutions suggested were: a.) different IMF's, b.) different rates of galactic wind and c.) different rates of infall. In Figure 1 we show graphically the solution b.) which refers to different rates of galactic wind obtained just by varying λ. Later on, Matteucci & Tosi (1985) tried the three solutions in a detailed numerical model and concluded that the most likely solution is b.), on the basis of reproducing the majority of the observed features in these galaxies, including the mass-metallicity relation.

Later on, Pilyugin (1993) forwarded the idea that the observed spread could be due to self-pollution of the HII regions, which do not mix efficiently with the surrounding medium, coupled with "enriched" or "differential" galactic winds, namely different chemical elements are lost at different rates. Other models in the following years (Marconi *et al.* 1994; Kunth *et al.* 1995; Bradamante *et al.* 1998) followed the suggestions of differential

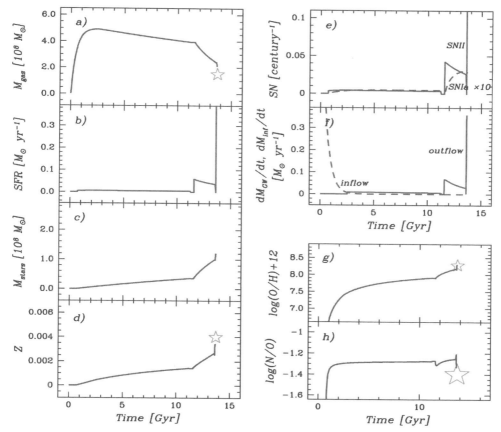

Figure 2. Comparison between data and model for NGC1569. The model (continuous lines) is from Romano *et al.* (2006), where the references to the data (stars) can also be found.

winds and introduced the novelty of the contribution to the chemical enrichment and energetics of the ISM by SNe of different Type (II, Ia and Ib).

2.1. *Galactic outflows*

Galactic outflows have been observed in some irregular dwarfs (e.g. Martin *et al.* 2002). More recently outflows have been observed with Chandra (Ott *et al.* 2005) in several star bursting dwarfs including IZw18 (see Aloisi this conference) and NGC1569 (see later). However, it is not entirely clear whether the outflowing material will eventually be lost from the potential well of the galaxy (wind) or it will cool and recollapse back into the galaxy (galactic fountain).

In models of chemical evolution of dwarf irregulars (e.g. Bradamante *et al.* 1998) the feedback effects are taken into account and the condition for the development of a wind is:

$$(E_{th})_{ISM} \geqslant E_{Bgas} \qquad (2.1)$$

namely, that the thermal energy of the gas is larger or equal to its binding energy.

The thermal energy of gas due to SN and stellar wind heating is:

$$(E_{th})_{ISM} = E_{th_{SN}} + E_{th_w}, \qquad (2.2)$$

but the contribution from stellar wind is negligible relative to that of SNe. Two parameters here play a fundamental role: the efficiency transfer of the SN energy and the stellar wind energy into the ISM. In other words, one has to assume that a fraction of the initial SN blast wave energy ($E_o = 10^{51}$ erg) is going to thermalize the ISM, and the same for stellar winds although they can be neglected. These efficiencies are unfortunately still largely unknown. In Bradamante *et al.* (1998) the SN energy transfer efficiency is assumed to be 3% but it depends on the physical conditions of the ISM and in some situation it can be 100% (see Recchi *et al.* 2001).

On the other hand, the binding energy of the gas, $E_{Bgas}(t)$, should be computed by assuming a dark matter halo. The typical model for a BCG has a luminous mass of $(10^8 - 10^9)M_\odot$, a dark matter halo ten times larger than the luminous mass and various values for the parameter describing the distribution of the dark matter halo (in the Bradamante *et al.* paper the dark matter is 10 times more diffuse than the luminous matter). It is found that the galactic wind in these galaxies develops easily but it carries out mainly metals so that the total mass lost in the wind is small. These results have been confirmed by more sophisticated chemo-dynamical models (see Recchi *et al.* 2001).

2.2. *A specific galaxy: NGC1569*

As an example we show the chemical evolution results for a local dwarf irregular NGC1569, which has been observed with the *Hubble Space Telescope* (HST) and its color-magnitude diagram derived. From this color-magnitude diagram it has been inferred the star formation history of this object, in particular that it suffered two major bursts in the last 2 Gyrs and low but continuous star formation before (Angeretti *et al.* 2005). Gas outflows have also been observed (Ott *et al.* 2005). Romano *et al.* (2006) modeled the chemical evolution of NGC1569 following the SFR suggested by the color-magnitude diagram. They assumed that the rate of mass outflow is proportional to the Type II and Ia SN

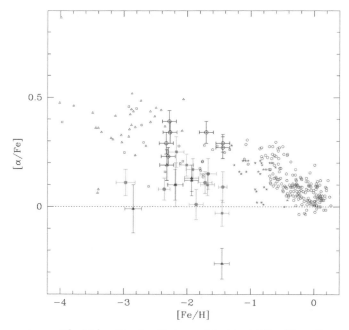

Figure 3. Comparison of [α/Fe] ratios in dSphs and in the Milky Way: figure from Shetrone *et al.* (2003). The blue dots represent Galactic stars whereas the others are the dSphs.

rates instead than to the SFR. This implies a more realistic treatment of the outflow relative to previous models where the wind stops when the SF stops. In this case, instead, the winds continue due to the energetic input of Type Ia SNe. In figure 2 we show the results for NGC1569 compared with the observations.

3. Dwarf Spheroidals

Dwarf spheroidal galaxies (dSphs) are small systems surrounding the Milky Way which are often indicated as the building blocks of the Galaxy in the framework of a hierarchical growth of structures. It is interesting to compare the abundance ratios, recently measured in an accurate way also in these systems, with those of the Milky Way. In Figure 3 we

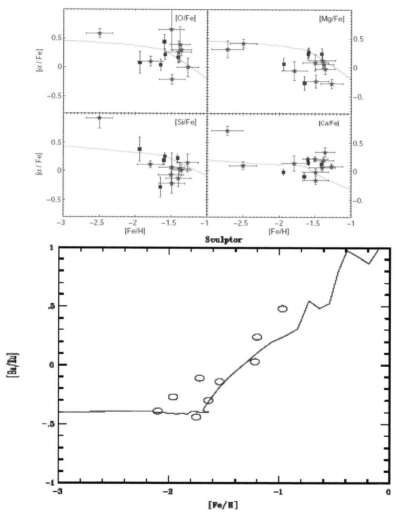

Figure 4. Upper figure: abundance ratios for the α-elements O, Mg, Si and Ca, as measured by Koch *et al.* (2008), shown as blue stars, in comparison with the model calculations from LMC06 (solid lines). Also plotted are the five data points from Shethrone *et al.* (2003) (black squares). Figure and the rest of data from Koch *et al.* (2008). Lower figure: measured and predicted [Ba/Eu] ratios for the dSph Sculptor. The model is represented by the solid line and it is from LMC06. Figure and data from Geisler *et al.* (2007).

show such a comparison, indicating that apart from a small overlap of the dSph data with the Galactic data at relatively low metallicities, the dSph data have a quite different behaviour, showing low [α/Fe] ratios at low metallicities. This behaviour resembles the one predicted for systems with slow star formation (see Matteucci 2001). In fact, the time-delay with which the bulk of Fe is restored by Type Ia SNe, when combined with different star formation rates, produces different behaviours for the [α/Fe] ratios: a fast star formation produces a longer plateau, whereas a slow star formation produces a smaller plateau and a steeper decline, relative to the solar vicinity. This induces to think that the different behaviour of the [α/Fe] ratios in dSphs has something to do with their SF histories. To check this point Lanfranchi & Matteucci (2004) and Lanfranchi *et al.* (2006, hereafter LMC06) have computed models for dSphs by adopting the SF histories suggested by the color-magnitude diagrams of these systems. Their models include also powerful galactic winds assumed to be proportional to the SFR, which devoid the dSphs of all the residual gas after star formation. As one can see in Figure 4, where model predictions and observations are shown, the model reproduces quite well the observations of Carina. This is due to the combination of a slow SFR and a strong wind which subtracts gas from the system thus further lowering the SFR. Always in Figure 4 we show the predictions of LMC06 for the [Ba/Eu] ratio in Sculptor and the agreement is also quite good. Finally, LMC06 reproduced very well the stellar metallicity distribution in Carina including the lack of stars with [Fe/H]<-3.0 (Helmi *et al.* 2006): this is due to the assumption that these galaxies do not evolve as a closed box but infall played a role in the early stages of their formation.

4. Conclusions

We have described the evolution of low metallicity dwarfs, in particular dwarf irregulars and dwarf spheroidals.

Our main conclusions can be summarized as:

• The main properties of dwarf irregulars can be well reproduced with metal enhanced galactic winds. The winds carry out mostly metals, with a small net loss of gas. The star formation in these systems could have proceeded either in small bursts or at very low level but continuously with bursts overimposed.

• The dSphs instead must have suffered huge losses of gas, which lasted for several Gyrs, coupled with a low star formation efficiency. Low star formation and strong winds allow us to reproduce most of their observed abundance ratios (α- and s- and r-process elements). DSphs seem to have evolved differently from the Milky Way, as shown by their different [α/Fe] and [s,r/Fe] ratios.

• The observed lack of stars with [Fe/H] <-3.0 dex in dSphs can be well explained if these galaxies formed by infall of gas on a timescale of ~ 0.5 Gyr. The problem is similar to the G-dwarf metallicity distribution in the solar vicinity.

• Objects similar to the local dwarf irregulars and dSphs are unlikely to have been the building blocks of the Milky Way.

References

Angeretti, L., Tosi, M., Greggio, L., Sabbi, E., Aloisi, A. & Leitherer, Claus 2005, *AJ*, 129, 2203
Ballero, S.K., Matteucci, F. & Chiappini, C. 2006, *New Astr.*, 11, 306
Bradamante, F., Matteucci, F. & D'Ercole, A. 1998, *A&A*, 337, 338
D'Ercole, A. & Brighenti, F. 1999, *MNRAS*, 309, 941
Geisler, D., Wallerstein, G., Smith, V. V. & Casetti-Dinescu, D. I. 2007, *PASP* 119, 939

optical depth is dependent on both the mass and the radius. For the elliptical galaxies, to match the large amount of reprocessing observed in the SEDs of the SCUBA galaxies in the SHADES dataset, the time spent initially by the young stars in the dense MCs was increased to a larger value than that found for local normal starforming galaxies.

Also investigated are the potential errors you could expect if instead of following the evolution of the dust component of the ISM, two common assumption are adopted. These *errors* are given as the difference between our fiducial models with the detailed dust treatment and the models using the two simplifications. They will only be correct for the models presented here and not for any real galaxy in particular. However the chemical evolution models have been thoroughly tested against observations in previous works and, in this work the SEDs have been shown to be broadly consistent with the SEDs of the morphological type modeled. In addition, in all cases, the models generated using the dust evolution of CPM08 fit the observations better than the models generated when the two assumptions are adopted (figures 1b, c and d). Therefore the errors presented in this work can be viewed as an example of the magnitude of errors that could be expected if the two assumptions are adopted in spiral, irregular and elliptical galaxies. In particular it has been shown that, for spiral galaxies, the errors that could be expected are small and decrease with the age of the galaxy (see figure 2a). This is unsurprising, since these assumptions have been developed based on observations of the Milky Way and normal local starforming galaxies. For irregular and elliptical galaxies, however, it has been shown that, potentially, the errors introduced by adopting the assumptions could be much larger (figures 2b and 2c). Of the assumptions it has been shown that, in all galactic types, the largest errors will be introduced if a relationship for the dust-to-gas ratio with the metallicity is adopted, instead of using a full dust evolution model to calculate this quantity. These errors would be largest in the IR part of the spectrum.

Acknowledgments

This work was supported through a Marie Curie studentship for the 6th Framework Research and Training network MAGPOP, contract number MRTN-CT-2004-503929.

References

Bauer, A. E., Drory, N., Hill, G. J., & Feulner, G. 2005, *ApJ*, 621, L89
Calura, F., Pipino, A., & Matteucci, F. 2008, *A&A*, 479, 669
Clements *et al.*, 2008, *MNRAS*, 387, 247
Dale, D. A. & Helou, G. 2002, *ApJ*, 576, 159
Dale D. A., Gil de Paz A., Gordon K. D., *et al.* 2007 *ApJ*, 655, 863
Eales S., Lilly S., Webb T., Dunne L., Gear W., Clements D., & Yun M., 2000, *AJ*, 120, 2244
Granato G. L. *et al.*, 2000, *ApJ*, 542, 710
Hammer, F., Flores, H., Elbaz, D., Zheng, X. Z., Liang, Y. C., & Cesarsky, C. 2005, A&A, 430, 115
Hauser, M. G. & Dwek, E. 2001, *ARA&A*, 39, 249
Iglesias-Pramo, J. *et al.* 2007 *ApJ*, 670, 279
Panuzzo, P., Granato, G. L., Buat, V., Inoue, A. K., Silva, L., Iglesias-Pramo, J., & Bressan, A 2007 *MNRAS*, 375, 640
Lilly, S. J., Eales, S. A., Gear, W. K. P., *et al.* 1999, *ApJ*, 518, 641
Popescu, C. C. & Tuffs, R. J., 2005 *AIPC*, 761, 155P
Silva, L., Granato, G. L., Bressan, A., & Danese, L., 1998, *ApJ*, 509, 103
Vega, O., Clemens, M. S., Bressan, A., Granato, G. L., Silva, L., & Panuzzo, P. 2008, *A&A*, 484, 631

Low-Metallicity Star Formation:
From the First Stars to Dwarf Galaxies
Proceedings IAU Symposium No. 255, 2008
L.K. Hunt, S. Madden & R. Schneider, eds.

About the Chemical Evolution of dSphs (and the peculiar Globular Cluster ω Cen)

Andrea Marcolini[1] and Annibale D'Ercole[2]

[1] Centre for Astrophysics, University of Central Lancashire,
Preston, Lancashire, PR1 2HE, United Kingdom
email: amarcolini@uclan.ac.uk

[2] INAF, Osservatorio Astronomico di Bologna,
via Ranzani 1, 40127 Bologna, Italy
email: annibale.dercole@bo.astro.it

Abstract. We present three dimensional hydrodynamical simulations aimed at studying the dynamical and chemical evolution of the interstellar medium (ISM) in isolated dwarf spheroidal galaxies (dSphs). This evolution is driven by the explosion of Type II and Type Ia supernovae, whose different contribution on both the dynamics and chemical enrichment is taken into account. Radiative losses are effective in radiating away the huge amount of energy released by SNe explosions, and the dSph is able to retain most of the gas allowing a long period ($\geqslant 2 - 3$ Gyr) of star formation, as usually observed in this kind of galaxies. We are able to reproduce the stellar metallicity distribution function (MDF) as well as the peculiar chemical properties of strongly O-depleted stars observed in several dSphs. The model also naturally predicts two different stellar populations, with an anti-correlation between [Fe/H] and velocity dispersion, similarly to what observed in the Sculptor and Fornax dSphs. These results derive from the inhomogeneous pollution of the SNe Ia, a distinctive characteristic of our model. We also applied the model to the peculiar globular cluster (GC) ω Cen in the hypothesis that it is the remnant of a formerly larger stellar system, possibly a dSph.

Keywords. hydrodynamics, methods: numerical, stars: abundances, ISM: evolution, ISM: abundances, galaxies: dwarf, (Galaxy:) globular clusters: individual (ω Cen)

1. Introduction

Due to their proximity, the galaxies of the Local Group (see Mateo 1995 and Geisler *et al.* 2007 for a review) offer a unique opportunity to study in details their structural, dynamical and chemical properties and to test different theories of galaxy formation.

Owing to their low metallicity and lack of neutral hydrogen, it was initially believed that dSphs are relatively simple objects whose ISM is completely removed by SN II explosions after a very short intense star formation period (e.g. Dekel & Silk 1986). Doubts about this picture come from the high resolution spectroscopy of several dSphs showing a wide range in metallicity. For instance, Shetrone *et al.* (2001) have observed stars in Draco and Ursa Minor with values of [Fe/H] in the range $-3 \leqslant$[Fe/H]$\leqslant -1.5$. The same authors also found that their observed dSphs have [α/Fe] abundances that are 0.2 dex lower than those of Galactic halo field stars at the same metallicity. This suggests that the bulk of the stars in these systems formed in gas self polluted by SNe II as well as SNe Ia and that the star formation (SF) must continue over a relatively long timescale in order to allow a sufficient production of iron by SNe Ia (Ikuta & Arimoto 2002, Lanfranchi & Matteucci 2004, Marcolini *et al.* 2006; but see also Recchi *et al.* 2007, Salvadori *et al.* 2008).

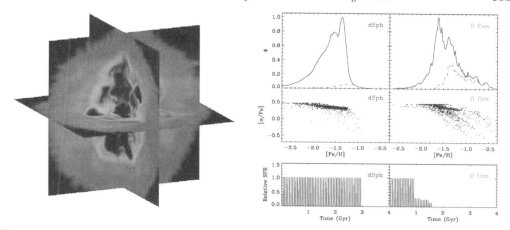

Figure 1. Left: distribution of the logarithm of the density in three orthogonal planes at ~ 400 Myr after the beginning of the simulation. Note the inner bubble carved by the SNe explosions and the shocks propagating outward. Right: [Fe/H] distribution of the models resembling the Draco dSph (left panels) and the ω Cen globular cluster (right panels) together with the corresponding [α/Fe]-[Fe/H] diagrams; the blue line and blue dots represent stars with [α/Fe]$\leqslant 0.2$ (i.e. affected by SNe Ia inhomogenous pollution, see text). The lower panels represent the assumed SFH for the two models.

A complex star formation history (SFH) is further suggested by several facts: *i*) isolated low mass dSphs such as Phoenix (Young et al. 2007) and Leo T (de Jong *et al.* 2008) were able to form stars up to 100 Myr ago; *ii*) the SFHs of dwarf galaxies are strongly dependent on their local environment, the fraction of passively evolving galaxies dropping from $\sim 70\%$ in dense environments, to zero in the rarefied field (Haines *et al.* 2007); *iii*) dwarf ellipticals and dSphs cluster around the dominant spirals galaxies, while gas rich star forming dwarf Irregulars are found at larger distances (van den Bergh 1994). These points highlight the role of the environment (tidal interaction/ram pressure stripping), and disfavors a scenario in which the evolution is due uniquely to internal processes.

Here we briefly report some results obtained by Marcolini et al. (2006, 2007, 2008) with a model which turns to be consistent with many properties of the Draco dwarf and, with minimal assumptions, to the chemical properties of the peculiar system ω Centauri (Marcolini *et al.* 2007), which is believed to be the remnant of an ancient dSph.

2. dSphs Model

Let us consider the reference model of Marcolini *et al.* (2006) which, although employed to study the general characteristics of dSphs, is tailored to explore the evolution of the Draco dSph. The simulation starts with the ISM in hydrostatic equilibrium in the extended dark matter halo potential well. The amount of initial gas corresponds to the cosmological baryonic fraction of the dark matter halo ($M_{ISM} = 0.18 M_{DM}$). The SFH is given *a priori*, assuming that stars form in a sequence of 50 instantaneous bursts separated in time by 60 Myr (see Fig. 1). We also assume that SNe II explode at a constant rate for 30 Myr after the occurrence of each starburst, while the SNe Ia rate follows the prescription of Matteucci & Recchi (2001). Each SN explosion is stochastically placed into the galaxy according to its radial probability $P(r) = M_\star(r)/M_{\star,tot}$, where $M_\star(r)$ and $M_{\star,tot}$ are the nowadays radial stellar mass profile and total stellar mass, respectively (cf. Marcolini *et al.* 2006 for more details).

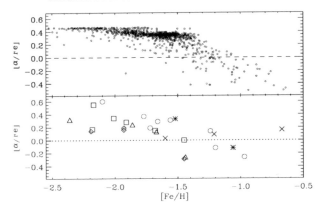

Figure 2. Abundance ratio [α/Fe] versus [Fe/H] for 500 sampled stars of the reference model at $t = 2$ Gyr, compared to a dataset of 28 dSph stars (Draco: triangles ; Fornax: crosses; Leo I: asterisks; Sextans: diamonds; Sculptor: circles; Ursa Minor: squares) collected from the literature (see Marcolini *et al.* 2008 for details).

Although the total energy released by the SNe II explosions is larger than the binding energy of the ISM, efficient radiative losses enable the galaxy to retain most of its gas, which thus remains available for the aforementioned prolonged SFH. The burst of SNe II associated with each stellar burst pushes the bulk of the ISM to the outskirt of the galaxy (see Fig. 1, left panel). Once the explosions cease (~30 Myr after each star burst episode), the ISM flows back towards the center of the galaxy; when the next burst occurs, the gas is pushed outwards again. This oscillatory behavior leads to a rather efficient and homogeneous pollution of the ISM by the SNe II ejecta. We note that as the SFH has been fixed *a priori*, it has no direct relation to the gas reservoir within the galaxy. However, stars do form during the quiescent periods between bursts, when the gas has "settled", and an *a posteriori* consistency for the SFH is recovered. For example a similar periodic ISM behaviour was recovered in similar low-mass galaxies simulations performed by Stinson *et al.* (2007). We point out that, while the observed dSphs are deprived of gas, the galaxy in the present model can not expel the bulk of its ISM by internal mechanisms. Following the discussion given in the Introduction, an external cause, such as the interaction with the Milky Way (Mayer *et al.* 2006), must be invoked at some evolutive stage of the galaxy to get rid of its ISM.

Given their lower rate, SNe Ia do not significantly affect the general hydrodynamical behavior of the ISM, but their role is relevant for the chemical evolution of the stars. Because of their longer evolutionary timescales, SN Ia progenitors created in previous starbursts continue to explode during the subsequent quiescent periods, when the gas is flowing back into the central region. During these periods the higher ambient gas density (together with the lower SNe Ia explosion rate) cause the SNe Ia remnants to be isolated from one another, forming chemically inhomogeneous regions (we refer to these regions as "SNe Ia pockets"). These pockets are "washed out" by successive phases of expansion and collapse of the ISM, due to the effects of SNe II, but new pockets form during the quiescent phases between consecutive starbursts. At odds with the ejecta of SNe II, SN Ia debris are rich in iron and deficient in α-elements. Thus, stars forming in the SN Ia pockets possess lower [α/Fe] and higher [Fe/H] ratios than those formed elsewhere.

2.1. *Comparison with Local dSphs*

As stressed above, the reference model discussed in Marcolini et al. (2006) was tailored to fit the Draco galaxy. In Fig. 1 we show the MDF of this model at the end of the

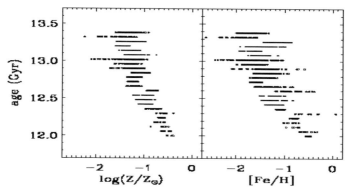

Figure 3. Age-Z and age-[Fe/H] distributions of 1000 sampled stars for the ω Cen model. Note the metallicity spread among coeval stars.

simulation (3 Gyr). The maximum value of the distribution occurring at [Fe/H]$\sim$-1.5 and the mean value of $\langle$[Fe/H]$\rangle = -1.65$ with a spread of $\sim$1.5 dex are compatible with observations (Bellazzini *et al.* 2002). The agreement is larger at the high metallicity tail of the distribution which, in our model, is shaped by the stars formed in the SN Ia pockets.

As discussed above, the SNe Ia are also responsible for the [α/Fe] spread present at larger values of [Fe/H] in the observed dSphs, as shown in Fig. 2 (lower panel). From this figure it is also apparent that the most metal poor stars have [α/Fe] $\simeq 0.5$, typical of pure SN II enrichment, while a decrement of ~ 0.2 dex in the plateau is achieved at higher metallicities. This drop occurs in our simulations after $t \sim 2.0 - 3.0$ Gyr to allow the cumulative effect of SNe Ia to be appreciable (top panel of Fig. 2). We thus conclude that, at least in our models, only a prolonged star formation history (> 2.0 Gyr) can account for the chemical differences between the Galactic halo and the dSphs as shown by Shetrone *et al.* (2001). Such a long SFH for dSphs is also found in the chemical models by Fenner *et al.* (2006) who are unable to reproduce the Ba/Y ratio unless stars formed over an interval long enough for the low-mass stars to pollute the ISM with *s*-elements.

The inhomogeneous pollution by SNe Ia discussed above is particularly important in the central galactic region, where the SN Ia rate is greater, and the density of the ambient gas is higher. This naturally accounts for a radial segregation of Fe-rich stars in the central regions of dSphs. As these stars in our model are also α-depleted, a similar radial segregation of α-poor stars should be observed to test our model. We finally stress that our model envisages a central depression in the radial distribution of the stellar velocity dispersion (Marcolini *et al.* 2008). This naturally entails two different stellar populations with an anti-correlation between [Fe/H] and velocity dispersion, which has been observed in the case of the Sculptor dSph (Tolstoy *et al.* 2004) and the Fornax dSph (Battaglia *et al.* 2006).

2.2. *ω Cen*

The stellar system ω Cen (NGC 5139) is unique among Galactic star clusters in terms of its structure, kinematics, and stellar content. It is the only known GC showing a clear [Fe/H] spread spanning the metallicity range $-1.6 <$[Fe/H]< -0.6 (Norris et al. 1996, Johnson *et al.* 2008). Recent photometric surveys have revealed the presence of multiple sequences in its color-magnitude diagram (CMD), indicating a complex star formation history (e.g. Sollima *et al.* 2005). It is suggested that ω Cen is the nucleus of a larger stellar system, possibly a dwarf galaxy, that lost most of its stars and gas in the interaction with the Milky Way $\sim$10 Gyr ago (e.g. Bekki & Norris 2006). In this framework Marcolini *et al.* (2007) focused on the evolution of the central region of their dSph model, where the inhomogeneous pollution of the SNe Ia is particularly effective.

In Fig. 1 the MDF within the inner 90 pc is plotted assuming a total SFH lasting ~ 1.5 Gyr as shown in the same Figure (see Marcolini *et al.* 2007 for more details and about the assumptions used for this model). The MDF shows a bimodal structure similar to that observed in ω Cen (e.g. Norris *et al.* 1996), with a maximum at [Fe/H] = -1.6 and a secondary peak at [Fe/H]~ -1.3 accounting for $\sim 25\%$ of the cluster's stellar content. Comparing the [α/Fe]-[Fe/H] diagram of this model with the one typical of dSphs (c.f. Fig. 1), it is possible to note the much larger number of α-depleted stars (see also the blue line in the MDF which represents stars with α/Fe $\leqslant 0.2$). The diagram is in reasonable agreement with the findings of Pancino *et al.* (2002) and Origlia *et al.* (2003), who find that while the metal-poor and intermediate-metallicity stellar populations of ω Cen have the expected α-element overabundance observed in halo and GC stars ($\langle[\alpha/\mathrm{Fe}]\rangle \simeq$ 0.3-0.4), the most metal-rich population ([Fe/H]~ -0.6) shows a significantly lower α-enhancement ($\langle[\alpha/\mathrm{Fe}]\rangle \simeq 0.1$).

Given the distinctive role of the SN Ia pollution, the [Fe/H] content of this system is not simply proportional to the metal content and a large spread in the age-metallicity relation is always present (see Fig. 3). These peculiarities have important consequences on the cluster CMD. In fact, they reduce by a factor of 40% the large (and still unexplained) helium overabundance usually invoked to account for the anomalous position of the blue main sequence observed in ω Cen.

References

Battaglia G. *et al.*, 2006, A&A, 459, 423

Bekki, K. & Norris, J. E., 2006, ApJL, 637, L109

Bellazzini M., Ferraro F. R., Origlia L., Pancino E., Monaco L., & Oliva E., 2002, AJ, 124, 3222

Dekel A. & Silk J., 1986, ApJ, 303, 39

de Jong *et al.*, 2008, astro-ph/08014027

Fenner, Y., Gibson, B. K., Gallino, R., & Lugaro, M., 2006, ApJ, 646, 184

Geisler, D., Wallerstein, G., Smith, V. V., & Casetti-Dinescu, D. I., 2007, PASP, 119, 939

Haines C., Gargiulo A., La Barbera F., Mercurio A., Merluzzi, & Busarello, 2007, MNRAS, 381, 7

Ikuta, C. & Arimoto, N., 2002, A&A, 391, 55

Johnson, C. I., Pilachowski, C. A., Simmerer, J., & Schwenk, D., astro-ph/08042607

Lanfranchi G. A. & Matteucci F., 2004, MNRAS, 351, 1338

Marcolini A., D'Ercole A., Brighenti F., & Recchi S., 2006, MNRAS, 371, 643

Marcolini A., Sollima A., D'Ercole A., Gibson B. K., & Ferraro F. R., 2007, MNRAS, 382, 443

Marcolini A., D'Ercole A., Battaglia G., & Gibson B. K, 2008, MNRAS, 386, 2173

Mateo M. L., 1998, ARA&A, 36, 435

Matteucci, F. & Recchi, S., 2001, ApJ, 558, 351

Mayer L., Mastropietro C., Wadsley J., Stadel J., & Moore B., 2006, MNRAS, 369, 1021

Norris J. E., Freeman K. C., & Mighell K. J., 1996, ApJ, 462, 241

Origlia L., Ferraro F. R., Bellazzini M., & Pancino E., 2003, ApJ, 591, 916

Pancino E., Pasquini L., Hill V., Ferraro F. R., & Bellazzini M., 2002, ApJ, 568, L101

Recchi, S., Theis, C., Kroupa, P., & Hensler, G, 2007, A%A, 470, 5

Salvadori, S., Ferrara, A., & Schneider, R., 2008, MNRAS, 386, 348

Shetrone M. D., Côté P., & Sargent W. L. W., 2001, ApJ, 548, 592

Sollima A., Ferraro F. R., Pancino E., & Bellazzini M., 2005, MNRAS, 357, 265

Stinson, G. S., Dalcanton, J. J., Quinn, T., Kaufmann, T., & Wadsley, J., 2007, ApJ, 667, 170

Tolstoy E., *et al.* 2004, ApJ, 617, L119

van den Bergh S., 1994, ApJ, 428, 617

Young L. M., Skillman E. D., Weisz D. R., & Dolphin A. E., 2007, ApJ, 659, 331

Low-Metallicity Star Formation:
From the First Stars to Dwarf Galaxies
Proceedings IAU Symposium No. 255, 2008
L.K. Hunt, S. Madden & R. Schneider, eds.

Young Star Clusters in the Small Magellanic Cloud: Impact of Local and Global Conditions on Star Formation

Elena Sabbi[1], Linda J. Smith[1,2], Lynn R. Carlson[3], Antonella Nota[1,4], Monca Tosi[5], Michele Cignoni[5], Jay S. Gallagher III[6], Marco Sirianni[1,4], and Margaret Meixner[1]

[1] Space Telescope Science Institute
3700 San Martin Drive, Baltimore, MD, 21218, USA
email: sabbi@stsci.edu

[2] University College London
London, UK

[3] Johns Hopkins University
Baltimore, MD, USA

[4] European Space Agency, Research and Scientific Support Department
Baltimore, MD, USA

[5] INAF-Osservatorio Astronomico di Bologna
Bologna, Italy

[6] University of Wisconsin
Madison, WI, USA

Abstract. We compared deep images acquired with the Advanced Camera for Surveys on board of the Hubble Space Telescope with mid-IR Spitzer Space Telescope images and University College London Echelle Spectrograph spectra of NGC 346 and NGC 602, two of the youngest star clusters in the Small Magellanic Cloud. Our multi-wavelength approach allowed us to infer very different origins for the clusters: while NGC 346 is likely the result of the hierarchical collapse of a giant molecular cloud, NGC 602 is probably the result of the collision and consequent interaction of two H I shells of gas.

Keywords. Magellanic Clouds, galaxies: star clusters, stars: formation, stars: mass function, stars: pre–main-sequence

1. Introduction

As the closest star forming dwarf galaxy ($\sim 60\,\mathrm{kpc}$), the Small Magellanic Cloud (SMC) represents an ideal laboratory for detailed studies of resolved stellar populations in this extremely common class of objects. In addition its present day metallicity (Z=0.004) and low dust content (1/30 the Milky Way) make the SMC the best local analog to the vast majority of late-type dwarfs.

Deep images acquired with the Advanced Camera for Surveys (ACS) on board of the Hubble Space Telescope (HST) provide excellent photometry well below the turn-off (TO) of the oldest stellar population in the SMC, allowing us to infer an accurate star formation history (SFH) over the entire Hubble time. Furthermore the high spatial resolution of ACS allowed us to spatially resolve even the densest star cluster in the SMC to probe the cluster formation and evolution in late-type dwarf galaxies.

The availability of multi-wavelength surveys from the radio band to the far-IR, combined with the simple kinematics of the galaxy allow us to identify the possible triggers of star formation (SF), while using high resolution spectroscopy we can study the effect of reduced stellar wind power on the early cluster evolution.

With the aim of understanding the impact of global and local conditions on early star cluster evolution we recently acquired deep F555W ($\sim$ V) and F814W ($\sim$ I) ACS Wide Field Channel (WFC) images of two of the youngest SMC star clusters: NGC 346 and NGC 602.

2. Photometric properties of the star clusters

NGC 346 is the most massive star forming region of the SMC, and is located in the bar of the SMC, a region characterized by high gas and stellar densities. The cluster contains half of the known O stars in the SMC, and is ionizing the bright nebula N66.

A first inspection of the NGC 346 m_{F555W} vs. $m_{F555W} - m_{F814W}$ color magnitude diagram (CMD, Fig. 1-*left panel*) reveals that different stellar populations are present in this area:

• Young stars. The bright ($12.5P \lesssim m_{F555W} \lesssim 22$) and blue ($0.3 \lesssim m_{F555W} - m_{F814W} \lesssim 0.4$) main sequence (MS), well visible to the upper left of the CMD, as well as the sequence of red ($1.5 \lesssim m_{F555W} - m_{F814W} \lesssim 2.2$) and faint ($m_{F555W} \lesssim 21$) pre-MS stars in the mass range $0.6 - 3 M_\odot$ are representative of a stellar population that formed between 3 and 5 Myr ago (Nota *et al.* 2006, Sabbi *et al.* 2007).

• Intermediate-age and old stars. An older stellar population is easily distinguishable in the CMD. Its rich MS extends from $m_{F555W} \simeq 22$ down to $m_{F555W} \simeq 26.5$. The evolved phases of this population are well delineated: the narrow subgiant branch (SGB) ($m_{F555W} \simeq 21.6$; $0.45 \lesssim m_{F555W} - m_{F814W} \lesssim 0.95$); the very well defined red giant branch (RGB) (with the brightest stars at $m_{F555W} \simeq 17.3$ and $m_{F555W} - m_{F814W} \simeq 0.65$), and the red clump (RC) ($m_{F555W} \simeq 19.5$). Isochrones fitting indicates that the majority of the older stars in this field formed in a single SF episode approximately 4.5 Gyr ago (Sabbi *et al.* 2007).

Figure 2 shows the present day mass function (PDMF) of NGC 346 (open circles) between 0.8 and 60 M$_\odot$. In this range of masses the slope of the NGC 346 PDMF $\Gamma = -1.43 \pm 0.18$ (Sabbi *et al.* 2008), in good agreement with the value derived by Salpeter (1955) for the solar neighborhood ($\Gamma_{Salpeter} = -1.35$).

NGC 602 is a small and bright star forming region, located between the SMC wing and the Magellanic bridge, a region characterized by low gas and dust content. Figure 1–right panel shows NGC 602 m_{F555W} vs. $m_{F555W} - m_{F814W}$ CMD. The most striking feature of this CMD is the rich population of pre-MS stars in the mass range 0.6–3 M$_\odot$ (Carlson *et al.* 2007). The TO near $m_{F555W} \simeq 22$, the SGC also at $m_{F555W} \simeq 22$, and the RC around $m_{F555W} \simeq 19.5$ belong likely to the SMC field stellar population that is present even in this low stellar density region. Isochron fitting indicates an age of $\sim$ 4 Myr for NGC 602 (but see the poster presented in this conference by Cignoni *et al.* for a detailed SFH history of this region).

We derived the PDMF of NGC 602 in the mass range 0.8–30 M$_\odot$ (Fig.2–filled ellipses). From the weighted least mean squares fit of the data we derived a slope $\Gamma = -1.25 \pm 0.22$ (Cignoni *et al.* 2008b) in excellent agreement with what we found in NGC 346.

Does the universality of the MF imply that the SF process is independent of both the local and global conditions?

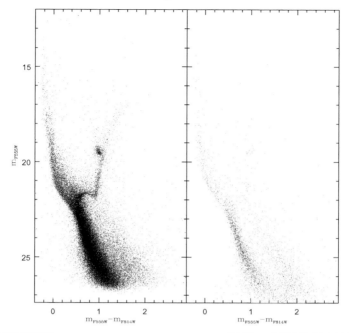

Figure 1. ACS/WFC m_{F555W} vs. $m_{F555W} - m_{F814W}$ CMDs of NGC 346 (*left panel*) and NGC 602 (*right panel*).

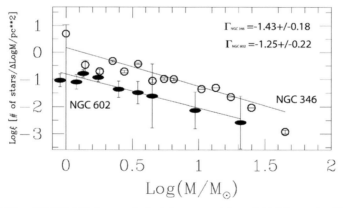

Figure 2. PDMFs of NGC 346 (open circles) between 0.8 and 60 $M_\odot$ and of NGC 602 (filled ellipses) between 0.8 and 30 $M_\odot$.

3. NGC 346 structure and kinematics

The high spatial resolution of ACS/WFC showed that the NGC 346 stars are organized in a number of small compact sub-clusters, which vary in density, dust content, and morphology, but, within the uncertainties of isochrone fitting, are likely coeval. The majority of these associations are still embedded in HII gas, and dust. Arcs and filaments of dust and gas depart from the most central sub-clusters and connect them with the external associations (Sabbi *et al.* 2007).

The sub-clusters are coincident with the CO clumps described by Rubio *et al.* (2000). Recent Spitzer Space Telescope (SST) mid-IR observations also revealed that all the sub-clusters contain one or more young stellar objects (YSOs) with ages younger than

1 Myr (Simon *et al.* 2007). By assuming a Salpeter IMF, Simon *et al.* (2007) calculated that more than 3000 M$_\odot$ have been formed in the last $\sim$ 1 Myr, concluding that 6% of the current SF in the SMC is taking place in NGC 346.

The fact that the stars are organized in sub-clusters, connected by filaments of dust and gas, and that star formation is still ongoing in the sub-clusters suggest that NGC 346 is the result of the collapse and subsequent fragmentation of an initial giant molecular cloud into multiple "seeds" of SF, and may represent a good observational match to the conditions predicted by the hierarchical fragmentation of a molecular cloud in a turbulent interstellar medium (Klessen & Burket 2000; Bonnell & Bate 2002; Bonnell *et al.* 2003). According to this model, the fragmentation of the cloud is due to supersonic turbulent motions present in the gas. The turbulence induces the formation of shocks in the gas and produces filamentary structures (Bate *et al.* 2003). The chaotic nature of the turbulence increases the density in the filamentary structures locally. When regions of high density become self-gravitating, they start to collapse to form stars. Simulations show that SF occurs simultaneously at several different locations in the cloud (Bonnell *et al.* 2003), as appears to be the case for NGC 346.

The formation of shocks in the gas, due to the initial supersonic turbulence, should rapidly remove kinetic energy from the gas (Ostriker *et al.* 2001). Therefore to further confirm the origin of NGC 346, we observed many sub-clusters with the University College London Echelle Spectrograph (UCLES) at the Anglo-Australian Telescope. We find an average radial velocity $v_{rad} = 165.2$ km/s, with a dispersion $\sigma = 1.9$ km/s, further supporting the hypothesis that NGC 346 is the result of a hierarchical collapse. Our spectroscopic analysis also indicates that the ionized gas is quiescent, with no evidence of large-scale motions, confirming that the mechanical feedback is reduced at low metallicity (see the discussion in Mokiem *et al.* 2007), and proving that supernovae have not yet exploded (Smith *et al.* – in preparation).

4. NGC 602 kinematics

In the ACS/WFC images NGC 602 appears as a small star cluster surrounded by two dusty arches, characterized by numerous dark "pillars of creations". SST observations revealed numerous class 0.5 and I YSOs on the verge of the pillars, indicating that SF propagated from the center of NGC 602, and a second generation of stars is currently forming in the periphery of the cluster (Carlson *et al.* 2007).

Even if the morphology of this association is reminiscent of an expanding bubble of gas, UCLES high-resolution spectra show very low velocity dispersion, excluding large-scale motions (Nigra *et al.* 2008). Another possible explanation for the NGC 602 morphology is that we are observing a cavity eroded by the OB UV stellar radiation.

NGC 602 and its associated H II region, N90, formed in a relatively isolated and diffuse environment. Its isolation from other regions of massive star formation and the relatively simple surrounding H I shell structure encouraged us to try to constrain the processes that may have led to its formation. Using the shell catalog derived from the 21 cm neutral hydrogen (H I) spectrum survey data (Staveley-Smith *et al.* 1997; Staminirovíc *et al.* 1999) we identified a distinct H I cloud component that is likely the progenitor cloud of the cluster and the H II region which probably formed in blister fashion from the cloud's periphery. A comparison between H I and H II kinematics suggests that star formation in NGC 602 was triggered by compression and turbulence associated with H I shell interaction $\sim$ 7 Myr ago (Nigra *et al.* 2008).

5. Conclusions

We have recently analyzed the two very young SMC star clusters NGC 346 and NGC 602, which are located in two regions characterized by very different gas and stellar densities. We derived the PDMF of the clusters over two orders of magnitude, further confirming its universality.

Both clusters contain noticeable populations of pre-MS stars and YSOs, indicating that in both cases SF is still ongoing and residual gas is still present. The ionized gas is quiescent, and we did not find evidence of stellar wind interaction, confirming the hypothesis that mechanical feedback is reduced at low metallicities.

Our multi-wavelength approach allowed us to infer very different origins for the clusters, suggesting that different local condition might deeply affect the formation and the evolution of star clusters since the earliest phases.

References

Bonnell, I. A. & Bate, M. R. 2002, MNRAS, 336, 659

Bonnell, I. A., Bate, M. R., & Vine, S. G. 2003, MNRAS, 343, 413

Cignoni, M. *et al.* 2008, Proceedings IAU Symposium No. 255, "Low-Metallicity Star Formation: From the Firts Stars to Dwarf Galaxies", L.K. Hunt, S. Madden, & R. Schneider, eds.

Cignoni, M. *et al.* 2008, AJ, (submitted)

Klessen, R. S. & Burkert, A. 2000, ApJS, 128, 287

Mokiem, M. R., *et al.* 2007, A&A, 473, 603

Nigra, L., Gallagher, J. S., III, Smith, L. J., Stanimirović, S., Nota, A., & Sabbi, E. 2008, PASP, (accepted for publication) astro-ph/0808.1033

Nota, A. *et al.* 2006, ApJ, 640, L29

Ostriker, E. C., Stone, J. M., & Gammie, C. F. 2001, ApJ, 546, 980

Rubio *et al.* 2000, A&A

Sabbi, E., Sirianni, M., Nota, A., Tosi, M., Gallagher, J., Meixner, M., Oey, M. S., Walterbos, R., Pasquali, A., Smith, L.J., & Angeretti, L. 2007, AJ, 133, 44

Sabbi, E., Sirianni, M., Nota, A., TOsi, M., Gallagher, J. Smith, L. J., Angeretti, L., Oey, M. S., Walterbos, R., & Pasquali, A. 2008, AJ, 135, 173

Simon, J. H., *et al.* 2007, ApJ, 670, 313

Stanimirović, S., Staveley-Smith, L., Dickey, J. M., Sault, R. J., & Snowden, S. L. 1999, MNRAS, 302, 417

Staveley-Smith, L., Sault, R. J., Hatzidimitriou, D., Kesteven, M. J., & McConnell, D. 1997, MNRAS, 289, 225

Low-Metallicity Star Formation:
From the First Stars to Dwarf Galaxies
Proceedings IAU Symposium No. 255, 2008
L.K. Hunt, S. Madden & R. Schneider, eds.

© 2008 International Astronomical Union
doi:10.1017/S1743921308024757

Modeling the ISM Properties of Metal-Poor Galaxies and Gamma-Ray Burst Hosts

Emily M. Levesque[1], Lisa J. Kewley[1], Kirsten Larson[1], Leonie Snijders[2]

[1]Institute for Astronomy, University of Hawaii
2680 Woodlawn Dr., Honolulu, HI 96822

[2]Leiden Observatory, Leiden University
P.O. Box 9513, 2300 RA Leiden, The Netherlands

Abstract. Recent research has suggested that long-duration gamma-ray bursts (LGRBs) occur preferentially in low-metallicity environments, but the exact nature of this correlation is currently a matter of intense debate. We use the newest generation of the Starburst99/Mappings code to generate an extensive suite of cutting-edge stellar population synthesis models, covering a wide range of physical parameters specifically tailored for modeling the ISM environments of metal-poor galaxies and LGRB host galaxies. With our models, we generate optical emission line diagnostics, which will allow us to examine the ISM properties and stellar populations of a variety of galaxy populations in unprecedented detail. While accurately modeling low-metallicity galaxies still poses a challenge to these models, future improvements to these grids will have profound consequences for our understanding of metal-poor galaxies, their ISM environments, and the nature of their role as the hosts of LGRBs.

Keywords. galaxies: ISM, galaxies: abundances, galaxies: stellar content

1. Introduction

Long-duration gamma-ray bursts (LRGBs) have long been considered excellent tools for probing star-formation in distant galaxies. In recent years, however, several studies have uncovered a connection between LGRBs and low-metallicity galaxies that could threaten their utility as unbiased tracers of star formation in our universe. Such studies find that these events' hosts lie below the standard mass-metallicity relation for dwarf galaxies (Stanek *et al.* 2006, Kewley *et al.* 2007), and are morphologically and chemically distinct from the hosts of burstless core-collapse SNe (Fruchter *et al.* 2006, Modjaz *et al.* 2008).

However, there are many authors that doubt the validity of this apparent low-metallicity bias, noting that it could be, for example, simply a by-product of a bias towards young stellar population age (Bloom *et al.* 2002, Berger *et al.* 2007). There are also arguments that such a bias should not affect the utility of LGRBs as star formation tracers at high redshift, where the mean metallicity is lower as a whole (Fynbo *et al.* 2006).

Understanding the relationship between LGRBs and low-metallicity environments is extremely important, as it could challenge the use of these events as tracers of star formation in normal galaxies. A metallicity bias would suggest that LGRBs are not the best means of probing early star formation, as they would be considerably less likely to occur in normal star-forming galaxies (Stanek *et al.* 2006).

This kind of detailed analyses of LGRB ISM environments requires the use of a comprehensive, detailed, and robust grid of stellar population synthesis models that can reproduce the observed spectra of metal-poor galaxies and LGRB hosts. With such models, the ISM properties of the galaxies can be probed in detail, and the similarities and differences between the two populations can be quantified and applied to our current understanding of LGRB progenitor models and metal-poor galaxy properties.

In this paper, we present an extensive suite of model galaxy spectra covering a wide range of physical parameters, generated using the newest generation of the Starburst99/Mappings III code. We demonstrate the application of these diagnostics to spectra of a variety of galaxy populations, examine the results of these comparisons, and discuss future work in this area.

2. Starburst99/Mappings III Model Grids

2.1. *Model Grid Parameters*

To model our sample of galaxies we have used the Starburst99 code (Leitherer *et al.* 1999, Vázquez & Leitherer 2005) to generate theoretical spectral energy distributions (SEDs), which in turn were used in Mappings III photoionization models to produce model galaxy spectra that could be compared to our observations.

Starburst99 is an evolutionary synthesis code that can be used to generate synthetic ionizing far-ultraviolet (FUV) radiation spectra as a function of metallicity, star formation history, and the age and evolution of the stellar populations. These populations are produced by use of model stellar atmospheres and spectra along with evolutionary tracks for massive stars. Starburst99 generates the final synthetic FUV spectrum as output, which is then taken as input by the Mappings III shock and photoionization code, originally developed by Binette *et al.* (1985) and most recently improved to include a more sophisticated treatment of dust (Groves *et al.* 2004). For a more detailed discussion of Mappings III see Dopita *et al.* (2000) and Kewley *et al.* (2001). The parameters taken as inputs by Starburst99 and Mappings III are illustrated in Figure 1.

With these codes we have computed a complete grid of plane-parallel isobaric photoionization models, ranging in age from 0 to 10 Myr in increments of 0.5 Myr. When generating our Starburst99/Mappings III stellar population synthesis models, we adopted a broad grid of parameters to facilitate comparison with a wide range of galaxy samples:

Star Formation History: We model both a zero-age instantaneous burst of star formation, with a fixed mass of $10^6 M_\odot$, and a continuous star formation history where the star formation rate (SFR) is constant at a rate of $1 M_\odot$ per year, starting from an initial time and assuming a stellar population that is large enough to render the fluctuating contributions from high-mass stars negligible.

Metallicity: We model the full range of metallicities available from the evolutionary tracks of the Geneva group, which includes five metallicities of $z = 0.001$, 0.004, 0.008, 0.02, and 0.04, where $Z_\odot = 0.02$

Evolutionary Tracks: We adopted the two evolutionary tracks of the Geneva group that are currently available in Starburst99: the Geneva "standard" (STD) mass loss tracks, and the Geneva "high" (HIGH) mass loss tracks. The STD evolutionary tracks were originally published in a series of papers by the Geneva group (Schaller *et al.* 1992; Schaerer *et al.* 1993a, 1993b; Charbonnel *et al.* 1993); the HIGH tracks, published in Meynet *et al.* (1994), include higher mass loss rates derived by doubling those adopted in the STD models. While many advances have since been made in our understanding of stellar physics, adopting these tracks in our stellar population synthesis models is still advisable. The STD mass loss tracks are the more applicable of the two when considering the effects of wind clumping on mass loss rates (Crowther *et al.* 2002), and the HIGH mass loss tracks produce a reasonable approximation of the mass loss rates resulting from

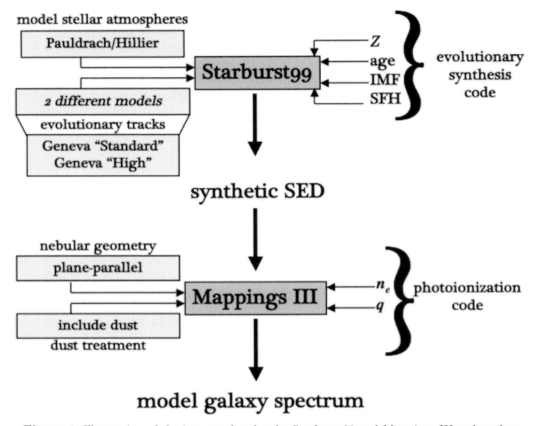

Figure 1. Illustration of the inputs taken by the Starburst99 and Mappings III codes when generating our model grids.

the effects of rotation, when surface mixing results in an earlier start of the WR phase (Meynet, private communication).

Ionization parameter: In Mappings III, the ionization parameter q (cm s^{-1}) is defined as the maximum velocity possible for an ionization front being driven by the local radiation field, and can be translated to a dimensionless ionization parameter by dividing by the speed of light (Dopita *et al.* 2000). In our model grid, we adopted seven different ionization parameters ($q = 1 \times 10^7, 2 \times 10^7, 4 \times 10^7, 8 \times 10^7, 1 \times 10^8, 2 \times 10^8$, and 4×10^8 cm/s).

Electron density: We adopt two different electron densities n_e for this work, of $n_e = 10$ and $n_e = 100$. We assume an isobaric density structure for these models, and thus n_e is specified by the dimensionless pressure/mean temperature ratio. For the remainder of this paper, we present results that adopt $n_e = 100$.

3. Comparison to Observations

3.1. *Galaxy Samples*

With the generation of these grids complete, we can now compare the results of our Starburst99/Mappings models to observed spectra from a variety of galaxy populations:

Metal-Poor Galaxies (MPGs): The MPG spectra were selected by Brown *et al.* (2008) from a larger survey of blue compact galaxies (BCGs). The galaxies in this sample are all categorized as MPGs: their metallicities, $12 + \log(O/H)$, range from 7.41 to 8.32 as

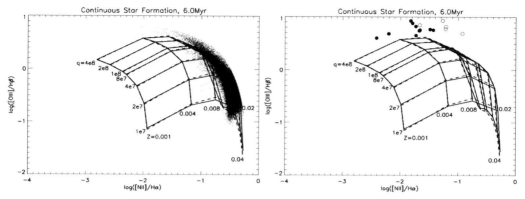

Figure 2. [NII]/Hα vs. [OIII]/Hβ emission line diagnostic grids for a 6.0 Myr continuous star formation model grid; an electron density of 100 is assumed. Left: the grid is in excellent agreement with the position of 60,920 SDSS star-forming galaxies (small points). Right: the agreement of the grid with the MPGs (filled circles), and LGRB hosts (open circles). We see that the models are in poor agreement with the MPG and LGRB host samples. In the electronic color version of these figures, lines of constant metallicity are shown in blue/green, while lines of constant ionization parameter are shown in red/yellow.

determined by the electron temperature (T_e) diagnostic, and from 7.66 to 8.25 using the R_{23} diagnostic put forth by Kewley & Dopita (2002) and refined by Kobulnicky & Kewley (2004). These galaxies inhabit the same region of the luminosity-metallicity plot for dwarf irregular galaxies as LGRB hosts, showing that both of these populations have distinctly low metallicites relative to their luminosities (for more information see Figure 6 of Brown *et al.* 2008, and discussion therein).

LGRB Host Galaxies: In addition to the metal-poor galaxies, we include a sample of galaxies that have hosted LGRBs, taken from the Gamma-ray burst Host Studies (GHostS) archive (Savaglio *et al.* 2006). We have selected a small sample of LGRB host galaxies that have accurate spectroscopic observations with sufficient spectral coverage and emission line detections.

Sloan Digital Sky Survey (SDSS) galaxies: Finally, to compare our models to the general local $(z \leqslant 0.3)$ galaxy population, we plot a sample of 60,920 star forming galaxies from SDSS; for more discussion of this sample see Kewley & Ellison (2008).

The MPG and SDSS galaxy fluxes were corrected for local extinction effects based on the Hα/Hβ emission line ratio, assuming the Balmer decrement for case B recombination (Hα/Hβ = 2.85 for T = 10^4 K and $n_e \sim 10^2$ - 10^4 cm^{-3}, following Osterbrock 1989) and the Cardelli *et al.* (1989) reddening law with the standard total-to-selective extinction ratio $R_V = 3.1$ (the line fluxes for the LGRB host galaxies were taken from the literature and had been previously dereddened).

3.2. *Emission Line Diagnostics Grids*

For our analyses, we have plotted the models in a series of emission line diagnostic grids, comparing a variety of line ratios selected to examine the evolution of specific properties such as metallicity and ionization parameter across the full parameter space of our models. In Figure 2, we compare the HIGH (solid line) and STD (dotted line) model grids to the position of these galaxy samples.

It is clear from examining these figures that, while the models agree with the position of the SDSS galaxies (Figure 2, left panel), they are in very poor agreement with the MPG and LGRB host samples (Figure 2, right panel). We conclude from our comparison of these observations with our model grid that, while the models can be applied quite effectively to the general galaxy population, further improvements are needed before they can be accurately used to determine the ISM properties of low-metallicity galaxy populations.

4. Discussion and Future Work

One potential means of improving our models concerns the new generation of the Geneva evolutionary tracks, which accommodate for the first time the effects of rotation on the stellar population (Vázquez *et al.* 2007). The effects of rotation are particularly critical at lower metallicities. Expanding our current grid of evolutionary models by include the rotating evolutionary tracks as they become available will help us to further probe the effect that these tracks have on the eventual outcome of the models, as well as take important strides towards improving these grids and making them applicable to lower-metallicity galaxy populations.

References

Berger, E., Fox, D. B., Kulkarni, S. R., Frail, D. A., & Djorgovski, S. G. 2007 *ApJ*, 660, 504
Binette, L., Dopita, M. A., & Tuohy, I. R. 1985 *ApJ*, 297, 476
Bloom, J. S., Kulkarni, S. R., & Djorgovski, S. G. 2002, *ApJ*, 121, 1111
Brown, W., Kewley, L. J., & Geller, M. J. 2008, *AJ*, 135, 92
Cardelli, J. A., Clayton, G. C., & Mathis, J. S. 1989, *ApJ*, 345, 245
Charbonnel, C., Meynet, G., Maeder, A., Schaller, G., & Schaerer, D. 1993, *A&AS*, 101, 415
Crowther, P. A., Dessart, K., Hillier, D. J., Abbott, D. B., & Fullterton, A. W. 2002, *A&A*, 392, 653
Dopita, M. A., Kewley, L. J., Heisler, C. A., & Sutherland, R. S. 2000, *ApJ*, 542, 224
Fruchter, A .S. *et al.* 2006, *Nature*, 441, 463
Fynbo, J. P. U., *et al.* 2006, *A&A*, 451, 47
Groves, B., Dopita, M., & Sutherland, R. 2004, *ApJS*, 153, 9
Kewley, L. J., Brown, W. R., Geller, M. J., Kenyon, S. J., & Kurtz, M. J. 2007, *AJ*, 133, 882
Kewley, L. J. & Dopita, M. A. 2002, *ApJS*, 142, 35
Kewley, L. J., Dopita, M. A., Sutherland, R. S., Heisler, C. A., & Trevena, J. 2001, *ApJ*, 556, 121
Kewley, L. J., & Ellison, S. 2008, arXiv:0801.1849
Kewley, L. J., Groves, B., Kauffman, G., & Heckman, T. 2006, *MNRAS*, 372, 961
Kobulnicky, H. A. & Kewley, L. J. 2004, *ApJ*, 617, 240
Kong, X. & Cheng, F. Z. 2002, *A&A*, 389, 845
Leitherer, C., *et al.* 1999, *ApJS*, 123, 3
Meynet, G., Maeder, A., Schaller, G., Schaerer, D., & Charbonnel, C. 1994, *A&AS*, 103, 97
Modjaz, M., Kewley, L. J., Kirshner, R. P., Stanek, K. Z., Challis, P., Garnavich, P. M., Greene, J. E., Kelly, P. L., Prieto, J. L. 2008, *AJ* 135, 1136
Osterbrock, D. 1989, *Astrophysics of gaseous nebulae and active galactic nuclei* (University Science Books)
Savaglio, S., Glazebrook, K., & Le Borgne, D. 2006, in: S. S. Holt, N. Gehrels, & J. A. Nousek (eds.), *American Institute of Physics Conference Series* (Publ. de l'Observatoire de Paris), p. 540-545
Schaerer, D., Charbonnel, C., Meynet, G., Maeder, A., & Schaller, G. 1993a, *A&AS*, 102, 339
Schaerer, D., Meynet, G., Maeder, A., & Schaller, G. 1993b, *A&AS*, 98, 523
Schaller, G., Schaerer, D., Meynet, G., Maeder, A. 1992, *A&AS*, 96, 269
Stanek, K. Z., *et al.* 2006, *Acta Astron.*, 56, 333
Vázquez, G. A. & Leitherer, C. 2005, *ApJ*, 621, 695
Vázquez, G. A., Leitherer, C., Schaerer, D., Meynet, G., & Maeder, A. 2007, *ApJ*, 663, 995

Low-Metallicity Star Formation:
From the First Stars to Dwarf Galaxies
Proceedings IAU Symposium No. 255, 2008
L.K. Hunt, S. Madden & R. Schneider, eds.

Dwarf galaxies and the Magnetisation of the IGM

Uli Klein[1]

[1]Argelander-Institut für Astronomie,
Auf dem Hügel 71, Bonn, Germany
email: uklein@astro.uni-bonn.de

Abstract. With the operation of LOFAR, a great opportunity exists to shed light on a problem of some cosmological significance. Diffuse radio synchrotron emission not associated to any obvious discrete sources as well as Faraday rotation in clusters of galaxies both indicate that the intergalactic or intracluster medium (IGM, ICM) is pervaded by a weak magnetic field, along with a population of relativistic particles. Both, particles and fields must have been injected into the IGM either by Active Galactic Nuclei (AGN) or by normal star-forming galaxies. Excellent candidates for the latter are starburst dwarf galaxies, which in the framework of hierarchical structure formation must have been around in large numbers. If this is true, one should be able to detect extended synchrotron halos of formerly highly relativistic particles around local starburst or post-starburst dwarf galaxies. With LOFAR, one should easily find these out to the Coma Cluster and beyond.

Keywords. acceleration of particles, magnetic fields, polarization, radiation mechanisms: non-thermal, ISM: jets and outflows, ISM: magnetic fields, galaxies: clusters: general, galaxies: dwarf, galaxies: starburst, radio continuum: galaxies

1. Introduction

Dwarf galaxies play a key role in the enrichment of the ICM or IGM, not only as far as heavy elements are concerned, but possibly also regarding the magnetisation. According to the standard bottom-up scenario of galaxy formation, primeval galaxies must have injected much of their (enriched) interstellar medium (ISM) into the IGM during the initial bursts of star formation, thereby "polluting" large volumes of intergalactic space because of their high number density. By the same token, the following two properties render dwarf galaxies potentially very efficient in injecting a relativistic plasma into their surroundings: first, they exist in large numbers, and second, they possess low escape velocities, making it easier to expel their interstellar gas, as compared to massive spiral galaxies. Such galactic winds are in fact seen in some prototypical low-mass galaxies in the local universe.

2. Outflows and the magnetisation of galaxy clusters

Galaxy clusters are known to be pervaded by a relativistic plasma, i.e. particles with (mildly) relativistic energy and magnetic fields. In the centres of some rich clusters, radio halos have been found, produced by either primary or secondary electrons (see Fig. 1, left). The problem with primary electrons is their relatively short lifetime, limited by

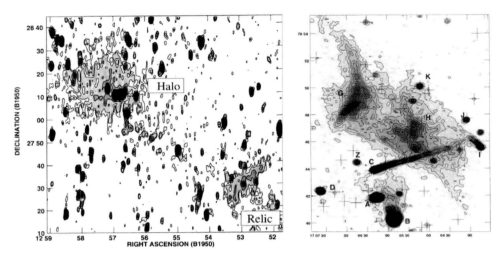

Figure 1. Diffuse radio continuum emission from clusters of galaxies: central radio halo and peripheral structure in the Coma Cluster (left, from Giovannini *et al.* 1993), and radio relic in A 2556 (right, from Röttgering *et al.* 1994).

synchrotron and inverse-Compton losses:

$$t_{1/2} = 1.59 \cdot 10^9 \cdot \frac{B^{1/2}}{B^2 + B^2_{cmb}} \left[\left(\frac{\nu}{\text{GHz}} \right) (1+z) \right]^{-1/2} , \qquad (2.1)$$

where

$$B_{cmb} = (1+z)^2 \mu G \qquad (2.2)$$

is the equivalent magnetic field strength of the cosmic microwave background. This latter equation results from the similar dependences of synchrotron and inverse-Compton losses,

$$\dot{E}_{synch} \propto -B^2 \cdot E^2 \qquad (2.3)$$

$$\dot{E}_{IC} \propto -u_{rad} \cdot E^2 , \qquad (2.4)$$

where u_{rad} is the energy density of the radiation field (which for the perfect black-body CMB radiation is $\propto T^4_{cmb}$). Eq. (2.1) then tells us that, for instance, with a magnetic field of $B = 1$ μG, particles radiating at 1.4 GHz will be rendered invisible after $t_{1/2} = 10^8$ yr in the local universe ($z = 0$). Hence, primary electrons require continuous injection. However, diffuse radio halos are never associated with any obvious "fresh" source, and one therefore has to invoke secondary electrons. These are produced by hadronic collisions of relativistic protons with the thermal gas in a pion-muon chain. Owing to their much larger mass, protons have synchrotron lifetimes exceeding a Hubble time. Finally, it should be pointed out that completely independent evidence for the magnetisation of (at least the) central regions of galaxy clusters comes from the observation of Faraday rotation (e.g. Clarke *et al.* 2001).

Clusters undergoing large-scale merging frequently exhibit so-called radio relics, mostly on their periphery (Fig. 1, right). In contrast to radio halos, these are significantly polarised, probably reflecting magnetic-field enhancement in the compression zones where the subclusters produce large-scale shockwaves during their mutual penetration. Naturally, this also provides an efficient acceleration mechanism. Particles that were formerly highly relativistic (electrons in particular) have cooled via synchrotron and

Session III

Explosive events in low-metallicity environments

Low-Metallicity Star Formation:
From the First Stars to Dwarf Galaxies
Proceedings IAU Symposium No. 255, 2008
L.K. Hunt, S. Madden & R. Schneider, eds.

Supernovae and their Evolution in a Low Metallicity ISM

Roger A. Chevalier

Department of Astronomy, University of Virginia,
P.O. Box 400325, Charlottesville, VA 22904, USA
email: rac5x@virginia.edu

Abstract. Observations of core collapse supernovae and their progenitors generally support expectations of increasing mass loss with increasing initial mass. Mass loss rates are expected to decline at lower metallicity, and there are prospects for directly testing this for the red supergiant progenitors of Type IIP supernovae. However, there are indications that mass loss rates for high mass early type stars may be overestimated and that there are mass loss mechanisms that do not decline at lower metallicity. In this case, there may be supernova emission from strong circumstellar interaction even at low metallicity. Although there is evidence for dust formation in freely expanding ejecta of supernovae, the quantities are relatively small. Another promising site of dust formation is the circumstellar interaction region, but this should occur in only a fraction of supernovae.

Keywords. Galaxies: abundances, stars: mass loss, supernovae: general, supernova remnants

1. Introduction

In recent years, the discovery of 100's of supernovae per year, as well as multiwavelength observations of these events, has clarified the landscape of supernova types. Mass loss during the stellar evolution is crucial for the type of supernova, and the mass loss properties near the time of the explosion can be investigated by multiwavelength observations. While the general supernova properties have been clarified for metallicities close to solar, there are puzzles regarding very massive stars and there remains considerable uncertainty in going to low metallicities.

In section 2, stellar evolution calculations of massive stars are considered. These calculations have become quite sophisticated, but still involve some assumptions that need to be examined. Observations bearing on these expectations are discussed in section 3. Dust formation by supernovae is discussed in section 4, and a concluding discussion in section 5.

2. Expectations from Massive Star Evolution

Evolutionary studies of single massive stars show that the properties of the supernova at the end of a star's life depend crucially on the mass loss leading up to the core collapse. At solar metallicity, estimates of mass loss as a function of stellar effective temperature and luminosity come from observations of stellar mass loss in the Galaxy. The result is that mass loss increases rapidly with mass, so stars with initial mass $\sim 9\ M_\odot$ end their lives with most of their H envelopes, while massive stars lose their H envelope by the time of the explosion. The transition to the loss of H envelope occurs at an initial mass $\sim 30\ M_\odot$ (Heger *et al.* 2003). Among the massive stars, at low metallicity there is also a transition to the point where the core becomes so massive that there is direct collapse to a black hole without an explosion, $\sim 40\ M_\odot$.

In going to low metallicity, the mass loss rates due to stellar winds are generally taken to have a dependence $\propto Z^n$ with $n \approx 0.5$ (e.g., Heger *et al.* 2003). For O stars and Wolf-Rayet stars there is some observational support for this dependence, but for red supergiants this is simply an estimate. Mass loss from red supergiants is likely to be a combination of hydrodynamic processes (e.g., pulsations or large convective cells) and radiation pressure on dust grains. The dependence on dust suggests a Z dependence, but the nature of the dependence is uncertain. The general implication of the metallicity dependence of mass loss rates is that at low metallicity, the fraction of stars that retain their H envelopes increases and one goes directly from supernovae with H envelopes to core collapses to black holes without normal supernovae.

One change among the supernovae in going to low metallicity is that the stars can end their lives as BSGs (blue supergiants, radii $< 100 \, R_\odot$), as opposed to RSGs (red supergiants, radii $500 - 1500 \, R_\odot$). In a study of $13 - 25 \, M_\odot$ stellar models, Chieffi *et al.* (2003) found that all the stars at $Z = 0.02$ (solar) ended as RSGs, but that at $Z = 0.00$ all the stars ended as BSGs. At intermediate metallicities, the more massive stars tend to explode as BSGs and the less massive ones as RSGs. The progenitor radius is important for the peak luminosity of a supernova because the expansion from a small radius reduces the radiative energy in the supernova. Chieffi *et al.* (2003) estimate that the events with BSG progenitors are 1.5 magnitudes fainter than those with RSG progenitors.

Stars with initial mass $\geqslant 60 \, M_\odot$ are subject to pulsational instability related to nuclear burning. Baraffe *et al.* (2001) find that at $Z = 0$, the instability is slow and relatively unimportant, but even a minute metallicity $\sim 10^{-6}$ is sufficent to give substantial mass loss by pulsations. Stars of somewhat higher mass, $140 - 260 \, M_\odot$, are subject to become pair instability supernovae. In these events, the core becomes so hot that pairs form, causing the adiabatic index to drop below $4/3$ and resulting in the collapse of the core region. The contraction leads to enhanced burning and explosion. At $140 \, M_\odot$ there is weak Si burning, an explosion energy of 3×10^{51} ergs, and synthesis of a trace amount of ^{56}Ni, while at $260 \, M_\odot$, the explosion energy is 100×10^{51} ergs, with the synthesis of up to $50 \, M_\odot$ of ^{56}Ni (Scannapieco *et al.* 2005). These events clearly have the potential to produce very luminous supernovae. However, if there is mass loss during the stellar evolution, the core does not grow to the mass necessary for the pair instability to occur. With standard mass loss rates, pair instability supernovae are not expected for metallicities near solar because of the mass loss (Heger *et al.* 2003).

These results are modified in the case of close binary stars. Binary interaction can result in the loss of the H envelope in cases where single star mass loss would not be able to bring it about. In addition, the merger of stars can lead to a more massive H envelope than would occur in the single star case.

In the standard view, stars with initial masses $\geqslant 40 \, M_\odot$ collapse to black holes. However, it is possible that a disk forms around the black hole and the energy generated by accretion is sufficient to blow up the star. This is a possible model for long duration gamma-ray bursts (GRBs) (Woosley 1993). A problem is that in the evolution of single stars, the core does not end up with enough angular momentum to form an accretion disk. This problem is alleviated by having relatively little mass loss during the evolution because the mass loss carries away stellar angular momentum. However, the supernovae observed to be associated with GRBs are of Type Ic, which do not have their H envelopes, implying mass loss. Yoon & Langer (2005) and Woosley & Heger (2006) suggested a possible solution to this problem, noting that a rapidly rotating main sequence star can completely burn because of rotational mixing, resulting in a Wolf-Rayet star even though there has been little mass loss. In order to avoid too much mass-loss, the mass loss rates must be much less than the commonly used values at solar metallicity.

3. Implications of Core Collapse Supernova Observations

Observations of the basic core collapse supernova types near solar metallicity, IIP (plateau light curve), IIL (linear light curve) and Ibc (no H), are consistent with the general theoretical expectations. Modeling the light curves of SNe IIP had long shown consistency with the explosion of RSGs at the ends of their lives. In recent years, there has been direct observations of the progenitors of SNe IIP, showing that they are the expected RSGs (Li *et al.* 2007 and references therein). In addition, radio and X-ray observations have shown evidence for the mass loss rates expected for RSGs at the ends of their lives (Chevalier *et al.* 2006). The mass loss rates are expected to depend both on the mass of the progenitor star and its metallicity. The mass can be estimated from the direct progenitor detection or from modeling the supernova light curve, and the metallicity from observations of the surrounding ISM. Although the data are not yet of sufficient quality to do this, it may eventually be possible to directly measure the relation between mass loss rate and metallicity for the supernova progenitors.

As noted above, at very low metallicity these stars should explode as BSGs. Such events should stand out by their light curves, but have not been observed, only part of which can be explained by the fainter magnitudes that these supernovae have (Cappellaro *et al.* 1997). The implication is that we are not observing supernovae in very metal poor regions. The only supernova that clearly exploded as a BSG was SN 1987A and, in that case, it is plausible that the reason was a binary merger (Podsiadlowski *et al.* 2007). The asymmetric circumstellar medium around SN 1987A supports the merger scenario. Although the Large Magellanic Cloud is somewhat metal poor, it is not sufficiently metal poor to regularly result in supernovae exploding as BSGs.

The supernova Types IIL, IIb, and Ibc are consistent with the expectation that as one goes to higher initial mass, there is less of a H envelope left at the time of the explosion. Binary interaction can also give rise to a low mass H envelope; evidence for such an action is clearest for SN 1993J, where the binary companion has apparently been detected in the postsupernova light (Maund *et al.* 2004).

A supernova type that has been more perplexing is the IIn, which is characterized by relatively narrow Hα emission and a blue continuum (Schlegel 1990). The luminosities observed to late times for some of these objects are indicative of high rates of mass loss $\dot{M} \sim 10^{-3}(v_w/10 \text{ km s}^{-1})$ M$_\odot$ yr^{-1} (e.g., Chugai & Chevalier 2006). In addition, P Cygni profiles observed in the Hα line indicate circumstellar velocities of 100's km s^{-1}. Smith & Owocki (2006) make the point that the mass loss rates and velocities are higher than those in the winds of RSGs, and are roughly compatible with observations of LBVs (luminous blue variables), which are thought to be stars with initial masses $\geqslant 40-50$ M$_\odot$ that do not go through a RSG phase. In the standard view of stellar evolution, these stars would go through an extended Wolf-Rayet phase after losing their H envelopes. The supernovae would then not interact directly with the dense mass loss. However, if the mass loss rate during the O star phase has been overestimated, it may be possible to have the supernova be nearly contemporaneous with the LBV phase. The cause of mass loss during the LBV phase is poorly understood; it may have to do with continuum opacity or with explosive events. Smith & Owocki (2006) note that these mechanisms do not have a metallicity dependence, so that the strong mass loss can occur even at low metallicity, in contrast to the standard view that going to low metallicity should lead to a lower circumstellar density at the time of the supernova. In this case, Type IIn supernovae could persist to low metallicity.

One development in the search for supernovae has been the implementation of 'blind' searches. Traditionally, supernova searches have targeted nearby luminous galaxies. In some of the new searches (e.g., the Texas Supernova Search (TSS) with the ROTSE-IIb telescope) a wide field of view is observed without targeting particular galaxies. Such searches have the potential to discover supernovae in small, typically metal poor, galaxies and luminous, distant, rare supernovae. These searches have found a number of Type IIn supernovae, although there are not sufficient statistics to comment on the rate of these events.

An interesting discovery from the TSS survey is the very luminous SN 2006gy (Smith *et al.* 2007; Ofek *et al.* 2007); the radiated energy in the first 200 days was $\sim 10^{51}$ ergs, comparable to the total mechanical energy of a normal supernova. One suggestion for the high luminosity is that the event was a pair instability supernova, with the creation of 22 $M_\odot$ of ^{56}Ni (Smith *et al.* 2007). The prediction of this model would be the presence of a late tail to the light curve, powered by radioactivity. Although the supernova has become very faint at optical wavelengths, it has become infrared bright (Smith *et al.* 2008b), which may require a radioactive power source. However, the metallicity of the host galaxy of SN 2006gy is roughly solar (Ofek *et al.* 2007), which is above the range expected for the occurrence of pair instability. This scenario requires an unexpected low rate of mass loss from the progenitor star. An alternative to radioactivity is power from shock interactions, but strong X-ray and radio emission, a signature of circumstellar interaction, has not been detected. However, absorption effects at high density could be the reason for the lack of emission (Ofek *et al.* 2007). Future multiwavelength observations should be able to discriminate between these possibilities.

Some of the best evidence for the expectation that GRBs have low metallicity progenitors has come from estimating the metallicity of the local environment of nearby GRBs and broad-lined Type Ic supernovae (Modjaz *et al.* 2008). The GRB related objects are systematically at lower metallicity, with the dividing line between them at [12+log(O/H)]= 8.5. However, if the solar O abundance is [12+log(O/H)]= 8.7, the metallicity of the most metal rich GRB event, GRB 980425/SN 1998bw, is only slightly below the solar value. Among more distant GRBs, there is some evidence indicating that GRB host galaxies have lower metallicity than other galaxies at a similar redshift (e.g., Fruchter *et al.* 2006).

There is not yet much observational information on supernovae in low metallicity galaxies. Based on broad Hα emission, Izotov *et al.* (2007) found possible evidence for Type IIP and IIn supernovae in metal poor blue compact dwarf galaxies. Interestingly, the possible Type IIn supernovae were only present in galaxies with [12+log(O/H)]$\leqslant$ 8.0, or 5 times lower than the solar value. However, the time dependence of these objects still needs to be demonstrated to show that they are indeed supernovae and not active galactic nuclei. Isotov (this meeting) found little time evolution in the Type IIn candidates over 5 years, implying that they are likely to be active galactic nuclei. The Type IIP candidates remain uncertain.

Prieto *et al.* (2008) have examined the relation between supernova type and host galaxy metallicity for > 100 supernovae found in the recent SDSS survey; the host metallicities span the range $\sim 0.1 - 2.7\ Z_\odot$. They find that the Type Ibc supernovae tend to occur in more metal rich galaxies than the Type II's, as expected from the general scenario of stellar evolution described in the previous section. However, the supernova that they find in the most metal poor galaxy ($Z \sim 0.05\ Z_\odot$) is a Type Ic, SN 2007bg. Interestingly, this supernova is a luminous radio source (A. Soderberg, private communication), which suggests that it has a dense circumstellar medium. Prieto *et al.* (2008) suggest a link to GRBs, but this supernova showed evidence for high velocity H and He absorption lines (Harutyunyan *et al.* 2007), which have not been seen in supernovae associated with GRBs. The supernova thus appears to be a normal broad-lined Type Ibc supernova, perhaps similar to SN 2003bg, that is well below the dividing line found by Modjaz *et al.* (2008).

4. Ejection of Dust by Supernovae

Dust formation in freely expanding supernova ejecta has long been predicted, but has been difficult to observe. SN 1987A provided a clear case of dust formation, as judged by a shift in the emission to infrared wavelengths and absorption of redshifted gas, but the amount of dust formed, $\sim 5 \times 10^{-4}$ $M_\odot$ (Ercolano *et al.* 2007), was small. Considering the available mass of refractory elements, the efficiency of dust formation was $\sim 10^{-3}$. There have been numerous infrared observations of more recent supernovae, but the dust emission can generally be interpreted in terms of circumstellar dust. One case with dust formation in the ejecta is the Type IIP SN 1999em (Elmhamdi *et al.* 2003), but, once again, the amount of dust is estimated to be relatively small. In the case of the Type IIP SN 2003gd, Sugerman *et al.* (2006) found evidence for 0.02 $M_\odot$ of dust, which would imply a relatively high efficiency of 0.1 for dust formation. However, Meikle *et al.* (2007) find that Sugerman *et al.* (2006) overestimated the amount of dust and that only 4×10^{-5} $M_\odot$ of newly condensed dust is needed. They attribute some of the infrared emission to an echo. They note that the absorption of redshifted gas estimated from line profiles is comparable to that observed in SN 1987A, so that a comparable amount of dust may be indicated.

In recent years, another possible site of newly formed dust has drawn attention. When there is supernova interaction with a dense circumstellar medium, cooling shock waves may occur in the shocked region, allowing the temperature to drop to the point where dust condensation can occur. Because of the high pressure in the interaction region, the gas densities can exceed 10^{10} cm^{-3}, which is higher than the density in the freely expanding ejecta. Pozzo *et al.* (2004) found evidence for this mechanism in SN 1998S, but an especially good case for dust formation in this situation is provided by SN 2006jc, where both a rising infrared flux and absorption in circumstellar line emission was observed (Smith *et al.* 2008a). However, Nozawa *et al.* (2008) found that although ~ 1.5 $M_\odot$ of dust might be formed according to their condensation calculations, the observational evidence indicated the formation of just $\sim 10^{-3}$ $M_\odot$ of dust. This mechanism should be important in supernovae where circumstellar interaction is especially strong, i.e. the Type IIn supernovae. Fox *et al.* (2008) recently argued for the importance of this mechanism for the infrared emission from the Type IIn SN 2005ip and suggested that it may be more generally applicable to SNe IIn.

The dust mass estimates given here are for warm dust ($T \geqslant 300$ K). There are no reliable limits on the amount of cool dust. However, in the radiation field of a supernova, the dust is expected to be warm unless it is shielded by optical depth effects.

In addition to young supernovae, it is possible to detect dust in supernova remnants. The Crab Nebula is thought to come from a star with an initial mass of $8 - 10$ $M_\odot$, based on its element abundances, so it is plausibly the result of a Type IIP supernova and provides a good case of comparison for the supernovae. *Spitzer* observations of the Crab show evidence for dust emission, and an estimated dust mass $\sim 10^{-3}$ $M_\odot$ (Temim *et al.* 2006), which is comparable to the supernova results. One can draw the conclusion that Type IIP supernovae are generally inefficient at forming dust.

5. Discussion

Our understanding of the lower mass core collapse supernovae is in good shape. The stars explode as RSGs that have had relatively little mass loss during their evolution. The mass loss at the time of the supernova can be investigated by multiwavelength observations and there are prospects for determining the variation of mass loss rate with metallicity.

For more massive stars, mass loss effects are strong and there is more uncertainty in the evolution. There are several indications that generally assumed hot star mass loss rates are overestimated: (1) Some Type IIn supernovae show evidence for circumstellar densities and velocities that are typical of LBVs. Ordinarily, LBVs would be expected to go through a Wolf-Rayet phase before becoming supernovae, but reduced mass loss may allow an explosion close to the LBV phase. (2) There is possible evidence for a pair instability supernova at a metallicity close to solar. Such an event requires a very massive star with a small amount of mass loss. (3) A plausible requirement for a GRB is a small amount of mass loss from the progenitor, so there is little loss of angular momentum during the stellar evolution. Although these considerations point to low mass loss rates, there are also requirements for high rates. The occurrence of SNe Ibc requires mass loss to end with a stripped star, unless the Wolf-Rayet star is formed by rotational mixing during the main sequence with little mass loss. The case of SN 2007bg is of special interest because because it is in a very low metallicity host and has radio evidence for a dense circumstellar medium. These objects point to another mechanism for driving the mass loss in some cases: binary interaction is a good candidate.

Observations of supernovae show clear evidence for the formation of dust in the ejecta, which is relevant to the origin of dust in the early universe. Observations of Type IIP supernovae indicate the formation of $\sim 10^{-3}$ $M_\odot$ of dust in the freely expanding ejecta. A similar amount of dust has been inferred in the Crab Nebula, which might have its origin in a Type IIP supernova. This amount is considerably less that the ~ 0.1 $M_\odot$ of dust per supernova needed to produce the dust that may be present in the early universe (Todini & Ferrara 2001). A possible solution is that most of the metals in the universe are not produced by Type IIP supernovae, but by supernovae with more massive progenitor stars.

Acknowledgements

I am grateful to the organizers for putting on a stimulating meeting in a very pleasant location. This research was supported in part by NASA grant NNG06GJ33G.

References

Baraffe, I., Heger, A., & Woosley, S. E. 2001, *ApJ*, 550, 890
Cappellaro, E., Turatto, M., Tsvetkov, D. Y., Bartunov, O. S., Pollas, C., Evans, R., & Hamuy, M. 1997, *A&A*, 322, 431
Chevalier, R. A., Fransson, C., & Nymark, T. K. 2006, *ApJ*, 641, 1029
Chieffi, A., Domínguez, I., Höflich, P., Limongi, M., & Straniero, O. 2003, *MNRAS*, 345, 111
Chugai, N. N., & Chevalier, R. A. 2006, *ApJ*, 641, 1051
Elmhamdi, A., *et al.* 2003, *MNRAS*, 338, 939
Ercolano, B., Barlow, M. J., & Sugerman, B. E. K. 2007, *MNRAS*, 375, 753
Fox, O., *et al.* 2008, *ApJ*, in preparation
Fruchter, A. S., *et al.* 2006, *Nature*, 441, 463
Harutyunyan, A., *et al.* 2007, *CBET*, 948, 1
Heger, A., Fryer, C. L., Woosley, S. E., Langer, N., & Hartmann, D. H. 2003, *ApJ*, 591, 288

Izotov, Y. I., Thuan, T. X., & Guseva, N. G. 2007, *ApJ*, 671, 1297

Li, W., Wang, X., Van Dyk, S. D., Cuillandre, J.-C., Foley, R. J., & Filippenko, A. V. 2007, *ApJ*, 661, 1013

Maund, J. R., Smartt, S. J., Kudritzki, R. P., Podsiadlowski, P., & Gilmore, G. F. 2004, *Nature*, 427, 129

Meikle, W. P. S., *et al.* 2007, *ApJ*, 665, 608

Modjaz, M., *et al.* 2008, *AJ*, 135, 1136

Nozawa, T., *et al.* 2008, *ApJ*, in press (arXiv:0801.2015)

Ofek, E. O., *et al.* 2007, *ApJ*, 659, L13

Podsiadlowski, P., Morris, T. S., & Ivanova, N. 2007, in: S. Immler, K. Weiler, & R. McCray (eds.), *Supernova 1987A: 20 Years After* (Melville, NY: AIP), p. 125

Pozzo, M., Meikle, W. P. S., Fassia, A., Geballe, T., Lundqvist, P., Chugai, N. N., & Sollerman, J. 2004, *MNRAS*, 352, 457

Prieto, J. L., Stanek, K. Z., & Beacom, J. F. 2008, *ApJ*, 673, 999

Scannapieco, E., Madau, P., Woosley, S., Heger, A., & Ferrara, A. 2005, *ApJ*, 633, 1031

Schlegel, E. M. 1990, *MNRAS*, 244, 269

Smith, N., & Owocki, S. P. 2006, *ApJ*, 645, L45

Smith, N., *et al.* 2007, *ApJ*, 666, 1116

Smith, N., Foley, R. J., & Filippenko, A. V. 2008a, *ApJ*, 680, 568

Smith, N., *et al.* 2008b, *ApJ*, in press (arXiv:0802.1743)

Sugerman, B. E. K., *et al.* 2006, *Science*, 313, 196

Temim, T., *et al.* 2006, *AJ*, 132, 1610

Todini, P. & Ferrara, A. 2001, *MNRAS*, 325, 726

Woosley, S. E. 1993, *ApJ*, 405, 273

Woosley, S. E. & Heger, A. 2006, *ApJ*, 637, 914

Yoon, S.-C. & Langer, N. 2005, *A&A*, 443, 643

Low-Metallicity Star Formation:
From the First Stars to Dwarf Galaxies
Proceedings IAU Symposium No. 255, 2008
L.K. Hunt, S. Madden & R. Schneider, eds.

First Stars – Type Ib Supernovae Connection

Ken'ichi Nomoto[1,2], Masaomi Tanaka[2], Yasuomi Kamiya[2], Nozomu Tominaga[3] and Keiichi Maeda[1]

[1]Institute for the Physics and Mathematics of the Universe, University of Tokyo, Kashiwa, Chiba 277-8568, Japan
email: `nomoto@astron.s.u-tokyo.ac.jp`

[2]Department of Astronomy, University of Tokyo, Bunkyo-ku, Tokyo 113-0033, Japan
[3]National Astronomical Observatory, Mitaka, Tokyo 113-0033, Japan

Abstract. The very peculiar abundance patterns observed in extremely metal-poor (EMP) stars can not be explained by conventional normal supernova nucleosynthesis but can be well-reproduced by nucleosynthesis in hyper-energetic and hyper-aspherical explosions, i.e., Hypernovae (HNe). Previously, such HNe have been observed only as Type Ic supernovae. Here, we examine the properties of recent Type Ib supernovae (SNe Ib). In particular, SN Ib 2008D associated with the luminous X-ray transient 080109 is found to be a more energetic explosion than normal core-collapse supernovae. We estimate that the progenitor's main sequence mass is $M_{MS} = 20 - 25 M_\odot$ with an explosion of kinetic energy of $E_K \sim 6.0 \times 10^{51}$ erg. These properties are intermediate between those of normal SNe and hypernovae associated with gamma-ray bursts. Therefore, such energetic SNe Ib could also make an important contribution to the chemical enrichment in the early Universe.

Keywords. Galaxy: halo, gamma rays: bursts, nuclear reactions, nucleosynthesis, abundances, stars: abundances, stars: Population II, supernovae: general

1. Metal Poor Stars – Hypernovae – GRB Connections

The abundance patterns of the extremely metal-poor (EMP) stars are good indicators of supernova (SN) nucleosynthesis, because the Galaxy was effectively unmixed at [Fe/H] < -3. Thus they could provide useful constraints on the nature of First Supernovae and thus First Stars.

The EMP stars are classified into three groups according to [C/Fe] (e.g., Hill, François, & Primas 2005, Beers & Christlieb 2005):

(1) [C/Fe] ~ 0, normal EMP stars ($-4 < $ [Fe/H] < -3);
(2) [C/Fe] $\gtrsim +1$, Carbon-enhanced EMP (CEMP) stars ($-4 < $ [Fe/H] < -3);
(3) [C/Fe] $\sim +4$, hyper metal-poor (HMP) stars ([Fe/H] < -5, e.g., HE 0107–5240,
Christlieb *et al.* 2002, Bessell & Christlieb 2005; HE 1327–2326, Frebel *et al.* 2005).
Table 1 summarizes other abundance features of various EMP stars. Many of these EMP stars have high [Co/Fe].

We have shown that such peculiar abundance patterns can not be explained by conventional normal supernova nucleosynthesis but can be reproduced by nucleosynthesis in hyper-energetic and hyper-aspherical explosions, i.e., Hypernovae (HNe) (e.g., Maeda *et al.* 2002, Maeda & Nomoto 2003, Tominaga *et al.* 2007, Tominaga 2007).

The abundance pattern of the Ultra Metal-Poor (UMP) star (HE 0557–4840: Norris *et al.* 2007) is shown in Figure 1 and compared with the HN ($E_{51} = 20$) and SN ($E_{51} = 1$) models of $25 M_\odot$ stars. The Co/Fe ratio ([Co/Fe]~ 0) requires a high energy explosion and the high [Sc/Ti] and [Ti/Fe] ratios require a high-entropy explosion. The HN model is

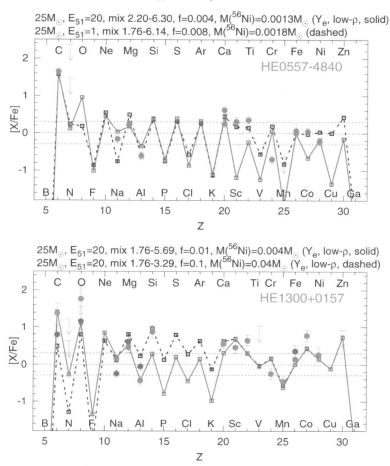

Figure 1. Comparisons of the abundance patterns between the mixing-fallback models and the UMP star HE0557–4840 (upper: Norris *et al.* 2007), and the CEP star HE1300+0157 (lower: Frebel *et al.* 2007).

in a good agreement with the abundance pattern of HE 0557–4840. The model indicates $M(^{56}\mathrm{Ni}) \sim 10^{-3} M_\odot$ being similar to faint SN models for CEMP stars.

The abundance pattern of the CEMP-no star (i.e., CEMP with no neutron capture elements) HE 1300+0157 (Frebel *et al.* 2007) is shown in Figure 1 (lower) and marginally reproduced by the hypernova model with $M_{\mathrm{MS}} = 25 M_\odot$ and $E_{51} = 20$. The large [Co/Fe] particularly requires the high explosion energy.

Previously, Hypernova-like explosions with $E_{51} > 10$ have been found only in Type Ic supernovae (SNe Ic), which are core-collapse supernovae characterized by the lack of hydrogen and helium.

Recently, several interesting Type Ib supernovae (SNe Ib) have been observed to show quite peculiar features. SNe Ib are another type of envelope-stripped core collapse SN but characterized by the presence of prominent He lines. Thus it is interesting to examine the explosion energy and other properties of SNe Ib in comparison with Hypernovae and normal SNe.

Here we present our analysis of peculiar SNe Ib 2008D and 2006jc.

Table 1. Metal-poor stars.

Name	[Fe/H]	Features	Reference
HE 0107–5240	−5.3	C-rich, Co-rich?, [Mg/Fe]∼ 0	Christlieb *et al.* 2002
HE 1327–2326	−5.5	C, O, Mg-rich	Frebel *et al.* 2005, Aoki *et al.* 2006
HE 0557–4840	−4.8	C, Ca, Sc, Ti-rich, [Co/Fe]∼ 0	Norris *et al.* 2007
HE 1300+0157	−3.9	C, Si, Ca,Sc,Ti, Co-rich	Frebel *et al.* 2007
HE 1424–0241	−4.0	Co,Mn-rich, Si,Ca,Cu-poor	Cohen *et al.* 2007
CS 22949–37	−4.0	C,N,O,Mg,Co,Zn-rich	Depagne *et al.* 2002
CS 29498–43	−3.5	C,N,O,Mg-rich, [Co/Fe]∼ 0	Aoki *et al.* 2004
BS 16934–002	−2.8	O,Mg-rich, C-poor	Aoki *et al.* 2007

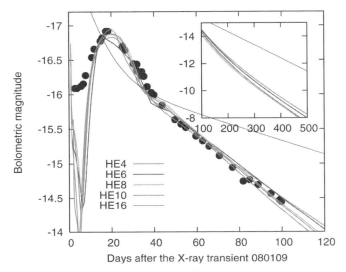

Figure 2. The pseudo-bolometric ($UBVRIJHK$) light curve (LC) of SN 2008D compared with the results of LC calculations with the models HE4 (red), HE6(blue), HE8 (green), HE10 (magenta) and HE16 (gray). The pseudo-bolometric LC is shown in filled (left) and open (right) circles. The thin black line shows the decay energy from ^{56}Ni and ^{56}Co [$M(^{56}$Ni$) = 0.07\ M_\odot$]. At late epochs, it is roughly equal to the optical luminosity under the assumption that γ-rays are fully trapped. The bolometric magnitude at $t \sim 4$ days after the X-ray transient is brighter by ~ 0.25 mag than that shown by other papers (Soderberg *et al.* 2008; Malesani et al. 2008; Modjaz *et al.* 2008b; Mazzali *et al.* 2008), which is shown by the thin arrow in the left panel.

2. Energetic Type Ib Supernova SN 2008D

SN 2008D was discovered as a luminous X-ray transient in NGC 2770. The X-ray emission of the transient reached a peak ~ 65 seconds, lasting ~ 600 seconds, after the observation started. The X-ray spectrum is soft, and no γ-ray counterpart was detected by the *Swift* BAT. The optical counterpart was discovered at the position of the X-ray transient, confirming the presence of a SN 2008D (see, e.g., Soderberg *et al.* 2008).

SN 2008D showed a broad-line optical spectrum at early epochs ($t \lesssim 10$ days, hereafter t denotes time after the transient, 2008 Jan 9.57 UT, Soderberg *et al.* 2008). However, the spectrum changed to that of a normal Type Ib SN, i.e., a SN with He absorption lines and without H lines (Modjaz *et al.* 2008). To date, the SNe associated with GRBs or XRFs are all Type Ic, i.e., SNe without H and He absorption.

The pseudo-bolometric ($UBVRIJHK$) light curve (LC) is compared with the theoretical models of He star models HE4, HE6, HE8, HE10, and HE16, whose masses are

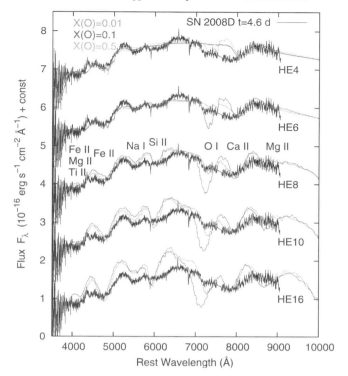

Figure 3. The spectrum of SN 2008D at $t = 4.6$ days from the X-ray transient (black line, Mazzali *et al.* 2008) compared with synthetic spectra (color lines). The spectra are shifted by 6.0, 4.5, 3.0, 1.5, 0.0 from top to bottom. The model spectra are reddened with $E(B−V) = 0.65$ mag. From top to bottom, the synthetic spectra calculated with HE4, H46, HE8, HE10 and HE16 are shown. The red, blue and green lines show the synthetic spectra with oxygen mass fraction $X(O) = 0.01$, 0.1, and 0.5, respectively. Since the synthetic spectra with $X(O) = 0.1$ for more massive models than HE4 already show too strong O I line, the spectra with $X(O) = 0.5$ are not shown for these models.

$M_\alpha = 4, 6, 8, 10$, and $16 M_\odot$, respectively. These He stars correspond to the main-sequence stellar masses of $M_{ms} \sim 15$, 20, 25, 30, and 40 $M_\odot$.

Since the timescale around the peak depends on both M_{ej} and E_K as $\propto \kappa^{1/2} M_{ej}^{3/4} E_K^{-1/4}$, where κ is the optical opacity (Arnett 1982), a specific kinetic energy is required for each model to reproduce the observed timescale. The derived set of ejecta parameters are $(M_{ej}/M_\odot,\ E_K/10^{51}\ \text{erg}) = (2.7, 1.1), (4.4, 3.7), (6.2, 8.4), (7.7, 13.0)$ and $(12.5, 26.5)$ for the case of HE4, HE6, HE8, HE10 and HE16, respectively. The ejected ^{56}Ni mass is $\sim 0.07 M_\odot$ in all models.

We have done a detailed theoretical study of emission from SN 2008D. The bolometric LC and optical spectra are modeled based on realistic progenitor models and the explosion models obtained from hydrodynamic/nucleosynthetic calculations (Tanaka *et al.* 2008).

We have found that HE4, HE10 and HE16 are not consistent with SN 2008D. Both HE6 and HE8 have a small inconsistency related to the boundary between the He-rich and O-rich layers. It seems that a model between HE6 and HE8 may be preferable.

We thus conclude that the progenitor star of SN 2008D had a He core mass $M_\alpha = 6 − 8 M_\odot$ prior to the explosion. This corresponds to a main sequence mass of $M_{MS} = 20 − 25 M_\odot$. We find that SN 2008D is an explosion with $M_{ej} = 5.3 \pm 1.0 M_\odot$ and $E_K = 6.0 \pm 2.5 \times 10^{51}$ erg. The mass of the central remnant is $1.6 − 1.8 M_\odot$, which is near the

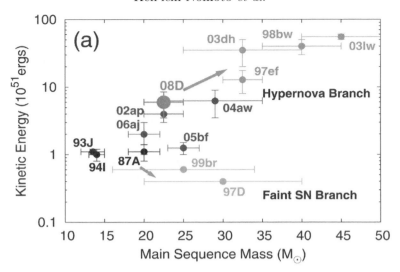

Figure 4. The kinetic explosion energy E as a function of the main sequence mass M of the progenitors for several supernovae/hypernovae. Hypernovae are the SNe with $E_{51} > 10$.

boundary mass between a neutron star and a black hole. Note that the error bars reflect just the uncertainty of the LC and spectral modeling.

Figure 4 shows the kinetic energy of the ejecta and the ejected ^{56}Ni mass as a function of the estimated main sequence mass for several core-collapse SNe (see, e.g., Nomoto *et al.* 2007). SN 2008D is shown by a red circle.

Comparison with other Type Ib SNe shown in Figure 4 is possible only for SN 2005bf although SN 2005bf is a very peculiar SN that shows a double peak LC with a very steep decline after the maximum, and increasing He line velocities (Anupama et al. 2005; Tominaga *et al.* 2005; Folatelli *et al.* 2006; Maeda et al. 2007). The LC of SN 2005bf is broader than that of SN 2008D, while the expansion velocity of SN 2005bf is lower than that of SN 2008D. These facts suggest that SN 2005bf is the explosion with a lower E_K/M_{ej} ratio.

The spectra of SN 2008D and SN 1999ex are very similar (Valenti et al. 2008b), while SN 2005bf has lower He velocities. The He lines in SN 1993J are very weak at this epoch. The Fe features at 4500-5000Å are similar in these four SNe, but those in SN 2005bf are narrower. Malesani *et al.* (2008) suggested that the bolometric LCs of SNe 1999ex and 2008D are similar. The similarity in both the LC and the spectra suggests that SN 1999ex is located close to SN 2008D in the $E_K - M_{MS}$ and $M(^{56}Ni) - M_{MS}$ diagrams.

Malesani *et al.* (2008) also pointed the similarity of the LCs of SNe 1993J and 2008D. But the expansion velocity is higher in SN 2008D (see, e.g. Prabhu *et al.* 1995). Thus, both the mass and the kinetic energy of the ejecta are expected to be smaller in SN 1993J. In fact, SN 1993J is explained by the explosion of a $4M_\odot$ He core with a small mass H-rich envelope (Nomoto *et al.* 1993; Shigeyama *et al.* 1994; Woosley *et al.* 1994).

3. Dust-Forming Type Ib Supernova SN 2006jc

Another recent event, SN Ib 2006jc, is characterized by dust formation in the ejecta as found from NIR and MIR observations.

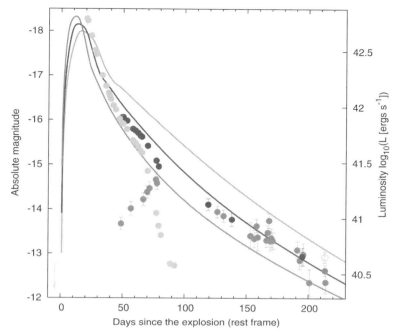

Figure 5. Comparison between the synthetic LCs for the models with $E_{51} = 5$ and $M_{ej} = 5.1 M_{\odot}$, $E_{51} = 10$ and $M_{ej} = 4.9 M_{\odot}$, and $E_{51} = 20$ and $M_{ej} = 4.6 M_{\odot}$, and the LCs of SN 2006jc ($L_{UV} + L_{opt}$, $L_{IR,est} (\nu < 3 \times 10^{14} Hz)$, L_{bol}, $L_{IR,hot} (\nu < 3 \times 10^{14} Hz)$, $L_{IR} (\nu < 3 \times 10^{14} Hz)$).

We present a theoretical model for Type Ib supernova (SN) 2006jc. We calculate the evolution of the progenitor star, hydrodynamics and nucleosynthesis of the SN explosion, and the SN bolometric LC. The synthetic bolometric LC is compared with the observed bolometric LC constructed by integrating the UV, optical, near-infrared (NIR), and mid-infrared (MIR) fluxes.

The progenitor is assumed to be as massive as $40 M_{\odot}$ on the zero-age main-sequence. The star undergoes extensive mass loss to reduce its mass down to as small as $6.9 M_{\odot}$, thus becoming a WCO Wolf-Rayet star. The WCO star model has a thick carbon-rich layer, in which amorphous carbon grains can be formed. This could explain the NIR brightening and the dust feature seen in the MIR spectrum. We suggest that the progenitor of SN 2006jc is a WCO Wolf-Rayet star having undergone strong mass loss and such massive stars are the important sites of dust formation. We derive the parameters of the explosion model in order to reproduce the bolometric LC of SN 2006jc by the radioactive decays: the ejecta mass $4.9 M_{\odot}$, hypernova-like explosion energy 10^{52} ergs, and ejected ^{56}Ni mass $0.22 M_{\odot}$.

We also calculate the circumstellar interaction and find that a CSM with a flat density structure is required to reproduce the X-ray LC of SN 2006jc. This suggests a drastic change of the mass-loss rate and/or the wind velocity that is consistent with the past luminous blue variable (LBV)-like event.

We have thus found SN Ib 2006jc is almost a HN-like energetic explosion. This is suggestive for the SN Ib contribution to the early enrichment in the Universe. Also dust formation in a WCO star seems to be quite important.

4. Concluding Remarks

We presented a theoretical model for SN 2008D associated with the luminous X-ray transient 080109. These models are tested against the bolometric LC and optical spectra. This is the first detailed model calculation for the Type Ib SN that is discovered shortly after the explosion.

The main sequence mass of the progenitor of SN 2008D is estimated to be $M_{\rm MS} = 20 - 25 M_\odot$, between normal SNe and GRB-SNe (or hypernovae). The kinetic energy of SN 2008D is also intermediate. Thus, SN 2008D is located between the normal SNe and the "hypernovae branch" in the $E_{\rm K} - M_{\rm MS}$ diagram (upper panel of Fig. 4). The ejected ^{56}Ni mass in SN 2008D ($\sim 0.07 M_\odot$) is similar to the ^{56}Ni masses ejected by normal SNe but much smaller than those in GRB-SNe.

These energetic SNe Ib, as indicated from both 2008D and 2006jc, could make an important contribution to the chemical enrichment in the early Universe, although the explosions are not as extreme as Hypernovae. Dust formation in WCO stars could also be important.

References

Aoki, W., *et al.* 2004, *ApJ* 608, 971
Aoki, W., *et al.* 2006, *ApJ* 639, 897
Aoki, W., *et al.* 2007, *ApJ* 660, 747
Bailyn, C.D., Jain, R.K., Coppi, P., & Orosz, J.A. 1998, *ApJ* 499, 367
Beers, T. & Christlieb, N. 2005, *ARA&A* 43, 531
Bessell, M. S. & Christlieb, N. 2005, in V. Hill *et al.* (eds.), *From Lithium to Uranium*, Proc. IAU Symposium No. 228 (Cambridge: Cambridge Univ. Press), p. 237
Cayrel, R., *et al.* 2004, *A&A* 416, 1117
Christlieb, N., *et al.* 2002, *Nature* 419, 904
Cohen, J.G., *et al.* 2007, *ApJ* 659, L161
Depagne, E., *et al.* 2002, *A&A* 390, 187
Frebel, A., *et al.* 2005, *Nature* 434, 871
Frebel, A., *et al.* 2007, *ApJ* 658, 534
Galama, T., *et al.* 1998, *Nature* 395, 670
Hill, V., François, P., & Primas, F. (eds.) 2005, *From Lithium to Uranium: Elemental Tracers of Early Cosmic Evolution*, Proc. IAU Symp. No. 228 (Cambridge: Cambridge Univ. Press)
Iwamoto, K., Mazzali, P. A., Nomoto, K., *et al.* 1998, *Nature* 395, 672
Iwamoto, N., Umeda, H., Tominaga, N., Nomoto, K., & Maeda, K. 2005, *Science* 309, 451
Maeda, K., Nakamura, T., Nomoto, K., *et al.* 2002, *ApJ* 565, 405
Maeda, K. & Nomoto, K. 2003, *ApJ* 598, 1163
Malesani, J., *et al.* 2006, *ApJ* 609, L5
Nomoto, K., *et al.* 2004, in C. L. Fryer (ed.), *Stellar Collapse* (Astrophysics and Space Science: Kluwer), p. 277 (astro-ph/0308136)
Nomoto, K., *et al.* 2006, *Nuclear Phys A* 777, 424 (astro-ph/0605725)
Nomoto, K., *et al.* 2007, *Nuovo Cimento* 121, 1207 (astro-ph/0702472)
Norris, J. E., *et al.* 2007, *ApJ* 670, 774
Tominaga, N., Tanaka, M., Nomoto, K., *et al.* 2005, *ApJ* 633, L97
Tominaga, N., Maeda, K., Umeda, H., Nomoto, K., Tanaka, *et al.* 2007, *ApJ* 657, L77
Tominaga, N., Umeda, H., & Nomoto, K. 2007, *ApJ* 660, 516
Tominaga, N. 2007, *ApJ* submitted (arXiv:0711.4815)
Umeda, H. & Nomoto, K. 2002, *ApJ* 565, 385
Umeda, H. & Nomoto, K. 2005, *ApJ* 619, 427
Woosley, S. E. & Bloom, J. S. 2006, *ARA&A* 44, 507

Low-Metallicity Star Formation:
From the First Stars to Dwarf Galaxies
Proceedings IAU Symposium No. 255, 2008
L.K. Hunt, S. Madden & R. Schneider, eds.

© 2008 International Astronomical Union
doi:10.1017/S1743921308024800

Supernova Nucleosynthesis in the early universe

Nozomu Tominaga[1], Hideyuki Umeda[2], Keiichi Maeda[3], Ken'ichi Nomoto[3,2] and Nobuyuki Iwamoto[4]

[1]Optical and Infrared Astronomy Division, National Astronomical Observatory, Mitaka, Tokyo, Japan
email: nozomu.tominaga@nao.ac.jp

[2]Department of Astronomy, School of Science, University of Tokyo, Bunkyo, Tokyo, Japan

[3]Institute for the Physics and Mathematics of the Universe, University of Tokyo, Kashiwa, Chiba, Japan

[4]Nuclear Data Center, Nuclear Science and Engineering Directorate, Japan Atomic Energy Agency, Tokai, Ibaraki, Japan

Abstract. The first metal enrichment in the universe was made by supernova (SN) explosions of population (Pop) III stars. The history of chemical evolution is recorded in abundance patterns of extremely metal-poor (EMP) stars. We investigate the properties of nucleosynthesis in Pop III SNe by comparing their yields with the abundance patterns of the EMP stars. We focus on (1) jet-induced SNe with various properties of the jets, especially energy deposition rates $[\dot{E}_{\rm dep} = (0.3 - 1500) \times 10^{51} {\rm ergs\ s^{-1}}]$, and (2) SNe of stars with various main-sequence masses $(M_{\rm ms} = 13 - 50 M_\odot)$ and explosion energies $[E = (1 - 40) \times 10^{51} {\rm ergs}]$. The varieties of Pop III SNe can explain the observations of the EMP stars: (1) higher [C/Fe] for lower [Fe/H] and (2) trends of abundance ratios [X/Fe] against [Fe/H].

Keywords. Galaxy: halo, gamma rays: bursts, nuclear reactions, nucleosynthesis, abundances, stars: abundances, stars: Population II, supernovae: general

1. Introduction

Long-duration γ-ray bursts (GRBs) have been found to be accompanied by luminous and energetic Type Ic supernovae [SNe Ic, called hypernovae (HNe)] (e.g. Galama *et al.* 1998). Although the explosion mechanism is still under debate, photometric observations (a "jet break", e.g. Frail *et al.* 2001) and spectroscopic observations (a nebular spectrum, e.g. Maeda *et al.* 2002) indicate that they are aspherical explosions with jet(s).

The aspherical explosions are indirectly suggested from the abundance patterns of extremely metal-poor (EMP) stars with [Fe/H] < -3† which are suggested to show nucleosynthesis yields of a single core-collapse SN (e.g. Beers & Christlieb 2005). C-enhanced EMP (CEMP) stars have been well explained by faint SNe with large fallback (Umeda & Nomoto 2005; Iwamoto *et al.* 2005; Nomoto *et al.* 2006; Tominaga *et al.* 2007b). Some CEMP stars show enhancement of Co and Zn (e.g. Depagne *et al.* 2002) that requires explosive nucleosynthesis under high entropy. In a *spherical* model, a high entropy explosion corresponds to a high energy explosion that inevitably synthesizes a large amount of ^{56}Ni, i.e., a bright SN (e.g. Woosley & Weaver 1995). This incompatibility will be solved if a faint SN is associated with a narrow jet within which a high entropy region is confined (Umeda & Nomoto 2005).

† Here $[{\rm A/B}] \equiv \log_{10}(N_{\rm A}/N_{\rm B}) - \log_{10}(N_{\rm A}/N_{\rm B})_\odot$, where the subscript $\odot$ refers to the solar value and $N_{\rm A}$ and $N_{\rm B}$ are the abundances of elements A and B, respectively.

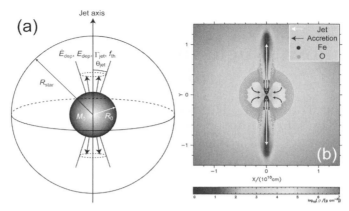

Figure 1. (a) Schematic picture of the jet-induced explosion. (b) Density structure of the $40\,M_\odot$ Pop III star explosion model of $\dot{E}_{\mathrm{dep},51} = 15$ at 1 sec after the start of the jet injection.

2. Models

We investigate the jet-induced explosions (e.g. Maeda & Nomoto 2003; Nagataki *et al.* 2006) of $40M_\odot$ Pop III stars (Umeda & Nomoto 2005; Tominaga *et al.* 2007b) using a two-dimensional special relativistic Eulerian hydrodynamic code (Tominaga 2007).

We inject the jets at a radius R_0, corresponding to an enclosed mass of M_0, and follow the jet propagation (Figs. 1ab). Since the explosion mechanism is unknown, the jets are treated parametrically with the following five parameters: energy deposition rate ($\dot{E}_{\mathrm{dep}}$), total deposited energy (E_{dep}), initial half angle of the jets (θ_{jet}), initial Lorentz factor (Γ_{jet}), and the ratio of thermal to total deposited energies (f_{th}).

We investigate the dependence of nucleosynthesis outcome on $\dot{E}_{\mathrm{dep}}$ for a range of $\dot{E}_{\mathrm{dep},51} \equiv \dot{E}_{\mathrm{dep}}/10^{51}\,\mathrm{ergs\,s^{-1}} = 0.3 - 1500$. The diversity of $\dot{E}_{\mathrm{dep}}$ is consistent with the wide range of the observed isotropic equivalent γ-ray energies and timescales of GRBs (e.g. Amati *et al.* 2007). Variations of activities of the central engines, possibly corresponding to different rotational velocities or magnetic fields, may well produce the variation of $\dot{E}_{\mathrm{dep}}$. We expediently fix the other parameters as $E_{\mathrm{dep}} = 1.5 \times 10^{52}\,\mathrm{ergs}$, $\theta_{\mathrm{jet}} = 15°$, $\Gamma_{\mathrm{jet}} = 100$, $f_{\mathrm{th}} = 10^{-3}$, and $M_0 = 1.4M_\odot$ ($R_0 \sim 900$ km) in the models.

The hydrodynamical calculations are followed until the homologously expanding structure is reached ($v \propto r$). The nucleosynthesis calculations are performed as post-processing with thermodynamic histories traced with marker particles that represent individual Lagrangian elements. In computing the jet composition, we assume that the jet initially has the composition of the accreted stellar materials.

3. Jet-induced Supernovae

3.1. *Fallback*

Figure 2a shows "accreted" regions for models with $\dot{E}_{\mathrm{dep},51} = 120$ and 1.5 ergs s^{-1}, where the accreted mass elements are initially located in the progenitor. The inner matter is ejected along the jet-axis but not along the equatorial plane. On the other hand, the outer matter is ejected even along the equatorial plane, since the lateral expansion of the shock terminates the infall as the shock reaches the equatorial plane.

The remnant mass (M_{rem}) is larger for lower $\dot{E}_{\mathrm{dep}}$. This stems from the balance between the ram pressures of the injecting jet (P_{jet}) and the infalling matter (P_{fall}). In order to inject the jet, P_{jet} should overcome P_{fall}. P_{jet} is determined by R_0, $\dot{E}_{\mathrm{dep}}$, θ_{jet}, Γ_{jet}, and

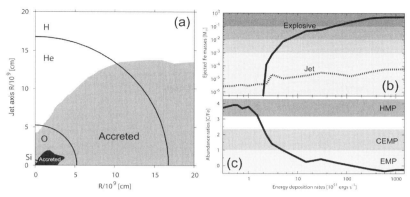

Figure 2. (a) Initial locations of the mass elements which are finally accreted for models with $\dot{E}_{\rm dep,51} = 120$ (*black*) and with $\dot{E}_{\rm dep,51} = 1.5$ (*gray*). (b) Ejected Fe mass (*solid line*: explosive nucleosynthesis products, *dotted line*: the jet contribution) as a function of the energy deposition rate. (c) Dependence of abundance ratio [C/Fe] on the energy deposition rate.

$f_{\rm th}$, thus being constant in time in the present models. On the other hand, $P_{\rm fall}$ decreases with time, since the density of the outer materials decreases following the gravitational collapse (e.g. Fryer & Mészáros 2003). For lower $\dot{E}_{\rm dep}$, $P_{\rm jet}$ is lower, so that the jet injection ($P_{\rm jet} > P_{\rm fall}$) is realized at a later time when the central remnant becomes more massive due to more infall. As a result, the accreted region and $M_{\rm rem}$ are larger for lower $\dot{E}_{\rm dep}$.

A model with lower $\dot{E}_{\rm dep}$ has larger $M_{\rm rem}$, higher [C/Fe], and smaller amount of Fe [$M({\rm Fe})$] because of the larger amount of fallback (Figs. 2bc, Tominaga *et al.* 2007a). The larger amount of fallback decreases the mass of the inner core relative to the mass of the outer layer. The fallback of the O layer also reduces $M({\rm Fe})$ because Fe is mainly synthesized explosively in the Si and O layers. The variation of $\dot{E}_{\rm dep}$ in the jet-induced explosions predicts that the variation of [C/Fe] corresponds to that of $M({\rm Fe})$.

3.2. *Comparison with the spherical supernova model*

The calculations of the jet-induced explosions show that the ejection of the inner matter is compatible with the fallback of the outer matter (Fig. 2a). This is consistent with the two-dimensional illustration of the mixing-fallback model (Fig. 3a) proposed by Umeda & Nomoto (2002). The mixing-fallback model has three parameters; initial mass cut [$M_{\rm cut}({\rm ini})$], outer boundary of the mixing region [$M_{\rm mix}({\rm out})$], and a fraction of matter ejected from the mixing region (f). The remnant mass is written as

$$M_{\rm rem} = M_{\rm cut}({\rm ini}) + (1 - f)[M_{\rm mix}({\rm out}) - M_{\rm cut}({\rm ini})]. \qquad (3.1)$$

The three parameters would relate to the hydrodynamical properties of the jet-induced explosion models, e.g. the inner boundary (M_0), the outer edge of the accreted region ($M_{\rm acc,out}$), and the width between the edge of the accreted region and the jet axis.

We calculate a model with $\dot{E}_{\rm dep,51} = 120$ and $M_0 = 2.3M_\odot$ ($R_0 \sim 3 \times 10^3$km) to compare the yield of the jet-induced SN explosion with that of the spherical SN explosion. The inner boundary, the outer edge of the accreted region, and the central remnant mass of the jet-induced explosion model are $M_0 = 2.3M_\odot$, $M_{\rm acc,out} = 12.2M_\odot$, and $M_{\rm rem} = 8.1M_\odot$. Its abundance pattern is well reproduced by the spherical SN model with the same main-sequence mass $M_{\rm ms} = 40M_\odot$ and the explosion energy $E = 3 \times 10^{52}$ ergs (Fig. 3b). The parameters of the mixing-fallback model are $M_{\rm cut}({\rm ini}) = 2.3M_\odot$,

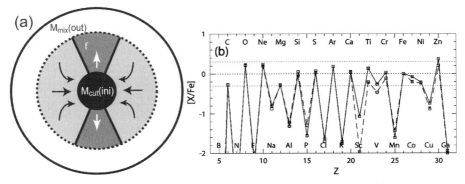

Figure 3. (a) Two-dimensional illustration of the mixing-fallback model. (b) Comparison of the abundance patterns of the jet-induced SN model with $\dot{E}_{\mathrm{dep},51} = 120$ and $M_0 = 2.3 M_\odot$ (*solid line*) and the mixing-fallback model (*dashed line*).

$M_{\mathrm{mix}}(\mathrm{out}) = 10.8 M_\odot$ and $f = 0.19$. The resultant M_{rem} (= $9.2 M_\odot$) is slightly larger than M_{rem} of the jet-induce SN model.

There, however, are some elements showing differences. The differences stem from the high-entropy explosion due to the concentration of the energy injection (e.g. Maeda & Nomoto 2003). The enhancements of [Sc/Fe] and [Ti/Fe] improve agreements with the observations. Such a thermodynamical feature of the jet-induced explosion model cannot be reproduced by the mixing-fallback model. A "low-density" modification might mimic the high-entropy environment (e.g. Umeda & Nomoto 2005; Tominaga *et al.* 2007b).

4. Trends with Metallicity

We calculate SN models with $M_{\mathrm{ms}} = 13 - 50 M_\odot$. The explosion energies are set to be consistent with the observations of present SNe (e.g. Tanaka *et al.* 2008). The SN yields are compared with the abundance patterns of the EMP stars that show certain trends of abundance ratios [X/Fe] with respect to [Fe/H] (Cayrel *et al.* 2004).

The observed abundance ratios against [Fe/H] are compared with yields of individual SN models and the Salpeter's IMF-integrated yield (Fig. 4). [Fe/H] of a next-generation star is determined by $M(\mathrm{Fe})$ and the swept-up H mass. Since the swept-up H mass is almost proportional to E of the SN, [Fe/H] of the next-generation stars are determined by a equation $[\mathrm{Fe/H}] = \log_{10}\left[\frac{M(\mathrm{Fe})}{M_\odot} \Big/ \left(\frac{E}{10^{51}\mathrm{ergs}}\right)^{6/7}\right] - C$ (Thornton *et al.* 1998), where C is assumed to be a constant value of 1.4. [Fe/H] of the IMF-integrated abundance ratios are assumed to be same as normal SN models ([Fe/H] ~ -2.6).

HNe explode with $E \gtrsim 10^{52}$ergs and eject large amount of ^{56}Ni ($\gtrsim 0.1 M_\odot$), while normal SNe explode with $E \sim 10^{51}$ergs and eject $\sim 0.07 M_\odot$ of ^{56}Ni. Therefore, according to the above equation, [Fe/H] of a next-generation star originated from a HN is lower than that of a next-generation star originated from a normal SN. The higher-energy explosion raises explosive nucleosynthesis under higher entropy and thus leads higher [Zn/Fe]. This accompaniment explains the observed trends of [Zn/Fe] against [Fe/H]. The trends of other elements are also reproduced by the variations of M_{ms} and E (Fig. 4).

5. Conclusion

We focus on two interesting properties observed in the abundance patterns of the metal-poor stars: (1) the higher [C/Fe] for lower [Fe/H] and (2) the trends of [X/Fe] against

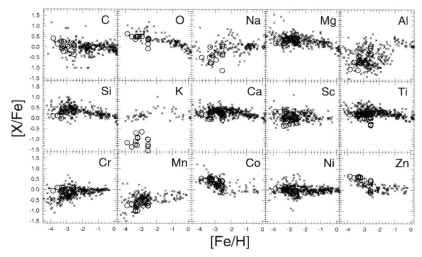

Figure 4. Comparison between the [X/Fe] trends of observed stars (e.g. Cayrel *et al.* 2004; Honda *et al.* 2004: *cross*) and SN models with the mixing-fallback model and applied the Y_e and "low-density" modifications (Tominaga *et al.* 2007b: *circles*).

[Fe/H]. The variations of the metal-poor stars are explained by the variations of SNe that contribute the metal enrichment of the early universe. Especially, (1) the variation of the energy deposition rates explains the tendency of [C/Fe] against [Fe/H] and (2) the variations of $M_{\rm ms}$ and E explain the trends of [X/Fe] against [Fe/H]. We propose that the abundance patterns of the metal-poor stars will provide additional constraints on the explosion mechanism of GRBs and SNe beyond the direct observations.

References

Amati, L., Della Valle, M., Frontera, F., *et al.* 2007, *A&A*, 463, 913

Beers, T.C. & Christlieb, N. 2005, *ARAA*, 43, 531

Cayrel, R., Depagne, E., Spite, M., *et al.* 2004, *A&A*, 416, 1117

Depagne, E., Hill, V., Spite, M., *et al.* 2002, *A&A*, 390, 187

Frail, D.A., Kulkarni, S. R., Sari, R., *et al.* 2001, *ApJ* (Letters), 562, L55

Fryer, C. & Mészáros, P. 2003, *ApJ* (Letters), 588, L25

Galama, T. J., Vreeswijk, P. M., van Paradijs, J., *et al.* 1998, *Nature*, 395, 670

Honda, S., Aoki, W., Kajino, T., *et al.* 2004, *ApJ*, 607, 474

Iwamoto, N., Umeda, H., Tominaga, N., Nomoto, K., & Maeda, K. 2005, *Science*, 309, 451

Maeda, K., Nakamura, T., Nomoto, K., *et al.* 2002, *ApJ*, 565, 405

Maeda, K. & Nomoto, K. 2003, *ApJ*, 598, 1163

Nagataki, S., Mizuta, A., & Sato, K. 2006, *ApJ*, 647, 1255

Nomoto, K., Tominaga, N., Umeda, H., *et al.* 2006, *Nucl. Phys. A*, 777, 424 (astro-ph/0605725)

Tanaka, M., Tominaga, N., Nomoto, K., *et al.* 2008, *ApJ*, submitted (arXiv:0807.1674)

Thornton, K., Gaudlitz, M., Janka, H.-Th., & Steinmetz, M. 1998, *ApJ*, 500, 95

Tominaga, N., Maeda, K., Umeda, H., *et al.* 2007a, *ApJ* (Letters), 657, L77

Tominaga, N., Umeda, H., & Nomoto, K. 2007b, *ApJ*, 660, 516

Tominaga, N. 2007, *ApJ*, submitted (arXiv:0711.4815)

Umeda, H. & Nomoto, K. 2002, *ApJ*, 565, 385

Umeda, H. & Nomoto, K. 2005, *ApJ*, 619, 427

Woosley, S. E. & Weaver, T. A. 1995, *ApJS*, 101, 181

Low-Metallicity Star Formation:
From the First Stars to Dwarf Galaxies
Proceedings IAU Symposium No. 255, 2008
L.K. Hunt, S. Madden & R. Schneider, eds.

© 2008 International Astronomical Union
doi:10.1017/S1743921308024812

Powerful explosions at $Z = 0$?

Sylvia Ekström[1], Georges Meynet[1], Raphael Hirschi[2] and André Maeder[1]

[1] Geneva Observatory, University of Geneva, Maillettes 51 - CH 1290 Sauverny, Switzerland

[2] University of Keele, Keele, ST5 5BG, UK

Abstract. Metal-free stars are assumed to evolve at constant mass because of the very low stellar winds. This leads to large CO-core mass at the end of the evolution, so primordial stars with an initial mass between 25 and 85 $M_\odot$ are expected to end as direct black holes, the explosion energy being too weak to remove the full envelope.

We show that when rotation enters into play, some mass is lost because the stars are prone to reach the critical velocity during the main sequence evolution. Contrary to what happens in the case of very low- but non zero-metallicity stars, the enrichment of the envelope by rotational mixing is very small and the total mass lost remains modest. The compactness of the primordial stars lead to a very inefficient transport of the angular momentum inside the star, so the profile of $\Omega(r)$ is close to $\Omega r^2 = \text{const}$. As the core contracts, the rotation rate increases, and the star ends its life with a fast spinning core. Such a configuration has been shown to modify substantially the dynamics of the explosion. Where one expected a weak explosion or none at all, rotation might boost the explosion energy and drive a robust supernova. This will have important consequences in the way primordial stars enriched the early Universe.

Keywords. stars: evolution, stars: rotation, stars: chemically peculiar, supernovae: general

1. The first stars

Population III (Pop III) stars occupy a key position in the history of the Universe, as a link between the pure H-He Universe at the beginning of time, and the metal-rich one we observe nowadays around us. Their evolution is important, because during their life they form the first heavy elements, but their death is also important, because this is the moment at which the newly synthesised elements are released in the surrounding medium, starting the chemical enrichment of the Universe.

While the first studies on the star formation at $Z = 0$ concluded that only very massive objects could form (Bromm *et al.* 2002; Abel *et al.* 2002), later findings show that the radiative feedback of the very first stars could lead to a second generation of metal-free stars with lower initial mass (Omukai and Yoshii 2003; O'Shea *et al.* 2005; Greif and Bromm 2006). Could these stars contribute to the chemical enrichment of the Universe? The consensual picture is that most of them are directly swallowed by a black hole.

2. Evolution at $Z = 0$

Since they are deprived of metals, the massive Pop III stars cannot rely on the CNO cycle to sustain their gravity. The *pp*-chains are not very sensitive in temperature and cannot completely halt the initial collapse. The stars continue their contraction until the central temperature is high enough to allow some carbon to be produced through the 3α reaction.

As a consequence, the main sequence (MS) occurs at much higher central temperature (T_c) and density (ρ_c), as shown in Fig. 1 (*left*). The great compactness implies that the

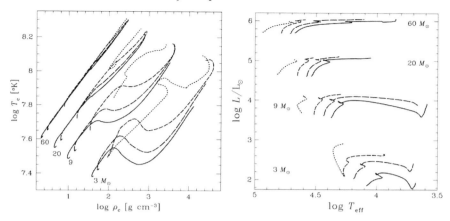

Figure 1. Evolution during the main sequence of massive stars at various metallicities: $Z = 0.020$ (solid line); $Z = 0.002$ (long-dashed line); $Z = 10^{-5}$ (short-dashed line); $Z = 0$ (dotted line). *Left:* $\log T_c - \log \rho_c$ diagram. *Right:* HR diagram.

zero-age main sequence (ZAMS) is shifted toward higher effective temperature ($T_{\rm eff}$), as shown in Fig. 1 (*right*). Without any metals, the envelope is transparent, so the stars remain in the blue side of the HR diagram throughout the whole MS.

The lack of metals has another consequence: the radiative winds are supposed to scale with the metallicity as $\dot{M} \propto (Z/Z_\odot)^\alpha$ with $\alpha = 0.5$ (Kudritzki and Puls 2000; Nugis and Lamers 2000; Kudritzki 2002) or 0.7-0.8 (Vink *et al.* 2001; Vink and de Koter 2005; Mokiem *et al.* 2007). At $Z = 0$, we thus expect no radiative winds, and most Pop III models are computed with the hypothesis of constant mass. In fact, Kudritzki (2002) shows that very low-metallicity stars close to the Eddington limit $L_{\rm Edd} = 4\pi cGM/\kappa$ (with κ the electron-scattering opacity and obvious meaning for the other quantities) are subject to a weak but non-zero mass loss. Vink and de Koter (2005) show that there is a flattening in the metallicity dependence of the WR winds below $Z \approx 10^{-3}$. Also Smith and Owocki (2006) show that continuum-driven eruptive mass ejections can occur at any metallicity, even $Z = 0$, leading to strong mass loss episodes like those observed in luminous blue variables (LBVs).

Even if the mass loss is non zero, it is in any case weak, so the Pop III stars end their life with a large helium core (M_α), as shown by Marigo *et al.* (2001). In such a case, the expected fate of the stars can deviate from the usual "SN explosion with neutron star or black hole remnant" seen at higher metallicity. Heger *et al.* (2003) determine the fate of massive stars at various metallicities as a function of their M_α at the pre-supernova (preSN) stage:

- a Type II SN: $M_\alpha < 9 \ M_\odot$
- a BH by fallback: $9 \ M_\odot \leqslant M_\alpha < 15 \ M_\odot$
- a direct BH: $15 \ M_\odot \leqslant M_\alpha < 40 \ M_\odot$
- a pulsational pair-instability followed by a SN with BH formation: $40 \ M_\odot \leqslant M_\alpha < 64 \ M_\odot$
- a pair-instability SN (PISN): $64 \ M_\odot \leqslant M_\alpha \leqslant 133 \ M_\odot$
- or a direct BH collapse: $M_\alpha > 133 \ M_\odot$

Relating M_α with the initial mass of the star in the case of $Z = 0$, this means that stars with a mass between 25 and 140 $M_\odot$ or above 260 $M_\odot$ will not contribute at all to the enrichment of the early Universe.

Could this picture be revised?

3. Rotation effects

Meynet *et al.* (2006) show that at low (but non-zero) metallicity, rotation drastically changes the evolution of massive stars. Two processes are involved:

(*a*) During the MS, because of the low radiative winds, the star loses very little mass and thus very little angular momentum. As the evolution proceeds, the stellar core contracts and spins up. If a coupling exists between the core and the envelope (*i.e.* meridional currents or magnetic fields), the surface may be accelerated up to the critical velocity and the star may experience a mechanical mass loss due to the centrifugal acceleration. The matter is launched into a decretion disc (Owocki 2005), which may be dissipated later by the radiation field of the star.

(*b*) Rotation induces an internal mixing that depends on the gradient of the rotational rate Ω. After H exhaustion, the core contracts and the envelope expands, stretching the Ω-gradient. The mixing becomes strong and enriches the surface in heavy elements. Rotation also favours a redward evolution after the MS, allowing the star to spend more time in the cooler part of the HR diagram, where mass loss is increased. The outer convective zone dives deep inside the star and dredges up freshly synthesised heavy elements. The surface metallicity is dramatically enhanced (by a factor of 10^6 for a 60 $M_\odot$ at $Z_{ini} = 10^{-8}$). The opacity of the envelope is increased, and the radiative winds may thus be drastically enhanced.

Meynet *et al.* (2006) show that at $Z = 10^{-8}$, while a non-rotating 60 $M_\odot$ model loses only 0.27 $M_\odot$, a similar model computed with an initial velocity $v_{ini} = 800$ km s^{-1} loses 36 $M_\odot$, *i.e.* 60 % of its initial mass.

Now what happens if the metallicity is strictly $Z = 0$?

Ekström *et al.* (2008a) show that the first process (the mechanical mass loss during the MS) is slightly reduced because the meridional circulation is weak, so the star almost evolves in a regime of local angular momentum conservation $\Omega r^2 = \text{const}$, and reaches the critical velocity late in its MS evolution. The second process (the enrichment of the surface) barely occurs at $Z = 0$. This can be explained by the fact that some He is burnt already during the MS: at H exhaustion, the star is hot enough to move on to core He burning without any structural readjustment, and the He burning occurs mainly in the blue side of the HR diagram. The stretching of the Ω-gradient does not occur, and the outer convective zone, if any, remains thin. Most of the post-MS evolution occurs with a surface metallicity much lower than 10^{-6}.

As shown in Fig. 2, most of the mass is lost during the MS, and it amounts at most to 10% of the initial mass.

4. Explosion

A direct consequence of the evolution close to the local angular momentum conservation is that the models keep a high angular momentum content in the core throughout their whole evolution. Figure 3 (*left*) shows that the angular velocity of the iron core of a 85 $M_\odot$ model at the pre-supernova stage is more than 5 orders of magnitude higher than the surface velocity. On the right panel of the same figure, we see that in the innermost parts, the specific angular momentum of the same model is higher than that needed to create an accretion disc around a rotating BH.

According to Nomoto *et al.* (2003), there are indications that the most energetic supernovae, the hypernovae (HN), could be due to stellar rotation. Rotation is supposed to boost the explosion energy of a collapsing star by various mechanisms:

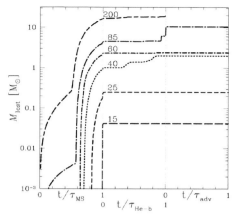

Figure 2. Mass lost during the evolution of Pop III models of various masses (Ekström *et al.* 2008a): 15 $M_\odot$ (long-dash line); 25 $M_\odot$ (short-dashed line); 40 $M_\odot$ (dotted line); 60 $M_\odot$ (dot-short dashed line); 85 $M_\odot$ (dot-long dashed line); 200 $M_\odot$ (short dash-long dashed line). The *x*-axis is a temporal axis, divided in three parts: the fraction of the MS evolution, the fraction of the core He burning phase, and the fraction of the rest of the advanced phases.

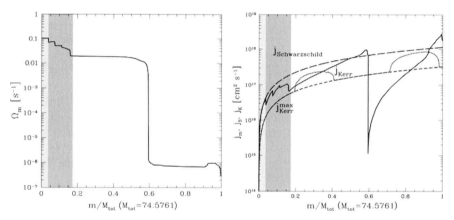

Figure 3. Pre-supernova structure of a Pop III 85 $M_\odot$ model. The light grey shaded area marks the border of the remnant, and the dark grey that of the CO core. *Left:* Ω profile. *Right:* specific angular momentum profile. $j_{\text{Schwarzschild}} = \sqrt{12}Gm/c$ (long-dashed line) and $j_{\text{Kerr}}^{\max} = Gm/c$ (short-dashed line) is the minimum specific angular momentum necessary for a non-rotating and a maximally-rotating black hole, respectively. j_{Kerr} (dotted line) is the minimum specific angular momentum necessary to form an accretion disc around a rotating black hole.

- it drives an anisotropy of the neutrinos sphere (Shimizu *et al.* 2001; Kotake *et al.* 2003),
- it lowers the critical neutrinos luminosity (the luminosity at which the stalled shock revives) (Yamasaki and Yamada 2005),
- it modifies the gravity (Burrows *et al.* 2005),
- it changes the geometry of the mass accretion, making it aspherical (Burrows *et al.* 2005),
- it generates vortices, which dredge up heat and increase the neutrinos luminosity (Burrows *et al.* 2005).

The models presented here are massive at the time of their death, they have heavy cores and they almost retained their entire envelope, but they have fast rotating iron cores, and therefore their final collapse could present a highly aspherical geometry. It would be extremely interesting to perform numerical simulations of the collapse of these rotating models so we could check whether the energy released in the collapse is sufficient to overcome the high gravitation of the star and drive a successful explosion. For the time being, we can only say that we expect rotating Pop III stars to undergo a stronger explosion than what is commonly admitted for non-rotating Pop III stars of this mass range.

5. Conclusions

The lack of metals affects strongly the stellar evolution at $Z = 0$. One of the main results is to reduce dramatically the radiative mass loss. This means that rotating stars will keep all their angular momentum until the end of their evolution. At the pre-supernova stage, they will present a fast-rotating iron core, so we expect the collapse to be highly aspherical. This may boost the explosion energy and allow a strong supernova explosion where only a weak one or none was expected.

Note that the inclusion of magnetic fields could alter these results by providing the core-envelope coupling which is lacking in Pop III stars. However, in that case, a strong mass loss would be expected (Ekström *et al.* 2008b), so the evolution of the star would anyway be strongly affected with regard to the consensual picture.

References

Abel, T., Bryan, G. L., and Norman, M. L. 2002, *Science* 295, 93

Bromm, V., Coppi, P. S., and Larson, R. B. 2002, *ApJ* 564, 23

Burrows, A., Walder, R., Ott, C. D., and Livne, E. 2005, *Nucl. Phys. A* 752, 570

Ekström, S., Meynet, G., Chiappini, C., Hirschi, R., and Maeder, A. 2008a, *A&A* (in press), arXiv:0807.0573

Ekström, S., Meynet, G., and Maeder, A. 2008b, in *IAU Symposium*, Vol. 250 of *IAU Symposium*, pp 209–216

Greif, T. H. and Bromm, V. 2006, *MNRAS* 373, 128

Heger, A., Fryer, C. L., Woosley, S. E., Langer, N., and Hartmann, D. H. 2003, *ApJ* 591, 288

Kotake, K., Yamada, S., Sato, K., and Shimizu, T. M. 2003, *Nucl. Phys. A* 718, 629

Kudritzki, R. P. 2002, *ApJ* 577, 389

Kudritzki, R.-P. and Puls, J. 2000, *ARAA* 38, 613

Marigo, P., Girardi, L., Chiosi, C., and Wood, P. R. 2001, *A&A* 371, 152

Meynet, G., Ekström, S., and Maeder, A. 2006, *A&A* 447, 623

Mokiem, M. R., de Koter, A., Vink, J. S., Puls, J., Evans, C. J., Smartt, S. J., Crowther, P. A., Herrero, A., Langer, N., Lennon, D. J., Najarro, F., and Villamariz, M. R. 2007, *A&A* 473, 603

Nomoto, K., Umeda, H., Maeda, K., Ohkubo, T., Deng, J., and Mazzali, P. A. 2003, *Nucl. Phys. A* 718, 277

Nugis, T. and Lamers, H. J. G. L. M. 2000, *A&A* 360, 227

Omukai, K. and Yoshii, Y. 2003, *ApJ* 599, 746

O'Shea, B. W., Abel, T., Whalen, D., and Norman, M. L. 2005, *ApJ* 628, L5

Owocki, S. 2005, in R. Ignace and K. G. Gayley (eds.), *The Nature and Evolution of Disks Around Hot Stars*, Vol. 337 of *ASPC*, p. 101

Shimizu, T. M., Ebisuzaki, T., Sato, K., and Yamada, S. 2001, *ApJ* 552, 756

Smith, N. and Owocki, S. P. 2006, *ApJ* 645, L45

Vink, J. S. and de Koter, A. 2005, *A&A* 442, 587

Vink, J. S., de Koter, A., and Lamers, H. J. G. L. M. 2001, *A&A* 369, 574

Yamasaki, T. and Yamada, S. 2005, *ApJ* 623, 1000

Low-Metallicity Star Formation:
From the First Stars to Dwarf Galaxies
Proceedings IAU Symposium No. 255, 2008
L.K. Hunt, S. Madden & R. Schneider, eds.

Wind anisotropy and stellar evolution

Cyril Georgy, Georges Meynet and André Maeder

Geneva Observatory, University of Geneva, Maillettes 51 - CH 1290 Sauverny, Switzerland

Abstract. Mass loss is a determinant factor which strongly affects the evolution and the fate of massive stars. At low metallicity, stars are supposed to rotate faster than at the solar one. This favors the existence of stars near the critical velocity. In this rotation regime, the deformation of the stellar surface becomes important, and wind anisotropy develops. Polar winds are expected to be dominant for fast rotating hot stars.

These polar winds allow the star to lose large quantities of mass and still retain a high angular momentum, and they modify the evolution of the surface velocity and the final angular momentum retained in the star's core. We show here how these winds affect the final stages of massive stars, according to our knowledge about Gamma Ray Bursts. Computation of theoretical Gamma Ray Bursts rate indicates that our models have too fast rotating cores, and that we need to include an additional effect to spin them down. Magnetic fields in stars act in this direction, and we show how they modify the evolution of massive star up to the final stages.

Keywords. stars: evolution, stars: magnetic fields, stars: mass loss, gamma rays: bursts

1. Effects of rotation on the stellar surface

Rotation has a strong influence on the stellar surface. Indeed, it adds a centrifugal component to the gravity, which modifies the shape of the surface, and various quantities such as the effective temperature T_{eff} and the mass loss flux. These effects can be derived from the von Zeipel theorem (von Zeipel 1924), which is originally valid for conservative cases of angular momentum distribution, and was treated in the more general case of the so-called "shellular rotation" by Maeder (1999). This theorem gives the relation between the local flux $\mathbf{F}$ of the star and the local effective gravity:

$$\mathbf{F} = -\frac{L(P)}{4\pi G M_\star(P)} \mathbf{g}_{\text{eff}} \left(1 + \zeta(\theta)\right) \tag{1.1}$$

where $L(P)$ is the luminosity on the isobar and $\mathbf{g}_{\text{eff}}$ the local effective gravity. The two remaining terms are given by

$$\zeta(\theta) = \left[\left(1 - \frac{\chi_T}{\delta}\right)\Theta + \frac{H_T}{\delta}\frac{\mathrm{d}\Theta}{\mathrm{d}r}\right] P_2(\cos(\theta)) \tag{1.2}$$

$$M_\star = M\left(1 - \frac{\Omega^2}{2\pi G \rho_{\text{m}}}\right) \tag{1.3}$$

Here, ρ_{m} is the internal average density, $M_\star$ represents the effective mass, modified by the rotation velocity Ω, $\chi = 4acT^3/(3\kappa\rho)$ is the thermal conductivity coefficient and χ_T is its partial derivative with respect to T, $\Theta = \frac{\tilde{\rho}}{\rho}$ is the ratio of the horizontal density fluctuation to the average density on the isobar (Zahn 1992). δ is the thermodynamic coefficient $\delta = -\left(\partial \ln \rho / \partial \ln T\right)_{P,\mu}$ and H_T is the temperature scale height. Generally the term $\zeta(\theta)$ is very small, and we can neglect it.

The total gravity at the surface of the star is given by $\mathbf{g}_{\mathrm{tot}} = \mathbf{g}_{\mathrm{eff}} + \mathbf{g}_{\mathrm{rad}}$ with $\mathbf{g}_{\mathrm{rad}} = \frac{\kappa \mathbf{F}}{c}$ as the term due to radiative forces and κ the total Rossland mean opacity. The Eddington limit in a rotating star is defined by the vanishing of $\mathbf{g}_{\mathrm{tot}}$ and we find the limiting flux

$$\mathbf{F}_{\mathrm{lim}} = -\frac{c}{\kappa} \mathbf{g}_{\mathrm{eff}} \, . \tag{1.4}$$

The Eddington factor at a given colatitude, which is given by the ratio of the local flux to the limiting flux, becomes (Maeder 1999)

$$\Gamma_\Omega(\theta) = \frac{L(P)}{L_{\mathrm{max}}} \quad \text{with} \quad L_{\mathrm{max}} = \frac{4\pi c G M}{\kappa(\theta)\,(1 + \zeta(\theta))} \left(1 - \frac{\Omega^2}{2\pi G \rho_{\mathrm{m}}} \right) \, . \tag{1.5}$$

A first interesting consequence is that the maximum luminosity of a star, given by $\Gamma_\Omega(\theta) = 1$, is lowered by rotation (compare L_{max} given above with the non-rotating one $L_{\mathrm{max,\,no\,rot}} = \frac{4\pi c G M}{\kappa}$).

Rotation has also a strong impact on the mass loss rate of the star. Maeder and Meynet (2000) found the following relation between the mass loss rate at a given rotational velocity $\dot{M}(\Omega)$ and the non-rotating one:

$$\frac{\dot{M}(\Omega)}{\dot{M}(0)} = \frac{(1 - \Gamma)^{\frac{1}{\alpha} - 1}}{\left[1 - \frac{\Omega^2}{2\pi G \rho_{\mathrm{m}}} - \Gamma \right]^{\frac{1}{\alpha} - 1}} \tag{1.6}$$

where $\Gamma = \frac{\kappa L}{4\pi c G M}$ is the "non-rotating" Eddington factor. We see that rotation will increase the total mass loss rate of the star.

Rotation has another effect on the mass loss rate: it is no longer isotropic at the surface of a rotating star, but becomes colatitude-dependent. Maeder and Meynet (2000) give the following local mass loss rate $\Delta \dot{M}$ per unit surface $\Delta\sigma$:

$$\frac{\Delta \dot{M}}{\Delta \sigma} \simeq (k\alpha)^{\frac{1}{\alpha}} \left(\frac{1 - \alpha}{\alpha} \right)^{\frac{1 - \alpha}{\alpha}} \left[\frac{L(P)}{4\pi G M_\star(P)} \right]^{\frac{1}{\alpha}} \frac{g_{\mathrm{eff}}}{(1 - \Gamma_\Omega(\theta))^{\frac{1}{\alpha} - 1}} \tag{1.7}$$

with κ and α the force multiplier parameters. We neglect here the small effect of $\zeta(\theta)$. Rotation favors mass loss through the terms $M_\star$ and Γ_Ω. Mass loss occurs preferentially where g_{eff} is small, $i.e$ at the poles. Then mass loss rate is thus varying as a function of the colatitude, producing the anisotropic wind phenomenon.

Figure 1 shows various effects of the rotation at the surface of a $20\,\mathrm{M}_\odot$ star at a metallicity of 10^{-5} and at 95% of the critical rotation velocity. We can first remark that the star becomes oblate, with an equatorial-to-polar radius ratio $\frac{R_{eq}}{R_{pol}} \simeq 1.3$. Then we note the variation of the effective temperature with respect to colatitude: T_{eff} at the pole is around $48000\,\mathrm{K}$, while it is only around $34000\,\mathrm{K}$ at the equator. Finally, we see that the mass loss flux is larger at the pole by a factor of ~ 3.8. This allows the star to lose 10% less angular momentum than if the same amount of mass was lost isotropically.

2. Models without magnetic field and rate of GRB

Let us briefly recall here the main assumptions of the so-called "collapsar" model for long-soft Gamma Ray Bursts (Woosley 1993). In order to produce such an event, the following conditions must be fulfilled:
- formation of a black hole;
- enough angular momentum in the stellar core in order to form an accretion disk around the BH;
- formation of a type Ic supernova (see Woosley & Bloom 2006).

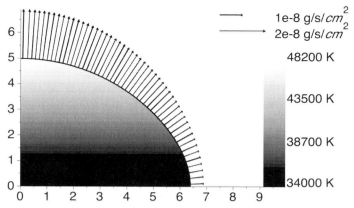

Figure 1. Effects of rotation on a $20\,M_\odot$ star at $Z = 10^{-5}$ and with $\frac{\Omega}{\Omega_{\text{crit}}} = 0.95$. The star is seen equator-on, and the axis are in $R_\odot$. The latitudinal variation of T_{eff} is shown (gray scale), and the arrows represents the masse flux at a given colatitude.

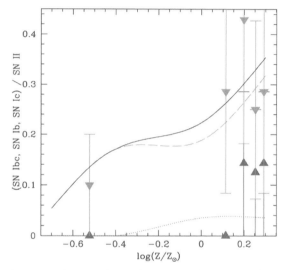

Figure 2. SNIb/SNII (solid line), SNIb/SNII (dotted line) and SNIc/SNII (dashed line) ratios. The triangles are observed SNIb/SNII ratio at various metallicities and the upside down gray triangles observed SNIc/SNII ratio (extracted from Prieto *et al.* (2008)).

These three points are consistent with a fast rotating massive star. It is thus interesting to study the evolution of the rate of type Ic supernovae. Figure 2 shows the relative rates for type Ib and type Ic SNe, computed with rotating models without a magnetic field. To distinguish between type Ib and type Ic, we use a criterion based on the amount of He ejected during the SN event: all models ejecting more than $0.55\,M_\odot$ of helium and no hydrogen are considered as type Ib, the models ejecting less than $0.55\,M_\odot$ of helium (and still no hydrogen) are considered to give birth to a type Ic SN event. The interesting curve for our purpose is the dashed one, representing the SN Ic / SN II ratio with respect to the metallicity. We see that the number fraction of type Ic supernova becomes higher at higher metallicity. This is in good agreement with the trend shown by the observed type Ic to type II SNe ratios given by Prieto *et al.* (2008). There is much observational evidence indicating that long-soft GRBs occur preferentially in metal-poor regions. For instance, Modjaz *et al.* (2008) find that GRB events appear at low metallicity: between $0.2 < Z/Z_\odot < 0.7$. Thus only type Ic events in metal poor regions (or part of them) can occur simultaneously with a GRB event.

Moreover, if we consider our models at metallicities that are compatible with the observed GRB range of metallicity, we see that all the models producing a type Ic SN keep enough angular momentum in the core to fulfilled the collapsar model conditions (see Hirschi *et al.* (2005)). We can thus determine the ratio of GRB event to the total number of core collapse SNe; we found that this rate is around 15% for metallicities between 0.4 and $0.7\,Z_\odot$. In comparison, Podsiadlowski *et al.* (2004) found an observational rate of $0.04\% - 8\%$, depending of the aperture angle of the bipolar jets produced during the GRB event. Our theoretical rate is therefore much larger than the observational one. These two facts lead to the conclusion that not all type Ic SNe produce a GRB. We have thus to find a way to reduce the number of GRB progenitor candidates, in order to reproduce the observational rate. One possibility is to introduce new ingredients in our models to extract more angular momentum from the core during the stellar life.

3. Models with magnetic field and wind anisotropy

Following Spruit (2002), we have included in our models the effect of magnetic field amplified at the expense of the excess energy in the shear. As noted just above, this produces a strong coupling between the differentially rotating layers, and tend to build a solid–body rotation profile. Contrarily to the previous models, where the rotational velocity is only weakly coupled between the surface and the core, models with magnetic field have a strong coupling, and thus, the loss of angular momentum due to mass loss is quickly transmitted to the core. This implies a strong extraction of angular momentum during the evolution, particularly when the mass loss is strong at the surface (e.g. during the Wolf–Rayet phases).

To explore the combined effects of magnetic field and wind anisotropy, we computed two $60\,M_\odot$ models at $Z = 0.002$ with $\Omega/\Omega_{\rm crit} = 0.75$, with and without the treatment of the wind anisotropy. With respect to the work by Meynet and Maeder (2007), we have improved significantly the treatment of the wind anisotropy, allowing this treatment to apply even when the critical velocity is reached and checking very carefully that the sum of the angular momentum remaining in the star and the angular momentum lost in the wind remains constant all over the evolution (see Fig. 3, bottom panel). As we shall see below, this improvements lead to effects which although still important are less pronounced as in Meynet and Maeder (2007).

For both models, the strong mixing induced by the magnetic field leads to the so-called "quasi-chemically-homogenous" evolution (see Yoon and Langer 2005 and Woosley and Heger 2006). Figure 3 (top panel) shows the evolution of the total angular momentum of the stellar interior during its evolution. We see that the anisotropic model (dashed curve) is slightly higher than the isotropic one (solid curve). This is due to the effect of wind anisotropy, as shown in the medium panel: the anisotropic model loses less angular momentum (around 7%), while the rotational velocity is high (first part of the evolution). Then, the star becomes a WR, and the high mass loss rate implies a strong breaking of the rotation: the surface velocity is sufficiently below the critical velocity for removing any anisotropy in the wind, and there is no more difference between these two models.

We see that when magnetic field is accounted for, the WR phase has a crucial influence on the total angular momentum kept in the star (and thus in the core, through the coupling produced by the magnetic field). Our model does not keep enough angular

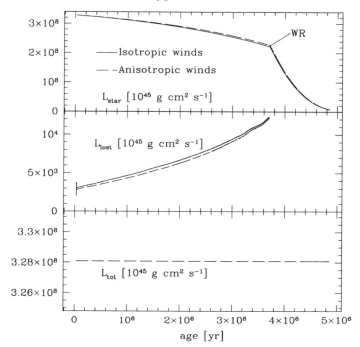

Figure 3. Top panel: evolution of the total angular momentum kept in the star (in units of $[10^{45}\,g\,cm^2\,s^{-1}]$). The dashed curve is the model with anisotropic wind, the solid curve for the isotropic one. Middle panel: angular momentum removed by wind at each time step, in the same units: dashed curve for anisotropic model, solid for isotropic one. Bottom panel: Sum of the angular momentum of the star and the integrated angular momentum removed by wind.

momentum for the collapsar model, and is thus not a good GRB progenitor candidate. However, models with very high initial rotational velocity would most probably develop larger differences between the iso– and anisotropic treatment. The same would be true for models with lower mass loss rate (less massive models, lower metallicity). We will address this point in a forthcoming paper.

References

Hirschi, R., Meynet, G. & Maeder, A. 2005, *A&A* 443, 581
Maeder, A. 1999, *A&A* 347, 185
Maeder, A. & Meynet, G. 2000, *A&A* 361, 159
Meynet, G. & Maeder, A. 2007, *A&A* 464, L11
Modjaz, M., Kewley, L., Kirshner, R. P., Stanek, K. Z., Challis, P., Garnavich, P. M., Greene, J. E., Kelly, P. L. & Prieto, J. L. 2008, *AJ* 135, 1136
Podsiadlowski, P., Mazzali, P. A., Nomoto, K., Lazzati, D. & Cappellaro, E. 2004, *ApJL* 607, L17
Prieto, J. L., Stanek, K. Z. & Beacom, J. F. 2008, *ApJ* 673, 999
Spruit, H. C. 2002, *A&A* 381, 923
von Zeipel, H. 1924, *MNRAS* 84, 665
Woosley, S. E. 1993, *ApJ* 405, 273
Woosley, S. E. & Bloom, J. S. 2006, *ARAA* 44, 507
Woosley, S. E. & Heger, A. 2006, *ApJ* 637, 914
Yoon, S.-C. & Langer, N. 2005, *A&A* 443, 643
Zahn, J.-P. 1992, *A&A* 265, 115

Low-Metallicity Star Formation:
From the First Stars to Dwarf Galaxies
Proceedings IAU Symposium No. 255, 2008
L.K. Hunt, S. Madden & R. Schneider, eds.

Low-Mass and Metal-Poor Gamma-Ray Burst Host Galaxies

Sandra Savaglio[1]

[1] Max-Planck Institute for Extraterrestrial Physics,
Giessenbachstr., PF 1312, D-85741, Garching bei München, Germany
email: savaglio@mpe.mpg.de

Abstract. Gamma-ray bursts (GRBs) are cosmologically distributed, very energetic and very transient sources detected in the γ-ray domain. The identification of their x-ray and optical afterglows allowed so far the redshift measurement of 150 events, from $z = 0.01$ to $z = 6.29$. For about half of them, we have some knowledge of the properties of the parent galaxy. At high redshift ($z > 2$), absorption lines in the afterglow spectra give information on the cold interstellar medium in the host. At low redshift ($z < 1.0$) multi-band optical-NIR photometry and integrated spectroscopy reveal the GRB host general properties. A redshift evolution of metallicity is not noticeable in the whole sample. The typical value is a few times lower than solar. The mean host stellar mass is similar to that of the Large Magellanic Cloud, but the mean star formation rate is five times higher. GRBs are discovered with γ-ray, not optical or NIR, instruments. Their hosts do not suffer from the same selection biases of typical galaxy surveys. Therefore, they might represent a fair sample of the most common galaxies that existed in the past history of the universe, and can be used to better understand galaxy formation and evolution.

Keywords. Galaxies: ISM, galaxies: abundances, galaxies: dwarf, galaxies: high-redshift, gamma rays: bursts, gamma rays: observations

1. Introduction

More than 40 years have passed since the discovery of the first gamma-ray burst (GRB), the most energetic explosions in the universe, by the US military satellite Vela. It took 30 years to demonstrate their cosmological origin (Metzger *et al.* 1997). Their γ-ray energies (typically 10^{51} ergs, emitted in less than a couple of minutes) emerge from a collimated jet in a core-collapse supernovae, or the merger of two compact objects (neutron stars or black holes).

Due to the highly transient nature of GRBs, their redshift, measured from the optical afterglow or the host galaxy, is known today for 150 objects only (Figure 1). Although GRBs are very rare (a rate of 1 event every 10^5 years in a galaxy is estimated, after correcting for the jet opening angle), they are so energetic that a few events are expected to be detectable from Earth every day. As shown by the discovery of high redshift events (the highest ever being GRB 050904 at $z = 6.3$; Kawai *et al.* 2006), GRBs offer the opportunity to explore the most remote universe, under extreme conditions, hard to observe using traditional tools.

The first scientific paper on GRBs dates from 6 years after the Vela discovery (Klebesadel, Strong & Olson 1973). In 1975, already 100 different theories where proposed, and today more than 5000 refereed papers on GRBs have been published. Before the 1997 discovery, theories on the energetic emission ranked from the impact of comets onto neutron stars (Harwit & Salpeter 1973) to collisions of chunks of antimatter with normal stars (Sofia & van Horn 1974). However, the most likely hypothesis was already

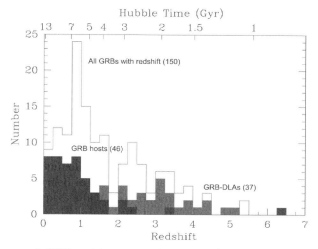

Figure 1. Histogram of GRBs with measured redshift (empty histogram 150 objects). The high-z and low-z filled histograms are the subsample of GRBs studied using optical afterglow spectra (GRB-DLAs) and multiband photometry of the host galaxies (GRB hosts), respectively.

postulated years earlier, when nothing was really known about GRBs: Colgate (1968) predicted γ-ray emission from supernovae in distant galaxies.

The curse and blessing of GRBs is their fast fading. The light curve is extremely steep and therefore very hard to catch. The emitted energy is so immense that it cannot last long. However, it allows us to see a hidden universe. One extreme case is the recent event GRB 080319B at $z = 0.937$ (Bloom *et al.* 2008), visible for a short time by naked eye (peak optical magnitude $m = 5.6$). This GRB was already hard to observe spectroscopically with the largest telescopes 7 hours after its discovery.

2. Studying small star-forming galaxies at low and high redshift with GRBs

GRB studies offer the opportunity to complement our partial view of the universe, and better understand the formation and evolution of galaxies in general. The typical GRB host is a small star-forming galaxy (Le Floc'h *et al.* 2003; Christensen *et al.* 2004; Prochaska *et al.* 2004). Galaxies with similar characteristics, but without known GRB, are dominating the universe. This effect can only have been more important in the past, because galaxies become bigger with time and not smaller.

High redshift galaxies have been explored for more than two decades using absorption lines in QSO spectra. The so-called damped Lyman-α systems (DLAs) probe the physical state of neutral gas in galaxies intersecting QSO sight lines. Although very efficient (QSOs are bright and detected up to $z = 6.4$), this technique does not allow, except for some special cases, the study of the emitting component (stars and hot gas) of the parent galaxies because of the strong glare of the background light.

The direct detection of a large number of high-z galaxies was possible with the advent of modern technologies of 10m-class and space telescopes. The Lyman-break technique (Steidel *et al.* 1996) revealed that massive and chemically evolved galaxies were already in place up to $z = 3$ (Erb *et al.* 2006). The difference with QSO-DLA galaxies (metal poor and small) is the effect of Lyman-break galaxies being the tip of the iceberg of the

Table 1. Overview of host diagnostics with GRB observations.

Diagnosis	GRB-DLA (AG spectroscopy)	GRB host (Int. spectroscopy)	GRB host (Photometry)
SFR	×	√√	√
Metallicity	√√√	√√	×
Dust extinction	√	√√	×
Dust depletion	√√√	×	×
Stellar mass	×	×	√√√
Age	×	√	√

whole galaxy population. The typical high-z galaxy is much fainter and metal poor, and also much harder to find. Much harder, unless a GRB event takes place.

The recent exploration of the SFR density of the universe for different stellar mass bins (Juneau *et al.* 2005) has shown that small galaxies (like GRB hosts) prevailed in terms of star-formation in the $z < 2$ universe. Moreover, the mass-metallicity relation (Tremonti *et al.* 2004) and its redshift evolution (Savaglio *et al.* 2005; Erb *et al.* 2006; Maiolino *et al.* 2008) have revealed that small galaxies have evolved more and over a longer time than big galaxies. Therefore, it is important to go deeper in terms of stellar mass to understand galaxy formation in the distant universe. GRB hosts can serve this purpose. Last but not least, using GRB observations it is possible to study the cold gas and the emitting stars and gas in the same galaxy.

Table 1 summarizes the physical parameters studied with GRBs. Observables are grouped in three categories: *i*) the spectroscopy of the optical afterglow (AG); *ii*) the integrated optical spectroscopy of the host galaxy; *iii*) the multi-band optical and NIR photometry of the host. The diagnosis that can (marked with √) or cannot (marked with ×) be investigated are listed. In this contribution we will describe most of them (excepting the age of the stellar population and the dust extinction in the host).

3. Spectroscopy of optical afterglows

Starting from the discovery of the first optical GRB afterglow (Metzger *et al.* 1997), it became clear that GRB spectra looked similar to QSO-DLAs (Savaglio, Fall & Fiore 2003), with the presence of strong absorption lines of neutral or singly ionized species. The main difference is that GRB strong absorption lines (GRB-DLAs) probe the gas in the GRB vicinity and inside the host galaxy, while most QSO-DLAs are physically unrelated to the background source.

GRB-DLAs have on average higher column densities of absorption lines than QSO-DLAs. This is shown in Figure 2, with the histograms of the neutral hydrogen and the singly ionized zinc and iron. The difference is even larger if one considers that the column density of a species is higher when the background source is external to the galaxy (as for QSO-DLAs) than when the background source is inside the galaxy (as for GRB-DLAs). In the former case, the line of sight crosses the entire galaxy, whereas in the latter only the gas in front of the stars is probed. The simplest situation would be that the GRB is at the center and therefore the observed column is a factor of two less than what a background source would see.

As of today, one or more column densities of elements is known for 37 GRB-DLAs (Figure 1; Savaglio 2006 and references therein). The larger column densities with respect to QSO-DLAs can be explained if GRB hosts are larger in size and/or the gas along the

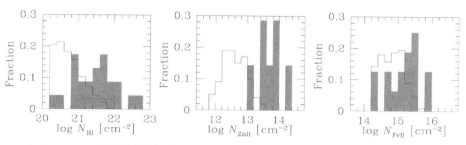

Figure 2. Fraction of QSO-DLAs (empty histograms) or GRB-DLAs (filled histograms) per column density bins for different ions (from left to right: HI, ZnII and FeII).

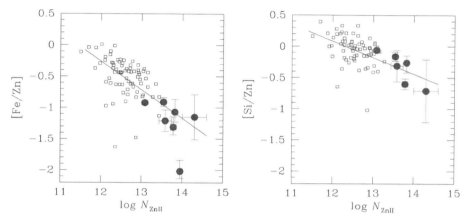

Figure 3. Iron-to-zinc and silicon-to-zinc relative abundances as compared to ZnII column density, for QSO-DLAs (open squares) and GRB-DLAs (filled dots). The straight lines show the linear correlations between quantities.

sight lines is denser than in QSO-DLAs. Moreover, from a close look at the column density distributions (Figure 2), it can be noticed that the difference between QSO-DLAs and GRB-DLAs is stronger for ZnII than for FeII. This can be an indication of a larger dust content in GRB-DLAs. In fact, zinc is little dust depleted, while a large fraction of iron is generally locked into dust grains (Savage & Sembach 1996); the fraction of iron depleted is larger for larger gas column densities.

This is apparent when comparing [Fe/Zn] and [Si/Zn] to the ZnII column densities (Figure 3). Silicon is mildly depleted in dust grains, more than iron and less than zinc. Negative and large values of [Fe/Zn] and [Si/Zn], together with large column densities of ZnII, indicate large columns of dust in GRB-DLA sight lines (Savaglio 2006). The continuous distribution of points in QSO-DLAs and GRB-DLAs in Figure 3 suggests that the larger column density of metals in GRB-DLAs is mostly the effect of something intrinsic to the gas (e.g. larger gas density) and not to the galaxies (e.g. larger gas clouds).

The detection of several heavy elements and HI allows to estimate the metal content in GRB-DLAs. Two main assumptions are generally made. First, the ionization correction is neglected (the gas is mostly in neutral form). Second, the dust depletion correction is estimated when more than one element, with different refractory properties, are detected. The metal content in GRB-DLAs at different redshifts can be used to trace the chemical evolution of the universe. This was done in the recent years by several authors (Berger *et al.* 2006; Fynbo *et al.* 2006; Savaglio 2006; Prochaska *et al.* 2007).

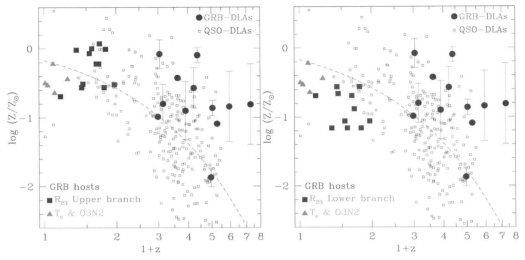

Figure 4. Metallicity as a function of redshift for 12 GRB-DLAs (filled circles), 17 GRB hosts (filled triangles and squares), and QSO-DLAs (small open squares). Left and right panels show the upper or lower branch solutions for a subsample of GRB hosts with R_{23} measurements, respectively (Savaglio *et al.* 2008). The dashed line is the fit for QSO-DLAs.

The most recent analysis of the metallicity as a function of redshift for 12 GRB-DLAs is shown in Figure 4. On average GRB-DLAs have larger metallicities than QSO-DLAs. No redshift evolution is detected in the redshift interval $2 < z < 6.3$. According to the approach followed by Prochaska *et al.* (2007), where several GRB-DLA metallicities are turned into lower limits, the difference is even larger.

The source of this significant difference in chemical enrichment is not well understood. Dust obscuration can play a role, as galaxies with dust content as large as those measured in GRB sight lines (Kann *et al.* 2006) would highly obscure background QSOs (Savaglio 2006). Another possibility is the different impact parameter in the two classes of absorbers. Using numerical simulations of star-forming galaxies (Sommer-Larsen & Fynbo 2008), Fynbo *et al.* (2008) proposed that the difference can be explained by a metallicity gradient in the absorbing galaxies. GRBs are associated with star-forming regions, which are likely located closer to the galaxy center (where metallicity is higher) than the randomly distributed QSO-DLAs.

4. GRB host galaxies

More than 50 galaxies associated with GRB events have been discovered so far. The largest public database GHostS (GRB Host Studies; www.grbhosts.info) lists many important observed parameters for most of them. These include multi-band photometry and optical emission-line fluxes. GHostS is also an interactive tool offering features such as SDSS and DSS sky viewing through Virtual Observatory services (Savaglio *et al.* 2007).

The typical GRB host (Figure 5) is an intermediate-redshift, sub-luminous, blue, star-forming galaxy (Le Floc'h *et al.* 2003; Christensen *et al.* 2004; Prochaska *et al.* 2004). The largest sample of GRB hosts studied so far contains 46 galaxies (Savaglio, Glazebrook, Le Borgne 2008). Its redshift distribution is shown in Figure 1. The average redshift and dispersion are $< z > = 1.0 \pm 1.0$. Among other parameters, Savaglio *et al.* (2008)

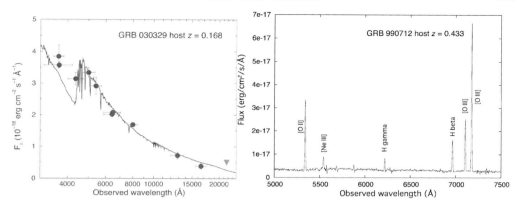

Figure 5. Two typical GRB host galaxies. *Left panel*: optical-NIR spectral energy distribution of the host of GRB 030329 (Gorosabel *et al.* 2005). The solid curve is the best-fit template with metallicity 1/5 solar. *Right panel*: optical integrated-light spectrum of the host of GRB 990712 (Küpcü Yoldaş *et al.* 2006).

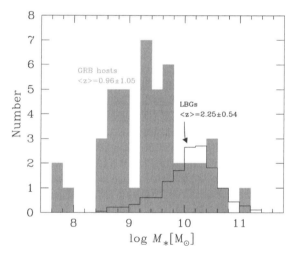

Figure 6. Stellar mass histogram of GRB hosts (filled histogram), mean redshift and dispersion $z = 0.96 \pm 1.05$. For comparison, the empty histogram represents LBGs (redshift interval $1.3 < z < 3$; Reddy *et al.* 2006) normalized to the GRB host histogram for $M_* > 10^{10}$ M$_\odot$.

measured the SFR and stellar mass of the sample. The histogram distribution of the latter is shown in Figure 6.

The specific star formation rate ($SSFR = SFR/M_*$) or its inverse, the growth time scale ($\rho_* = M_*/SFR$) as a function of redshift is shown in Figure 7. This shows how active a galaxy is. In particular ρ_* gives the time that a galaxy needs to form the observed stellar mass, if the measured SFR has been constant over its entire life. In Figure 7 we also mark the age of the universe (Hubble time) as a function of redshift. Galaxies above the line can be considered quiescent. Most GRB hosts are below the line, half of them have SSFR above 0.8 Gyr^{-1}, or $\rho_* < 800$ Myr.

Metallicities in the HII regions of GRB hosts are measured with different methods, using emission-line fluxes. Results for 17 GRB hosts at $z < 1.0$ are shown in Figure 4

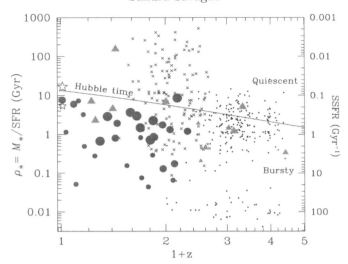

Figure 7. Growth time scale $\rho_* = M_*/SFR$ (left y-axis) or specific star formation rate SSFR (right y-axis) as a function of redshift. Filled circles and triangles are GRB hosts with SFRs measured from emission lines and UV luminosities, respectively. Small, medium and large symbols are hosts with $M_* \leqslant 10^{9.0}$ $M_\odot$, $10^{9.0}$ $M_\odot < M_* \leqslant 10^{9.7}$ $M_\odot$, and $M_* > 10^{9.7}$ $M_\odot$, respectively. The curve is the Hubble time as a function of redshift, and indicates the transition from bursty to quiescent mode for galaxies. Crosses are field galaxies at $0.5 < z < 1.7$ (Juneau *et al.* 2005; Savaglio *et al.* 2005). Dots are LBGs at $1.3 \lesssim z \lesssim 3$ (Reddy *et al.* 2006). The big and small stars at zero redshift represent the Milky Way and the Large Magellanic Cloud, respectively.

(Savaglio *et al.* 2008). The 5 metallicities derived using the electron temperature T_e and O3N2 methods, give a more reliable result. The other metallicities, mainly measured with the notoriously problematic R_{23} calibrator, which gives two solutions (lower and upper branch; Kewley & Ellison 2008), are more uncertain. Either solution, combined with all the other metallicities (including those in GRB-DLAs), does not reveal a redshift evolution of metallicities in GRB hosts in the interval $0 < z < 6.3$ (Figure 4).

5. Discussion

The main, still unsolved, question about GRB hosts is whether they represent a fair sample of the whole star-forming galaxy population, or they are a distinct population of galaxies. GRB hosts are generally small, star-forming galaxies detected at any redshift, up to $z = 6.3$. The mean stellar mass (measured mainly for the low-z subsample) is a few times above 10^9 $M_\odot$. That is the stellar mass of the Large Magellanic Cloud. The mean SFR is 5 times higher than the LMC (Savaglio *et al.* 2008). Metallicities of the hosts, measured from absorption lines in the afterglow spectra at $z > 2$ or from emission lines in the host spectra at $z < 1.0$, do not indicate a clear redshift evolution (Figure 4), with values mainly between solar and 1/10 solar. This suggests that the chemical enrichment is not a parameter characterizing the GRB host population.

It is clear that small star-forming galaxies are the most common galaxies that existed in the entire history of the universe. For instance, the star formation history of the universe, studied for different galaxy stellar-mass bins, indicates that small galaxies dominated for redshift $z < 2$, where massive galaxies experienced a fast decline (Juneau *et al.* 2005). GRB hosts are low stellar-mass star-forming galaxies and can provide a very efficient way to understand galaxy formation and evolution in the most active phase of the universe. Future missions will give the possibility to study them in much larger quantities.

Acknowledgements

We thank the conference organizers for the outstanding organization. The author thanks Karl Glazebrook and Damien Le Borgne for the very fruitful and long-lasting collaboration.

References

Berger, E., Penprase, B. E., Cenko, S. B., Kulkarni, S. R., Fox, D. B., Steidel, C. C., & Reddy, N. A. 2006, ApJ, 642, 979

Bloom, J. S., *et al.*, 2008, ArXiv e-prints, 803, arXiv:0803.3215

Christensen, L., Hjorth, J., & Gorosabel, J. 2004, A&A, 425, 913

Colgate, S. A. 1968, Canadian Journal of Physics, 46, 476

Erb, D. K., Shapley, A. E., Pettini, M., Steidel, C. C., Reddy, N. A., & Adelberger, K. L. 2006, ApJ, 644, 813

Fynbo, J. P. U., *et al.*, 2006, A&A, 451, L47

Fynbo, J. P. U., Prochaska, J. X., Sommer-Larsen, J., Dessauges-Zavadsky, M., & Møller, P. 2008, ArXiv e-prints, 801, arXiv:0801.3273

Gorosabel, J., *et al.*, 2005, A&A, 444, 711

Harwit, M., & Salpeter, E. E. 1973, ApJ, 186, L37

Juneau, S., *et al.*, 2005, ApJ, 619, L135

Kann, D. A., Klose, S., & Zeh, A. 2006, ApJ, 641, 993

Kawai, N., *et al.*, 2006, Nature, 440, 184

Kewley, L. J. & Ellison, S. L. 2008, ApJ, 681, 1183

Klebesadel, R. W., Strong, I. B., & Olson, R. A. 1973, ApJ, 182, L85

Küpcü Yoldaş, A., Greiner, J., & Perna, R. 2006, A&A, 457, 115

Le Floc'h, E., *et al.*, 2003, A&A, 400, 499

Maiolino, R., *et al.*, 2008, ArXiv e-prints, 806, arXiv:0806.2410

Metzger, M. R., Djorgovski, S. G., Kulkarni, S. R., Steidel, C. C., Adelberger, K. L., Frail, D. A., Costa, E., & Frontera, F. 1997, Nature, 387, 878

Prochaska, J. X., *et al.*, 2004, ApJ, 611, 200

Prochaska, J. X., Chen, H.-W., Dessauges-Zavadsky, M., & Bloom, J. S. 2007, ApJ, 666, 267

Reddy, N. A., Steidel, C. C., Erb, D. K., Shapley, A. E., & Pettini, M. 2006, ApJ, 653, 1004

Savage, B. D. & Sembach, K. R. 1996, ARA&A, 34, 279

Savaglio, S. 2006, New Journal of Physics, 8, 195

Savaglio, S., *et al.*, 2005, ApJ, 635, 260

Savaglio, S., Fall, S. M., & Fiore, F. 2003, ApJ, 585, 638

Savaglio, S., Budavári, T., Glazebrook, K., Le Borgne, D., Le Floc'h, E., Chen, H.-W., Greiner, J., & Yoldas, A. K. 2007, The Messenger, 128, 47

Savaglio, S., Glazebrook, K., & Le Borgne, D. 2008, ArXiv e-prints, 803, arXiv:0803.2718

Sofia, S. & van Horn, H. M. 1974, ApJ, 194, 593

Sommer-Larsen, J. & Fynbo, J. P. U. 2008, MNRAS, 385, 3

Steidel, C. C., Giavalisco, M., Pettini, M., Dickinson, M., & Adelberger, K. L. 1996, ApJ, 462, L17

Tremonti, C. A., *et al.*, 2004, ApJ, 613, 898

Low-Metallicity Star Formation:
From the First Stars to Dwarf Galaxies
Proceedings IAU Symposium No. 255, 2008
L.K. Hunt, S. Madden & R. Schneider, eds.

The Luminosity Function of Long Gamma-Ray Burst and their rate at $z \geqslant 6$

R. Salvaterra[1], S. Campana[1], G. Chincarini[1,2], T.R. Choudhury[3], S. Covino[1], A. Ferrara[4], S. Gallerani[5], C. Guidorzi[1], and G. Tagliaferri[1]

[1]INAF, Osservatorio Astronomico di Brera, via E. Bianchi 46, I-23807 Merate (LC), Italy

[2]Università degli Studi di Milano Bicocca, Piazza della Scienza 3, I-20126 Milano, Italy

[3]Institute of Astronomy, Madingley Road, Cambridge CB3 0HA, UK

[4]SISSA/International School for Advanced Studies, Via Beirut 4, I-34100 Trieste, Italy

[5]Institute of Physics, Eötvös University, Pázmány P. s. 1/A, 1117 Budapest, Hungary

Abstract. We compute the luminosity function (LF) and the formation rate of long gamma ray bursts (GRBs) in three different scenarios: i) GRBs follow the cosmic star formation and their LF is constant in time; ii) GRBs follow the cosmic star formation but the LF varies with redshift; iii) GRBs form preferentially in low–metallicity environments. We then test model predictions against the *Swift* 3-year data, showing that scenario i) is robustly ruled out. Moreover, we show that the number of bright GRBs detected by *Swift* suggests that GRBs should have experienced some sort of luminosity evolution with redshift, being more luminous in the past. Finally we propose to use the observations of the afterglow spectrum of GRBs at $z \geqslant 5.5$ to constrain the reionization history, and then applied our method to the case of GRB 050904.

Keywords. gamma rays: bursts, stars: formation, cosmology: observations, (galaxies:) inter-galactic medium

1. Introduction

Long Gamma Ray Bursts (GRBs) are powerful flashes of high–energy photons occurring at an average rate of a few per day throughout the Universe up to very high redshift (the current record is $z = 6.29$). The energy source of a long GRB is believed to be associated with the collapse of the core of a massive star (see Mészáros 2006 for a review). One of the main goals of the *Swift* satellite (Gehrels *et al.* 2004) is to tackle the key issue of the GRB luminosity function (LF). Unfortunately, although the number of GRBs with good redshift determination has been largely increased by *Swift*, the sample is still too poor (and bias dominated) to allow a direct measurement of the LF. We use here the *Swift* 3-year data to constraint the GRB LF and its evolution (Salvaterra & Chincarini 2007, Salvaterra *et al.* 2008b). Moreover, we show a possible use of GRBs detected at $z \geqslant 5.5$ to study the history of reionization (Gallerani *et al.* 2008).

2. Model description

The observed photon flux, P, in the energy band $E_{\min} < E < E_{\max}$, emitted by an isotropically radiating source at redshift z is

$$P = \frac{(1+z) \int_{(1+z)E_{\min}}^{(1+z)E_{\max}} S(E)dE}{4\pi d_L^2(z)}, \tag{2.1}$$

where $S(E)$ is the differential rest–frame photon luminosity of the source, and $d_L(z)$ is the luminosity distance. To describe the typical burst spectrum we adopt the functional form proposed by Band *et al.* (1993), i.e. a broken power–law with a low–energy spectral index α, a high–energy spectral index β, and a break energy $E_b = (\alpha - \beta)E_p/(2+\alpha)$, with $\alpha = -1$ and $\beta = -2.25$ (Preece *et al.* 2000). In order to broadly estimate the peak energy of the spectrum, E_p, for a given isotropic–equivalent peak luminosity, $L = \int_{1\,\mathrm{keV}}^{10000\,\mathrm{keV}} ES(E)dE$, we assumed the validity of the correlation between E_p and L (Yonetoku *et al.* 2004).

Given a normalized GRB LF, $\phi(L)$, the observed rate of bursts with $P_1 < P < P_2$ is

$$\frac{dN}{dt}(P_1 < P < P_2) = \int_0^\infty dz \frac{dV(z)}{dz} \frac{\Delta\Omega_s}{4\pi} \frac{\Psi_{\mathrm{GRB}}(z)}{1+z} \int_{L(P_1,z)}^{L(P_2,z)} dL'\phi(L'), \qquad (2.2)$$

where $dV(z)/dz$ is the comoving volume element†, $\Delta\Omega_s$ is the solid angle covered on the sky by the survey, and the factor $(1+z)^{-1}$ accounts for cosmological time dilation. Finally, $\Psi_{\mathrm{GRB}}(z)$ is the comoving burst formation rate. In this work, we assume that the GRB LF is described by a power law with an exponential cut–off at low luminosities, i.e. $\phi(L) \propto (L/L_{\mathrm{cut}})^{-\xi} \exp(-L_{\mathrm{cut}}/L)$.

We consider three different scenarios: i) **no evolution model**, where GRBs follow the cosmic star formation and their LF is constant in time; ii) **luminosity evolution model**, where GRBs follow the cosmic star formation but the LF varies with redshift; iii) **density evolution model**, where GRBs form preferentially in low–metallicity environments. In the first two cases, the GRB formation rate is simply proportional to the global SFR, i.e. $\Psi_{\mathrm{GRB}}(z) = k_{\mathrm{GRB}}\Psi_\star(z)$. We use here the recent determination of the SFR obtained by Hopkins & Beacom (2006), slightly modified to match the observed decline of the SFR with $(1+z)^{-3.3}$ at $z \gtrsim 5$ suggested by recent deep–field data (Stark *et al.* 2006). For the luminosity evolution model, we also assume that the cut–off luminosity in the GRB LF varies as $L_{cut} = L_0(1+z)^\delta$. Finally, for density evolution case, the GRB formation rate is obtained by convolving the observed SFR with the fraction of galaxies at redshift z with metallicity below Z_{th} using the expression computed by Langer & Norman (2006). In this scenario, $L_{cut} = \mathrm{const} = L_0$.

3. Results

The free parameters in our model are the GRB formation efficiency k_{GRB}, the cut–off luminosity at $z = 0$, L_0, and the LF power index ξ. We optimized the value of these parameters by χ^2 minimization over the observed differential number counts in the 50–300 keV band of BATSE (Stern *et al.* 2000). We find that it is always possible to find a good agreement between models and data. Moreover, we can reproduce also the 3–year differential peak flux count distribution in the 15-150 keV *Swift* band without changing the best fit parameters (Salvaterra & Chincarini 2007). We then check the resulting redshift distributions in the light of the *Swift* 3–year data, focusing on the large sample of GRBs detected at $z > 2.5$ and $z > 3.5$ (Fig. 1 panels a & b). The no evolution model is ruled out by the number of certain high-z GRBs. This result is robust since it does not depend on the assumed SFR at high-z nor on the faint–end of the GRB LF. In conclusion, the existence of a large sample of bursts at $z > 2.5$ in the *Swift* 3-year data imply that GRBs have experienced some kind of evolution, being more luminous or more common in the past (Salvaterra & Chincarini 2007).

† We adopted the 'concordance' model values for the cosmological parameters: $h = 0.7$, $\Omega_m = 0.3$, and $\Omega_\Lambda = 0.7$.

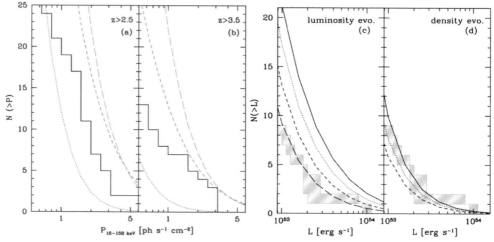

Figure 1. Panels a & b: cumulative number of GRBs at $z > 2.5$ (a) and at $z > 3.5$ (b) as a function of the photon flux P. Dotted line refers to the no evolution model, short dashed to the luminosity evolution model ($\delta = 1.5$) and long-dashed to the density evolution model ($Z_{th} = 0.1\ Z_\odot$). The number of sources detected by *Swift* in three years is shown as solid histogram. Note that the observed detections are lower limits, since many high–z GRBs can be missed by optical follow–up searches. A field of view of 1.4 sr for *Swift* is adopted. **Panels c & d:** cumulative number of luminous GRBs detected by *Swift* in three years, shown with the histogram, as function of the isotropic equivalent peak luminosity, L. Shaded area takes into account the errors on the determination of L. Note that the data are to be considered as lower limits of the real number of *Swift* detections. For pure luminosity evolution models (panel c): solid line is for $\delta = 3$, dotted line for $\delta = 2.5$, short–dashed line for $\delta = 2$, and long–dashed for $\delta = 1.5$. For pure density evolution models (panel d): Solid line is for $Z_{th} = 0.1\ Z_\odot$, dotted line is for $Z_{th} = 0.2\ Z_\odot$, and short–dashed line is for $Z_{th} = 0.3\ Z_\odot$.

In order to discriminate between luminosity and density evolution models, we compute the number of luminous GRBs, i.e. bursts with isotropic peak luminosity $L \geqslant 10^{53}$ erg s^{-1} in the 1-10000 keV band (Salvaterra et al. 2008b). We compare model predictions with the number of bright bursts detected by *Swift*. Conservatively, our data sample contains only bursts with a good redshift measurement and whose peak energy was well constrained by *Swift* itself or other satellites (such as HETE-2 or Konus-Wind). We stress here that this number represents a lower limit on the real number of bright GRBs detected, since some luminous bursts without z and/or E_p can be present in the *Swift* catalog. Results for the pure luminosity (density) evolution models are plotted in the panel c (d) of Fig. 1. Data are shown with the histogram where the shaded area takes into account errors on the determination of L. We find that models involving pure luminosity evolution requires $\delta \gtrsim 1.5$ to reproduce the number of known bright GRBs. On the other hand, models in which GRB formation is confined in low–metallicity environments fall short to account for the observed bright GRBs for $Z_{th} > 0.1$. Assuming $Z_{th} = 0.1\ Z_\odot$, the model reproduces the observed number of bright GRBs, taking also into account the errors in the determination of L. This means that essentially all bright bursts present in the 3-year *Swift* catalog have a measured redshift and well constrained peak energy. So, although this model cannot be discarded with high confidence, the available data indicate the need of some evolution in the GRB LF even for such a low value of Z_{th}. For $Z_{th} = 0.3\ Z_\odot$, as required by collapsar models (MacFadyen & Woosley 1999), only ∼6 bursts with $L \geqslant 10^{53}$ erg s^{-1} should have been detected in three years, highly underpredicting the number of *Swift* sure identifications. Thus, pure density evolution models, where the GRB LF is constant with redshift, are ruled out by the number of bright GRBs. In conclusion, available data suggest that GRBs have experienced some luminosity evolution with cosmic time.

4. GRBs from the reionization epoch

We can now compute a robust lower limit on the number of bursts detectable by *Swift* at very high z. Assuming a trigger threshold $P \geqslant 0.4\,\mathrm{ph\,s^{-1}\,cm^{-2}}$, at least $\sim 5 - 10\%$ of detected GRBs should lie at $z \geqslant 5$, with $> 1 - 3$ GRB yr^{-1} at $z \geqslant 6$. These numbers double by lowering the *Swift* trigger threshold by a factor of two (Salvaterra *et al.* 2008a).

High-z GRBs are a useful and unique tool to study the Universe near and beyond the reionization epoch. Gallerani *et al.* (2008) have studied the possibility of constraining the reionization history using the statistics of the dark portions (gaps) produced by intervening neutral hydrogen along the line of sight (LOS) in the afterglow spectra of GRB at $z \geqslant 5.5$. Two reionization models, both consistent with available observations of the high-z Universe, are considered: *(i) early reionization model* (ERM) where $z_{reion} \sim 7$ and *(ii) late reionization model* (LRM) where $z_{reion} \sim 6$. Suppose now that a GRB at redshift z_{GRB} is observed at a given flux level in the J band, F_{J}. We can then ask what is the probability that the largest of the dark gaps in its afterglow spectrum is found within a given width range. The results are shown in Fig. 2 for two different width ranges; the left (right) panels refer to the ERM (LRM) case. The isocontours correspond to a probability of 15%, 30%, 45%, and 60%. We find that the two models populate the $(z_{\mathrm{GRB}}, F_{\mathrm{J}})$ plane in a very different way. In particular, for largest gaps in the 40–80 Å range, the highest probability is obtained for fainter afterglows in the ERM than for the LRM. For largest gaps in the range 80–120 Å, the probability is in general higher in the LRM with respect to the ERM. Note that, in the ERM, only a few spectra should contain the largest gap in this range for $F_{\mathrm{J}} \gtrsim 10 - 40\ \mu\mathrm{Jy}$. Fig. 2 allows a straightforward comparison between data

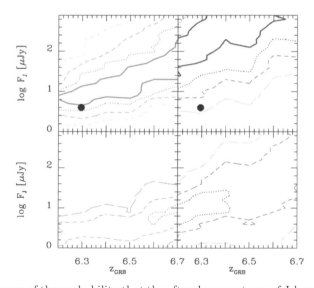

Figure 2. Isocontours of the probability that the afterglow spectrum of J-band flux F_{J} associated with a GRB at redshift z_{GRB}, contains the largest gap in the range 40–80 Å (top panels) and in the range 80–120 Å (bottom panel). The left (right) panel shows the results for the ERM (LRM). The isocontours correspond to probability of 15% (long dashed line), 30% (short dashed line), 45% (dotted line), and 60% (solid line). The black point indicates the position in the $(z_{\mathrm{GRB}}, F_{\mathrm{J}})$ plane of GRB 050904.

and model results. It is then natural to apply this procedure to GRB 050904 (black filled circle in Fig. 2). The probability to find the largest gap of 65 Å is $> 45\%$ in the ERM, i.e. almost half of the LOS contains the largest gap in the range 40–80 Å for a burst with the redshift and flux of GRB 050904. Such probability drops for the LRM to $\sim 15\%$ clearly indicating that in this case the GRB 050904 observation represents a much rarer event. Although a large sample of high-z GRBs is required before we conclude that a model in which reionization was complete at $z \sim 7$ is favored by the data, the discriminating power of the proposed method is already apparent.

This kind of analysis requires high signal-to-noise, high-resolution spectra of GRB afterglow spectra at $z \geqslant 5.5$ obtained with the largest ground-based telescopes soon after the burst detection. To avoid wasting observing time, we developed a very effective strategy to spot reliable $z \geqslant 5$ candidates on the basis of promptly available information provided by *Swift* (Campana *et al.* 2007, Salvaterra *et al.* 2007). The selection criteria adopted are: long burst observed durations ($T_{90} \gtrsim 60$ s), faint γ-ray photon fluxes ($P \lesssim 1\,\mathrm{ph\,s^{-1}\,cm^{-2}}$), and no optical counterpart in the V and bluer filters of UVOT ($V \gtrsim 20$). We tested our selection procedure against the last ~ 2 years of *Swift* data showing that our method is very efficient and clean (i.e. no low-z interloper is present in the sample).

5. Conclusions

We have tested different formation and evolution scenarios for long GRB against the 3-year *Swift* dataset. We found that *Swift* data strongly rule out models in which GRBs follow the cosmic star formation with a constant LF over time. In particular, the number of bright GRBs suggests that GRBs should have experienced some sort of luminosity evolution with cosmic time, being more luminous in the past. Finally we have shown that GRBs at $z \geqslant 5.5$ can be use to constrain the reionization history and we applied our method to the case of GRB 050904 at $z = 6.29$.

References

Band D.L. *et al.*, 1993, *ApJ*, 413, 281
Campana *et al.*, 2007, *A&A*, 464, L25
Gallerani, S., Salvaterra, R., Ferrara, A., & Choudhury T.R. 2008, *MNRAS*, 388, L84
Gehrels, N., *et al.*, 2004, *ApJ*, 611, 1005
Hopkins A. M. & Beacom J. F. 2006, *ApJ*, 651, 142
Langer L. & Norman C. A. 2006, *ApJ*, 638, L63
MacFadyen, A., & Woosley, S. 1999, *ApJ*, 524, 262
Mészáros, P. 2006, *Reports of Progress in Physics*, 69, 2259
Preece R. D., Briggs M. S., Mallozzi R. S., Pendleton G. N., Paciesas W. S., & Band D. L. 2000, *ApJS*, 126, 19
Salvaterra, R. & Chincarini, G., 2007, *ApJ*, 656, L49
Salvaterra, R., Campana, S., Chincarini, G., Tagliaferri, G., Covino, S. 2007, *MNRAS*, 380, L45
Salvaterra, R., Campana, S., Chincarini, G., Covino, S., Tagliaferri, G. 2008a, *MNRAS*, 385, 189
Salvaterra, R., Guidorzi, C., Campana, S., Chincarini, G., Tagliaferri G. 2008b, *MNRAS* submitted, arXiv:0805.4104
Stark D. P., Bunker A. J., Ellis R. S., Eyles L. P., Lacy M. 2007, *ApJ*, 659, 84
Stern B. E., Tikhomirova Y., Stepanov M., Kompaneets D., Berezhnoy A., Svensson R. 2000, *ApJ*, 540, L21
Yonetoku D., Murakami T., Nakamura T., Yamazaki R., Inoue A. K., Ioka K. 2004, *ApJ*, 609, 935

Low-Metallicity Star Formation:
From the First Stars to Dwarf Galaxies
Proceedings IAU Symposium No. 255, 2008
L.K. Hunt, S. Madden & R. Schneider, eds.

The Star Formation History of the GRB 050730 Host Galaxy

Francesco Calura[1]

[1]Dipartimento di Astronomia, Universita' di Trieste, via G. B. Tiepolo 11, 34131
TRIESTE - ITALY
email: `fcalura@oats.inaf.it`

Abstract. The long GRB 050730 observed at redshift $z \sim 4$ allowed the determination of the elemental abundances for a set of different chemical elements. We use detailed chemical evolution models taking into account also dust production to constrain the star formation history of the host galaxy of this long GRB. For the host galaxy of GRB 050730, we derive also some dust-related quantities and the specific star formation rate, namely the star formation rate per unit stellar mass. Finally, we compare the properties of the GRB host galaxy with those of Quasar Damped Lyman Alpha absorbers.

Keywords. Gamma rays: bursts, Galaxies: high-redshift, Galaxies: abundances, ISM: general

1. Introduction

Gamma-ray bursts (GRBs) afterglows provide insight into the interstellar medium (ISM) of galaxies during the earliest stages of their evolution. In several cases, thanks to the GRB afterglows it has been possible to study the dust content, the star formation rates and the stellar mass of the GRB host (Bloom *et al.* 1998; Savaglio *et al.* 2008). In a few cases, the determination of their chemical abundance pattern has been possible (Savaglio *et al.* 2003; Vreeswijk *et al.* (2004); Prochaska *et al.* 2007).

Chen *et al.* (2005) reported on the chemical abundances for the damped Lyman Alpha system (DLA) associated with the host galaxy of GRB 050730. Their analysis showed that this gas was metal poor with modest depletion. These results were subsequently expanded and tabulated by Prochaska *et al.* (2007) (hereafter P07).

In this paper, we aim at determining the star formation history of a GRB host galaxy by studying the chemical abundances measured in the afterglow spectrum of GRB 050730. For this purpose, we use a detailed chemical evolution model. Our aim is to constrain the star formation rate and the age of the host galaxy of GRB 050730, and possibly to expand our study to other systems. The plan of this paper is as follows. In Section 2, we briefly introduce the chemical evolution model. In Section 3, we present and discuss our main results.

2. Chemical evolution models including dust

The chemical evolution model used here to derive the star formation history of the GRB 050730 host galaxy is similar to that developed by Bradamante *et al.* (1998) for dwarf irregular galaxies. Since some chemical species studied in this paper are refractory (Fe, Ni), in the chemical evolution model we include also a treatment of dust production and destruction, based on the work by Calura, Pipino & Matteucci (2008, hereafter CPM08).

We use the model for dwarf irregular galaxies since several observational investigations provided strong evidence that most of the GRBs originate in gas rich, star-forming sub-luminous ($L < L_*$) galaxies with relatively low metallicities ($Z < Z_\odot$) (Bloom *et al.* 1998; Prochaska *et al.* 2004). The dwarf galaxy is assumed to form by means of a continuous infall of pristine gas until a mass M_{tot} is accumulated. The evolution of dwarf irregular galaxies is characterized by a continuous star formation history, characterized by a low star formation efficiency ($\nu \leqslant 0.1 Gyr^{-1}$). The star formation rate $\psi(t)$ is directly proportional the gas fraction $G(t)$ at the time t, according to the Schmidt law:

$$\psi(t) = \nu G(t). \tag{2.1}$$

Supernovae (SNe) Ia and II are responsible for the onset of a galactic wind, when the thermal energy of the ISM exceeds its binding energy, which is related to the presence of a dark matter halo in which the galaxy is embedded (for more details, see Bradamante *et al.* 1998, Lanfranchi & Matteucci 2003). The binding energy of the gas is influenced by assumptions concerning the presence and distribution of dark matter (Matteucci 1992). A diffuse ($R_e/R_d = 0.1$, where R_e is the effective radius of the galaxy and R_d is the radius of the dark matter core) and massive ($M_{dark}/M_{Lum} = 10$) dark halo has been assumed for each galaxy. The time at which the wind develops depends on the assumed star formation efficiency. In general, the higher the star formation (SF) efficiency, the earlier the wind develops. The models used throughout this paper are characterized by very low star formation efficiencies and by a young age, much lower than the times of the onset of the galactic winds. For these reasons, the galactic winds have no effect on the main results obtained in this paper.

Chemical enrichment from various types of stars is properly taken into account. The stellar yields are mainly from Woosley & Weaver (1995) for massive stars, from Meynet & Maeder (2002) for low and intermediate mass stars (LIMS) and from Iwamoto *et al.* (1999) for Type Ia SNe.

We assume a Salpeter initial mass function (IMF). We assume a cosmological model characterized by $\Omega_m = 0.3$, $\Omega_\Lambda = 0.7$ and a Hubble constant $H_0 = 70 \, \mathrm{km \, s^{-1} \, Mpc^{-1}}$.

2.1. *Dust production and destruction*

Two elements studied in these paper are refractory: Ni and Fe. To model their gas phase abundances, we need to take into account dust production and destruction.

The model for dust evolution used in this paper is described in detail in CPM08. Here, we summarize its main features. For the refractory chemical element labeled i, a fraction δ_i^{SW}, δ_i^{Ia}, and δ_i^{II} is incorporated in dust grains by low and intermediate mass stars, type Ia SNe, and type II SNe, respectively. These quantities are the dust condensation efficiencies of the element i in various stellar objects. Here we assume $\delta_i^{SW} = \delta_i^{Ia} = \delta_i^{II} \equiv \delta_i = 0.1$. This choice is motivated by recent mid-infrared observations of one supernova (Sugerman *et al.* 2003), which provided an upper limit of $\delta_i^{II} \leqslant 0.12$. The value assumed here is further supported by theoretical studies of the local dust cycle (Edmunds 2001).

Dust grains are usually destroyed by the propagation of SN shock waves in the warm/ionised interstellar medium (Jones *et al.* 1994). If $G_{dust,i}$ is the fraction of the element i locked into dust and G is the gas fraction, the destruction rate is calculated as $G_{dust,i}/\tau_{destr}$, where τ_{destr} is the dust destruction timescale, calculated as:

$$\tau_{destr} = (\epsilon M_{SNR})^{-1} \frac{G}{R_{SN}}. \tag{2.2}$$

(McKee 1989). $M_{SNR} = 1300 M_\odot$ is the mass of the interstellar gas swept up by the SN remnant. R_{SN} is the total SNe rate, i.e. the sum of the rates of Type Ia and Type II SNe.

Unless otherwise specified, we assume that no dust accretion, occurring mainly in cold molecular clouds, is taking place in the GRB host galaxy. Our choice is motivated by the fact that very little molecular H is observed in local dwarfs, with molecular-to-atomic gas fractions of $\sim 10\%$ or lower (Clayton *et al.* 1996). Our assumption is further supported by the lack of H_2 absorption lines observed in the spectra of GRB afterglows (Whalen *et al.* 2008). In the presence of intense SF, molecular clouds are likely to rapidly dissolve, allowing very little dust accretion to occur.

2.2. *Results and discussion*

The abundance ratios between two elements formed on different timescales can be used as "cosmic clocks" and provide us with information on the roles of LIMS and SNe in the enrichment process (Matteucci 2001). In particular, the study of abundance ratios such as $[\alpha/Fe]$ and $[N/\alpha]$† is quite helpful, since the α-elements (O, S, Mg) are produced on short timescales by Type II SNe, whereas the Fe-peak elements and nitrogen are produced on long timescales by Type Ia SNe and low and intermediate-mass stars, respectively. As shown by Dessauges-Zavadsky *at al.* (2004; 2007, hereafter D07), the simultaneous study of the abundance ratios between different elements as functions of a metallicity tracer (such as [S/H] or [Fe/H]) may be used to constrain the star formation history of a given system, whereas the study of the abundance ratios versus redshift can be used to constrain the age of the system.

† All the abundances between two different elements X and Y are expressed as $[X/Y] = log(X/Y) - log(X/Y)_\odot$, where (X/Y) and $(X/Y)_\odot$ are the ratios between the mass fractions of X and Y in the ISM and in the sun, respectively. We use the set of solar abundances as determined by Grevesse *et al.* (2007).

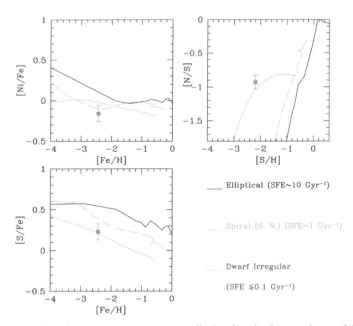

Figure 1. Observed abundance ratios versus metallicity for the host galaxy of GRB 050730 as reported by P07 (solid squares with error bars). The solid line, dashed line and dotted line are predictions computed by means of a chemical evolution model for an elliptical galaxy, a spiral galaxy and a dwarf irregular galaxy, respectively, and without including dust depletion (see text for further details).

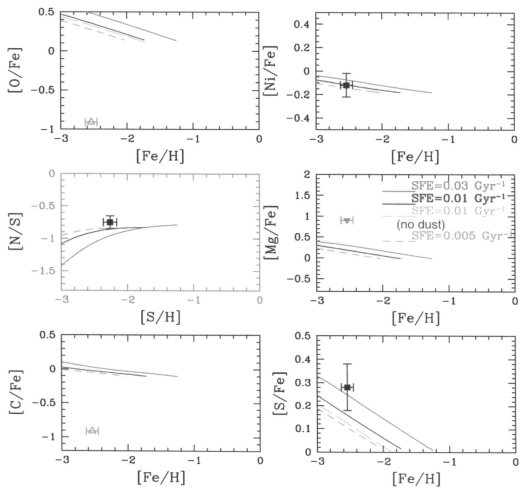

Figure 2. Observed abundance ratios versus metallicity for the host galaxy of GRB 050730 as derived by P07 (solid squares with error bars). The triangles stand for upper or lower limits. The thick lines represent models with increasing SF efficiency, with the lowest curves having the lowest SF efficiency values. The thin dashed lines do not include dust depletion. The red box (N/S vs S/H) is used for abundance ratios between non-refractory elements.

In Fig. 1, we show the predicted evolution of [N/Fe], [Ni/Fe] and [S/Fe] as a function of various metallicity tracers, such as [Fe/H] or [S/H], for various star formation histories describing galaxies of different morphological types: an elliptical galaxy (solid lines), a spiral galaxy (more precisely, a model for the Solar Neigbourhood, dashed lines, see CPM08 for a detailed model description) and a dwarf irregular galaxy (dotted lines), compared with the values observed for GRB 050730 DLA (solid squares with error bars). Fig 1 shows clearly that the most likely progenitor for the host galaxy of GRB 050730 is a dwarf irregular galaxy, and how the elliptical and spiral model appear indadequate to describe the abundance pattern observed by P07 in the GRB 050730 host. In the following, we will use the dwarf irregular galaxy model to constrain the parameters of the SF history and the age of the host galaxy GRB 050730.

The next step is to constrain the main star formation history parameter of the host galaxy of GRB 050730, i.e. its star formation efficiency ν. We search for the model which

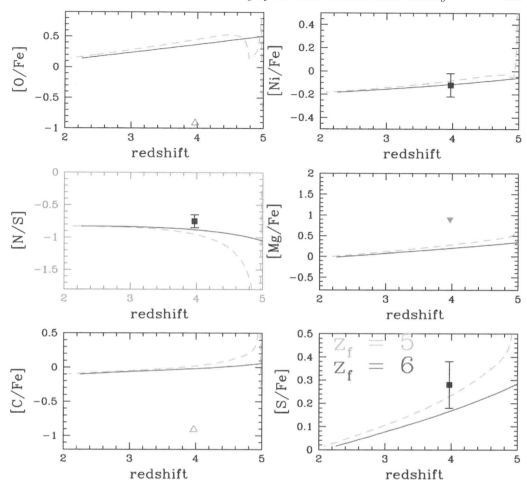

Figure 3. Observed abundance ratios versus redshift for the host galaxy of GRB 050730 as derived by P07 (solid squares with error bars). The triangles stand for upper or lower limits. The solid (dashed) lines represent a redshift of formation $z_f = 6$ ($z_f = 5$). The thin dashed lines do not include dust depletion. The red box (N/S vs redshift) is used for abundance ratios between non-refractory elements.

best reproduces the observed abundances. In figure 2, we show the predicted evolution of several abundance ratios for various models, characterized by different SF efficiencies. The model which best reproduces the observed abundances is that characterized by a SF efficiency $\nu = 0.01$ Gyr^{-1}.

Once we have determined the star formation efficiency of the best model, the following step is to constrain the age of the GRB host galaxy. To perform this task, we study the abundance ratios vs redshift (Fig. 3). The abundances are best reproduced at an age of ~ 0.4 Gyr, corresponding to a redshift of formation $z_f \sim 5$. At this age, the specific star formation rate is ~ 5 Gyr^{-1}, the dust-to-gas ratio is $\sim 10^{-6}$, in agreement with the upper limit of $8 \cdot 10^{-6}$ as derived observationally by P07. The dust-to-metals ratio is 0.04, roughly consistent with the upper limit of 0.1 measured by P07.

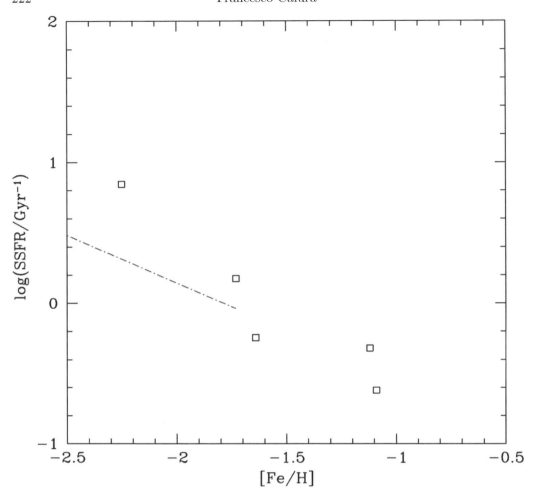

Figure 4. Predicted evolution of the SSFR vs [Fe/H] for the best model for the host galaxy of GRB 050730. The open squares are the values determined for 5 QSO DLAs in the sample by D07.

Finally, in Fig. 4, we compare the predicted SSFR vs [Fe/H] obtained for the best model for the host galaxy of GRB 050730, with SSFR derived for a sample of QSO DLAs by D07. The evolution of the SSFR of the GRB host is compatible with the values found for QSO DLAs, possibly indicating similar chemical evolution paths. However, this may be the result of a coincidence, since in principle GRB DLAs are expected to represent denser and more metal-enriched galactic regions than QSO DLAs. By extending the method presented here to other systems, it will be possible to shed light on possible analogies and differences between QSO and GRB DLAs.

Acknowledgements

It is a pleasure to thank my collaborators, i.e. Jason Prochaska, Mirka Dessauges-Zavasky and Francesca Matteucci. I acknowledge financial contribution from contract ASI-INAF I/016/07/0.

References

Bloom, J. S.; Djorgovski, S. G.; Kulkarni, S. R.; Frail, D. A., 1998, ApJ, 507, L25

Bradamante, F., Matteucci, F., D'Ercole, A. 1998, A&A, **337**, 338

Calura, F.; Pipino, A.; Matteucci, F., 2008, A&A, 479, 669 (CPM08)

Clayton, Geoffrey C., Green, J., Wolff, Michael J., Zellner, Nicolle E. B., Code, A. D., Davidsen, Arthur F., WUPPE Science Team, HUT Science Team, 1996, ApJ, 460 313

Dessauges-Zavadsky, M.; Calura, F.; Prochaska, J. X.; D'Odorico, S.; Matteucci, F., 2004, A&A, 416, 79

Dessauges-Zavadsky, M.; Calura, F.; Prochaska, J. X.; D'Odorico, S.; Matteucci, F., 2007, A&A, 470, 431 (D07)

Edmunds, M. G., 2001, MNRAS, 328, 223

Grevesse, N.; Asplund, M.; Sauval, A. J., 2007, SSRv, 130, 105

Jones, A. P., Tielens, A. G. G. M., Hollenbach, D. J., McKee, C. F., 1994, ApJ, 433, 797

Iwamoto, K.; Brachwitz, F.; Nomoto, K.; Kishimoto, N.; Umeda, H.; Hix, W. R.; Thielemann, F.-K., 1999, ApJS, 125, 439

Lanfranchi, G., Matteucci, F., 2003, MNRAS, 345, 71

Matteucci, F., 1992, ApJ, 397, 32

Matteucci, F., 2001, *The chemical evolution of the Galaxy*, Astrophysics and space science library, Volume 253, Dordrecht: Kluwer Academic Publishers

McKee C. F., 1989, in Allamandola L. J., Tielens A. G. G. M., eds, Interstellar Dust, Proc. IAU Symposium 135. Kluwer, Dordrecht, p. 431

Meynet, G., Maeder, A., 2002, A&A, 381, L25

Prochaska, J. X.; Bloom, J. S.; Chen, H.-W.; Hurley, K. C.; Melbourne, J.; Dressler, A.; Graham, J. R.; Osip, D. J.; Vacca, W. D., 2004, ApJ, 611, 200

Prochaska, J. X.; Chen, H.-W.; Dessauges-Zavadsky, M.; Bloom, J. S., 2007, ApJ, 666, 267

Savaglio, S., Fall, S. M., Fiore, F., 2003, ApJ, 585, 638

Savaglio, S., Glazebrook, K., Le Borgne, D., 2008, ApJ, submitted , arXiv0803.271

Sugerman, B. E. K., *et al.*, 2006, Science, 313, 196

Vreeswijk, P. M., *et al.*, 2004, A&A, 419, 927

Whalen, D., Prochaska, J. X., Heger, A., Tumlinson, J., 2008, ApJ, submitted

Woosley, S.E., Weaver, T.A., 1995, ApJS, 101, 181

Session IV

Dust and gas as seeds for metal-poor star formation

Low-Metallicity Star Formation:
From the First Stars to Dwarf Galaxies
Proceedings IAU Symposium No. 255, 2008
L.K. Hunt, S. Madden & R. Schneider, eds.

Dust and Gas as Seeds for Metal-Poor Star Formation

Deidre A. Hunter

Lowell Observatory, 1400 West Mars Hill Road, Flagstaff, Arizona, USA
email: dah@lowell.edu

Abstract. I address the issue of dust and gas as seeds for metal-poor star formation by reviewing what we know about star formation in nearby dwarf galaxies and its relationship to the gas and dust. I (try to) speculate on the extent to which processes in nearby galaxies mimic star formation in the early universe.

Keywords. galaxies: dwarf, galaxies: ISM, galaxies: evolution

1. Star formation and gas

Star formation in dwarf galaxies (dIm) is occurring at low, and even very low, average atomic gas densities. The gas densities are sufficiently low that star formation is at best marginally unstable to large scale gravitational instabilities; moreover, the instability criterion does not predict where stars form (models: Safronov 1960, Toomre 1964, Quirk 1974; dwarf observations: Hunter & Plummer 1996, Meurer *et al.* 1996, van Zee *et al.* 1997, Hunter *et al.* 1998, Rafikov 2001, Leroy *et al.* 2008). (See, for example, Figure 1, right top.) This suggests that cloud formation in dwarfs is more difficult and inefficient than in the inner parts of spirals (Dong *et al.* 2003, Li *et al.* 2005a) and that other, more local, processes dominate (Elmegreen & Hunter 2006). Furthermore, we expect the difficulty in forming stars to increase as the galaxy mass decreases (Li *et al.* 2005b). So, just what do we know about star formation in dwarfs and its relationship to the (atomic) gas?

Star formation is a local process.

Since large-scale spontaneous gravitational instabilities are not dominant, other—local—processes must be important in tiny galaxies. That this is the case is seen observationally: star formation is occurring in HI clouds or complexes even where the *average* gas density is "too low" (Hunter & Gallagher 1985, 1986; Phillipps *et al.* 1990; van der Hulst *et al.* 1993; Taylor *et al.* 1994; van Zee *et al.* 1997; Meurer *et al.* 1998; de Blok & Walter 2006). For example, in NGC 2366 the HI peaks associated with star-forming regions are close to the Toomre critical density even though the average gas density in the surroundings is well below it (see Figure 2; Hunter *et al.* 2001a). Star formation in outer disks, where the gas is highly sub-critical, must certainly be a local process (Roye & Hunter 2000, Komiyama *et al.* 2003, Parodi & Binggeli 2003).

Not all HI is equal.

However, not all of the HI is equal when it comes to star formation. Braun (1997) found that 60–90% of HI is in cool filaments in spirals and dwarfs (but see Usero *et al.*, in prep), and it is the cool HI that we expect to be associated with new star formation. In fact, Young & Lo (1996) found that 20% of the HI in Leo A is in a component with a low velocity dispersion, and similarly for a small sample of other dwarfs (Young *et al.* 2003, Begum *et al.* 2006b). Even little Leo T, the recently discovered dIm in the Local Group with an M_V of only -7 (Irwin *et al.* 2007), has a cool HI component ($\sigma \sim 2\,\mathrm{km/s}$;

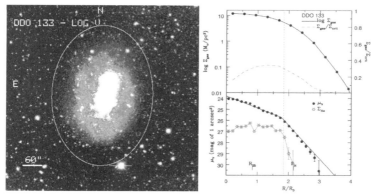

Figure 1. *Left*: V-band image of the dIm DDO 133. *Right, top*: Gas surface density of DDO 133 and $\Sigma_{gas}/\Sigma_{crit}$. The gas density is much lower than the Toomre critical density. *Right, bottom*: V-band and Hα surface brightness profiles. The V profile is steeper in the outer disk.

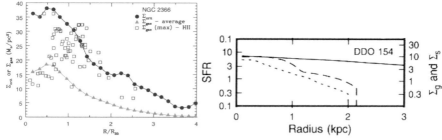

Figure 2. *Left*: HI of dIm NGC 2366. The gas peaks in star-forming regions are close in density to the Toomre critical density even though the average surface density is much lower. Adapted from Hunter *et al.* (2001a). *Right*: Example of star formation rate profile (*long dashed line*) following the average surface density of older stars (*short dashed line*) better than it follows the average gas surface density (*solid line*). Adapted from Hunter *et al.* (1998).

Ryan-Weber *et al.* 2008). In NGC 6822, de Blok & Walter (2006) concluded that the cool gas is more important in determining the star formation *locally* than the total HI, although the relationship between cool HI and star formation is not deterministically simple (Young *et al.* 2003, Begum *et al.* 2006b). Thus, the immediate reservoir of gas for cloud formation may not be as extensive as the integrated HI mass would indicate.

Star formation follows the older stars.

The azimuthally-averaged star formation rate follows the stellar surface density of the older stars with radius better than it follows the gas surface density in most dwarfs (see Figure 2). So, star formation appears to be tied to existing stars (Hunter & Gallagher 1985, Brosch *et al.* 1998, Hunter *et al.* 1998, Stewart *et al.* 2000, Parodi & Binggeli 2003, Hunter & Elmegreen 2004, Leroy *et al.* 2008), with some time averaging (10^8 yrs, Andersen & Burkert 2000). In many galaxies, this is true even in the far outer disk.

Stars blow holes, and porosity has consequences.

One way for star formation to be tied to the presence of older stars is through star-induced star formation (Öpik 1953; Gerola *et al.* 1980; Comins 1983). Star formation changes the interstellar medium (ISM; Martin 1997, Mühle *et al.* 2005). Massive stars have strong winds and then die as supernovae, so where concentrations of massive stars form, large holes can be blown in the ISM (Puche *et al.* 1992, Martin 1998, Walter & Brinks 1999, Parodi & Binggeli 2003, Cannon *et al.* 2005, de Blok & Walter 2006). These holes are surrounded by higher density shells and in these shells star-forming clouds can

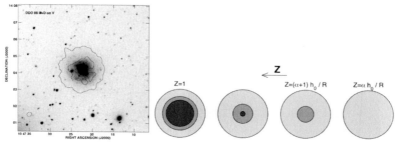

Figure 3. *Left*: V image of the Sm DDO 88 with contours of integrated HI emission (600×520 pc beam; Simpson *et al.* 2005). Stars have blown a 3-kpc diameter gas ring. *Right*: Sketch of molecular clumps as a function of metallicity from Bolatto *et al.* (1999). As the metallicity goes down, the molecular core traced by CO shrinks and the PDR grows.

develop (Mori *et al.* 1997, van Dyk *et al.* 1998). A beautiful example is Constellation III in the LMC (Dopita *et al.* 1985). There, stars that formed in the interior 12–16 Myr ago blew a 1.4-kpc size hole in the gas (Dolphin & Hunter 1998). More recently ($\sim$6 Myrs) stars have formed in the higher density shell surrounding the hole.

The action of the massive stars can create a swiss-cheese morphology in the ISM, and some dwarfs have lots of holes in their gas. In fact the filling factor in holes can be very high, exceeding that in M31 by factors of 3–800 (Oey *et al.* 2001; Bagetakos *et al.*, in prep). The rearrangement of the ISM can also be severe in tiny galaxies, and without shear, structures can last a long time. In DDO 88, for example, the HI looks like a giant doughnut and the ring, located at $1/2R_H$, contains 30% of the total HI mass (see Figure 3; Simpson *et al.* 2005). The massive stars may even blow the gas out of the galaxy, and completely away if the total mass is small enough, about 10^7 M$_\odot$ (Martin 1998, Mac Low & Ferrara 1999, Ferrara & Tolstoy 2000, but see Ott *et al.* 2005).

Not all holes have to be made by massive stars. There are holes in outer disks and low density HI regions without obvious young stars (LMC: Kim *et al.* 1999; Holmberg II: Rhode *et al.* 1999, Stewart *et al.* 2000, Bureau & Carignan 2002; DDO 154: Hoffman *et al.* 2001), and models show that these could be the result of turbulence and other processes such as thermal and gravitational instabilities (Wada *et al.* 2000, Piontek & Ostriker 2004, Dib & Burkert 2005, but see Sánchez-Salcedo 2001).

But what are the consequences of this porosity to star formation? Although secondary star formation can form in the shells, porosity in general may also regulate the star formation process by making it harder to form more stars—like a speed bump to star formation (Silk 1997, Scalo & Chappell 1999, Martin 1999, Parodi & Binggeli 2003). This is partly because the holes allow UV photons to heat the gas and reduce the cold gas supply (Silk 1997). In addition, one can imagine that the porosity would make it hard to form multiple generations beyond the second one. For example, in the LMC, the HI is so filamentary that it is hard to see how the second generation in the shell of Constellation III is going to have much of a chance to form a third generation (see the HI map of Kim *et al.* 1999). For the galaxy as a whole, which process—secondary star formation or regulating further star formation—wins or the roles they both play aren't clear.

One consequence of this on-going rearrangement of the ISM is that star formation must move around the galaxy on kpc scales. And this is what is observed (Payne-Gaposhkin 1974; Hodge 1969, 1980; Hunter *et al.* 1982; Hunter & Gallagher 1985, Hunter & Gallagher 1986, Schombert *et al.* 2001). These statistical sorts of fluctuations across a galaxy can be severe in dwarfs just because they are small. For example, you could end up with a galaxy that is blue in one half, where star formation took place recently, and red in the other half, where older stars dominate (Hunter & Elmegreen 2006).

There is evidence for turbulence compression.

There is evidence for another local process taking place in dwarfs—turbulence compression. First, the ISM is structured into clouds of all sizes (that is, a fractal) whose distributions resemble those of compressible turbulence. This is seen in the LMC (Elmegreen *et al.* 2001), the SMC (Stanimirovic *et al.* 1999), and other nearby dwarfs (Westpfahl *et al.* 1999, Begum *et al.* 2006a). Second, other distributions are also consistent with sampling a fractal turbulent gas, including the HII region luminosity function (Kingsburgh & McCall 1998, Youngblood & Hunter 1999), Hα probability distribution function (Hunter & Elmegreen 2004), V-band Fourier Transform power spectra (Willett *et al.* 2005), and star cluster mass functions (e.g., Hunter *et al.* 2003). In Holmberg II, for example, Dib & Burkert (2005) see correlated structure in the ISM for scales less than 6 kpc and argue for a turbulence driver that acts on scales of order 6 kpc.

Turbulence can heat the gas and make it harder to form stars, thereby regulating star formation (Struck & Smith 1999). But it can also bring gas together and create density enhancements that become self-gravitating if the gas is dense enough for the gravitational potential energy to exceed the energy in turbulent motions (Krumholz & McKee 2005).

Stars form in outer disks.

Stars have formed in outer disks in the past, and FUV surface brightness profiles indicate that, in many dwarfs, young stars also exist far into the outer disk (Hunter *et al.*, in prep). Ultra-deep imaging shows that the stellar disks continue to surface brightnesses of even 30 mag/arcsec^2 in V with no end in sight (see Figure 1). Thus, stars have formed in extreme low-density conditions in dwarf outer disks, and this region poses a particularly difficult test of our understanding of the cloud/star formation process.

In outer disks, we often find complex stellar surface brightness profiles (see Figure 1). About 24% of a survey of 94 dIm show a stellar surface brightness profile that becomes suddenly steeper at about 2 disk scale lengths (Hunter & Elmegreen 2006). This kind of break is also seen in spirals (see, for example, Pohlen *et al.* 2002, Kregel & van der Kruit 2004) and in disks at redshifts of $0.6 < z < 1.0$ (Pérez 2004). People have historically called this a "truncation" because they thought they were seeing the end of the disk, but I think it is better referred to as a break because it represents a change rather than an ending. The difference between spirals and dwarfs is that the break occurs much closer to the center of the galaxy in dwarfs.

What happens to the star formation process at the break? Different theoretical approaches predict such breaks: 1) Andersen & Burkert (2000) suggest that self-regulated evolution within a confining dark halo will lead naturally to exponential density profiles that are somewhat flatter in the central regions. 2) Schaye (2004) argues that star formation occurs where there is cold gas, triggering gravitational instabilities. The threshold gas column density for this transition is around 3–10×10^{20}/cm^2. Tiny Leo T has a peak column density of 7×10^{20}/cm^2 and it has formed stars as recently as 200 Myr ago (Ryan-Weber *et al.* 2008, Irwin *et al.* 2007). And star formation is found where locally the gas density is greater than 4–6×10^{20}/cm^2 in a sample of low luminosity dwarfs (Hunter *et al.* 2001a, Begum *et al.* 2006b). The break, according to Schaye, occurs where the *average* gas density drops below this threshold. 3) Li *et al.* (2005b) determined from hydrodynamic simulations that there should be a sharp drop in the star formation rate at 2× the disk scale length and that stars are more important than gas in destabilizing dwarf disks. 4) Elmegreen & Hunter (2006) suggest that inner disks are dominated by large-scale gravitational instabilities while in outer disks local processes driven by a low level of bulk motions of the gas form clouds with densities high enough *locally* to form

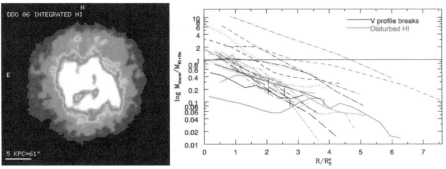

Figure 4. *Left*: Integrated HI map of dIm DDO 86 (1.4×1.3 kpc beam). Circled HI peaks at $5R_D$ have maxima of $5 \times 10^{20}/\mathrm{cm}^2$, within expected thresholds for star formation (Schaye 2004), even though the average Σ_{gas} there is $3 \pm 1 \times 10^{20}/\mathrm{cm}^2$. *Right*: M_{gas}/M_{stars} for dwarfs. Dwarfs become increasingly gas-rich with radius, implying a decreasing cloud formation efficiency.

stars (see Figure 4). The profile break is the transition from the dominance of large-scale processes to the dominance of local processes. This requires a low level of motion of the gas continuing into the outer disk, as is observed. One mechanism for producing local density enhancements is the magnetorotational instability, where the angular velocity decreases outward and magnetic fields are present. Piontek & Ostriker (2005) predict that turbulent velocity dispersions go up, reaching a quasi-steady plateau, as the average gas density goes down. This results in departures from thermal equilibrium and local density variations that might be particularly useful in outer disks. 5) Roškar *et al.* (2008) suggest that the break forms in a (spiral) disk within 1 Gyr and that the break moves outward as the gas cools. The break in their model corresponds to a rapid drop in the star formation rate associated with a drop in the cool gas surface density relative to the Toomre critical density. 6) Bournaud *et al.* (2007) suggest the outer exponential is built from tidal debris. All but the last model agree that the break corresponds to a change in the star formation rate due to changes in the physical conditions within the gas.

There is a wide variety of gas surface density profiles.

One thing all dwarfs have in common is that the ratio of the mass in stars to mass in gas changes steadily with radius. Dwarfs are usually gas dominated even in the centers, but become more so with radius. This implies a steady decrease in the large-scale cloud formation efficiency with radius (see Figure 4; Leroy *et al.* 2008).

The wide variety of gas surface density fall-offs with radius is striking. Some profiles are relatively flat (in terms of log of the surface density), some drop precipitously, and others decrease steadily. So, just what is the role of the gas in determining the nature of the stellar disk, especially outer disks? To explore the relationship of the HI profiles to properties of the stellar disk, I fit Sersic functions to HI surface density profiles of 19 dwarf galaxies: $\log \Sigma_{gas}(R) = (\log \Sigma_{gas})_0 - 0.434(R/R_0)^{1/n}$. Dwarfs are fit well with $n \leqslant 1$; a higher n value means the gas is more centrally concentrated.

There are a few trends between the way the gas falls off with radius and the star-forming characteristics of the galaxies (see Figure 5). First, the star formation activity is more centrally concentrated in galaxies where the gas is more centrally concentrated. Second, the central gas density is more important in determining the integrated star formation rate than the details of the gas profile; a higher central gas density correlates with a higher integrated star formation rate. Third, galaxies with breaks in their stellar surface brightness profiles tend to have gas profiles that are less centrally concentrated (lower n) and have lower central gas densities, but there are similar galaxies without

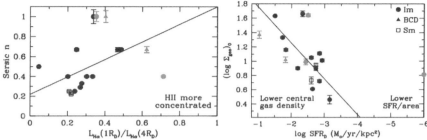

Figure 5. Results of fitting gas surface density profiles with Sersic functions. *Left*: Sersic n vs. $L_H\alpha(1R_D)/L_H\alpha(4R_D)$, a measure of the central concentration of the star formation activity. A higher n means a higher degree of central concentration of HI. The star formation activity is more centrally concentrated in galaxies where the HI is also more centrally concentrated. *Right*: The integrated star formation rate normalized to πR_D^2 SFR_D correlates with the central gas density, not the shape of the Σ_{gas} profile.

x

optical breaks as well. So, all in all, the gas does seem to play a role in determining the stellar disk, but the effects are not as dramatic as the variation in gas profiles.

No dark galaxies.

Almost all puddles of gas must form stars according to Taylor & Webster (2005). In their models the gas in galaxies with more than 10^6 M$_\odot$ in gas and stars becomes unstable and forms stars until there is a stellar radiation field to warm the ISM and stabilize it. The key to instability is efficient H_2 cooling. Similarly, Warren *et al.* (2007) suggest that isolated galaxies with shallow dark matter potentials will produce the minimum quantity of stars needed to stabilize the gas disk. Taylor and Webster suggest that local dwarfs represent the minimum rates of self-regulated star formation. Leo T must define the extreme. It has only 3×10^5 M$_\odot$ of HI and a peak column density of 7×10^{20}/cm^2, but it has formed stars in the past 200 Myrs (Irwin *et al.* 2007, Ryan-Weber *et al.* 2008).

Starbursts happen.

Lee *et al.* (2007) find evidence that the star formation histories of dwarfs with $M_B > -15$ are more erratic than those of brighter dwarfs. Specifically, the equivalent width of Hα becomes more scattered. They argue that at this brightness level there is some change in the star formation process, or at least in the regulatory nature of the process. One possibility is that feedback from massive stars has a larger negative impact, and the evolution becomes bursty or at least "gaspy" (Marconi *et al.* 1995, Schombert *et al.* 2001). This emphasizes the need to look for trends across the dwarf galaxy class.

Stinson et al. (2007) have simulated the collapse of dwarf galaxies. They find that supernovae feedback can quench star formation as the gas collapses. Gas flows out in a hot halo, cools, and then forms stars again. The result is episodic bursts of star formation. On the other hand, Li *et al.* (2005a) argue that direct feedback from starbursts is minor because most of the energy is deposited above the disk, not in it.

Blue Compact Dwarfs (BCDs) have, or are, undergoing enhanced star formation at some level. Furthermore, star formation has migrated to the center within the last Gyr (Noguchi 1988, Hunter & Elmegreen 2004, but see van Zee *et al.* 2001). The HI is more centrally concentrated (Chamaraux 1977, Taylor *et al.* 1994, Meurer *et al.* 1998, van Zee *et al.* 2001), and the distribution of HI has changed with time (Simpson & Gottesman 2000). Some have unusually large HII regions (Youngblood & Hunter 1999), and many have peculiar HI velocity fields. Sometimes there is some object nearby to blame for the starburst, but not always (c.f., Taylor 1997; Stil & Israel 1998; Méndez & Esteban 2000; Campos-Aquilar & Moles 1991; Telles & Terlevich 1995; Telles & Maddox 2000; Simpson *et al.* 2008, in prep). Could some BCDs be advanced dwarf–dwarf mergers or dwarf and even-smaller-dwarf mergers (so there won't be a tidal tail, which we don't see)? BCDs may be showing us the frequency of dwarf–dwarf interactions.

2. Star formation and dust

The relationship between the gas and star formation is obvious since stars form out of gas clouds. The connection between dust and star formation is less direct. Dust affects the heating cycle in the ISM, and this affects the ability of the ISM to form cold, dense clouds that can form stars. The presence of even a small amount of dust can make a difference to the ability to form H_2 clouds (Schaye 2004). Thus, the low dust content *should* have consequences to the star formation process in dwarfs.

The dust and metal contents are low.

Although there are embedded star-forming regions in dwarfs and even dust in outer disks (e.g., Hinz *et al.* 2006, Walter *et al.* 2007), the dust-to-gas ratio is lower in dwarfs than in spirals by factors of 2–25 (Hunter *et al.* 1989; Lisenfeld *et al.* 2002; Galliano *et al.* 2003; Böttner *et al.* 2003; Cannon *et al.* 2006a,b; Leroy *et al.* 2007; Walter *et al.* 2007), and the ratio falls with metallicity (Lisenfeld & Ferrara 1998). However, there is some evidence for large quantities of cold dust that we generally miss (Madden *et al.* 2006).

One consequence of the low dust content, coupled with high porosity in some dwarfs, is long sight-lines for stellar UV radiation. This heats the dust and gas in the diffuse ISM and on the surfaces of molecular clouds. This results in warmer FIR dust temperatures (Hunter *et al.* 1989, Walter *et al.* 2007), and is especially an issue in starbursts (Madden 2000). It also means deeper penetration of star-forming gas clouds. This results in large photodissociation regions (PDRs) and small molecular cores in the star-forming clouds (Poglitsch *et al.* 1995, Madden *et al.* 1997, Hunter *et al.* 2001b). When the molecular gas is detected, we usually find that stars are forming in giant molecular clouds, just like in spirals (Wilson & Reid 1991, Ohta *et al.* 1992, Wilson 1994, Taylor *et al.* 1999, Hunter *et al.* 2001b, Leroy *et al.* 2006), but the clouds in dwarfs have a different proportion of PDR and core. Bolatto *et al.* (1999) (Figure 3) suggest that the molecular core in a typical cloud, as traced by CO, shrinks and the PDR grows with decreasing metallicity. At very low metallicities a cloud could be entirely PDR. And yet, the products of the star formation process—HII region luminosity functions, cluster mass functions, and stellar initial mass functions—appear to be normal. Even the star formation efficiency within low metallicity clouds appear to be normal: In NGC 6822's Hubble V star forming region the star formation efficiency is a nominal 10% (Israel *et al.* 2003).

The dust is different.

In dwarfs the mix of PAHs, large and small grains, and silicate and carbon grains is different than in spirals (Madden 2000, Hunter *et al.* 2001b, Engelbracht *et al.* 2005, Hogg *et al.* 2005, Madden *et al.* 2006, Hunter & Kaufman 2007, Rosenberg *et al.* 2008). Low metallicity ISMs are a hostile environment for PAHs: PAHs, as well as small grains, are destroyed by intense stellar radiation fields (Bolatto *et al.* 2000, 2007; Madden 2000; Hunter *et al.* 2001b; Galliano *et al.* 2003; but see Lisenfeld *et al.* 2002), and the presence of PAH emission decreases as the hardness of the radiation field increases (Madden *et al.* 2006). PAH emission relative to small grain emission correlates with metallicity (Engelbracht *et al.* 2005, Madden *et al.* 2006, Draine *et al.* 2007; Rosenberg *et al.* 2008; but see Walter *et al.* 2007). However, there is more PAH emission relative to starlight in higher star formation systems (Hunter *et al.* 2006, Jackson *et al.* 2006). So, there are differences in the nature of the dust in dwarfs compared to spirals, but does it make any difference? There is some evidence that FIR spectral energy distributions vary among

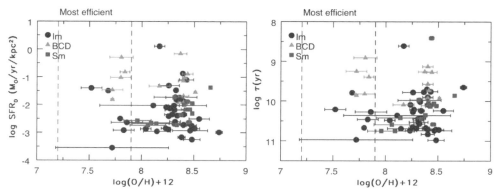

Figure 6. O/H vs. the $\mathrm{SFR}/\pi R_D^2$ (*left*) and vs. $\tau = M_{gas}/\mathrm{SFR}$ (*right*). The models of Spaans & Norman (1997) suggest that over the indicated metallicity range star formation is most efficient. The galaxies with the high star formation rates are BCDs, which may have other things going on, so we are not seeing an obvious effect. We need a better way to examine this prediction.

dwarfs, but this variation may not correlate with metallicity or the intensity of the stellar radiation field (Kiuchi *et al.* 2004).

The lower dust content should affect the star formation process.

But does it? Dust traces star formation, but to what extent does the lower dust content and different relative dust components in dwarfs *affect* the star formation process? Weak cooling of the gas should result in less efficient cloud formation because cooling takes longer (Dib & Burkert 2005). According to Schaye (2004), we can also expect a higher threshold for star formation when the metallicity is lower. So one would think that a lower dust and metal content would mean that a galaxy would have a harder time forming cold clouds. This would be a particular problem in regions of lower gas density such as the outer disk. Calura *et al.* (2008) show how dust production and destruction in dwarfs is tied to the galaxy's star formation history. But, the converse must also be true. The cooling problem would suggest that dwarfs have star formation histories in which the star formation *rate* increases at first and then levels off after the ISM reaches some metallicity. So, we expect the lower dust content to affect the star formation process and star formation histories, but the connection hasn't been shown observationally.

3. Initiating star formation

So, what do we need to form stars: high enough gas density and cool enough gas temperatures. Spaans & Norman (1997) have argued that a multi-phase ISM emerges when the metallicity reaches $0.02 Z_\odot$, and this is very important to the efficiency of star formation. Locally, the lowest metallicity star-forming galaxy has an oxygen abundance that is $1/40 \ Z_\odot$ (SBS0335-052, Izotov *et al.* 2005). Increasing metallicity enhances atomic and molecular cooling; increasing dust content absorbs part of the stellar radiation field and boosts H_2, and life is good (for star formation). Things go sour when the metallicity reaches $0.1 Z_\odot$ and ambipolar diffusion for clouds is no longer important. Then the ease of star formation should go down. (See Figure 6).

But before the first stars form and there are no metals, it would seem that cloud formation by normal means would be difficult (Li *et al.* 2005a). However, we have heard at this meeting that a single generation of very massive stars forming in the initial dark matter mini-haloes can raise the metallicity of the gas to 10^{-3}–$10^{-3.5} Z_\odot$ and that star formation enters the realm of quasi-normal above this critical metallicity. So, a freshly

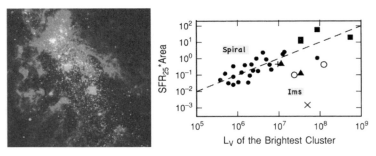

Figure 7. *Left*: Super star clusters (*blue*) and ionized gas (*red*) in the starburst dwarf NGC 1569. Images are from *HST* observations presented by Hunter *et al.* (2000). *Right*: V-band luminosity of the brightest star cluster in spiral and dIm galaxies vs. integrated galactic star formation rate. The brightest clusters in dIm galaxies are too bright for the galaxies' star formation rates. Adapted from Billett *et al.* (2002).

formed dwarf galaxy would not be starting from zero metallicity. Even so, this threshold is well below the value of $0.02Z_\odot$ where Spaans & Norman (1997) argue star formation becomes particularly easy.

I gather that some of the "small" galaxies in the early universe were dense objects—denser than dwarfs, and they became the central regions of today's spirals (Mao, Mo, & White 1998). Forming clouds in these units wouldn't be particularly difficult because the gas density was so high. But, maybe to get things going in dwarf galaxies in the early universe (but after the first stars have polluted the gas a little), you need an especially high gas density (for a dwarf). Then once the metallicity is raised to something like $0.02Z_\odot$, cloud formation becomes more efficient and can take place at lower gas densities.

How might we get the gas density up (temporarily) in (at least some) dwarfs? One way is through dwarf–dwarf interactions. Interactions, even minor ones, can have a big impact on a dwarf's ability to form stars (Dong *et al.* 2003). We see that some nearby dwarfs have formed very massive star clusters that are too big for the number of other clusters that are present (see Figure 7; Billett *et al.* 2002). These systems generally also show evidence for an external perturbation. Stellar bar structures also facilitate star formation in that barred dwarfs tend to have higher integrated star formation rates, and these can form in an interaction (Hunter & Elmegreen 2006). So, one could imagine the first dwarfs interacting and making massive star clusters or at least getting the high gas density necessary to begin forming stars.

But can a dwarf–dwarf interaction/merger make another dwarf, such as we see today? That this might be possible is suggested by the Local Group dIm WLM: conditions in this galaxy were right to form a globular cluster very early in life (Hodge *et al.* 1999) and yet today it looks like a typical dIm galaxy (Kepley *et al.* 2007). Many BCDs may be (minor?) dwarf–dwarf mergers. Once the current starburst is over in a BCD, would it be distinguishable from a regular dIm? In BCDs the HI tends to be more centrally concentrated, but Simpson & Gottesman (2000) argue that a starburst could move much of the HI back out to the edges of the optical galaxy. This scenario would produce HI ring structures, and such are seen in some dIm today (c.f. Figure 3, Simpson *et al.* 2005). We need simulations of dwarf–dwarf interactions in order to see what gas densities might be achieved and what the resulting galaxy would look like for different types of encounters.

4. Relation to star formation in early star-forming units

So, if I wanted to look at present-day star formation processes that might be applicable to star formation in freshly formed dwarf galaxies, where would I look? First, look at outer disks. Dwarfs are making star-forming clouds in outer disks where gas densities are quite low on average. Star formation out there must be pretty inefficient, but it does take place. So, outer disks are probably the place to look to understand the drivers for star formation in the smallest, lowest density objects in the early universe.

Second, look at dwarfs with a range of properties including metallicity, dust-to-gas ratios, and luminosities. We don't see any 10^{-3}–$10^{-3.5} Z_\odot$ metallicity galaxies or dust-free dwarfs today. But, nearby dwarfs do allow us to see trends with these properties down to the most extreme objects like Leo T that are forming stars nearby.

Third, look at BCDs as examples of externally induced star formation proceses that may have been especially important in the early universe.

Acknowledgements

DAH gratefully acknowledges funding from the US National Science Foundation through grant AST-0707563 and comments by B. Elmegreen, J. Cannon, and E. Brinks.

References

Andersen, R.-P. & Burkert, A. 2000, *ApJ*, 531, 296
Begum, A., Chengalur, J. N., & Bhardwaj, S. 2006a, *MNRAS*, 372, 33
Begum, A., *et al. MNRAS*, 365, 1220
Billett, O. H., Hunter, D. A. & Elmegreen, B. G. 2002, *AJ*, 123, 1454
Bolatto, A. D., Jackson, J. M., & Ingalls, J. G. 1999, *ApJ*, 513, 275
Bolatto, A. D., Jackson, J. M., Wilson, C. D., & Moriarty-Schieven, G. 2000, *ApJ*, 532, 909
Bolatto, A. D., *et al.* 2007, *ApJ*, 655, 212
Böttner, C., Klein, U., & Heithausen, A. 2003, *A&A*, 408, 493
Braun, R. 1997, *ApJ*, 484, 637
Bournaud, F., Elmegreen, B. G., & Elmegreen, D. M. 2007, *ApJ*, 670, 237
Brosch, N., Heller, A., & Almoznino, E. 1998, *ApJ*, 504, 720
Bureau, M. & Carignan, C. 2002, *AJ*, 123, 1316
Calura, F., Pipino, A., & Matteucci, F. 2008, *A&A*, 479, 669
Campos-Aguilar, A., & Moles, M. 1991, *A&A*, 241, 358
Cannon, J. M., *et al.* 2005, *ApJ*, 630, L37
Cannon, J. M., *et al.* 2006a, *ApJ*, 647, 293
Cannon, J. M., *et al.* 2006b, *ApJ*, 652, 1170
Chamaraux, P. 1977, *A&A*, 60, 67
Comins, N. E. 1983, *ApJ*, 266, 543
de Blok, W. J. G., & Walter, F., 2006, *AJ*, 131, 363
Dib, S. & Burkert, A. 2005, *ApJ*, 630, 238
Dolphin, A. E. & Hunter, D. A. 1998, *AJ*, 116, 1275
Dong, S., Lin, D. N. C., & Murray, S. D. 2003, *ApJ*, 596, 930
Dopita, M. A., Mathewson, D. S., & Ford, V. L. 1985, *ApJ*, 297, 599
Draine, B. T., *et al.* 2007, *ApJ*, 663, 866
Elmegreen, B. G. & Hunter, D. A. 2006, *ApJ*, 636, 712
Elmegreen, B. G., Kim, S., & Staveley-Smith, L. 2001, *ApJ*, 548, 749
Engelbracht, C. W., *et al.* 2005, *ApJ*, 628, 29
Ferrara, A. & Tolstoy, E. 2000, *MNRAS*, 313, 291
Galliano, F., *et al.* 2003, *A&A*, 407, 159
Gerola, H., Seiden, P. E., & Schulman, L. S. 1980, *ApJ*, 242, 517
Hinz, J., Misselt, K., Rieke, G., Smith, P. Blaylock, M., & Gordon, K. 2006, *ApJ*, 651, 874
Hodge, P. 1980, *ApJ*, 156, 847
Hodge, P. 1969, *ApJ*, 241, 125
Hodge, P., Dolphin, A., Smith, T., & Mateo, M. 1999, *ApJ*, 521, 577
Hogg, D. W., *et al.* 2005, *ApJ*, 624, 162

Hoffman, G. L., Salpeter, E. E., & Carle, N. J. 2001, *AJ*, 122, 2428

Hunter, D. A. & Elmegreen, B. G. 2004, *AJ*, 128, 2170

Hunter, D. A. & Elmegreen, B. G. 2006, *ApJS*, 162, 49

Hunter, D. A., Elmegreen, B. G., & Baker, A. L. 1998, *ApJ*, 493, 595

Hunter, D. A., Elmegreen, B. G., Dupuy, T. G., & Mortonson, M. 2003, *AJ*, 126, 1836

Hunter, D. A., Elmegreen, B. G., & Martin, E. 2006, *AJ*, 132, 801

Hunter, D. A., Elmegreen, B. G., & van Woerden, H. 2001a, *ApJ*, 556, 773

Hunter, D. A. & Gallagher, J. S. 1985, *ApJS*, 58, 533

Hunter, D. A. & Gallagher, J. S. 1986, *PASP*, 98, 5

Hunter, D. A., Gallagher, J. S., & Rautenkranz, D. 1982, *ApJS*, 49, 53

Hunter, D. A., Gallagher, J. S., Rice, W. L., & Gillett, F. C. 1989, *ApJ*, 336, 152

Hunter, D. A. & Kaufman, M. 2007, *AJ*, 134, 721

Hunter, D. A., O'Connell, R., Gallagher, J., & Smecker-Hane, T. 2000, *AJ*, 120, 2383

Hunter, D. A. & Plummer, J. D. 1996, *ApJ*, 462 732

Hunter, D. A., *et al.* 2001, *ApJ*, 553, 121

Irwin, M. J., *et al.* 2007, *ApJ*, 656, 13

Israel, F. P., Baas, F., Rudy, R. J., Skillman, E. D., & Woodward, C. E. 2003, *A&A*, 397, 87

Izotov, Y. I., Thuan, T. X., & Guseva, N. G. 2005, *ApJ*, 632, 210

Jackson, D. C., *et al.* 2006, *ApJ*, 646, 192

Kepley, A., Wilcots, E., Hunter, D., & Nordgren, T. 2007, *AJ*, 133, 2242

Kim, S., Dopita, M. A., Staveley-Smith, L., & Bessell, M. S. 1999, *AJ*, 118, 2797

Kingsburgh, R. L. & McCall, M. L.1998, *AJ*, 116, 2246

Kiuchi, G., Ohta, K., Sawicki, M., & Allen, M. 2004, *AJ*, 128, 2743

Komiyama, Y., *et al.* 2003, *ApJ*, 590, 17

Kregel, M. & van der Kruit, P. C. 2004, *MNRAS*, 355, 143

Krumholz, M. R. & McKee, C. F. 2005, *ApJ*, 630, 250

Lee, J. C., Kennicutt, R. C., Funes, J. G., S. J., Sakai, S., & Akiyama, S. 2007, *ApJ*, 671, L113

Leroy, A., Bolatto, A., Walter, F., & Blitz, L. 2006, *ApJ*, 643, 825

Leroy, A., Bolatto, A., Stanimirovic, S., Mizuno, N., Israel, F., & Bot, C. 2007, *ApJ*, 658, 1027

Leroy, A., *et al.* 2008, *AJ*, submitted

Li, Y., Mac Low, M.-M., & Klessen, R. S. 2005a, *ApJ*, 620, L19

Li, Y., Mac Low, M.-M., & Klessen, R. S. 2005b, *ApJ*, 626, 823

Lisenfeld, U., Israel, F. P., Stil, J. M., & Sievers, A. 2002, *A&A*, 382, 860

Lisenfeld, U. & Ferrara, A. 1998, *ApJ*, 496, 145

Mac Low, M.-M. & Ferrara, A. 1999, *ApJ*, 513, 142

Madden, S. C. 2000, *NewAR*, 44, 249

Madden, S. C., Galliano, F., Jones, A. P., & Sauvage, M. 2006, *A&A*, 446, 877

Madden, S. C., Poglitsch, A., Geis, N., Stacey, G. J., & Townes, C. H. 1997, *ApJ*, 483, 200

Mao, S., Mo, H. J., & White, S. D. M. 1998, *MNRAS*, 297, L71

Marconi, G., Tosi, M., Greggio, L., & Focardi, P. 1995, *AJ*, 109, 173

Martin, C. L. 1997, *ApJ*, 491, 561

Martin, C. L. 1998, *ApJ*, 506, 222

Martin, C. L. 1999, *ApJ*, 513, 156

Méndez, D. I. & Esteban, C. 2000, *A&A*, 359, 493

Meurer, G. R., Carignan, C., Beaulieu, S. F., & Freeman, K. C. 1996, *AJ*, 111, 1551

Meurer, G. R., Staveley-Smith, L, & Killeen, N. E. B.1998, *MNRAS*, 300, 705

Mori, M., Yoshii, Yl, Tsujimoto, T., & Nomoto, K. 1997, *ApJ*, 478, 21

Mühle, S., Klein, U., WIlcots, E. M., & Hüttemeister, S. 2005, *AJ*, 130, 524

Noguchi, M. 1988, *A&A*, 201, 37

Oey, M. S., Clarke, C. J., & Massey, P. 2001, in: De Boer, K. S., Dettmar, R.-J., & Klein, U. (eds.), *The Magellanic Clouds and other dwarf galaxies* (Aachen: Shaker-Verlag), p.181

Ohta, K., Sasaki, M., Yamada, T., Saito, M., & Nakai, N. 1992, *PASJ*, 44, 585

Öpik, E. J. 1953, *IrAJ*, 2, 219

Ott, J., Fabian, W., & Brinks, E. 2005, *MNRAS*, 358, 1453

Parodi, B. R. & Binggeli, B. 2003, *A&A*, 398, 501

Payne-Gaposhkin, C. 1974, *Smithsonian Contrib. Astrophys.*, No. 16

Pérez, I. 2004, *A&A*, 427, L17

Phillipps, S., Edmunds, M. G., & Davies, J. I. 1990, *MNRAS*, 244, 168

Piontek, R. A. & Ostriker, E. C. 2004, *ApJ*, 601, 905

Piontek, R. A. & Ostriker, E. C. 2005, *ApJ*, 629, 849

Poglitsch, A., *et al.* 1995, *ApJ*, 454, 293

Pohlen, M., Dettmar, R.-J., Lütticke, R., & Aronica, G. 2002, *A&A*, 392, 807

Puche, D., Westpfahl, D., Brinks, E., & Roy, J.-R. 1992, *AJ*, 103, 1841

Quirk, W. J. 1972, *ApJ*, 176, L9

Rafikov, R. R. 2001, *MNRAS*, 323, 445

Rhode, K. L., Salzer, J. J., Westpfahl, D. J., & Radice, L. A. 1999, *AJ*, 118, 323

Rosenberg, J. L., *et al.* 2008, *ApJ*, 674, 814

Roškar, R., *et al.* 2008, *ApJ*, 675, 65

Roye, E. W. & Hunter, D. A. 2000, *AJ*, 119, 1145

Ryan-Weber, E. V., *et al.* 2008, *MNRAS*, 384, 535

Safronov, V. S. 1960, *Ann d'Ap*, 23, 979

Sánchez-Salcedo, F. J. 2001, *ApJ*, 563, 867

Scalo, J. & Chappell, D. 1999, *MNRAS*, 310, 1

Schaye, J. 2004, *ApJ*, 609, 667

Schombert, J. M., McGaugh, S. S., & Eder, J. A. 2001, *AJ*, 121, 2420

Silk, J. 1997, *Apj*, 481, 703

Simpson, C. E., & Gottesman, S. T. 2000, *AJ*, 120, 2975

Simpson, C. E., Hunter, D. A., & Knezek, P. M. 2005, *AJ*, 129, 160

Spaans, M. & Norman, C. A. 1997, *ApJ*, 483, 87

Stanimirovic, S., *et al.* 1999, *MNRAS*, 302, 417

Stewart, S. G., *et al.* 2000, *ApJ*, 529, 201

Stil, J. M. & Israel, F. P. 1998, *A&A*, 337, 64

Stinson, G. S., Dalcanton, J. J., Quinn, T., Kaufmann, T., & Wadsley, J. 2007, *ApJ*, 667, 170

Struck, C. & Smith, D. C. 1999, *ApJ*, 527, 673

Taylor, C. L. 1997, *ApJ*, 480, 524

Taylor, C. L., Brinks, E., Pogge, R. W., & Skillman, E. D. 1994, *AJ*, 107, 97

Taylor, C. L., Hüttemeister, S., Klein, U., & Greve, A. 1999, *A&A*, 349, 424

Taylor, E. N. & Webster, R. L. 2005, *ApJ*, 634, 1067

Telles, E. & Maddox, S. 2000, *MNRAS*, 311, 307

Telles, E. & Terlevich, R. 1995, *MNRAS*, 275, 1

Toomre, A. 1964, *ApJ*, 139, 1217

van der Hulst, J. M., *et al.* 1993, *AJ*, 106, 548

van Dyk, S. D., Puche, D., & Wong, T. 1998, *AJ*, 116, 2341

van Zee, L., Haynes, M. P., Salzer, J. J., & Broeils, A. H. 1997, *AJ*, 113, 1618

van Zee, L., Salzer, J. J., & Skillman, E. D. 2001, *AJ*, 122, 121

Wada, K., Spaans, M., & Kim, S. 2000, *ApJ*, 540, 797

Walter, F. & Brinks, E. 1999, *AJ*, 118, 273

Walter, F., *et al.* 2007, *ApJ*, 661, 102

Warren, B. E., Jerjen, H., & Koribalski, B. S. 2007, *AJ*, 134, 1849

Westpfahl, DJ., Coleman, P. H., Alexander, J., & Tongue, T. 1999, *AJ*, 117, 868

Willett, K. W., Elmegreen, B. G., & Hunter, D. A. 2005, *AJ*, 129, 2186

Wilson, C. D. 1994, *ApJ*, 434, 11

Wilson, C. D. & Reid, I. 1991, *ApJ*, 366, 11

Young, L. M. & Lo, K. Y. 1996, *ApJ*, 462, 203

Young, L. M., van Zee, L., Lo, K. Y., Dohm-Palmer, R. C., & Beierle, M. 2003, *ApJ*, 592, 111

Youngblood, A. J. & Hunter, D. A. 1999, *ApJ*, 519, 55

Low-Metallicity Star Formation:
From the First Stars to Dwarf Galaxies
Proceedings IAU Symposium No. 255, 2008
L.K. Hunt, S. Madden & R. Schneider, eds.

Interstellar Chemistry: Radiation, Dust and Metals

Marco Spaans[1]

[1]Kapteyn Astronomical Institute, University of Groningen,
P.O. Box 800, 9700 AV, Groningen, the Netherlands
email: spaans@astro.rug.nl

Abstract. An overview is given of the chemical processes that occur in primordial systems under the influence of radiation, metal abundances and dust surface reactions. It is found that radiative feedback effects differ for UV and X-ray photons at any metallicity, with molecules surviving quite well under irradiation by X-rays. Starburst and AGN will therefore enjoy quite different cooling abilities for their dense molecular gas. The presence of a cool molecular phase is strongly dependent on metallicity. Strong irradiation by cosmic rays ($>200\times$ the Milky Way value) forces a large fraction of the CO gas into neutral carbon. Dust is important for H_2 and HD formation, already at metallicities of $10^{-4} - 10^{-3}$ solar, for electron abundances below 10^{-3}.

Keywords. astrochemistry, ISM: molecules, dust, early universe

1. Introduction

In the study of primordial chemistry, and subsequently the formation of the first stars, it is crucial to understand the ability of interstellar gas to cool through atomic and molecular emissions and to collapse. Furthermore, atomic and molecular species can be used to probe the ambient conditions, such as density and temperature, under which stars form. The basic questions posed by this contribution are: What sets the abundances of atoms and molecules that cool the gas? What is the role of radiation, metallicity and dust in molecule formation?

A number of different chemical processes are relevant to this effect:

Ion-molecule reactions: $A^+ + BC \rightarrow AB^+ + C$;
Neutral-neutral reactions: $A + BC \rightarrow AB + C$;
Dissociative recombination: $AB^+ + e^- \rightarrow A + B$;
Radiative recombination: $A^+ + e^- \rightarrow A + h\nu$;
Radiative association: $A + B \rightarrow AB + h\nu$;
Ionization: $A + CR/UV/X\text{-ray} \rightarrow A^+ + e^-$;
Dissociation: $AB + UV \rightarrow A + B$;
Charge transfer: $A^+ + B \rightarrow A + B+$;
Grain surface reactions: $Grain + A + B \rightarrow Grain + AB$.

In any chemical network, the above reactions play an important role. For example, the charge transfer between H^+ and O, followed by reactions with H_2 to H_3O^+, and dissociative recombination with e^-, leads to species like OH and H_2O (following certain branching ratios). Similarly complex routes exist for CO. In any case, many species are typically joined through different chemical routes. Thus, it is not trivial to construct concise chemical networks if one wants to include important molecules such as CO and

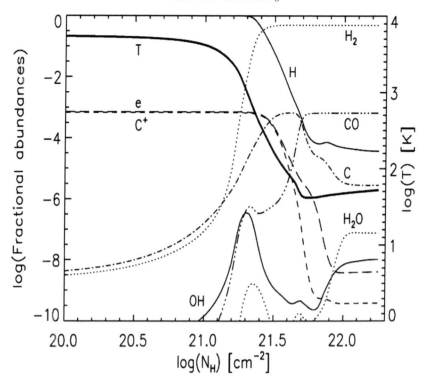

Figure 1. PDR model typical of a modest starburst inside a dwarf galaxy, at a density of 10^5 cm^{-3}.

H_2O†. Of course, in the limit of low metallicity, chemistry simplifies. Basically, no metals implies no molecules except for H_2 and HD (and a few minor species). Still, even small amounts of metals and dust ($\sim 10^{-4}$ solar) can be crucial to the efficient formation of species such as H_2, HD, CO, H_2O and many others, which is the purpose of this contribution.

2. Radiation

The impact of radiation is denoted by UV and X-ray dominated regions (PDRs and XDR, respectively). These are regions where photons dominate the thermal and chemical balance of the gas. Examples are O & B stars (HII regions), active galactic nuclei (AGN), and T Tauri stars. In PDRs, the radiation field comprises photons with energies $6 < E < 13.6$ eV. Heating is provided by photo-electric emission from dust grains and cosmic rays, while cooling proceeds through fine-structure emission lines like [OI] 63, 145 μm and [CII] 158 μm as well as emissions by H_2, CO and H_2O rotational and vibrational lines. As a rule of thumb, a 10 eV photon penetrates about 1/2 mag of dust.

In XDRs photon energies $E > 0.3$ keV are considered. Heating is provided by X-ray photo-ionizations that lead to fast electrons and Coulomb heating as well as H

† It is important to realize that water can be quite an important heating agent in the presence of a warm infrared background, like $T > 50$ K dust or a $z > 15$ CMB (Spaans & Silk 2000).

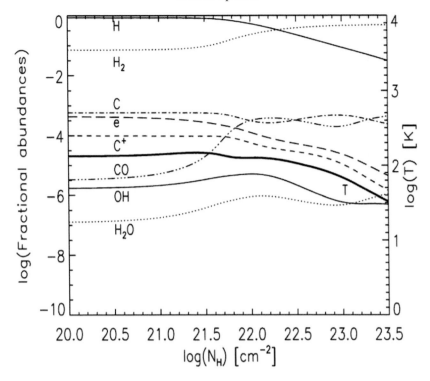

Figure 2. XDR model typical of gas at a few hudred pc from a Seyfert nucleus, at a density of 10^5 cm^{-3}.

and H_2 vibrational excitation followed by UV emission (Ly $alpha$, Lyman-Werner); H_3^+ recombination heating can be important as well. Cooling is provided by [FeII] 1.26, 1.64 μm; [OI] 63 μm; [CII] 158 and [SiII] 35 μm emission lines as well as thermal H_2 rotational and vibrational emissions and gas-dust cooling. Typically, a 1 keV photon penetrates 10^{22} cm^{-2}, because cross sections scale as $E^{-(2-3)}$. Figures 1 and 2 show typical examples of a PDR and XDR. Note the fact that molecules have an easier time surviving in an XDR, for the same impinging flux by energy (Meijerink & Spaans 2005), because molecular photo-dissociation cross sections peak in the UV. Furthermore, the heating efficiciency in XDRs can be 10-50%, while it is at most 1% in PDRs.

In Figure 3 it is shown that neutral carbon is an excellent mass tracer (as good as CO) under cosmic ray irradiations that exceed Milky Way values by more than factor of 100 (Meijerink *et al.* 2007). Also note that the collisional coupling between warm gas on cool dust grains can dominate the gas cooling for modest metallicities in PDRs and XDRs.

3. Metals

As the metallicity decreases, one finds smaller molecular clouds and the atomic cooling dominates by mass (Bolatto *et al.* 1999, Roellig *et al.* 2006). The occurrence of a multi-phase medium (Wolfire *et al.* 1995) depends strongly on metallicity and pressure

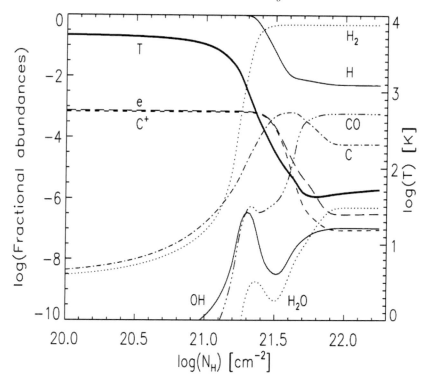

Figure 3. PDR model typical of gas irradiated by cosmic rays from a supernova rate of 2 per year, at a density of 10^5 cm^{-3}.

(Spaans & Norman 1997). Figure 4 shows that only a single interstellar phase occurs for metallicities below a percent of solar. At the same time, there is a region between 1 and 10 % of solar metallcity where star formation is most efficient. The reason is that these modest metallicities allow for efficient cooling without any line trapping (optical depth effects).

At metallicities well below 1% of solar, cooling is dominated by H_2 and HD emissions, which allow cooling down to $\sim$100 K only. This is illllustrated in Figure 5, where the strengths of the first two pure rotational H_2 lines are compared to the CO line spectral energy distribution, for a system with a 10^5 $M_\odot$ black hole accreting at Eddington. One can clearly see that both low metallicities and strong irradiation favor H_2 as the main coolant.

This is pertinent to the study of pop III.1 and III.2 star formation (e.g., Abel *et al.* 2002, Bromm *et al.* 2001; and contributions by Schneider, Ferrara and Tan in this volume). The relative contributions of molecular cooling depicted in Figure 5 show that the upcoming ALMA telescope will be able to see primordial systems that are growing a massive black hole, at redshifts of $z = 10 - 20$ (Spaans & Meijerink 2008).

Metallicity-dependent cooling is quite important for the collapse of gas clouds and the properties of the initial mass function. In particular, LTE effects and line trapping impact the effective equation of state (the thermodynamics) of the gas. This is further

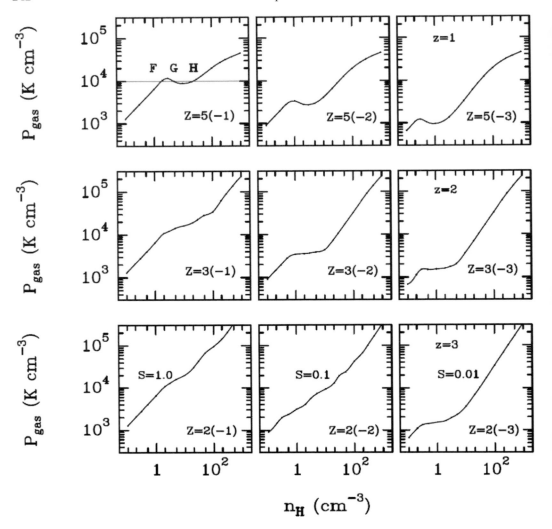

Figure 4. Phase diagrams for interstellar gas as a function of metallicity and background star formation rate.

discussed in detail, using the FLASH code, in the contribution of Hocuk (this volume). He finds that the level of fragmentation is a strong function of rotational energy and metallicity.

Furthermore, metals and molecules other than H_2 and HD impact the formation of structure in collapsing primordial systems. Detailed studies, using the Enzo code, that include a complete gas-phase chemistry and formation of H_2 and HD on dust grains, is presented by Aykutalp (this volume). She finds that pre-enrichment of young galaxies strongly lowers the Jeans mass of the gas clouds they contain.

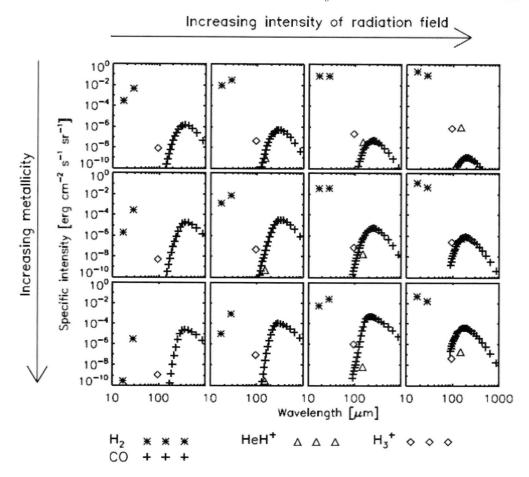

Figure 5. Relative contributions from CO (metal-rich) and H$_2$ (metal-poor) gas that is irradi-ated by a hard spectrum of primordial galaxy; as a function of metallicity (10^{-3} to 10^{-1}/top to bottom) and impinging flux (0.1 to 100 erg cm^{-2} s^{-1}/left to right). All panels are for a density of 10^5 cm^{-3}.

4. Dust

The formation of H$_2$ and HD on dust grains depends strongly on their surface prop-erties, as indicated in Figure 6. Hydrogen atoms can be weakly bound through van der Waals forces (physi-sorption) or strongly bound through covalent bonds (chemi-sorption). The advantage of the latter bond is that it allows atoms to bind to the surface even for dust temperatures well in excess of 100 K. The hydrogen atoms either thermally hop at high dust temperatures or tunnel at low ($\sim$10 K) temperatures. A comparison with the gas phase formation of H$_2$ through the H$^-$ route (see Figure 7) indicates that dust processes dominate H$_2$ formation for metallicities $>10^{-3.5}$ solar and electron abundances below 10^{-3} (Cazaux & Spaans 2004).

The formation of HD benefits above 10^{-3} solar as well, as long as the gas den-sity is above $10^{4.5}$ cm^{-3} and the electron abundance below 10^{-3} (Cazaux & Spaans

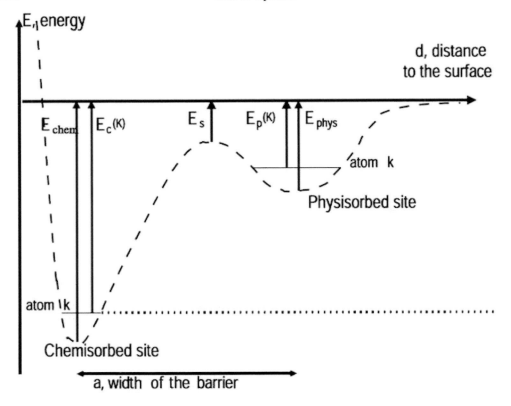

Figure 6. Typical grain surface characterization (Cazaux & Tielens 2004).

2008, in preparation). In this, the deuterium atom is more massive that atomic hydrogen, i.e., it is less mobile and more strongly bound to the dust grain (up to higher temperatures).

5. Conclusions

Radiative feedback effects differ for UV and X-ray photons at any metallicity, with molecules surviving quite well under irradiation by X-rays. Starburst and AGN will therefore enjoy quite different cooling abilities for their dense molecular gas. The presence of a cool molecular phase is strongly dependent on metallicity. Strong irradiation by cosmic rays ($>200\times$ the Milky Way value) forces a large fraction of the CO gas into neutral carbon. Dust is important for H_2 and HD formation, already at metallicities of $10^{-4} - 10^{-3}$ solar.

Finally, one should always solve the equations of statistical equilibrium to distinguish properly between the excitation, radiation and kinetic temperature of a system. I.e., the thermodynamic floor set by the CMB is only a hard one if the density is high enough (larger than the critical density of a particular transition) to drive collisional de-excitation.

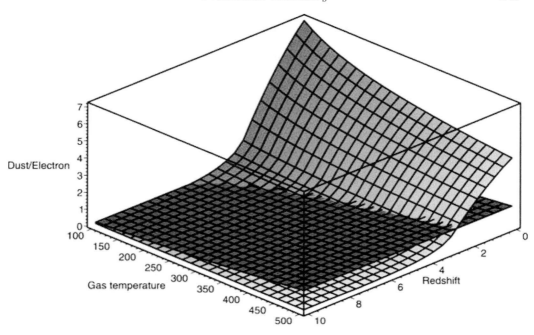

Figure 7. Comparison between the gas phase and dust surface formation routes for H_2.

References

Abel, T., Bryan, G.L., & Norman, M.L., 2002, Science, 295, 93

Bolatto, A.D., Jackson, J.M. & Ingalls, J.G., 1999, ApJ, 513, 275

Bromm, V., Ferrara, A., Coppi, P.S., & Larson, R.B., 2001, MNRAS, 328, 969

Cazaux, S. & Spaans, M., 2004, ApJ, 611, 40

Cazaux, S. & Tielens, A.G.G.M., 2004, ApJ, 604, 222

Meijerink, R. & Spaans, M., 2005, A&A, 436, 397

Meijerink, R., Spaans, M. & Israel, F.P., 2007, A&A, 461, 793

Roellig, M., Ossenkopf, V., Jeyakumar, S., Stutzki, J. & Sternberg, A., 2006, A&A, 451, 917

Spaans, M. & Meijerink, R., 2008, ApJ, 678, L5

Spaans, M. & Silk, J., 2000, ApJ, 538, 115

Spaans, M. & Norman, C.A., 1997, ApJ, 483, 87

Wolfire, M.G., Hollenbach, D., McKee, C.F., Tielens, A.G.G.M. & Bakes, E.L.O., ApJ, 443, 673

Low-Metallicity Star Formation:
From the First Stars to Dwarf Galaxies
Proceedings IAU Symposium No. 255, 2008
L.K. Hunt, S. Madden & R. Schneider, eds.

Dust Evolution from Nearby Galaxies: Bridging the Gap Between Local Universe and Primordial Systems

Frédéric Galliano[1]

[1]Department of Astronomy, University of Maryland, College Park, MD 20742, USA
email: galliano@astro.umd.edu

Abstract. This paper presents the results of a study aimed at understanding the evolution of the dust properties, as a function of both the environmental conditions and the metal enrichment of the system. I first review the peculiar dust properties of dwarf galaxies, and discuss attempts to understand their origin. Then, I discuss the evolution of the PAH and dust abundances, constrained by the UV-to-radio SED of nearby galaxies, comparing the properties of low-metallicity environments and more evolved systems. I discuss the long term evolution of dust in galaxies, comparing the grain production by various stellar progenitors to their destruction by SN blast waves and in H II regions. Finally, I will show how these models explain the paucity of PAHs in low-metallicity environments.

Keywords. stars: AGB, supernovae: general, ISM: abundances, ISM: dust, ISM: extinction, galaxies: dwarf, galaxies: evolution, galaxies: high-redshift, galaxies: starburst, infrared: galaxies

1. Introduction

Understanding the variations of dust properties with the age and physical conditions of a system has become crucial for comprehending galaxy evolution. Indeed, the wide database of infrared observations collected by the *Infrared Space Observatory* and the *Spitzer Space Telescope*, with an unprecedented quality, provides valuable observational constraints of the elementary evolutionary processes playing a role in the interstellar medium (ISM). How do dust abundances vary with metallicity? What processes affect the composition and size distribution of dust grains? How is the dust distributed throughout the various phases of the ISM? Answers to these fundamental questions are necessary to address the following higher level outstanding issues. How is the infrared spectral energy distribution (SED) of a whole star forming galaxy related to its physical conditions and star formation history? What were the ISM properties of the first galaxies formed?

These questions can be addressed from different points of view. The local approach consists in observing an individual region or object (SN II, AGB star, shock, PDR, etc.), in order to derive detailed information on a few particular processes. The difficulty of this approach is that such regions, accessible to current observatories, are not very numerous. On the opposite, the global approach – that I am going to discuss in this paper – consists in observing an entire galaxy, in order to derive average properties. The observations are easier in this case, although the difficulty of this approach is that all the individual processes are mixed, and not easily separable.

Nearby dwarf galaxies constitute extremely useful laboratories to address these questions, since they can be considered as snapshots of galaxy evolution at early epochs of their aging, due to their low elemental enrichment. In that sense, dwarf galaxies are also important objects to understand protogalaxies. Although dwarf galaxies are not rigorously identical to primordial systems, their study provides insights on the interplay between star formation and the ISM in low-metallicity environments. Moreover, the ISM of dwarf galaxies experiencing massive star formation is subjected to extreme conditions of radiation and numerous supernovae (SN) blast waves, providing unique constraints on the impact of these processes on the ISM.

This paper summarizes an original study aimed at understanding the origin of the weakness of the mid-IR bands in low-metallicity environments.

2. The Peculiar Infrared Properties of Low-Metallicity Environments

To begin with, dwarf galaxies do not constitute an homogeneous population of objects – they are a category by default. This category roughly encompasses systems with metal abundances lower than solar $(12 + \log(O/H)_\odot \simeq 8.8)$, as a consequence of the size-metallicity relation (e.g. Kunth & Östlin 2000). Therefore, it is very difficult to characterize them by an ensemble of well-defined properties. Alternatively, we are compelled to describe how their properties differ from normal metallicity systems and to derive trends between these properties.

The first infrared SEDs of dwarf galaxies were provided by *IRAS* (Hunter *et al.* 1989), and showed what optical studies suggested: a lower dust-to-gas mass ratio compared to the Galaxy. It was not until the times of the *Infrared Space Observatory* that the mid-IR properties and their spatial distribution could be studied in details. It appeared that starbursting dwarf galaxies presented similarities with giant Galactic H II regions: lack of polycyclic aromatic hydrocarbons (PAH) and steep rising mid-IR continuum (Thuan *et al.* 1999; Madden *et al.* 2006). Self-consistent modeling of the UV-to-mm SED of nearby blue compact galaxies by Galliano *et al.* (2003, 2005) confirmed these views and showed that the difference in observed properties could be explained by a systematic difference in intrinsic grain properties. In this model, the dust grains are eroded and fragmented by the numerous SN shock waves sweeping the ISM, accounting for the peculiar shape of their Magellanic-like extinction curves and their IR SED. In addition, this study reported the presence of a submillimetre emission excess which can be attributed to very cold, shielded dust $(T \lesssim 10 \text{ K})$. The significant amount of dust potentially hidden in these dense clumps, challenges our comprehension of dust evolution and ISM structure.

The exceptional sensitivity of *Spitzer* brought to this field tremendous advances. The spatial variations of the IR properties of several low-metallicity systems (like NGC 6822 Cannon *et al.* 2006) as well as the Magellanic clouds (Meixner *et al.* 2006; Bolatto *et al.* 2007) have been studied. *Spitzer* also allowed better statistics by reaching fainter sources (Engelbracht *et al.* 2005; Wu *et al.* 2006). In particular, it became clear that there was a trend between the strength of the aromatic features and the metallicity.

The believed carriers of these aromatic features, the PAHs, are known to be destroyed in H II regions by the hard UV radiation (e.g. Cesarsky *et al.* 1996). Madden *et al.* (2006) proposed that, due to the lower opacity of the ISM in dwarf galaxies – which is a consequence of the lower metal abundance – the hard UV photons prenetrate deeply and destroy the PAHs on larger scale, than in normal metallicity systems. This interpretation is based on the correlation of left panel of Fig. 1. On the other hand, O'Halloran *et al.* (2006) proposed that the PAHs could be widely destroyed by the numerous shock waves that sweep these galaxies (right panel of Fig. 1). However, observations of individual supernovae indicate that not only the PAHs, but also the carriers of the underlying continuum are destroyed by the shock (e.g. Reach *et al.* 2002).

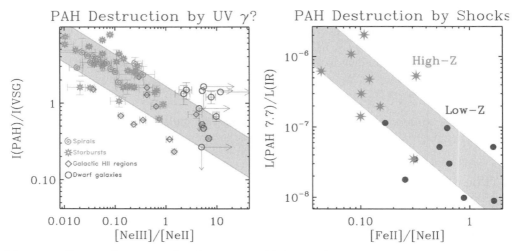

Figure 1. Two explanantions for the paucity of PAHs in low-metallicity environments. **Left:** correlation of the PAH strength with the $[\text{Ne\,III}]_{15.56\mu m}/[\text{Ne\,II}]_{12.81\mu m}$ ratio (Madden *et al.* 2006). The PAH intensity is normalised to the intensty of the very small grain (VSG) continuum. The $[\text{Ne\,III}]/[\text{Ne\,II}]$ ratio is very sensitive to hard UV photons, and provides therefore a tracer of ionizing stellar populations. **Right:** correlation of the PAH strength with the $[\text{Fe\,II}]_{25.99\mu m}/[\text{Ne\,II}]_{12.81\mu m}$ ratio (O'Halloran *et al.* 2006). The iron being mostly depleted into dust, its observation in the gas phase is presumed by the authors to be an indication of its sputtering by SN\,II shock blasts. Thus, O'Halloran *et al.* (2006) consider the $[\text{Fe\,II}]/[\text{Ne\,II}]$ ratio to be a shock tracer.

These explanations are based on observed feature intensities, which depend both on the PAH abundance and on their irradiation. To understand the origin of these relations and interpret them correctly, the derivation of the actual PAH abundance is needed.

3. Global SED Modeling of Nearby Galaxies

Derivation of the PAH abundances in galaxies from their observed mid-IR spectral features is complicated by the nature of their emission mechanism. PAHs are stochastically-heated by the interstellar radiation field (ISRF) and consequently only a fraction of their population is copiously emitting IR radiation at any given time. Correcting for the mass of PAHs too cold to radiate at mid-IR wavelengths requires detailed modeling of their stochastic heating process, and therefore detailed knowledge of the ISRF. The situation is further complicated by the fact that a significant fraction of the mid-IR emission originates from hot dust in H\,II regions radiating at the equilibrium dust temperature. Deriving the abundance of PAHs and other dust species in a galaxy requires therefore careful modeling of the stellar population that produces the ISRF that heats the dust in the diffuse ISM and the ionizing radiation that heats the dust in H\,II regions.

We constructed a self-consistent model for the evolution of the stellar populations and the composition of the ISM in a sample of 35 nearby galaxies, with metallicities ranging from 1/50 to $\simeq 2$ $Z_\odot$. The star formation history comprises of two distinct components: (1) a global continuous mode of star formation which is used to calculate the chemical evolution with a closed-box model, and the evolution of the stellar radiation using the PÉGASE stellar population code (Fioc & Rocca-Volmerange 1997); and (2) an "instantaneous burst" of star formation (age $\lesssim 10$ Myr). This short-lived burst does not contribute significantly to the metallicity of the gas, and is tailored to fit the observed UV, optical, radio, and the IR emission from dust in H\,II regions. The continuous star formation component provides a self-consistent picture of the evolution of stellar colors and metallicity as a function of age in all the galaxies in the sample. The relative contribution of these two components, as well as the age of the galaxy are fully constrained by the UV-to-radio continuum emission. Details of the models are presented in Galliano *et al.* (2008a).

The following describes the steps used to decompose the observed galactic SEDs into the various stellar and ISM emission components (see Fig. 2).

(*a*) We first decompose the radio continuum into free-free and synchrotron emission in order to constrain the emission measure from the galaxy.

(*b*) The resulting free-free continuum, together with observations of the mid-IR continuum between ≃5 and ≃60 μm are used to constrain the gas density and reradiated energy from the H II regions. We assume that any existing PAHs are destroyed inside the H II regions, and use a simple radiative transfer model (Galliano *et al.* 2008a) to calculate the absorbed radiation that is emitted as free-free emission from the gas, and thermal IR emission from the dust in the H II region.

(*c*) Optical and near-IR broadband emission are used to constrain the radiation escaping from the H II regions and that from the non-ionizing stars. They comprise the ISRF that heats the dust in PDRs.

(*d*) The observed far-IR/submm SED constrains the dust emission from the PDRs. The PAH-to-dust mass fraction is constrained by the detailed fit of the features seen on the mid-IR spectrum.

(*e*) Globally, the stellar luminosity absorbed by the gas and dust phases of the ISM is equal to the total reradiated and escaping power from the galaxy.

4. Dust Evolution in Galaxies

The model described in Sect. 3, when applied to the observations of each galaxy in our sample, provides the variation of PAH and dust abundances in galaxies, as a function of metallicity. Similar trends of dust abundances with metallicity have also been presented by Draine *et al.* (2007) and Engelbracht *et al.* (2008). However, Draine *et al.* (2007) used a much simpler SED without constraints from the mid-IR spectroscopy. Engelbracht *et al.* (2008) did not estimate the PAH masses. None of them provide a consistent interpretation of these trends.

To interpret these trends, we have developed the following dust evolution model (Galliano *et al.* 2008a, Fig. 3).

(*a*) We consider a closed box model. The star formation rate (SFR) follows the Kennicutt (1998) law, relating the surface densities of the gas to the surface density of the SFR. We adopt a Salpeter IMF.

(*b*) At any time, we follow the evolution of stars of different masses, until they release their elements into the ISM. Massive stars, evolving into supernovae have a short lifetime of a few Myr and therefore inject their elements promptly, while low-mass stars evolve to their post-AGB phase after ≃400 Myr.

(*c*) For each stellar progenitor, the elements are combined to form various types of grains (carbon, silicates, titanium oxides, etc.).

(*d*) A fraction of the elements locked-up in the grains is returned to the gas phase, when sputtered by a shock wave. We parametrize the dust destruction efficiency by SN II blast waves by considering that all the dust is destroyed around a supernovae, within a volume determined by the mass of gas $\langle m_{\rm ISM} \rangle$ it encloses. Therefore, the dust destruction is proportional to the supernova rate. We vary $\langle m_{\rm ISM} \rangle$ between 0 $M_\odot$ (no destruction)

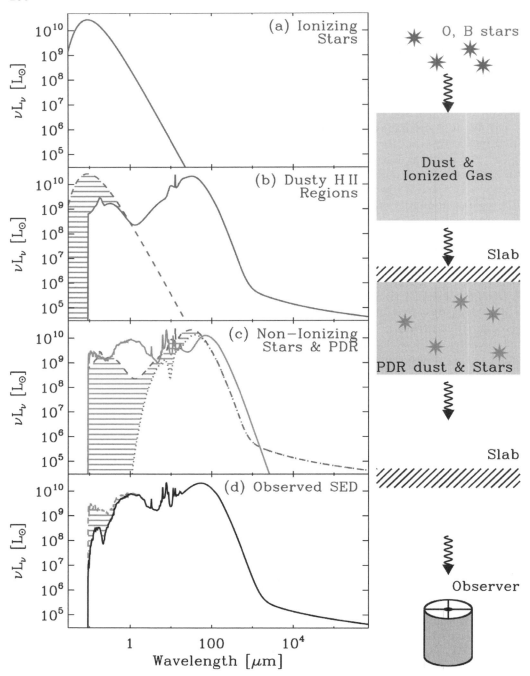

Figure 2. Schematic illustration of the of the model used to decompose the observed galactic SED into the various stellar and ISM emission components (Galliano *et al.* 2008a). **Panel (a)**: SED of the ionizing stars (purple curve). **Panel (b)**: IR spectrum of the dust in the H II regions (green curve). **Panel (c)**: Stellar emission and emission from dust residing in the neutral ISM (red curve). **Panel (d)**: The observed SED after passing through a slab of internal extinction (black curve). In each figure, the hatched region represents the emission from the previous panel that is absorbed by the relevant phase depicted in the panel.

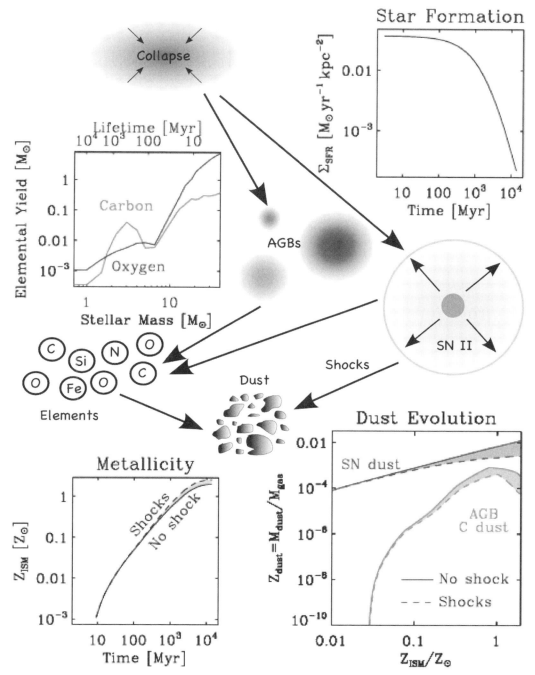

Figure 3. Flow chart of the dust evolution model. **(1)** We start from the collapse of a cloud, following the star formation rate given in the top right panel. **(2)** Stars of different masses and lifetimes are produced at any time. When these stars die, they release their elements in the ISM (top left panel). **(3)** These elements contribute to the metallicity of the gas (bottom left panel). A fraction of the elements are combined to form different kinds of dust species (bottom right panel). **(4)** Some of these grains are destroyed by the SN II blast waves.

and 300 $M_{\odot}$ (Galactic value). The two bottom plots of Fig. 3 demonstrate the effects of this parameter.

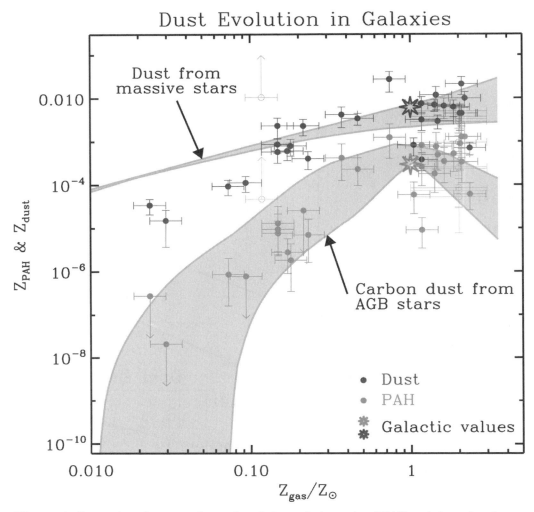

Figure 4. Comparison between observed and theoretical trends of PAH and dust abundances with metallicity (Galliano *et al.* 2008a). Z_{PAH} and Z_{dust} are respectively the PAH- and dust–to-gas mass ratios and Z_{gas} is the metallicity of the gas. The error bars are the results of the SED modeling. The grey filled curves are the results from the dust evolution model.

Fig. 4 shows the comparison between the observed trends of PAH and dust abundances with metallicity and the theoretical dust evolution in galaxies.

First, the observed PAH trend is in good agreement with the trend of carbon dust produced in AGB stars. Indeed, AGB stars are believed to be the principal source of PAHs. This trend provides a natural explanation of the paucity of PAHs in low-metallicity environments: due to their long lifetime ($\simeq$400 Myr), AGB stars begin to contribute to the injection of PAHs in the ISM, when the galaxy is already evolved. Therefore, there is a delay between the dust and the PAH contributions, and this delay translates into metallicity.

Second, the trend of the other dust components is in good agreement with the production of grains by SN II, down to $\simeq 1/10$ $Z_\odot$. At very low-metallcities, the observed dust-to-gas mass ratios are systematically lower than the SN II dust. A significant mass of cold dust could have been overlooked because these galaxies lack submillemeter data. Another possible origin of this disagreement could be that the star formation history of these galaxies is not continuous as in our dust evolution model.

The SN II yield (mass of dust formed by an average SN II) of this model is relatively high (consistent with the $\simeq 1$ $M_\odot$ SN^{-1} derived from observations of distant quasars; Dwek *et al.* 2007), compared to estimates on individual SN II ($\simeq 0.05$ $M_\odot$ SN^{-1}; Rho *et al.* 2008). This discrepancy between the global and the local approaches is one the main challenge in dust evolution, nowadays. The *Herschel* satellite, by extensively observing submillimeter wavelengths, will provide important constraints on the dust masses, and help address this issue.

References

Bolatto, A. D., Simon, J. D., Stanimirović, S., *et al.*, 2007, *ApJ*, 655, 212

Cannon, J. M., Walter, F., Armus, L., *et al.*, 2006, *ApJ*, 652, 1170

Cesarsky, D., Lequeux, J., Abergel, A., *et al.*, 1996, *A&A*, 315, L305

Draine, B. T., Dale, D. A., Bendo, G., *et al.*, 2007, *ApJ*, 663, 866

Dwek, E., Galliano, F., & Jones, A. P. 2007, *ApJ*, 662, 927

Engelbracht, C. W., Gordon, K. D., Rieke, G. H., *et al.*, 2005, *ApJL*, 628, L29

Engelbracht, C. W., Rieke, G. H., Gordon, K. D., *et al.*, 2008, *ApJ*, 678, 804

Fioc, M. & Rocca-Volmerange, B. 1997, *A&A*, 326, 950

Galliano, F., Dwek, E., & Chanial, P. 2008a, *ApJ*, 672, 214

Galliano, F., Madden, S. C., Jones, A. P., Wilson, C. D., & Bernard, J.-P. 2005, *A&A*, 434, 867

Galliano, F., Madden, S. C., Jones, A. P., *et al.*, 2003, *A&A*, 407, 159

Galliano, F., Madden, S. C., Tielens, A. G. G. M., Peeters, E., & Jones, A. P. 2008b, *ApJ*, 679, 310

Hunter, D. A., Gallagher, III, J. S., Rice, W. L., & Gillett, F. C. 1989, *ApJ*, 336, 152

Kennicutt, Jr., R. C. 1998, *ApJ*, 498, 541

Kunth, D. & Östlin, G. 2000, *A&A Rev.*, 10, 1

Madden, S. C., Galliano, F., Jones, A. P., & Sauvage, M. 2006, *A&A*, 446, 877

Meixner, M., Gordon, K. D., Indebetouw, R., *et al.*, 2006, *AJ*, 132, 2268

O'Halloran, B., Satyapal, S., & Dudik, R. P. 2006, *ApJ*, 641, 795

Reach, W. T., Rho, J., Jarrett, T. H., & Lagage, P.-O. 2002, *ApJ*, 564, 302

Rho, J., Kozasa, T., Reach, W. T., *et al.*, 2008, *ApJ*, 673, 271

Thuan, T. X., Sauvage, M., & Madden, S. 1999, *ApJ*, 516, 783

Wu, Y., Charmandaris, V., Hao, L., *et al.*, 2006, *ApJ*, 639, 157

Low-Metallicity Star Formation:
From the First Stars to Dwarf Galaxies
Proceedings IAU Symposium No. 255, 2008
L.K. Hunt, S. Madden & R. Schneider, eds.

Evolution of newly formed dust in Population III supernova remnants and its impact on the elemental composition of Population II.5 stars

Takaya Nozawa[1,2]**, Takashi Kozasa**[1]**, Asao Habe**[1]**, Eli Dwek**[3]**,
Hideyuki Umeda**[4]**, Nozomu Tominaga**[5]**, Keiichi Maeda**[2,6]**, and
Ken'ichi Nomoto**[2,4,7]

[1]Department of Cosmosciences, Graduate School of Science, Hokkaido University,
Sapporo 060-0810, Japan
email: tnozawa@mail.sci.hokudai.ac.jp
[2]Institute for the Physics and Mathematics of the Universe, University of Tokyo,
Kashiwa, Chiba 277-8568, Japan
[3]Laboratory for Astronomy and Solar Physics, NASA Goddard Space Flight Center,
Greenbelt, MD 20771, USA
[4]Department of Astronomy, School of Science, University of Tokyo,
Bunkyo-ku, Tokyo 113-0033, Japan
[5]Division of Optical and Infrared Astronomy, National Astronomical Observatory of Japan,
Mitaka, Tokyo 181-8588, Japan
[6]Max-Planck-Institut für Astrophysik, 85741 Garching, Germany
[7]Research Center for the Early Universe, School of Science, University of Tokyo,
Bunkyo-ku, Tokyo 113-0033, Japan

Abstract. We investigate the evolution of dust formed in Population III supernovae (SNe) by considering its transport and processing by sputtering within the SN remnants (SNRs). We find that the fate of dust grains within SNRs heavily depends on their initial radii $a_{\rm ini}$. For Type II SNRs expanding into the ambient medium with density of $n_{\rm H,0} = 1\,{\rm cm}^{-3}$, grains of $a_{\rm ini} < 0.05\,\mu{\rm m}$ are detained in the shocked hot gas and are completely destroyed, while grains of $a_{\rm ini} > 0.2\,\mu{\rm m}$ are injected into the surrounding medium without being significantly destroyed. Grains with $a_{\rm ini} = 0.05$–$0.2\,\mu{\rm m}$ are finally trapped in the dense shell behind the forward shock. We show that the grains piled up in the dense shell enrich the gas up to 10^{-6}–$10^{-4}\,Z_\odot$, high enough to form low-mass stars with 0.1–1 $M_\odot$. In addition, [Fe/H] in the dense shell ranges from -6 to -4.5, which is in good agreement with the ultra-metal-poor stars with [Fe/H] < -4. We suggest that newly formed dust in a Population III SN can have great impact on the stellar mass and elemental composition of Population II.5 stars formed in the shell of the SNR.

Keywords. dust, extinction, supernovae: general, hydrodynamics, shock waves, stars: abundances, stars: chemically peculiar, methods: numerical

1. Introduction

The first dust in the universe plays critical roles in the subsequent formation processes of stars and galaxies. Dust grains provide additional pathways for cooling of gas in metal-poor molecular clouds through their thermal emission and formation of H_2 molecules on the surface (e.g., Cazaux & Spaans 2004). In particular, the presence of dust decreases the value of the critical metallicity to 10^{-6}–$10^{-4}\,Z_\odot$ (Omukai *et al.* 2005; Schneider *et al.* 2006; Tsuribe & Omukai 2006), where the transition of star formation mode from massive Population III stars to low-mass Population II stars occurs. Since absorption and thermal emission by dust grains strongly depend on their composition, size distribution, and amount, it is essential to clarify the properties of dust in the early epoch of the universe, in order to elucidate the evolutionary history of stars and galaxies.

Dust grains at redshift $z > 5$ are considered to have been predominantly produced in supernovae (SNe). Theoretical studies have predicted that dust grains of 0.1–2 $M_\odot$ and 10–60 $M_\odot$ are formed in the ejecta of primordial Type II SNe (SNe II, Todini & Ferrara 2001; Nozawa *et al.* 2003) and pair-instability SNe (PISNe, Nozawa *et al.* 2003; Schneider *et al.* 2004), respectively. However, the newly formed dust is reprocessed via sputtering in the hot gas swept up by the reverse and forward shocks that are generated by the interaction between the SN ejecta and the surrounding medium. Thus, the size and mass of the dust can be greatly modified before being injected into the interstellar medium (ISM, Bianchi & Schneider 2007; Nozawa *et al.* 2007).

Here we present the results of the calculations for the evolution of newly formed dust within Population III SN remnants (SNRs), based on the dust formation model by Nozawa *et al.* (2003). We investigate the transport of dust and its processing by sputtering in the shocked hot gas, and report the size and amount of dust injected from SNe into the ISM. It is also shown that a part of the surviving dust grains are piled up in the dense SN shell formed behind the forward shock and can enrich the gas in the dense shell up to 10^{-6}–10^{-4} $Z_\odot$. We suppose that newly condensed dust in the SN ejecta has significant influence on the elemental abundances of Population II.5 stars, that is, the second-generation stars formed in the dense shell of Population III SNRs.

2. Evolution of Dust in Population III SNRs

We first briefly describe the models for the evolution of dust within SNRs. The time evolution of a SNR is numerically solved by assuming spherical symmetry. We adopt the hydrodynamic models of Population III SNe II with progenitor masses of $M_{pr} = 13$, 20, 25, and 30 $M_\odot$ and explosion energy of 10^{51} ergs by Umeda & Nomoto (2002) for the initial condition of the gas in the ejecta. For the ambient medium, we consider a uniform medium with a hydrogen number density of $n_{H,0} = 0.1$, 1, and 10 cm^{-3}. For the model of dust in the He core, we adopt the dust grains formed in the unmixed ejecta by Nozawa *et al.* (2003). Treating dust grains as test particles, we calculate the destruction and dynamics of dust by taking into account the size distribution as well as the spatial mass distribution of each grain species.

The results of the calculations are shown in Figure 1. Figure 1*a* shows the trajectories of C, Mg$_2$SiO$_4$, and Fe grains within the SNR for $M_{pr} = 20$ $M_\odot$ and $n_{H,0} = 1$ cm^{-3}, and Figure 1*b* shows the time evolution of their radii relative to the initial ones. The positions of the forward shock, the reverse shock, and the surface of the He core are depicted by the thick solid lines in Figure 1*a*. Initially, newly formed dust grains are expanding with the cool gas in the ejecta, and thus they undergo neither gas drag nor processing by sputtering. However, once they intrude into the hot gas swept up by the reverse shock penetrating into the ejecta, they acquire the high velocities relative to the gas and are eroded by kinetic and/or thermal sputtering. The evolution of dust grains after colliding with the reverse shock heavily depends on their initial radii a_{ini} and compositions.

For example, C grains formed in the outermost region of the He core encounter the reverse shock at 3650 yr. Since the deceleration by the gas drag is more efficient for smaller grains, C grains with $a_{ini} = 0.01$ μm quickly slow down. These small grains are eventually trapped in the hot gas ($\geqslant 10^6$ K) generated from the passage of the reverse and forward shocks and are completely destroyed by thermal sputtering. C grains of $a_{ini} = 0.1$ μm reduce their sizes by sputtering but cannot be completely destroyed. These grains

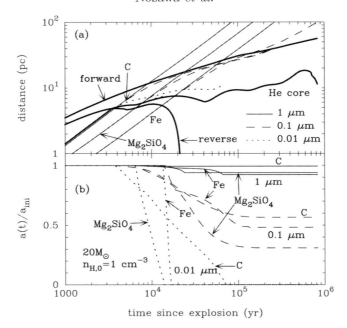

Figure 1. (*a*) Spatial evolution of C, Mg_2SiO_4, and Fe grains within the SNR for $M_{pr} = 20\ M_\odot$ and $n_{H,0} = 1\ cm^{-3}$ and (*b*) the temporal evolution of their radii relative to the initial values. The evolution of dust with $a_{ini} = 0.01$, 0.1, and 1 μm is shown by the dotted, dashed, and solid lines, respectively. The thick solid lines in (*a*) indicate the positions of the forward shock, the reverse shock, and the surface of the He core.

are finally captured in the dense SN shell formed behind the forward shock at $\sim 2 \times 10^5$ yr, where the gas temperature is too low ($<10^5$ K) to erode the dust grains by thermal sputtering. C grains with $a_{ini} = 1\ \mu m$, which are not efficiently decelerated by the gas drag, can pass through the forward shock front and are injected into the ambient medium without being significantly destroyed.

Mg_2SiO_4 grains, which are formed in the O-rich layer, collide with the reverse shock at about 6000 yr, but the dependence of their evolution on the initial radius is the same as for C grains. On the other hand, Fe grains formed in the innermost region of the ejecta hit the reverse shock after 13000 yr, and its 0.1 μm-sized grains are injected into the ambient medium because of the high bulk density.

As shown above, the small dust grains formed in the SN ejecta are predominantly destroyed by sputtering within the SNR. Thus, the size distribution of the surviving dust is dominated by larger grains, compared to the sizes at formation. Note that the evolution of dust within SNRs and thus the resulting size distribution of dust does not depend on the progenitor mass considered here, because their explosion energies are the same and the time evolution of the gas temperature and density within SNRs are similar. On the other hand, the ambient gas density strongly affects the evolution of dust in SNRs. The higher density in the ambient medium results in the higher density of the shocked gas and causes the efficient erosion and deceleration of dust due to more frequent collisions with the hot gas. Therefore, the initial radius below which dust is completely destroyed increases with increasing the ambient gas density and is 0.01, 0.05, and 0.2 μm for $n_{H,0} = 0.1$, 1, and 10 cm^{-3}, respectively. As a result, the total mass of the surviving dust is smaller for the higher ambient density and ranges from 0.01 to 0.8 $M_\odot$ for $n_{H,0} = 10$ to 0.1 cm^{-3}, depending on the size distribution of dust formed in each SN.

Table 1. Metallicities, [Fe/H], and abundances of C, O, Mg, and Si relative to Fe in the dense shell of primordial SN II remnants for various ambient gas densities.

M_{pr} ($M_\odot$)	$\log(Z/Z_\odot)$	[Fe/H]	[C/Fe]	[O/Fe]	[Mg/Fe]	[Si/Fe]
		$n_{\mathrm{H},0} = 0.1\,\mathrm{cm}^{-3}$				
13	-5.89	-6.43	-0.274	-0.699	-0.230	1.92
20	-5.44	-5.20	0.117	-0.595	0.034	0.410
25	-5.55	-5.90	1.11	-1.42	-0.500	-0.552
30	-5.33	-5.56	0.566	-0.043	0.739	0.866
		$n_{\mathrm{H},0} = 1\,\mathrm{cm}^{-3}$				
13	-4.72	-5.15	1.11	-0.555	-0.459	1.01
20	-4.68	-5.53	0.992	0.585	1.16	1.87
25	-4.79	-5.23	1.09	-0.412	0.407	0.989
30	-4.60	-5.11	0.797	0.242	1.09	1.26
		$n_{\mathrm{H},0} = 10\,\mathrm{cm}^{-3}$				
13	-4.40	-4.13	0.284	-2.54	-3.89	0.599
20	-4.09	-4.92	0.946	-2.15	-1.80	2.14
25	-3.91	-5.10	1.60	0.122	0.232	2.34
30	-3.84	-5.11	-0.207	0.375	-1.23	2.66

3. Metallicities and Elemental Abundances of Population II.5 Stars

In this section we discuss the influence of dust on the elemental composition of Population II.5 stars that are expected to form in the dense shell of Population III SNRs (Mackey *et al.* 2003; Salvaterra *et al.* 2004; Machida *et al.* 2005). As shown in the last section, the dust grains surviving the destruction but not injected into the ISM are piled up in the dense SN shell after 10^5–10^6 yr. This implies that the elemental composition of these piled-up grains can play an important role in the elemental abundances of Population II.5 stars. Furthermore, the existence of dust in the shell may enable the formation of stars with solar mass scales through its thermal emission if the gas is enriched to the critical metallicities (Omukai *et al.* 2005; Schneider *et al.* 2006; Tsuribe & Omukai 2006). Thus, we calculate the metallicities and metal abundances in the dense shell based on the elemental composition of piled-up grains, and compare with the observed abundance patterns of low-mass hyper-metal-poor (HMP) and ultra-metal-poor (UMP) stars.

Table 1 summarizes the calculated metallicities and elemental abundances in the dense shell of SN II remnants as a function of the ambient gas density. It should be noted that the metallicity in the shell ranges from 10^{-6} to 10^{-4} $Z_\odot$, which is considered to cause the formation of stars with 0.1–1 $M_\odot$. In addition, most of the calculated [Fe/H] abundances are in the range of -6 to -4.5, which is in good agreement with those for HMP and UMP stars. We also plot in Figure 2a the abundances of C, O, Mg, and Si relative to Fe in the shell for $n_{\mathrm{H},0} = 0.1$ and $1\,\mathrm{cm}^{-3}$. We can see that the calculated abundances of Mg and Si showing 1–100 times overabundances are consistent with the observations of HMP and UMP stars. Because the elemental composition of dust piled up in the shell can reproduce the abundance patterns of Fe, Mg, and Si in HMP and UMP stars, it is considered that the transport of dust separated from metal-rich gas within SNRs can be attributed to the elemental compositions of HMP and UMP stars, if they are Population II.5 stars.

However, as can be seen from Figure 2a, no model considered here can reproduce 10^2–10^4 times excesses of C and O observed in HMP stars. One of the reasons is that

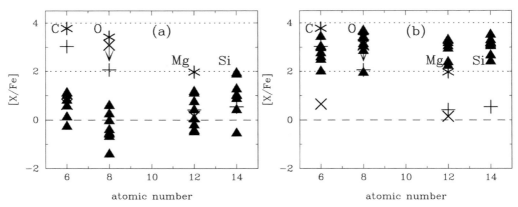

Figure 2. Abundances of C, O, Mg and Si relative to Fe in the dense shell of SN II remnants for $n_{H,0} = 0.1$ and $1\,cm^{-3}$ (*filled triangles*); (*a*) derived from the elemental composition of the grains piled up in the shell, and (*b*) derived from the elemental composition of the piled-up grains and the gas outside the innermost Fe layer. For observational data of HMP and UMP stars, the 3-D corrected abundances are adopted and are denoted by plus (HE0107-5240 with [Fe/H] = -5.62, Collet *et al.* 2006), asterisk (HE1327-2326 with [Fe/H] = -5.96, Frebel *et al.* 2008), and cross (HE0557-4840 with [Fe/H] = -4.75, Norris *et al.* 2007).

in the calculation we assumed that the metal-rich gas in the SN ejecta does not mix with the gas in the shell. Then we examine the abundance patterns in the shell by assuming that besides the piled-up grains, the gas outside the innermost Fe layer in the ejecta is incorporated into the shell. The results are shown in Figure 2*b*. In this case we can reproduce the very large overabundances of C and O, but the excesses of Mg and Si are too large ($\geqslant$100 times) to agree with the observations. However, it could be possible to reproduce the abundance patterns of refractory elements observed in HMP stars unless the Si-Mg-rich layer is mixed into the shell. Unfortunately, it is still being debated to what extent the gas in the ejecta can mix into the SN shell when Population II.5 stars form. Nevertheless, we can conclude that newly formed dust in a Population III SN can have great impact on the stellar mass and metal abundance of Population II.5 stars, if the metal-rich gas is not significantly incorporated into the dense gas shell.

Acknowledgements

    This work has been supported in part by a Grant-in-Aid for Scientific Research from the Japan Society for the Promotion of Sciences (18104003, 19740094).

References

Bianchi, S. & Schneider, R. 2007, *MNRAS*, 378, 973
Cazaux, S. & Spaans, M. 2004, *ApJ*, 611, 40
Collet, R., Asplund, M., & Trampedach, R. 2006, *ApJ*, 644, L121
Frebel, A., *et al.*, 2008, *ApJ*, accepted [arXiv:0805.3341]
Machida, M. N., *et al.*, 2005, *ApJ*, 622, 39
Mackey, J., Bromm, V., & Hernquist, L. 2003, *ApJ*, 586, 1
Norris, J. E., *et al.*, 2007, *ApJ*, 670, 774
Nozawa, T., *et al.*, 2003, *ApJ*, 598, 785

Nozawa, T., *et al.*, 2007, *ApJ*, 666, 955
Omukai, K., *et al.*, 2005, *ApJ*, 626, 627
Salvaterra, R., Ferrara, A., & Schneider, R. 2004, *New A*, 10, 113
Schneider, R., Ferrara, A., & Salvaterra, R. 2004, *MNRAS*, 351, 1379
Schneider, R., *et al.*, 2006, *MNRAS*, 369, 1437
Todini, P. & Ferrara, A. 2001, *MNRAS*, 325, 726
Tsuribe, T. & Omukai, K. 2006, *ApJ*, 642, L61

Low-Metallicity Star Formation:
From the First Stars to Dwarf Galaxies
Proceedings IAU Symposium No. 255, 2008
L.K. Hunt, S. Madden & R. Schneider, eds.

Dust properties and distribution in dwarf galaxies

Ute Lisenfeld[1,2], Monica Relaño[1,2], José Vílchez[3], Eduardo Battaner[1] and Israel Hermelo[1]

[1] Universidad Granada, Spain
email: ute@ugr.es, battaner@ugr.es

[2] Institute of Astronomy, University of Cambridge, UK
email: mrelano@ast.cam.ac.uk

[3] Instituto de Astrofísica de Andalucía, Granada, Spain
email: jvm@iaa.es

Abstract. We present a study of the extinction, traced by the Balmer decrement, in HII regions in the dwarf galaxies NGC 1569 and NGC 4214. We find that the large-scale extinction around the most prominent HII regions in both galaxies forms a shell in which locally the intrinsic extinction can adopt relatively high values ($A_V = 0.8 - 0.9$ mag) despite the low metallicity and thus the low overall dust content. The small-scale extinction (spatial resolution $\sim$0.3") shows fluctuations that are most likely due to variations in the dust distribution. We compare the distribution of the extinction to that of the dust emission, traced by *Spitzer* emission at 8 and 24μm, and to the emission of cold dust at 850μm. We find in general a good agreement between all tracers, except for the 850μm emission in NGC 4214 which is more extended than the extinction and the other emissions. Whereas in NGC 1569 the dust emission at all wavelengths is very similar, NGC 4214 shows spatial variations in the 24-to-850μm ratio.

We furthermore compared the 24μm and the extinction-corrected Hα emission from HII regions in a sample of galaxies with a wide range of metallicities and found a good correlation, independent of metallicity. We suggest that this lack of dependence on metallicity might be due to the formation of dust shells with a relatively constant opacity, like the ones observed here, around ionizing stars.

Keywords. ISM: dust, ISM: extinction, galaxies: ISM, galaxies: individual (NGC 1569), galaxies: individual (NGC 4214), galaxies: dwarf

---

1. Introduction

Interstellar dust can be studied via its emission and also via the extinction that it causes. Each method has its advantages and difficulties. The most common way of obtaining the extinction of the light coming from HII regions is based on the comparison of the Hα and Hβ recombination line fluxes. A major advantage of this method is the high spatial resolution achieved, determined by the resolution of the optical images. It is, however, difficult to derive the distribution of the dust from extinction maps because the relative distribution of the dust and the gas plays a major role and because maps of the Hα/Hβ ratio are biased towards low-extinction regions.

Alternatively, the dust can be studied via its emission, which depends on the amount of dust, the type of grains and the dust temperature determined by the interstellar radiation field (ISRF). Although very different models for the interstellar dust exist (see, e.g., Zubko *et al.* 2004 and references therein), they generally need to include three types of grains: big grains that are in thermal equilibrium with the ISRF, very small grains (VSGs) that are stochastically heated and are necessary to explain the mid-infrared emission

Low-Metallicity Star Formation:
From the First Stars to Dwarf Galaxies
Proceedings IAU Symposium No. 255, 2008
L.K. Hunt, S. Madden & R. Schneider, eds.

The Gas Phase in a Low Metallicity ISM

Elias Brinks[1], Se–Heon Oh[2], Ioannis Bagetakos[1], Frank Bigiel[3], Adam Leroy[3], Antonio Usero[1], Fabian Walter[3], W. J. G. de Blok[4], and Robert C. Kennicutt, Jr.[5]

[1] Centre for Astrophysics Research, University of Hertfordshire, Hatfield AL10 9AB, UK
[2] Research School of Astronomy & Astrophysics, The Australian National University, Mount Stromlo Observatory, Cotter Road, Weston Creek, ACT 2611, Australia
[3] Max–Plank–Institut für Astronomie, Königstuhl 17, 69117, Heidelberg, Germany
[4] Univ. of Cape Town, Dept. of Astronomy, Private Bag X3, Rondebosch 7701, South Africa
[5] Institute of Astronomy, University of Cambridge, Madingley Road, Cambridge CB3 0HA, UK

Abstract. We present several results from our analysis of dwarf irregular galaxies culled from The HI Nearby Galaxy Survey (THINGS). We analyse the rotation curves of two galaxies based on "bulk" velocity fields, i.e. velocity maps from which random non–circular motions are removed. We confirm that their dark matter distribution is best fit by an isothermal halo model. We show that the star formation properties of dIrr galaxies resemble those of the outer parts of larger, spiral systems. Lastly, we study the large scale (3–D) distribution of the gas, and argue that the gas disk in dIrrs is thick, both in a relative, as well as in an absolute sense as compared to spirals. Massive star formation through subsequent supernova explosions is able to redistribute the bulk of the ISM, creating large cavities. These cavities are often larger, and longer–lived than in spiral galaxies.

Keywords. galaxies: structure, galaxies: spiral, galaxies: ISM

1. Introduction

Dwarf irregular (dIrr) galaxies are known to be gas rich. Whereas their larger spiral cousins have gas fractions of order 10%, dIrr galaxies routinely have 50% of their baryonic mass in the gas phase. Although many dIrr galaxies are actively forming stars, and therefore must harbour at least some gas in molecular form, this phase of the ISM is elusive. Cold H_2, which makes up the bulk of GMCs, is hard to detect directly so CO emission is used as a proxy. However, the heavy element abundance of dwarfs is low which implies that the tracer molecule is scarce. Also, as a result of the low metallicity, their dust content is low which means that ionising radiation can penetrate deeper into dense clouds, at least until densities are high enough for self–shielding to become important. The upshot of all this is that no CO detection has been reported from any dwarf with a metallicity of $12 + \log[O/H] \lesssim 7.9$ (Taylor *et al.* 1998; Bolatto *et al.* 2008). Therefore, most of what we know about the gas phase in a low metallicity ISM comes from observations of the neutral, atomic phase, i.e., from H I studies, H II and dust being only minor constituents.

H I in dIrr galaxies generally extends well beyond R_{25}, the radius at an isophotal level of 25th mag arcsec^{-2}. The most extreme cases known are DDO 154 (Hoffman *et al.* 2001) and NGC 3741 (Begum *et al.* 2005). Their rotation is dominated by solid body rotation, most rotation curves barely starting to flatten by the time the last measured point is reached (de Blok *et al.* 2008). Even in their very centres, and contrary to what is found in spiral galaxies, they are dark-matter (DM) dominated (Carignan & Freeman 1988; de Blok *et al.* 2008). As their luminosity decreases, the gas fraction becomes proportionally more important to the point that at faint absolute luminosities the gas mass dominates

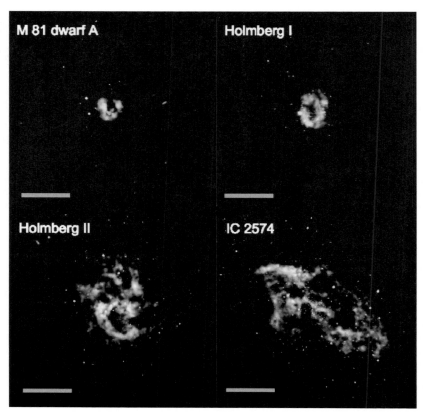

Figure 1. Some examples of dIrr galaxies, all members of the M 81 group, observed as part of THINGS. The H I emission is colour–coded in blue, the older stellar population is assigned an orange hue. Purple is a measure of the recent star formation activity and is a linear combination of the FUV flux measured with *GALEX*, and *Spitzer* 24 μm emission. The green bar measures 10 kpc. The intrinsic resolution of $\sim 6''$ at the distance of the galaxies depicted here corresponds to ~ 100 pc.

the baryonic matter budget. There seems to be a lower mass limit for H I in dwarfs of $M_{H\,I} \sim 5 \times 10^6 - 10^7$ M$_\odot$ which sets a limit to the lowest DM haloes which are able to capture and retain baryons (Taylor & Webster 2005). Even the lowest mass H I clouds have an optical counterpart; dark galaxies, i.e., DM haloes with gas but no stars, have yet to be found.

The observed velocity dispersions of H I in dwarfs is on average 6–9 km s^{-1}, but several studies have reported two components, a broad ubiquitous component of 9 km s^{-1} and a more narrow component, thought to trace cool, dense gas showing a velocity dispersion of ~ 4 km s^{-1} (Young & Lo 1996, 1997; Young *et al.* 2003; de Blok & Walter 2006). Lastly, 50% of field dwarf galaxies, i.e. those not obviously being a satellite of a larger system, either have an equal mass or lower mass dwarf companion (Taylor 1997).

Several groups have embarked on surveys of large numbers of dwarf galaxies to extend the results listed above and put them on a statistically more secure footing. Recent studies based on GMRT observations have been reported by Begum *et al.* (2006) and Begum *et al.* (2008). Two new Legacy Surveys have been awarded time at the NRAO†

† The National Radio Astronomy Observatory is a facility of the National Science Foundation operated under cooperative agreement by Associated Universities, Inc.

Very Large Array (VLA), i.e. ANGST by Ott *et al.*, and LITTLE THINGS by Hunter and collaborators.

The sample described in this contribution was observed as part of THINGS, The H I Nearby Galaxy Survey (Walter *et al.* 2008), and consists of 11 LSB and dIrr galaxies. The survey was carried out with the VLA in its B–, C–, and D–configuration, resulting in maps at a spatial resolution of 6″. The velocity resolution is $5.2\,\mathrm{km\,s^{-1}}$ or better and the observations reach typical 1σ column density sensitivities of $4 \times 10^{19}\,\mathrm{cm^{-2}}$ at 30″ resolution. Full details regarding the observations and data reduction can be found in Walter *et al.* (2008). Most of the galaxies in THINGS were drawn from the *Spitzer*

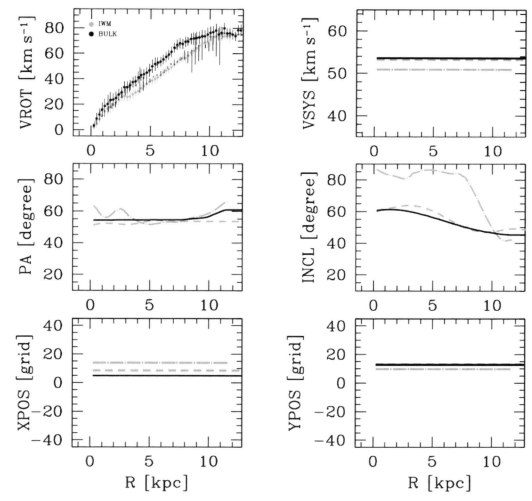

Figure 2. Tilted–ring analysis of the galaxy IC 2574 based on the bulk velocity field (top left panel; filled circles). The gray dashed lines in the other five panels are used as initial values for a tilted–ring fit to the bulk velocity field. The solid black lines show the adopted rotation curves of IC 2574 using the bulk velocity field. The fit to the bulk velocity field is compared with that based on the IWM velocity field (top left panel; filled grey circles). The gray long dash–dotted lines represent the fits of the geometrical parameters to the IWM velocity field. The large difference in inclination between IWM and bulk velocity fields is clearly evident in the panel labeled INCL and this results in a significant difference ($> 14\,\mathrm{km\,s^{-1}}$) in rotational velocity (figure taken from Oh *et al.* 2008).

Infrared Nearby Galaxies Survey (SINGS; Kennicutt *et al.* 2003), a multi-wavelength project designed to study the properties of the dusty ISM in nearby galaxies and most are also part of the *GALEX* (Galaxy Evolution Explorer) Nearby Galaxy Survey (Gil de Paz *et al.* 2007).

Fig. 1 shows a mosaic of false colour composites for four dIrr galaxies (see figure caption for details). In the sections which follow, we will highlight some of the papers which are being produced by us, notably the mass distribution and DM content of dIrr galaxies, their star formation (SF) characteristics, and the structure and morphology of their ISM.

2. Kinematics and Mass models

In the paper by de Blok *et al.* (2008) we present a rotation curve analyis of 19 THINGS galaxies. These are the highest quality H I rotation curves available to date for a large sample of nearby galaxies, spanning a wide range of H I masses and luminosities. Having said that, in the case of dwarf galaxies, motion due to gas streaming along a bar or oval distortion, or expanding shells around the site of OB associations, can give rise to quasi–random non–circular velocity components which correspond to an appreciable fraction of the rotational velocity. Ordinarily, the intensity–weighted mean velocities (IWM), or velocity fields, are affected by all these components. If one wishes to study the underlying mass distribution, one has to remove these non–circular components from the velocity field. Oh *et al.* (2008) devised a novel method to do just that and retrieve the underlying "bulk" velocity field (see their paper for details). Fig. 2 illustrates the resulting rotation curve based on the bulk velocity field and compares it with the one derived "classically", i.e., based on the IWM velocity field. The largest difference in the fitted parameters is in the inclination, which in turn translates to a considerable difference in the rotation curve and hence the inferred mass distribution.

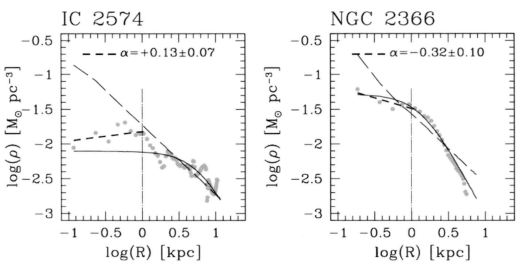

Figure 3. The derived mass density profiles of IC 2574 and NGC 2366. Long dashed and solid lines show the NFW halo model and the pseudo–isothermal halo model, respectively. Vertical long dash–dotted lines correspond to a radius of 1 kpc. The filled gray circles represent the dark matter density profile derived from the bulk rotation velocity. The inner slope of the derived dark matter density profile is denoted by α and is the result of a least squares fit (short dashed lines) to data points at radii less than 1 kpc. The measured inner slopes of the mass density profiles of IC 2574 and NGC 2366 are shown in the panels (figure taken from Oh *et al.* 2008).

Now that we have a more reliable rotation curve, less affected, if at all, by random non–circular motions, we follow the method described in de Blok & Bosma (2002) to determine the slope of the inner component of the mass density profile, $\rho \sim R^{\alpha}$. We measure the slopes, α, in plots of $\log(\rho)$ versus $\log(R)$ of the inner parts ($R < 1.2\,$kpc) of IC 2574 and NGC 2366 using a least squares fit and find the values of the slopes to be $\alpha = +0.13 \pm 0.07$ for IC 2574 and $\alpha = -0.32 \pm 0.10$ for NGC 2366, respectively. This is shown in Fig. 3 and is in good agreement with the earlier result of $\alpha = -0.2 \pm 0.2$ by de Blok & Bosma (2002) for a larger sample of LSB galaxies. These flat slopes imply that the dark matter distributions of IC 2574 and NGC 2366 are well characterized by a

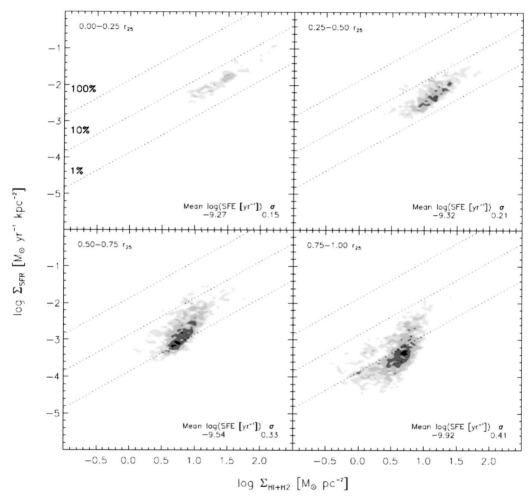

Figure 4. Variations of $\log \Sigma_{\mathrm{SFR}}$ versus $\log \Sigma_{\mathrm{H\,I+H_2}}$ with radius in spiral galaxies. The results for 7 spiral galaxies are plotted together in these diagrams. Green, orange, red, and magenta cells show contours of 1, 2, 5, and 10 independent data points per 0.05 dex–wide cell. Diagonal dotted lines show lines of constant SFE, indicating the level of SFR needed to consume 1%, 10% and 100% of the gas reservoir (including helium) in 10^8 years. Thus, the lines also correspond to constant gas depletion times of, from top to bottom, 10^8, 10^9, and 10^{10} yr. The data are plotted in 4 radius bins: $0.0 - 0.25\,R_{25}$, $0.25 - 0.5 R_{25}$, $0.5 - 0.75\,R_{25}$, and $0.75 - 1.0\,R_{25}$ (figure taken from Bigiel *et al.* 2008).

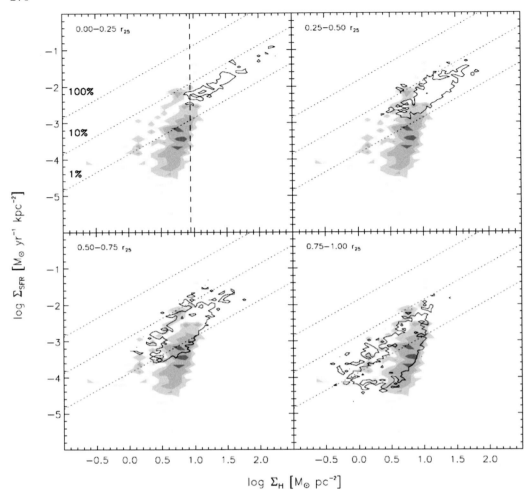

Figure 5. As Fig. 4 but for the dwarf galaxies in our sample. All four panels show the same distribution for the dIrrs. Note that the SFR density is plotted against Σ_{HI} (see text). Plotted over the dwarf distribution is the lowest contour for the spirals from the corresponding panel in Fig. 4. Thus each panel compares the distribution for the dwarfs to that in spiral galaxies from a particular radial range. The best agreement is seen in the bottom right panel, in which the black contour shows data from $0.75-1.0R_{25}$ in spiral galaxies. The vertical dashed line in the top left panel drawn at $9\,M_\odot\,pc^{-2}$ locates the total gas density above which the ISM is predominantly molecular in spiral galaxies (figure taken from Bigiel *et al.* 2008).

sizeable constant-density core, for which we expect $\alpha = 0$. This is in sharp contrast with the steep slope of $\alpha = -1$ predicted by the NFW profile (Navarro, Frenk & White 1996).

3. Star Formation Law

Following Kennicutt (1989, 1998) we investigate the relation between the gas and star-formation rate (SFR) surface density. This is usually expressed in terms of a power law relation $\Sigma_{SFR} \sim (\Sigma_{gas})^N$. The value of N has been found to cover a wide range in value, from 0.9–3.5 (see Bigiel *et al.* 2008, for a summary). Kennicutt (1998) finds a value of $N = 1.40 \pm 0.15$ when plotting the disk–averaged SFR against the total gas density (H I+H$_2$) in a sample of 61 nearby normal spiral and 36 infrared–selected starburst galaxies. Using THINGS we are now able to investigate the relation between SFR density and gas density on a pixel by pixel basis, the pixels being chosen to be at a common linear resolution of 750 pc, and to extend this relation to dwarf galaxies, thus probing the low–mass, low–metallicity regime. We use a linear combination of *GALEX* FUV and *Spitzer* 24 µm emission to determine the SFR (details and justification are discussed in Leroy *et al.* 2008).

We plot in Fig. 4 log Σ_{SFR} (the log of the SFR density) against log $\Sigma_{H I+H_2}$ (see figure caption for details) and we do this for 4 annuli: $0.0-0.25\,R_{25}$, $0.25-0.5 R_{25}$, $0.5-0.75\,R_{25}$, and $0.75-1.0\,R_{25}$. We show in Bigiel *et al.* (2008) that when plotting log Σ_{SFR} versus log Σ_{H_2} we find a power law relation with $N = 1.0 \pm 0.2$ across our sample of spiral galaxies. We interpret this as indication that H$_2$ forms stars at a constant efficiency in spirals. The average molecular gas depletion time is $\sim 2 \times 10^9$ years. We interpret the linear relation and constant depletion time over this range as evidence that stars are forming in GMCs with approximately uniform properties and that Σ_{H_2} may be more a measure of the filling fraction of giant molecular clouds than changing conditions in the molecular gas. And because the ISM in the central regions of galaxies tends to be largely molecular, the data in the two top panels of Fig. 4, corresponding to the central and hence molecule dominated ISM, reflect this result. The bottom panels, in contrast, show the relation in the outer parts of galaxies where atomic gas prevails. This is where we see that the power law relation breaks down.

In Fig. 5 we present a very similar graph for the dwarf galaxies in our sample. Because, as mentioned earlier, H$_2$ is difficult to detect and assuming that, if anything, the ISM in dIrr galaxies is predominantly atomic, we used log $\Sigma_{H I}$ as a proxy for the total gas content of dwarfs. In each of the four panels we show the same (i.e., all) independent data points for the dwarfs as coloured contours. What changes in this figure is the contour which corresponds to the lowest contour for the spirals from the corresponding panel in Fig. 4. What we learn from this is that the star formation properties of low–metallicity, dIrr galaxies resemble those encountered in the outskirts of spiral galaxies. In the outer regions of spirals, and therefore similarly in dwarfs, HI dominates and the SF is less efficient, the efficiency decreasing monotonically with radius.

4. H I Supershells

The superior resolution and sensitivity of THINGS reveals a wealth of structure in the ISM of gas–rich systems as can be appreciated from Fig. 1. For example, Holmberg II and IC 2574 are riddled with holes in their H I distribution. These features are common in disk galaxies, including dwarfs. Studies of the LMC (Kim *et al.* 1999) by several groups quite convincingly show that many of the H I holes are cavities in the ISM filled with hot, X–ray emitting gas (Dunne *et al.* 2001, and references therein). The origin of this gas is thought to be a result of Type II supernovae from massive stars which formed in super star clusters (SSCs) or OB associations (Oey & Clarke 1997).

Bagetakos *et al.* (2008) have detected more than 1000 holes in a total of 20 galaxies selected from THINGS, in both spirals as well as in dwarfs. This is the first time that the same detection technique is applied using such a large data set of uniformly high quality. The sizes of the H I holes range from about 100 pc (our resolution limit) to 2000 pc. Their expansion velocities vary from 5 to 35 km s^{-1}. We estimate their ages at $6 - 150$ Myr and their energy requirements, based on a simple single blast approximation (Chevalier 1974), at $10^{50}-10^{53}$ erg. The kinetic energy deposited is compatible with it being independent of galaxy type: an OB association is to first order oblivious to the nature of its host galaxy.

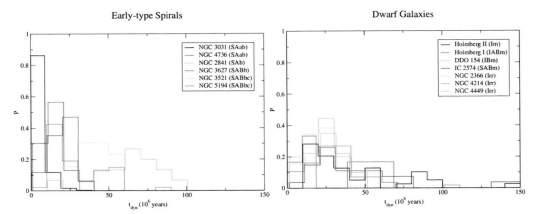

Figure 6. The number distributions of the kinematic ages of the H I holes divided into 2 different groups of galaxies: early–type spirals are pictured in the left panel and dwarf galaxies on the right. The y–axis shows the relative number distribution, P, the x–axis corresponds to the kinematic age.

In most galaxies, H I holes are found all the way out to the edge of the H I disk. Assuming that holes are the result of massive star formation we estimate the star formation rate and find that it correlates with values obtained by other SF tracers, corroborating the assertion that they are the result of the evolution of massive stars.

We find that the kinematic ages of the holes show a trend in the sense that holes in early–type spirals are younger than in dwarfs. This is illustrated in Fig. 6. This is probably due to a combination of factors: holes in spirals tend to be distorted and destroyed by the passage of spiral density waves and through shear which sets an upper limit to their age. No such mechanisms occur in dwarfs, which rotate more or less as solid bodies and lack spiral arms. Hence holes in dwarfs tend to survive for longer. We confirm that H I holes in dwarfs tend to be larger than in spiral galaxies (Brinks *et al.* 2002; Walter & Brinks 1999; Puche *et al.* 1992). This we ascribe to the fact that the mass volume density in the plane of a dIrr is lower than that of a larger spiral. Therefore, for similar observed H I velocity dispersions, the gas disk in a dwarf will be thicker, both in a relative as well as absolute sense. Because we also find that the amount of energy input is similar in dIrrs and in spirals, H I holes can grow to much larger diameters in dwarfs whereas they would suffer blow–out in the disks of spirals.

It has been argued that it would be easier for dwarf galaxies to lose a large fraction of their ISM as a result of feedback from massive star formation (Ferrara & Tolstoy 2000; Silich & Tenorio–Tagle 2001). Ott *et al.* (2005) investigated this in a sample of actively star forming dwarf galaxies. They reached the conclusion that SF activity as witnessed today in dIrrs is indeed capable of blowing enriched gas into their haloes which is then in principle able to escape. However, any extended low density envelope of material may delay this outflow on time–scales exceeding those of the cooling time of the hot gas.

References

Bagetakos, I., Brinks, E., Walter, F., de Blok, W. J. G., Rich, J. W., Usero, A., & Kennicutt, R.C., Jr. 2008, *AJ* (submitted)

Begum, A., Chengalur, J. N., & Karachentsev, I. D. 2005, *A&A*, 433, L1

Begum, A., Chengalur, J. N., Karachentsev, I. D., Kaisin, S. S., & Sharina, M. E. 2006, *MNRAS*, 365, 1220

Begum, A., Chengalur, J. N., Karachentsev, I. D., Sharina, M. E., & Kaisin, S. S. 2008, *MNRAS*, 386, 1667

Bigiel, F., Leroy, A., Walter, F., Brinks, E., de Blok, W. J.G., Madore, B., & Thornley, M.D. 2008, *AJ* (submitted)

Bolatto, A. D., Leroy, A. K., Rosolowsky, E., Walter, F., & Blitz, L. 2008, astro–ph/0807.0009

Brinks, E., Walter, F., & Ott, J. 2002, *ASP Conf. Proc.*, 275, 57

Carignan, C. & Freeman, K. C. 1988 *ApJ Lett.*, 332, L33

Chevalier, R. A. 1974, *ApJ*, 188, 501

de Blok, W. J. G. & Bosma, A. 2002, *A&A*, 385, 816

de Blok, W. J. G. & Walter, F. 2006, *AJ*, 131, 363

de Blok, W. J. G., Walter, F., Brinks, E., Trachternach, C., Oh, S.–H., & Kennicutt, R.C., Jr. 2008, *AJ* (accepted)

Dunne, B. C., Points, S. D., & Chu, Y.–H. 2001, *ApJS*, 136, 119

Ferrara, A. & Tolstoy, E. 2000, *MNRAS*, 313, 291

Gil de Paz, A., *et al.* 2007, *ApJS*, 173, 185

Hoffman, G. L., Salpeter, E. E., & Carle, N. J. 2001, *AJ*, 122, 2428

Kennicutt, R. C. 1989, *ApJ*, 344, 685

Kennicutt, R. C. 1998, *ApJ*, 498, 541

Kennicutt, R. C., Jr., *et al.* 2003, *PASP*, 115, 928

Kim, S, Dopita, M. A., Staveley–Smith, L.,& Bessell, M. S. 1999 *AJ*, 118, 2797

Leroy, A., Walter, F., Brinks, E., Bigiel, F., de Blok, W. J. G., Madore, B., & Thornley, M.D. 2008 *AJ* (accepted)

Navarro, J. F., Frenk, C. S., & White, S. D. M. 1996, *ApJ*, 462, 563

Oh, S.–H., de Blok, W. J. G., Walter, F., Brinks, E., & Kennicutt, R. C., Jr. 2008, *AJ* (accepted)

Ott, J., Walter, F., & Brinks, E. 2005, *MNRAS*, 358, 1453

Oey, M. S. & Clarke, C. J. 1997, *MNRAS*, 289, 570

Puche, D., Westpfahl, D., Brinks, E., & Roy, J.–R. 1992, *AJ*, 103, 1841

Silich, S. A. & Tenorio–Tagle, G. 2001, *ApJ*, 552, 91

Taylor, C. L. 1997, *ApJ*, 480, 524

Taylor, C. L., Kobulnicky, H. A., & Skillman, E. D. 1998 *AJ*, 116, 2746

Taylor, E. N. & Webster, R. L. 2005 *ApJ*, 634, 1067

Walter, F. & Brinks, E. 1999, *AJ*, 118, 273

Walter, F., Brinks, E., de Blok, W. J. G., Bigiel, F., Kennicutt, R. C., Jr., Thornley, M. D., & Leroy, A. 2008, *AJ* (accepted)

Young, L. M. & Lo, K. Y. 1996, *ApJ*, 462, 203

Young, L. M. & Lo, K. Y. 1997, *ApJ*, 490, 710

Young, L. M., van Zee, L., Lo, K. Y., Dohm–Palmer, R. C., & Beierle, M. E. 2003, *ApJ*, 592, 111

Low-Metallicity Star Formation:
From the First Stars to Dwarf Galaxies
Proceedings IAU Symposium No. 255, 2008
L.K. Hunt, S. Madden & R. Schneider, eds.

© 2008 International Astronomical Union
doi:10.1017/S1743921308024939

The Resolved Properties of Extragalactic Giant Molecular Clouds

Alberto D. Bolatto[1], Adam K. Leroy[2], Erik Rosolowsky[3], Fabian Walter[2], and Leo Blitz[4]

[1] Department of Astronomy, University of Maryland, College Park, MD 20742, USA
email: bolatto@astro.umd.edu

[2] Max-Planck-Institut für Astronomie, D-69117 Heidelberg, Germany

[3] Department of Mathematics, Statistics, and Physics, University of British Columbia at Okanagan, Kelowna, B.C. V1V 1V7, Canada

[4] Department of Astronomy, University of California, Berkeley, CA94720, US A

Abstract. Giant molecular clouds (GMCs) are the major reservoirs of molecular gas in galaxies, and the starting point for star formation. As such, their properties play a key role in setting the initial conditions for the formation of stars. We present a comprehensive combined inteferometric/single-dish study of the resolved GMC properties in a number of extragalactic systems, including both normal and dwarf galaxies. We find that the extragalactic GMC properties measured across a wide range of environments, characterized by the Larson relations, are to first order remarkably compatible with those in the Milky Way. Using these data to investigate trends due to galaxy metallicity, we find that: 1) these measurements are not in accord with simple expectations from photoionization-regulated star formation theory; 2) there is no trend in the virial CO-to-H_2 conversion factor on the spatial scales studied; and 3) there are measurable departures from the Galactic Larson relations in the Small Magellanic Cloud — the object with the lowest metallicity in the sample — where GMCs have velocity dispersions that are too small for their sizes. We will discuss the stability of these clouds in the light of our recent far-infrared analysis of this galaxy, and will contrast the results of the virial and far-infrared studies on the issue of the CO-to-H_2 conversion factor and what they tell us about the structure of molecular clouds in primitive galaxies.

Keywords. ISM: clouds, galaxies: ISM, galaxies: dwarf

1. Introduction

There is an emerging body of evidence suggesting that the formation of Giant Molecular Clouds (GMCs) provides the regulating step in transforming gas into stars in galaxies (Leroy *et al.* 2008; see also Elias Brinks's contribution in these proceedings). Moreover, GMC properties set the initial conditions for protostellar collapse, and likely play a determining role in setting the initial stellar mass function (McKee & Ostriker 2007). However, there is a dearth of data on the properties of GMCs in other galaxies, particularly low metallicity galaxies. Here we introduce a systematic study of GMCs across dwarf galaxies, and compare their properties with those measured in large galaxies and in the Milky Way. The full analysis is discussed by Bolatto *et al.* (2008).

Studies of GMCs in the Milky Way (Solomon *et al.* 1987; Heyer *et al.* 2004) find that GMCs are in approximate virial equilibrium, and obey uniform scaling relations commonly known as Larson laws (Larson 1981). These originate in the compressible supersonic magnetohydrodynamic turbulence (also known as Burger's turbulence) observed in the interstellar medium (Elmegreen & Scalo 2004).

2. Results and Discussion

How do extragalactic GMCs compare with those in the Galaxy? We have conducted a systematic study of spatially resolved extragalactic GMCs in galaxies in the Local Group and beyond, using a combination of interferometric and (for the Magellanic Clouds) single-dish CO observations. All observations have been analyzed in the same manner, using the `CPROPS` algorithm described by Rosolowsky & Leroy (2007).

Overall, we observe that the relationship between size, velocity dispersion, and luminosity observed for extragalactic GMCs in our sample are consistent with those determined in the Milky Way (Figure 1). This result underscores that the Galactic Larson relations provide a remarkably good description of CO-bright Giant Molecular Clouds independent of their environment, at least in the range of environments explored by this study (our lowest metallicity galaxy, as well as the lowest metallicity galaxy in which CO emission has been reliably detected, is the Small Magellanic Cloud).

Although the Larson relations are approximately Galactic there are some significant departures. GMCs in dwarf galaxies tend to be slightly larger than GMCs in the Milky Way, M 31, or M 33 for a given CO luminosity or velocity dispersion. The largest departures occur in the SMC. GMCs in dwarf galxies have average surface densities that are typically a factor of two under those of the Galaxy ($\Sigma_{\mathrm{GMC}} \approx 170 \; \mathrm{M}\odot \; \mathrm{pc}^{-2}$ for the Milky Way; Solomon *et al.* 1987).

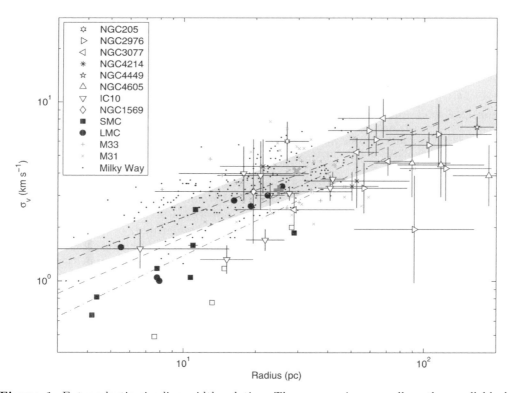

Figure 1. Extragalactic size-line width relation. The gray region as well as the small black points correspond to Solomon *et al.* (1987). The measurements for each galaxy are shown with the corresponding error bars. For the Magellanic Clouds, white and black symbols correspond to CO $1-0$ and CO $2-1$ measurements respectively. The large symbols for M31 and M33 correspond to their averages. The blue dashed and dot-dashed lines illustrate fits to all galaxies and dwarf galaxies only, respectively (see Bolatto *et al.* 2008 for details).

In the case of the Small Magellanic Cloud and probably also for IC 10, however, it seems that this explanation is not entirely viable since the departures are much too large and central cloud extinctions would be much too low. We suggest this is indirect evidence for large molecular envelopes faint in CO. We will come back to this point in a moment.

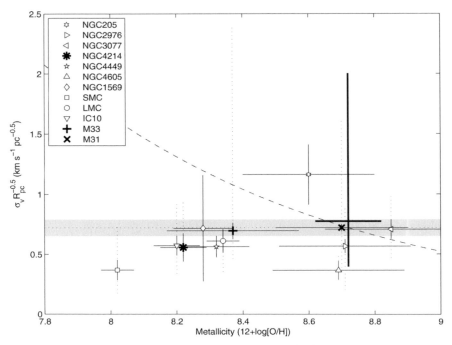

Figure 2. Galaxy averages of surface density vs. metallicity for GMCs in our sample. Filled lines indicate mean uncertainty, dotted lines indicate full observed range. The thick lines correspond to the Milky Way (full range shown in the vertical). The dashed line shows the expectation from photoinization-regulated star formation theory (McKee 1989).

Figure 2 shows that our data do not show evidence for the increase in GMC surface density with decreasing metallicity predicted by photoionization-regulated star formation theory (McKee 1989). Figure 3 shows that we do not see a systematic clear increase in the ratio of the virial to luminous mass (defined as the molecular mass obtained by using the Galactic CO-to-H_2 conversion factor) as a function of metallicity in our galaxies. In other words, we see an approximately constant CO-to-H_2 conversion factor in these GMCs. In fact, in the range of environments probed (that is, normal and dwarf non-starburst galaxies) the properties of resolved CO-bright GMCs are very uniform. The departure of the Small Magellanic Cloud from the Galactic ratio in Figure 3, for example, is entirely ascribable to the fact that the GMCs in this galaxy tend to be considerably smaller than GMCs in the Milky Way and the relation between virial and luminous mass is weakly dependent on mass (Solomon *et al.* 1987).

These results stand in contrast to analyses that use the far-infrared (either dust continuum or [CII] line emission) to trace molecular gas in low metallicity environments (e.g., Madden *et al.* 1997; Leroy *et al.* 2007). Studies of that type find a large increase in the CO-to-H_2 conversion factor and large parcels of CO-faint molecular gas. We suggest that the far-infrared and CO observations can be simultaneously understood if we assume that bright CO emission is only associated with the density peaks in metal-poor environments. Observations that resolve these density peaks find, as we show here, prop-

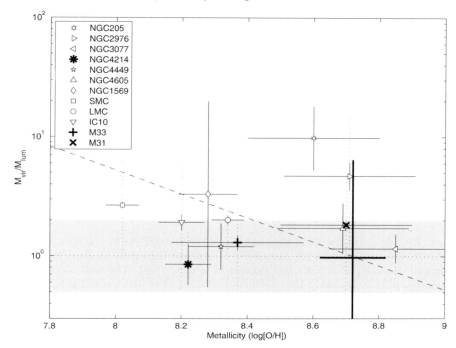

Figure 3. CO-to-H$_2$ conversion factor, obtained as the ratio of virial over luminous mass, versus metallicity for different galaxies in our sample. The gray line indicates the range in the Milky Way. The dashed line is the naive expectation Z^{-1}. Error bars are as described in Figure 2.

erties similar to Milky Way GMCs (see Heyer & Brunt 2004). They miss, however, large envelopes of CO-faint molecular gas. These envelopes would be CO-bright in objects of higher metallicity.

Such envelopes may contain much of the molecular gas as low metallicities (e.g., Leroy *et al.* 2007). Do they participate in star-formation? In the case of the Small Magellanic Cloud, for example, that appears necessary to preserve a normal star formation efficiency. It is unclear, however, how translucent gas may collapse and give rise to stars. Far-infrared observations of spectral transitions as well as continuum with high spatial resolution will be invaluable to understand the structure of the molecular gas in these enviroments.

References

Bolatto, A. D., Leroy, A. K., Rosolowsky, E., Walter, F., & Blitz, L. 2008, *ApJ*, in the press (arXiv0807.0009)

Elmegreen, B. G. & Scalo, J. 2004, *ARA&A*, 42, 211

Heyer, M. H. & Brunt, C. M., *ApJ*, 615, L45

Larson, R. B. 1981, *MNRAS*, 194, 809

Leroy, A., Bolatto, A., Stanimirovic, S., Mizuno, N., Israel, F., & Bot, C. 2007, *ApJ*, 658, 1027

Leroy, A., Walter, F., Brinks, E., Bigiel, F., de Block, W. J. G., Madore, B., & Thornley, M. D. 2008, *AJ*, in the press

Madden, S. C., Poglitsch, A., Geis, N., Stacey, G. J., & Townes, C. H. 1997, *AJ*, 483, 200

McKee, C. F. 1989, *ApJ*, 345, 782

McKee, C. F. & Ostriker, E. C. 2007, *ARA&A*, 45, 565

Rosolowsky, E. & Leroy, A. 2006, *PASP*, 118, 590

Solomon, P. M., Rivolo, A. R., Barrett, J., & Yahil, A. 1987, *ApJ*, 319, 730

Low-Metallicity Star Formation:
From the First Stars to Dwarf Galaxies
Proceedings IAU Symposium No. 255, 2008
L.K. Hunt, S. Madden & R. Schneider, eds.

Is there any pristine gas in nearby starburst galaxies?

Vianney Lebouteiller[1] and Daniel Kunth[2]

[1] Center for Radiophysics and Space Research, Cornell University, Space Sciences Building,
Ithaca, NY 14853-6801, USA
email: vianney@isc.astro.cornell.edu

[2] Institut dAstrophysique de Paris, UMR7095 CNRS, Universit Pierre & Marie Curie, 98 bis
boulevard Arago, 75014 Paris, France
email: kunth@iap.fr

Abstract. We derive the chemical composition of the neutral gas in the blue compact dwarf (BCD) Pox 36 observed with FUSE. Metals (N, O, Ar, and Fe) are underabundant as compared to the ionized gas associated with H II regions by a factor ~ 7. The neutral gas, although it is not pristine, is thus probably less chemically evolved than the ionized gas. This could be due to different dispersal and mixing timescales. Results are compared to those of other BCDs observed with FUSE. The metallicity of the neutral gas in BCDs seems to reach a lower threshold of $\sim 1/50\,Z_\odot$ for extremely-metal poor galaxies.

Keywords. ISM: abundances, (ISM:) HII regions, galaxies: abundances, galaxies: dwarf, galaxies: starburst, galaxies: ISM, ultraviolet: ISM, ultraviolet: galaxies

1. Introduction

Within the chemical downsizing scenario, in which massive galaxies are the first to form stars at a high rate (e.g., Cen & Ostriker 1999), dwarf galaxies can remain chemically unevolved since their formation. It is thus possible that some dwarf galaxies in the nearby Universe still contain pristine gas (Kunth & Sargent 1986; Kunth *et al.* 1994). An important step in understanding the process of metal enrichment of galaxies is to study all the gaseous phases involved in the gas mixing cycle, in particular the neutral phase.

Metals from the neutral gas can be observed through resonant lines in the far-ultraviolet (FUV). Blue compact dwarfs (BCDs) are ideal targets because they display large amounts of H I gas (Thuan & Martin 1981), and because the massive stars provide strong FUV continuum. The FUSE telescope (Moos *et al.* 2000) allows the observation of absorption-lines of H I together with many metallic species such as N I, O I, Si II, P II, Ar I, and Fe II. The neutral gas chemical composition was derived in several BCDs, IZw18 (Aloisi *et al.* 2003; Lecavelier *et al.* 2004), NGC1705 (Heckman *et al.* 2001), Markarian 59 (Thuan *et al.* 2002), IZw36 (Lebouteiller *et al.* 2004), NGC625 (Cannon *et al.* 2004), and SBS0335-052 (Thuan *et al.* 2005). Results showed that the neutral gas of BCDs has already been enriched with metals up to an amount of $\gtrsim 1/50\,Z_\odot$. This "threshold" metallicity value could represent a minimal amount due to starburst episodes. The second most important result is that the metallicity of the neutral gas is systematically lower than that of the ionized gas, implying that the neutral phase has been probably less processed.

We analyzed the blue compact dwarf galaxy Pox 36 which, because of its low H I column density, shows absorption-lines weak enough to be safely considered as unsaturated. Results are compared to those of the other BCDs.

2. Overview

The spectral continuum is provided by the UV-bright massive stars in the galaxy. Absorption-lines from species along the line of sight are superimposed on the continuum. The line of sight intersects the ISM from the Milky Way and the ISM from Pox 36 itself. The redshift of Pox 36 makes it possible to separate easily the Local absorption system from the intrinsic one. Absorption lines from the Milky Way are easily identified at an almost null radial velocity.

The absorption system corresponding to the neutral ISM in Pox 36 is detected at a velocity of $v_n = 1058 \pm 10\,\mathrm{km\,s^{-1}}$. We do not detect any H_2 lines. The radial velocity inferred from far-UV absorption lines is smaller than the value of the ionized gas as probed by optical emission lines, being consistent with a different origin of the two gaseous phases. More surprisingly, v_n is also smaller than the velocity derived from the 21 cm H I line. However, it must be stressed that the regions probed are different in terms of extent (because of the beam size) and in terms of depth (because of dust extinction). Hence the comparison of the velocity inferred from the optical, radio, and FUV must be interpreted with care. Instead, we take the opportunity given by the wealth of spectral features provided in the FUSE wavelength range to compare the relative velocities of the various galaxy components. In addition to tracers of the neutral gas, we have indeed access to the warm photoionized gas through the S III line at 1012.49 Å (giving $v_i = 1102 \pm 20\,\mathrm{km\,s^{-1}}$) and to the stars through the C III photospheric line at 1175.6 Å ($v_{stars} = 1082 \pm 20\,\mathrm{km\,s^{-1}}$). Hence using consistent data (only limited by the wavelength calibration), we find that the neutral gas is most likely indeed blue-shifted compared to the stars and to the ionized gas.

3. Stellar contamination and H I column density

The FUV spectrum of most of the BCDs and giant H II regions studied so far with FUSE is dominated by O stars. This is indicative of a starburst age younger than $\sim$ 10 Myr (Robert *et al.* 2003). Because of the high temperature, hydrogen is mostly ionized, and only weak H I photospheric lines can be observed which contribute to the − already saturated − core of the interstellar H I absorption line. Hence FUV spectra toward young starbursts allow a precise determination of the interstellar H I content. Older starbursts ($\gtrsim 10\,\mathrm{Myr}$) are dominated by B stars. Such stars have overall a more complex FUV spectrum than O stars because the degree of ionization is lower. For this reason, photospheric H I lines become prominent (Valls-Gabaud 1993). As an illustration, the H I lines from the Lyman series in the BCD Markarian 59 show "V-shaped" wings typical of B star photospheres (Thuan *et al.* 2002). Pox 36 shows a similar pattern. While Thuan *et al.* (2002) circumvented the problem by considering an artificial continuum to fit the profile of the H I lines, we decided to model the stellar absorption.

In order to reproduce the synthetic spectral continuum of Pox 36, we used the TLUSTY models for O and B stars (Lanz & Hubeny 2003; 2007) which provide detailed FUV absorption spectra. We choose to model the stellar spectrum of Pox 36 with a single stellar population. Since low-order Lyman lines (Lyα, Lyβ) are characterized by strong interstellar damping wings, interstellar H I dominates the global shape. On the other hand, cool stars contribute to prominent "V-wings" that are strongly constrained by the observed profile, especially of the high-order Lyman lines. We conclude that the stellar population contributing to the stellar H I profile is probably narrow around a given stellar type, in this case B0.

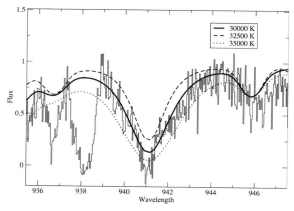

Figure 1. The Lyϵ line profile is dominated by stellar "V-wings" and gives a strong constraint on the photosphere temperature of the dominant stellar population. Models are drawn for a metallicity of $1/1000\,Z_\odot$, but the profile of stellar H I lines is unchanged for varying metallicities.

The wings of the stellar H I lines are strongly dependent on temperature. The best constraint is given by the high-order line of the Lyman serie since the interstellar contribution is saturated with no damping wings. Given the redshift of Pox 36 and the contamination by metallic lines from the Milky Way, Lyϵ is the cleanest H I line to constrain the temperature. The line profile gives a temperature of 30 000 K (Fig. 1), corresponding to a B0 class population.

Lyβ has prominent interstellar damped wings and can be used to constrain the interstellar H I column density if the stellar temperature is $\gtrsim$ 30 000 K, which is the case in Pox 36. After removing the stellar absorption component, we find $\log N$ (H I) = 20.28 ± 0.06. The uncertainty is only statistical and does not include systematic errors on the stellar model. We estimate a total conservative error of ± 0.3 dex.

Is this determination in agreement with the column density inferred from radio observations? Assuming an optically thin case and a distance of 20 Mpc, the integrated H I radio flux gives a mass of $M(\mathrm{H}) = 8.1 \times 10^8\,\mathrm{M}_\odot$. If the H I distribution is uniform, the average H I column density would be $\log N(\mathrm{H\ I}) \sim 20.7$. It must be stressed that the radio observation probes deeper lines of sight than those observed in the FUV. Hence, we could expect the H I column density measured in absorption to be roughly around half this determination, i.e., $\log N(\mathrm{H\ I}) \sim 20.4$. In addition, radio observations probe warm and cold neutral gas, while FUSE could observe mostly cold gas. As a conclusion, we estimate that the interstellar H I column density in Pox 36 is $\log N(\mathrm{H\ I}) = 20.3 \pm 0.4$ using Lyβ with conservative error bars.

4. Abundances

We derived metal column densities by fitting their profiles with Voigt profiles. We made the usual assumption of a single line of sight intersecting a medium with uniform properties (namely radial velocity, turbulent velocity, and column densities). The method to derive column densities from line profiles with FUSE is thoroughly discussed in Hébrard *et al.* (2002). Profiles are adjusted through a minimization of the χ^2 between the model and the observations. All the lines could be fit using simple Voigt profiles.

Abundances were then derived from column densities. It is the usual assumption to estimate the abundance of an element using the primary ionization state. We expect to find all elements with ionization potentials larger than that of hydrogen (13.6 eV) as

Table 1. Chemical abundances.

Element (tracer)	log (X/H)	$[\mathrm{X/H}]_n^1$	$[\mathrm{X/H}]_i^2$
N (N I)	-6.10 ± 0.42	-1.88 ± 0.42	-1.20 ± 0.06
O (O I)	-4.72 ± 0.41	-1.38 ± 0.41	-0.50 ± 0.05
Si (Si II)	-5.24 ± 0.47	-0.75 ± 0.47	...
P (P II)	-7.52 ± 0.47	-0.88 ± 0.47	...
Ar (Ar I)	-7.12 ± 0.47	-1.30 ± 0.47	-0.42 ± 0.06
Fe (Fe II)	-6.32 ± 0.42	-1.77 ± 0.42	-1.62 ± 0.11

Notes:
[1] Abundance in the neutral gas. [X/H] is defined as log (X/H) - log $(\mathrm{X/H})_\odot$, where log $(\mathrm{X/H})_\odot$ is the solar abundance. Solar abundances are from Asplund *et al.* (2005).
[2] Abundance in the ionized gas.

neutral atoms in the H I gas. This is the case for N, O, and Ar. On the other hand, Si, P, and Fe are mostly found as single-charged ions with negligible fractions of neutral atoms. Abundances are presented in Table 1.

It is clear that the neutral gas in Pox 36 is not pristine; it has already been enriched with heavy elements. This was expected since Pox 36 is more metal-rich than IZw18 and SBS0335-052 where no pristine gas has been found either (Lecavelier *et al.* 2004; Aloisi *et al.* 2003; Thuan *et al.* 2005).

Given its radial velocity, it seems that the neutral gas of Pox 36 is pushed away from the stellar clusters and their associated ionized gas. This is consistent with the gas being pushed by supernovae-driven shocks. The neutral gas cannot be enriched by the star-formation episode since we would expect a metallicity at least equal to or higher than that in the ionized gas of the H II regions. On the other hand, it is still possible that some of the atomic gas lies in front, being almost pristine. This would dilute the chemical abundances by adding H I without any metallic counterpart.

The final results (Table 1) show that N is underabundant by a factor ≈ 5 in the neutral gas of Pox 36 as compared to the ionized gas, while both O and Ar are underabundant by a factor ≈ 8. The relative agreement between the deficiency of N, O, and Ar in the neutral gas indicates that abundances of N and Ar are probably well determined, with only little ionization corrections required. Oxygen and argon are the best metallicity tracers available in the ionized gas, and agree with a metallicity of $\approx 1/3\,\mathrm{Z}_\odot$ in this phase. Considering these 2 elements, we find that the metallicity in the neutral gas should be $\approx 1/22\,\mathrm{Z}_\odot$, i.e., a factor ~ 7 below the value in the ionized gas. Our result in Pox 36 confirms those obtained for the other BCDs of the FUSE sample.

5. BCDs

We plot in Fig. 2 the elemental abundance [X/H] in the ionized gas and in the neutral gas of BCDs observed with FUSE. It can be seen that N and O (and Ar, not shown here) show identical trends, with a hint of a plateau at low metallicity and a positive correlation at higher metallicities. Part of the dispersion for [O/H] is probably due to saturation effects which could not be avoided in all cases. Part of the dispersion for [N/H] and [Ar/H] is probably due to ionization corrections. Since the trend is similar for N, O, and Ar, it seems that the corrections are on the same order for all the objects (probably within 0.5 dex), or that they are in fact negligible, as suggested by results of Pox 36.

A caveat exists in that the presence of foreground H I gas with no metals could dilute the metal abundances in the neutral gas. Therefore, except if the dilution factor is identical for all the sources, the presence of the plateau and of the correlation suggests that there is no significant dilution by metal-free H I gas.

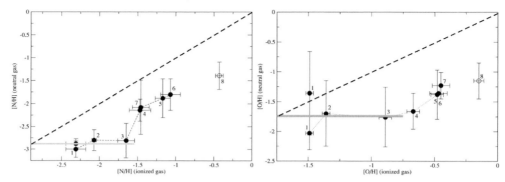

Figure 2. The abundances of nitrogen (left) and oxygen (right) in the ionized gas and in the neutral gas are compared. Labels: (1) IZw18, (2) SBS0335-052, (3) IZw36, (4) Mark 59, (5) Pox 36, (6) NGC 625, (7) NGC 1705, (8) NGC 604/M33. The dashed line indicates the 1:1 ratio.

At low-metallicities ($\lesssim 1/8\,Z_\odot$), there may be a plateau for which abundances in the neutral gas are not correlated with those in the ionized gas. The BCDs in this plateau are IZw18, SBS0335-052, and IZw36. There is a strong need for analyzing BCDs with similar metallicities to better characterize the plateau. The enrichment associated with this plateau cannot be due to population III stars but rather by multiple starburst episodes.

At higher metallicities $\gtrsim 1/5\,Z_\odot$, the abundances of the neutral gas and ionized gas are correlated. The global offset is around ~ 0.9 dex, i.e., almost a factor 10. This offset applies to objects with metallicities as large as solar. Moreover, it seems to apply also in the neutral gas of spiral galaxies (M 33/NGC 604; Lebouteiller *et al.* 2006) and not only to BCDs. We might be seeing here the effect of delayed enrichment of metals in the neutral gas.

References

Aloisi, A., Savaglio, S., Heckman, T. M., *et al.* 2003, ApJ, 595, 760
Cannon, J. M., McClure-Griffiths, N. M., Skillman, E. D., & Côté, S. 2004, ApJ, 607, 274
Cen, R. & Ostriker, J. P. 1999, ApJL, 519, L109
Hébrard, G., Lemoine, M., Vidal-Madjar, A., *et al.* 2002, ApJS, 140, 103
Heckman, T. M., Sembach, K. R., Meurer, G. R., *et al.* 2001, ApJ, 554, 1021
Kunth, D. & Sargent, W. L. W. 1986, ApJ, 300, 496
Kunth, D., Lequeux, J., Sargent, W. L. W., & Viallefond, F. 1994, A&A, 282, 709
Lanz, T. & Hubeny, I. 2003, ApJS, 146, 417
Lanz, T. & Hubeny, I. 2007, ApJS, 169, 83
Lebouteiller, V., Kunth, D., Lequeux, J., *et al.* 2004, A&A, 415, 55
Lebouteiller, V., Kunth, D., Lequeux, J., Aloisi, A., Désert, J.-M., Hébrard, G., Lecavelier Des
    Étangs, A., & Vidal-Madjar, A. 2006, A&A, 459, 161
Lecavelier des Etangs, A., Désert, J.-M., Kunth, D., *et al.* 2004, A&A, 413, 131
Moos, H. W., Cash, W. C., Cowie, L. L., *et al.* 2000, ApJL, 538, L1
Robert, C., Pellerin, A., Aloisi, A., *et al.* 2003, ApJS, 144, 21
Thuan, T. X. & Martin, G. E. 1981, ApJ, 247, 823
Thuan, T. X., Lecavelier des Etangs, A., & Izotov, Y. I. 2002, ApJ, 565, 941
Thuan, T. X., Lecavelier des Etangs, A., & Izotov, Y. I. 2005, ApJ, 621, 269
Valls-Gabaud, D. 1993, ApJ, 419, 7

Session V

Metal-poor IMFs, stellar evolution, and star-formation histories

Low-Metallicity Star Formation:
From the First Stars to Dwarf Galaxies
Proceedings IAU Symposium No. 255, 2008
L.K. Hunt, S. Madden & R. Schneider, eds.

© 2008 International Astronomical Union
doi:10.1017/S1743921308024964

The Metal-Poor IMF, Stellar Evolution, and Star Formation Histories

Evan D. Skillman[1]

[1]Astronomy Department, University of Minnesota, 116 Church St. SE, Minneapolis, MN 55455
email: skillman@astro.umn.edu

Abstract. I present an introduction to three important subjects relevant to low metallicity star formation: the IMF, stellar evolution, and star formation histories. I will draw on observations from the LCID (Local Cosmology from Isolated Dwarfs) project to illustrate some of these topics.

Keywords. stars: evolution, galaxies: evolution, galaxies: dwarf, (galaxies:) Local Group

1. Introduction

I was asked by the organizing committee to give an introductory lecture to the "IMF, Stellar Evolution, and Star Formation Histories," section of the conference program. Obviously a comprehensive overview of these three important topics is not possible here. Instead I settled on a few comments which I hoped might provoke some thought. I entertained myself by including references to work by all of the section's speakers in the talk, but some of those may have not been propagated through to this write-up.

2. The Initial Mass Function (IMF)

"Testing the Universality of the IMF remains as our primary challenge for the coming decade" Kennicutt (1998).

This quote comes from a review by Kennicutt, which provided the lead-off talk to a conference dedicated to the study of the IMF. The bulk of that talk focused on the importance of the IMF to just about every possible problem in astrophysics. Although ten years old, I highly recommend the talk as a good starting point for understanding the importance of the IMF and the biggest challenges in its study.

2.1. *Do we expect the IMF to change with metallicity?*

It is likely that the IMF is a reflection of the ISM from which the stars are born. Thus, characterizing the properties of the ISM at low-metallicity is vital to our understanding of the star formation process, and, accordingly, galaxy formation and evolution.

The chemical composition and content of the ISM in low-metallicity systems are fundamentally different than those of massive metal-rich galaxies. The dust and molecular gas content of galaxies are commonly assumed to scale with metallicity, with the most metal-poor systems being essentially devoid of dust, and showing a relative under-abundance of molecular gas. In massive galaxies, dust and molecular gas play a crucial role in the formation of stars, and yet, many low mass systems are actively forming stars in the assumed relative absence of dust and molecular gas.

Theoretical arguments indicate that the nature of the ISM in galaxies changes as a function of metallicity. For example, Norman & Spaans (1997) and Spaans & Norman (1997) emphasize a "phase transition" which occurs in the range of $Z \approx 0.03 - 0.1 \, Z_\odot$. In their models, the ISM at lower metallicities is characterized by a single temperature, and the familiar multiphase medium with stable cool and warm components does not appear until the metallicity has sufficiently increased. Thus, at lower metallicities, star formation is suppressed with obvious implications for the evolution of dwarf galaxies. There are also several theoretical studies which show that the character of photodissociation regions changes significantly as a function of metallicity, e.g., van Dishoeck & Black (1988), Maloney & Black (1988), Lequeux et al. (1994), Bolatto et al. (1999), Röllig et al. (2006).

Observational evidence is accumulating which supports the impression that the character of the ISM is different at lower metallicities. In the more than two decades since the earliest observations in dwarf galaxies by Elmegreen et al. (1980) and Tacconi & Young (1987), CO has not been detected in any galactic environment with $12+\log(\text{O/H})$ $< \sim 8.0$. Taylor et al. (1998) have claimed that there is a strong transition below Z ≈ 0.1 $Z_\odot$, where CO emission drops dramatically relative to galaxy properties such as size and luminosity. The lack of CO detections at lower metallicities was recently confirmed in a large survey for CO in dwarf galaxies by Leroy et al. (2005) and a very deep observation of the extremely metal-poor galaxy I Zw 18 ($12+\log(\text{O/H}) = 7.17$, Skillman & Kennicutt (1993)) by Leroy et al. (2007). There is on-going debate about the metallicity dependence of the conversion from CO emission to H_2 mass, see e.g., Cohen et al. (1988), Elmegreen (1989), Rubio et al. (1993), Wilson (1995), Verter & Hodge (1995), Arimoto et al. (1996), Bolatto et al. (2003), Leroy et al. (2006), Bolatto et al. (2008), and references therein, with the result that the H_2 masses of low metallicity dwarf galaxies are still not well known. Nonetheless, the complete absence of CO emission in dwarf galaxies showing a strong presence of star formation indicates that something is different at these lower metallicities.

Another indication that the nature of the ISM changes at low metallicity comes from the recent surveys measuring the diffuse 8 μm emission from hot (~ 350 K) dust and PAHs, i.e., Engelbracht et al. (2005), Hogg et al. (2005), Jackson et al. (2006), Rosenberg et al. (2006). Engelbracht et al. (2005) showed that an abrupt change in the properties of the ISM occurs at a metallicity 1/3 to 1/5 of the solar value. They showed that the ratio of the 8 to 24 μm luminosities dropped significantly in galaxies below this metallicity, which was attributed to the weakening of the PAH features at low-metallicity. Jackson et al. (2006) showed that this *transition metallicity* coincides with a drop off of other ISM components, namely the molecular gas content, and also that the slope of the diffuse 8 μm luminosities below this metallicity continues to very low metallicity (over one dex below the Engelbracht et al. (2005) transition metallicity).

Although the paucity of diffuse 8 μm emission at low-metallicity is well established, the Spitzer IRAC 8 μm imaging programs that have been so efficient at detecting this emission have had the major drawback that they do not distinguish between emission from the 7.7 μm PAH feature and the continuum emission from hot dust grains. Cannon et al. (2006) showed that one region in the dwarf starburst galaxy NGC 1705 that emitted strongly at 8 μm was dominated by emission from PAHs, while another region emitted only hot dust continuum radiation with no detected PAH emission. The absence of PAHs can be a result of the hard radiation fields that are pervasive in low-metallicity starburst systems destroying PAHs, see Madden et al. (2006), although it is not clear why in some cases the PAHs survive, e.g., Cannon et al. (2006). What is even less well understood are the relative contributions of PAH and hot-dust continuum emission in typical quiescent dwarf galaxies, which have very different radiation fields than starburst galaxies.

PAH emission is typically observed along the outer edges of PDRs, e.g., Giard et al. (1994), Helou et al. (2004), where they are thought to be transiently heated by single high-energy photons, c.f., Leger & Puget (1984). However, Lemke et al. (1998) observed PAH emission in a region of the Galaxy illuminated by only the typical interstellar radiation field of the solar neighborhood. Thus, it is clear that strong radiation fields *do* excite PAHs into emission, but it is not yet clear if they are *required* to excite them. Smith *et al.* (2007) recently published a comprehensive spectroscopic study of PAH emission in 59 galaxies over an impressive range of host galaxy metallicities and galaxy morphological types. They found a dependence of the relative PAH band strengths on host galaxy metallicity with the longer wavelength PAH bands being relatively weaker at lower metallicity, which was attributed to the inability of large PAHs to survive in low-metallicity environments.

Finally, the abundance of dust has long been known to depend strongly on the metallicity of the environment. Although the abundance of dust becomes low at low metallicities, recent observations have shown some evidence for dust even in the lowest metallicity environments known. Cannon *et al.* (2002) find $\sim 10^3$ $M_\odot$ of dust in the most metal-deficient galaxy known, I Zw 18. A similar abundance of dust is found in the second most metal-poor galaxy SBS 0335$-$052, see Dale *et al.* (2001), Houck *et al.* (2004). While the results are encouraging, it is important to note that almost all of the available information on the properties of dust at very low metallicities comes from either reddening or UV absorption observations. A comparatively unexplored area of the low-metallicity ISM is the far-IR emission from dust. This spectral region radiates the bulk of the bolometric luminosity from normal galaxies, e.g., Calzetti *et al.* (2000), and allows one to trace the locations of dust with much better precision than is available with other techniques. Previous far-IR investigations have typically excluded dwarf systems due to their lower surface brightness compared to more massive galaxies. Nonetheless, the majority of Local Group dIrr galaxies were detected by IRAS, c.f., Melisse & Israel (1994), Lisenfeld & Ferrara (1998).

In Marco Spaans' talk, he showed models where a phase transition occurred at metallicities between 1% and 0.1% of the solar value. Most of the above evidence indicates that major changes happen in the ISM at metallicities which are closer to 10% of the solar value. During later discussions at the conference, Marco indicated that the transition metallicity is a strong function of the ISM porosity, so a more porous ISM will experience the phase transition at higher metallicities (than the models that he showed). Thus, it may be possible that the observed changes could be interpreted within his phase transition model.

2.2. *Is there evidence for a metallicity dependence of the upper IMF cut-off?*

At metallicities of roughly the solar value, several studies suggest an upper limit to the IMF of $\approx$ 120 - 200 $M_\odot$, e.g., Weidner & Kroupa (2004), Oey & Clarke (2005), Figer (2005), but see also Melena *et al.* (2008). Despite a rather convincing figure in Figer (2005), this result is not without controversy, but it does provide a good start for this discussion.

The reason that there is controversy, even within our own galaxy, is that there are several criteria that an observation of a cluster must meet in order to be decisive on this point. First, the cluster must be young enough that the most massive stars have not yet evolved away. Second, the cluster must be massive enough that the upper IMF is completely populated. Third, the spread in the formation ages of the stars must be less than the lifetimes of the most massive stars. It is not clear that there are any known clusters which fulfill these three criteria without question. Additionally, luminosities and spectra must be converted into indicative masses with confidence. Some would argue that spectra are required for every star in order to meet this criterion, making this type of study very expensive in terms of telescope time, even in our own galaxy.

Given the above, is there any hope of making similar measurements at significantly lower metallicities? Specifically, can we make this type of observation at metallicities below where we might expect a phase transition in the ISM? In fact, this type of observation in an environment with a metallicity less than 10% of the solar value becomes considerably more difficult solely due to distance. The *nearest* low-metallicity, star forming environments (below one-tenth solar, where one might expect a putative phase change in the ISM) - the isolated dwarf irregular galaxies of the Local Group, e.g., Mateo (1998) - are ten times further away than the Magellanic Clouds.

Are there other ways to probe the upper IMF beside a strict stellar census in a cluster? From α/Fe ratios in stars, Tolstoy *et al.* (2003) suggest that they are seeing evidence for a truncated IMF in dSphs. However, Koch *et al.* (2008) point to the important roles of metal-enhanced winds and inhomogeneous mixing in the interpretation of α/Fe ratios, and question the uniqueness of the Tolstoy *et al.* hypothesis in interpreting the observations. As more observations of precision relative abundances become available in the dSphs, it may be possible to identify trends in the plots of X/Y vs. Fe/H to constrain the possible physical processes which give rise to these trends. From the small sample that Tolstoy *et al.* (2003) had to work with, the suggestion of a truncated IMF must be regarded as speculative, but the quantity of the data available is increasing rapidly.

There is really quite a bit at stake here. For example, Garnett (2002) has demonstrated that the effective oxygen yields of star forming dwarf galaxies increase with the mass of the galaxy. This is usually interpreted as due to the effect of metal enhanced winds flowing out of the small gravitational potential wells of the smallest galaxies. However, a metallicity dependent upper IMF truncation would have an identical effect.

2.3. *Is the upper IMF slope metallicity dependent?*

Leaving the upper mass cut-off, is it possible to constrain the slope of the upper IMF? Historically, the answer has been that there is no firm evidence for departures from the "universal" IMF, but this is also observationally very difficult. Naively, one would think that one can measure very large numbers of stars in nearby galaxies, and thus, be able to constrain the IMF. In practice, this becomes very difficult because of the degeneracy between the star formation history and the IMF. This is discussed in great detail in Miller & Scalo (1979), and, more recently, in Elmegreen & Scalo (2006).

In Fig. 1, from a study of NGC 5128 by Rejkuba *et al.* (2004), one can see that for color-magnitude diagrams of photometric depths typical of nearby galaxy studies that the constraints on the IMF are relatively weak. In the figure it is clear that reasonable fits to the main sequence luminosity function are possible over a large range in IMF slope.

If a large range of IMFs are consistent with the data typically observed in nearby galaxies, does that weaken the conclusions one derives from the study of the color-magnitude diagrams? The silver lining here is that the reconstructed star formation rate is not strongly dependent on the assumption of the IMF. Figure 6 of Angeretti *et al.* (2005) shows an example where the calculated star formation rate of NGC 1569 changes by less than a factor of two over a range in the IMF slope from $\sim$1 to $\sim$2.5. Of course, when the assumed IMF exceeds $\sim$2.7, the star formation begins to shoot up dramatically. Nonetheless, it is comforting that a reasonably large range in the assumed IMF results in rather stable calculated star formation rates.

3. Stellar Evolution

"Massive stars played a key role in the early evolution of the Universe ... It is therefore very important to understand their evolution." Hirschi *et al.* (2008)

"We hope to encourage observers to provide stronger observational constraints where they are needed .." Gallart *et al.* (2005)

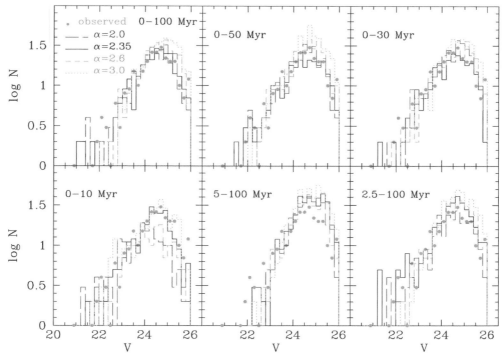

Figure 1. The observed Main Sequence (MS) luminosity function (dots) is compared with synthetic luminosity functions for different SFHs for $Z = 0.008$ and four different IMF slopes ($\alpha = 2.0$, 2.35, 2.6, and 3.0) as indicated in the legend. Synthetic luminosity functions correspond to models calculated adopting the appropriate distance, reddening and a constant SFR in the interval that is indicated in each panel (beginning-end of the burst). In the observed and simulated luminosity functions color and magnitude cuts are applied, and only stars bluer than $U - V = -0.7$ and brighter than $V = 26$ are compared (figure and caption from Rejkuba *et al.* 2004).

The first quote prompted my observation that the astronomical community is still dangerously low in the resources allocated to research on stellar evolution. Practically everything that we study is dependent to some degree on the results of stellar evolution models. As this modelling continues to explore new physical parameters (e.g., rotation, winds) the hope is that the models approach better representations of nature. We heard of great progress from new models in the talks by Leitherer and Ekström.

In this section I review some of my own experience with using the results of stellar evolution models, and close with what I believe is a promise to follow-on to the encouragement given in the second quote.

3.1. *Using HST to map recent star formation in dwarf galaxies: testing our knowledge of the star formation process*

My closest connection to stellar evolution research is my use of stellar evolution models to recreate recent star formation histories of nearby dwarf galaxies. Here I review the power of combining *Hubble Space Telescope* (HST) photometry with models of stellar evolution to produce spatially resolved recent star formation histories of galaxies with time resolution of order 30 Myr over the last few hundred Myr.

The first HST color magnitude diagram of the stars in the nearby dwarf irregular galaxy Sextans A by Dohm-Palmer *et al.* (1997) revealed for the first time a clear separation between the brightest main sequence stars and the blue helium burning (BHeB) stars - intermediate and high mass stars that have evolved beyond the main sequence, ignited helium burning in their cores, and have migrated back to the blue side of the color magnitude diagram. The separation of the main sequence stars and the BHeB was made possible by the high angular resolution of the HST (reducing photometric errors due to blends) and the low metallicity of Sextans A which leads to low differential reddening.

These BHeB stars afford a special opportunity to study the recent star formation histories of nearby galaxies. Because the position of a BHeB star in the CMD represents a unique age (as opposed to, for example, the main sequence or the red giant branch where a single position can correspond to a large range of ages), one can convert the BHeB luminosity function directly into a star formation history (SFH) (with the assumption of a universal initial mass function). The main limiting factor of this technique is that the position of the BHeB stars blends into the red clump at an age between 0.5 and 1 Gyr. Because one knows the positions of the BHeB stars in the galaxy, then one can produce a spatially resolved SFH (i.e., it is possible to produce "movies" of the recent star formation).

Translating the BHeB star luminosity function into a SFH depends on the accuracy of the stellar evolution models. We have different lines of evidence that the stellar evolution models provide an excellent guide to this stage of evolution. The first line came from the early HST observations themselves. Although the position of the blueward extension of the stellar evolution tracks is a strong function of both metallicity and stellar mass, excellent agreement between observations and models was found by choosing the stellar evolution tracks for the metallicity determined from the HII region abundances for Sextans A. This is a very important point. There are no low metallicity, young stellar clusters in the Milky Way galaxy or the Magellanic Clouds which allow the stellar evolution modelers to calibrate their codes in this regime (the oxygen abundance in Sextans A is a factor 3 lower than in the Small Magellanic Cloud). When these stars are first observed in the extragalactic context, the agreement with models can be taken as confirmation of a prediction.

Dohm-Palmer *et al.* (2002) used deeper HST photometry of Sextans A to compare recent star formation histories recovered from both the main-sequence stars and the BHeB stars for the last 300 Myr. The excellent agreement between these independent star formation rate (SFR) calculations is a resounding confirmation for the legitimacy of using the BHeB stars to calculate the recent SFR. Dolphin *et al.* (2003) derived the recent SFH of Sextans A from the entire CMD and found good agreement with that derived from the BHeB stars alone.

Additionally, Dohm-Palmer & Skillman (2002) used all of the HST observations of Sextans A in order to compare the ratio of blue to red supergiants. This ratio provides an observational constraint on the relative lifetimes of these two phases that is a sensitive test for convection, mass-loss, and rotation parameters. Analyzing the ratio as a function of age, or, equivalently, mass eliminates the confusion of unknown star formation histories (as in previous studies of this ratio). The functional form of the observed ratio matches the model extremely well with an offset of roughly a factor of 2 (and the offset is seen as support of the latest models which include rotation). Given these tests of reliability of the stellar models, we feel confident that using the BHeB stars to construct recent star formation histories is well justified.

Dohm-Palmer *et al.* (1998) employed HST photometry of four nearby dwarf irregular galaxies (Sextans A, Leo A, Pegasus, and GR 8) and derived recent star formation

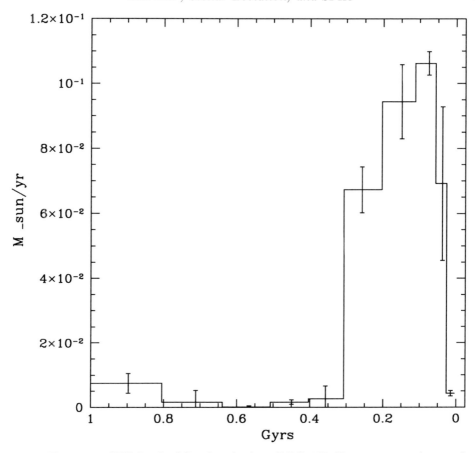

Figure 2. The recent SFH for the M81 dwarf galaxy DDO 165. Here we see an intense burst of star formation that ended just recently. The burst lasted for roughly 200 Myr.

histories for these galaxies. At the time, the surprising result was the lack of bursts or episodes of enhanced star formation. With time bins of only 25 Myr for the last 500 Myr, with the possible exception of enhanced star formation rates in the last 50 Myr for Sextans A and Leo A, all four galaxies are best described as nearly constant star formation rates. In retrospect, perhaps the lack of truly zero star formation rates is as surprising as the lack of periods of enhanced star formation.

3.2. HST ACS observations of M81 Group dwarfs

My collaborators and I have been able to continue this work through an HST program to study the star forming dwarf galaxies of the M81 Group. Studies of the impact of star formation ('feedback') on the properties of a galaxy are of fundamental importance to understanding galaxy evolution. One crucial aspect in these studies is a precise census of the recent star formation in a galaxy. We have recently obtained HST ACS observations of a sample of M81 Group dwarfs which allow deriving spatially resolved star formation histories with a time resolution of roughly 30 Myr over the last 500 Myr, see Weisz *et al.* (2008). Our sample comprises nine galaxies in the M81 group which is host to a wide diversity of dwarf star forming galaxies. They span ranges of 6 magnitudes in luminosity, 1000 in current star formation rate, and 0.5 dex in metallicity. The ACS observations allow us to directly observe the strength and spatial relationships of all of the star formation in these galaxies in the last 500 Myr. We can then quantify the star formation and measure (1) the fraction of star formation that is triggered by feedback, (2) the fraction of star formation that occurs in clusters and associations, and (3) to what degree star formation is governed by the feedback from star formation. The ACS observations will be complemented with high–quality ancillary data collected by our team for all galaxies (e.g., Spitzer, UV/optical/NIR, VLA HI). This will enable us to construct prescriptions of how star formation and feedback depend on metallicity, size, gas content, and current star formation rates in galaxies. As an example of the recent star formation history reconstructions, Fig. 2 shows the recent SFH of DDO 165, revealing a rather dramatic burst of star formation in the recent past.

In working on this new sample of galaxies, one of the things that we have discovered is that the stellar models do not follow the tracks of the BHeB stars nearly as well as in our studies of the low metallicity Local Group galaxies. To first order, the tracks predict the positions well; the trends predicted by changing metallicities are seen, but the detailed shapes in the color magnitude diagrams show some discrepancies. In the spirit of the Gallart *et al.* quotation, we will provide the stellar modellers with new constraints on this phase of stellar evolution. Finally, I note that the HST ACS Nearby Galaxy Survey (ANGST, PI Dalcanton) has provided new, high quality observations of a larger sample of nearby galaxies, so these constraints will grow in quantity.

4. Star Formation Histories

"The lack of a comprehensive scenario for the formation and evolution of dwarf spheroidal galaxies contrasts with the large amount of available data for these nearby Local Group satellites," Salvadori *et al.* (2008), and see also de Rijcke *et al.* (2005)

My work on star formation histories of galaxies has been almost exclusively confined to dwarf galaxies, which I will review here. Although the quote above sets a rather harsh view, I believe that significant progress has been made in the last decade.

4.1. Lifetime star formation histories of dwarf galaxies

The nearby dwarf galaxies of the Local Group are a unique probe of the chemical and structural evolution of the universe over its entire history. Interpretation of the fossil evidence contained within these systems provides direct and quantitative constraints on their star formation histories (SFHs) and broader evolution (e.g., Mateo 1998). However, dwarf galaxy evolution is dependent upon both local and cosmic environmental factors. It is therefore of tremendous importance to disentangle these effects for the nearby galaxies so that key results obtained locally can be correctly interpreted in a wider cosmological context. For example, it is important to understand whether the SFH of a dwarf galaxy is affected more by the radiation field of a nearby massive galaxy or by the diffuse background radiation field from the epoch of cosmic reionization.

The position — morphology relation seen for the lowest-mass systems in the Local Group indicates that local environmental factors can be as important as the generic cosmic structure formation framework; gas-poor, pressure-supported dwarf galaxies (dwarf spheroidal, dSph) are preferentially found as satellites to the MW and M31, whereas gas-rich rotating systems (dwarf irregular, dIrr) are preferentially found in isolated locales, e.g., van den Bergh (1994a). Additionally, the closest MW dSph companions, like Draco and Ursa Minor have only old stellar populations with ages $\geqslant 10$ Gyr, whereas those at a distance of 100 kpc or greater, like Fornax or Leo I have prominent young and intermediate-age stellar populations see, e.g., van den Bergh (1994b).

Detailed dynamical modeling of the orbital evolution of dwarf satellites provides clues to the role of local environment in dwarf galaxy evolution, c.f., Mayer *et al.* (2001a), Mayer *et al.* (2001b), Mayer *et al.* (2006). "Tidal stirring," which is a combination of tidal effects and disk shocking, can remove most of the gas from a dwarf galaxy and transform rotationally supported systems into pressure supported systems. Essentially, this will convert a dIrr galaxy into a dSph galaxy. The tidal stripping of both baryonic and dark matter from satellites is also known to occur, and while this process alone is unlikely to change the morphology of a system from a dIrr to a dSph, it will still change the observable properties of the dwarf, as spectacularly demonstrated by the Sagittarius dwarf galaxy. Finally, there is some evidence for a link between tidal interactions and star formation, such that tides are able to trigger star formation. All of these effects suggest that the observed properties of dwarf galaxies should be a complex function of their spatial location and orbital properties within the group where they are observed.

I am part of a cycle 14 HST program (LCID or Local Cosmology from Isolated Dwarfs) in which five isolated, Local Group dwarfs were observed to depths below the oldest main sequence turnoffs. Matteo Monelli's contribution to these proceedings gives a nice overview of the project and its goals. In this program, Cole *et al.* (2007) have uncovered one dwarf (Leo A) with evidence of suppression of star formation at early ages — resembling the expected signature of cosmic reionization. The dwarf irregular galaxy also shows evidence of delayed star formation, as seen in an earlier WFPC2 study by Skillman *et al.* (2003), although the new ACS observations allow for better time resolution and smaller errors on the star formation rates. The isolated dSph galaxies, Cetus and Tucana, show almost exclusively old stars attributed to an episode of star formation in the first few billion years of the age of the universe. The one complication with their color magnitude diagrams is the presence of a blue straggler population. It is not clear whether these can be interpreted as very low level star formation at later ages, or the result of primordial binary stars as discussed in Momany *et al.* (2007) and Mapelli *et al.* (2007). Finally, the "transition" galaxies, LGS-3 and Phoenix, are intermediate between with regard to star formation histories. They both show predominantly early star formation, but continue to form stars at a significantly level up to the present.

4.2. *IMF constraints from LCID observations*

Because of the exceptional quality of the LCID photometry, I was curious to see what constraints could be gained on the IMFs of the observed dwarfs. Sebastian Hidalgo ran a number of experiments, fitting the detailed color magnitude diagrams with different values of the IMF slope and then determining the χ^2 for the best possible fits. We concentrated these tests on LGS-3 and IC 1613. Because of the low level of recent star formation in LGS-3, the IMF slope could be changed over a large range (from zero to 5) with only very small changes in the χ^2 (1.05 to 1.2). This clearly showed that the best constraints on the IMF will come from galaxies with strong star formation at all ages. For IC 1613 we found that the χ^2 was much more sensitive to the IMF, bounding the slope to values between one and three. Given the quality of the photometry, this loose constraint may come as a disappointment. On the positive side, for the range of IMF values which gave satisfactory χ^2 values, the star formation histories appear nearly identical. There is a small degree of transferring early star formation for late star formation (as expected), but the overall trend with time remains preserved.

5. Summary

I'm not sure that a summary is appropriate for a introductory talk which is really just a collection of thoughts. However, at the conference I did emphasize that the Local Group and nearby dwarf galaxies can provide us with star formation histories for a significant number of galaxies sampling a range of environments. The range of metallicities in these nearby galaxies is large. In fact, some of them (e.g., Leo A and Sag DIG) are comparable to the low metallicity of I Zw 18, see, Skillman *et al.* (1989), Saviane *et al.* (2002), van Zee *et al.* (2006). This large range in metallicity in nearby galaxies allows us to search for metallicity dependences in the star formation process. The lifetime star formation histories that are available from the HST allow us to identify the physical processes that determine a galaxy's current evolutionary state (which we refer to as morphology). Interestingly, we now have very clear evidence that the small dwarf irregular galaxies have had significant delays in their star formation relative to the dSphs.

Acknowledgements

Finally, I would like to thank the conference organizers for a fantastic experience and my many collaborators for their efforts.

References

Angeretti, L., Tosi, M., Greggio, L., Sabbi, E., Aloisi, A., & Leitherer, C. 2005, *AJ*, 129, 2203
Arimoto, N., Sofue, Y., & Tsujimoto, T. 1996, *PASJ*, 48, 275
Bolatto, A. D., Jackson, J. M., & Ingalls, J. G. 1999, *ApJ*, 513, 275
Bolatto, A. D., Leroy, A., Israel, F. P., & Jackson, J. M. 2003, *ApJ*, 595, 167
Bolatto, A. D., Leroy, A. K., Rosolowsky, E., Walter, F., & Blitz, L. 2008, ArXiv e-prints, 807, arXiv:0807.0009
Calzetti, D., Armus, L., Bohlin, R. C., Kinney, A. L., Koorneef, J., & Storchi-Bergmann, T. 2000, *ApJ*, 533, 682
Cannon, J. M., Skillman, E. D., Garnett, D. R., & Dufour, R. J. 2002, *ApJ*, 565, 931
Cannon J. M., *et al.* 2006, *ApJ*, 647, 293
Cohen, R. S., Dame, T. M., Garay, G., Montani, J., Rubio, M., & Thaddeus, P. 1988, *ApJL*, 331, L95
Cole, A. A., *et al.* 2007, *ApJL*, 659, L17
Dale, D. A., Helou, G., Neugebauer, G., Soifer, B. T., Frayer, D. T., & Condon, J. J. 2001, *AJ*, 122, 1736
de Rijcke, S., Michielsen, D., Dejonghe, H., Zeilinger, W. W., & Hau, G. K. T. 2005, *A&A*, 438, 491
Dohm-Palmer, R. C. & Skillman, E. D. 2002 *AJ* 123, 1433
Dohm-Palmer, R. C., Skillman, E. D., Saha, A., Tolstoy, E., Mateo, M., Gallagher, J.S. Hoessel, J., Chiosi, C., & Dufour, R. J. 1997, *AJ* 114, 2514
Dohm-Palmer, R. C., Skillman, E. D., Gallagher, J. S., Tolstoy, E., Mateo, M., Dufour, R. J., Saha, A., Hoessel, J., & Chiosi, C. 1998, *AJ* 116, 1227
Dohm-Palmer, R. C., Skillman, E. D., Mateo, M., Saha, A., Dolphin, A., Tolstoy, E., Gallagher, J. S., & Cole, A. A. 2002, *AJ* 123, 813
Dolphin, A. E., *et al.* 2003, *AJ*, 126, 187
Elmegreen, B. G. 1989, *ApJ*, 338, 178
Elmegreen, B. G., Morris, M., & Elmegreen, D. M. 1980, *ApJ*, 240, 455
Elmegreen, B. G. & Scalo, J. 2006, *ApJ*, 636, 149
Engelbracht, C. W., Gordon, K. D., Rieke, G. H., Werner, M. W., Dale, D. A., & Latter, W. B. 2005 *ApJ*, 628, L29
Figer, D. F. 2005, *Nature*, 434, 192
Gallart, C., Zoccali, M., & Aparicio, A. 2005, *ARAA*, 43, 387
Garnett, D. R. 2002, *ApJ*, 581, 1019

Giard, M., Bernard, J. P., Lacombe, F., Normand, P. , & Rouan, D. 1994, *A&A*, 291, 239

Helou, G. *et al.* 2004, *ApJS*, 154, 253

Hirschi, R., Chiappini, C., Meynet, G., Maeder, A., & Ekström, S. 2008, IAU Symposium, 250, 217

Hogg, D. W., *et al.* 2005, *ApJ*, 624, 162

Houck, J. R. *et al.* 2004, *ApJS*, 154, 211

Jackson, D. C., Cannon, J. M., Skillman, E. D., Lee, H., Gehrz, R. D., Woodward, C. E., & Polomski, E. 2006, *ApJ*, 646, 192

Kennicutt, R. C., Jr. 1998, The Stellar Initial Mass Function (38th Herstmonceux Conference), 142, 1

Koch, A., Grebel, E. K., Gilmore, G. F., Wyse, R. F. G., Kleyna, J. T., Harbeck, D. R., Wilkinson, M. I., & Wyn Evans, N. 2008, *AJ*, 135, 1580

Leger, A. & Puget, J. L. 1984, *A&A*, 137, L5

Lemke, D., Mattila, K., Lehtinen, K., Laureijs, R. J., Liljestrom, T., Leger, A., & Herbstmeier, U. 1998, *A&A*, 331, 742

Lequeux, J., Le Bourlot, J., Des Forets, G. P., Roueff, E., Boulanger, F., & Rubio, M. 1994, *A&A*, 292, 371

Leroy, A., Bolatto, A. D., Simon, J. D., & Blitz, L. 2005, *ApJ*, 625, 763

Leroy, A., Bolatto, A., Walter, F., & Blitz, L. 2006, *ApJ*, 643, 825

Leroy, A., Cannon, J., Walter, F., Bolatto, A., & Weiss, A. 2007, *ApJ*, 663, 990

Lisenfeld, U. & Ferrara, A. 1998, *ApJ*, 496, 145

Madden, S. C., Galliano, F., Jones, A. P., Sauvage, M. 2006, *A&A*, 446, 877

Maloney, P. & Black, J. H. 1988, *ApJ*, 325, 389

Mapelli, M., Ripamonti, E., Tolstoy, E., Sigurdsson, S., Irwin, M. J., & Battaglia, G. 2007, *MNRAS*, 380, 1127

Mateo, M. L. 1998, *ARAA*, 36, 435

Mayer, L., Governato, F., Colpi, M., Moore, B., Quinn, T., Wadsley, J., Stadel, J., & Lake, G. 2001, *ApJL*, 547, L123

Mayer, L., Governato, F., Colpi, M., Moore, B., Quinn, T., Wadsley, J., Stadel, J., & Lake, G. 2001, *ApJ*, 559, 754

Mayer, L., Mastropietro, C., Wadsley, J., Stadel, J., & Moore, B. 2006, *MNRAS*, 369, 1021

Melena, N. W., Massey, P., Morrell, N. I., & Zangari, A. M. 2008, *AJ*, 135, 878

Melisse, J. P. M. & Israel, F. P. 1994, *A&AS*, 103, 391

Miller, G. E. & Scalo, J. M. 1979, *ApJS*, 41, 513

Momany, Y., Held, E. V., Saviane, I., Zaggia, S., Rizzi, L., & Gullieuszik, M. 2007, *A&A*, 468, 973

Norman, C. A. & Spaans, M. 1997, *ApJ*, 480, 145

Oey, M. S. & Clarke, C. J. 2005, *ApJL*, 620, L43

Rejkuba, M., Greggio, L., & Zoccali, M. 2004, *A&A*, 415, 915

Rosenberg, J. L., Ashby, M. L. N., Salzer, J. J., & Huang, J.-S. 2006, *ApJ*, 636, 742

Röllig, M., Ossenkopf, V., Jeyakumar, S., Stutzki, J., & Sternberg, A. 2006, *A&A*, 451, 917

Rubio, M., Lequeux, J., & Boulanger, F. 1993, *A&A*, 271, 9

Salvadori, S., Ferrara, A., & Schneider, R. 2008, *MNRAS*, 386, 348

Saviane, I., Rizzi, L., Held, E. V., Bresolin, F., & Momany, Y. 2002, *A&A*, 390, 59

Skillman, E. D., Kennicutt, R. C., & Hodge, P. W. 1989, *ApJ*, 347, 875

Skillman, E. D. & Kennicutt, R. C., Jr. 1993, *ApJ*, 411, 655

Skillman, E. D., Tolstoy, E., Cole, A. A., Dolphin, A. E., Saha, A., Gallagher, J. S., Dohm-Palmer, R. C., & Mateo, M. 2003, *ApJ*, 596, 253

Smith, J. D. T., *et al.* 2007, *ApJ*, 656, 770

Spaans, M. & Norman, C. A. 1997, *ApJ*, 483, 87

Tacconi, L. J. & Young, J. S. 1987, *ApJ*, 322, 681

Taylor, C. L., Kobulnicky, H. A., & Skillman, E. D. 1998, *AJ*, 116, 2746

Tolstoy, E., Venn, K. A., Shetrone, M., Primas, F., Hill, V., Kaufer, A., & Szeifert, T. 2003, *AJ*, 125, 707

van den Bergh, S. 1994, *AJ*, 107, 1328

van den Bergh, S. 1994, *ApJ*, 428, 617

van Dishoeck, E. F. & Black, J. H. 1988, *ApJ*, 334, 771

van Zee, L., Skillman, E. D., & Haynes, M. P. 2006, *ApJ*, 637, 269

Verter, F. & Hodge, P. 1995, *ApJ*, 446, 616

Weidner, C. & Kroupa, P. 2004, *MNRAS*, 348, 187

Weisz, D. R., Skillman, E. D., Cannon, J.M., Dolphin, A. E., Kennicutt, R. C., Jr., Lee, J., & Walter, F. 2008, *ApJ*, in press.

Wilson, C. D. 1995, *ApJ*, 448, L97

Low-Metallicity Star Formation:
From the First Stars to Dwarf Galaxies
Proceedings IAU Symposium No. 255, 2008
L.K. Hunt, S. Madden & R. Schneider, eds.

Stellar Evolution in the Early Universe

Raphael Hirschi[1,2], **Urs Frischknecht**[3], **F.-K. Thielemann**[3], **Marco Pignatari**[1,6], **Cristina Chiappini**[4,5], **Sylvia Ekström**[4], **Georges Meynet**[4], **& André Maeder**[4]

[1]Astrophysics group, Keele University, Lennard-Jones Lab., Keele, ST5 5BG, UK
email: r.hirschi@epsam.keele.ac.uk

[2]IPMU, University of Tokyo, Kashiwa, Chiba 277-8582, Japan

[3]Dept. of Physics & Astronomy, University of Basel, CH-4056, Basel, Switzerland

[4]Observatoire Astronomique de l'Université de Genève, CH-1290, Sauverny, Switzerland

[5]Osservatorio Astronomico di Trieste, Via G. B. Tiepolo 11, I-34131 Trieste, Italia

[6]JINA, University of Notre Dame, Notre Dame, IN 46556, USA

Abstract. Massive stars played a key role in the early evolution of the Universe. They formed with the first halos and started the re-ionisation. It is therefore very important to understand their evolution. In this paper, we describe the strong impact of rotation induced mixing and mass loss at very low metallicity (Z). The strong mixing leads to a significant production of primary ^{14}N, ^{13}C and ^{22}Ne. Mass loss during the red supergiant stage allows the production of Wolf-Rayet stars, type Ib,c supernovae and possibly gamma-ray bursts (GRBs) down to almost $Z = 0$ for stars more massive than 60 $M_\odot$. Galactic chemical evolution models calculated with models of rotating stars better reproduce the early evolution of N/O, C/O and ^{12}C/^{13}C. We calculated the weak s-process production induced by the primary ^{22}Ne and obtain overproduction factors (relative to the initial composition, $Z = 10^{-6}$) between 100-1000 in the mass range 60–90 $M_\odot$.

Keywords. Stars: mass loss, stars: Population II, stars: rotation, stars: supernovae, stars: Wolf-Rayet, Galaxy: evolution

1. Introduction

Massive stars ($M \gtrsim 10\ M_\odot$) started forming about 400 millions years after the Big Bang and ended the dark ages by re-ionising the Universe. They therefore played a key role in the early evolution of the Universe and it is important to understand the properties and the evolution of the first stellar generations to determine the feedback they had on the formation of the first cosmic structures. It is unfortunately not possible to observe the first massive stars because they died a long time ago but their chemical signature can be observed in low mass halo stars (called EMP stars), which are so old and metal poor that the interstellar medium out of which these halo stars formed are thought to have been enriched by one or a few generations of massive stars. Since the re-ionisation, massive stars have continuously injected kinetic energy (via various types of supernovae) and newly produced chemical elements (by both hydrostatic and explosive burning and s and r processes) into the interstellar medium of their host galaxies. They are thus important players for the chemo-dynamical evolution of galaxies. Most massive stars leave a remnant at their death, either a neutron star or a black hole, which often leads to the formation of pulsars or X-ray binaries.

The evolution of stars is governed by three main parameters, which are the initial mass, metallicity (Z) and rotation rate. The evolution is also influenced by the presence of magnetic fields and of a close binary companion. For massive stars around solar metallicity, mass loss plays a crucial role, in some cases removing more than half of the initial mass. Internal mixing, induced mainly by convection and rotation also significantly affect the evolution of stars. The properties of non-rotating low-Z stars are summarised in Sect. 2 of Hirschi *et al.* (2008). In short, at low Z, stars are more compact, usually have weaker winds (see however Pulstilnik *et al.* 2008) and are more massive (Bromm & Loeb 2003, Schneider *et al.* 2006). The fate of non-rotating massive single stars at low Z is summarised in Heger *et al.* (2003) and several groups have calculated the corresponding stellar yields (Heger & Woosley 2002, Chieffi & Limongi 2004, Tominaga *et al.* 2007). In this paper, after discussing the impact of rotation induced mixing and mass loss at low Z, we present the implications for the nucleosynthesis and for galactic chemical evolution in the context of extremely metal poor stars.

2. Rotation, internal mixing and mass loss

Massive star models including the effects of both mass loss and especially rotation better reproduce many observables around solar Z. For example, models with rotation allow chemical surface enrichments already on the main sequence (MS), whereas without the inclusion of rotation, self-enrichments are only possible during the red super-giant (RSG) stage (Heger & Langer 2000, Meynet & Maeder 2000). Rotating star models also better reproduce the ratio of WR to O type stars and also the ratio of type Ib+Ic to type II supernovae as a function of metallicity compared to non-rotating models, which underestimate these ratios (see contribution by Georgy in this volume and Meynet & Maeder 2005). The models at very low Z presented here use the same physical ingredients as the successful solar Z models. The value of $300 \ \mathrm{km\,s^{-1}}$ used as the initial rotation velocity at solar metallicity corresponds to an average velocity of about $220 \ \mathrm{km\,s^{-1}}$ on the main sequence (MS) which is close to the average observed value. One of the first surveys of OB stars (and in particular their surface velocities) was obtained by Fukuda (1982) and the most recent surveys are listed in Meynet *et al.* (2008). It is unfortunately not possible to observe very low Z massive stars and measure their rotational velocity since they all died a long time ago. The higher observed ratio of Be to B stars in the Magellanic clouds compared to our Galaxy (Maeder *et al.* 1999, Martayan *et al.* 2007) could point to the fact that stars rotate faster at lower metallicities. Also a low-Z star having the same ratio of surface velocity to critical velocity, v/v_{crit} (where v_{crit} is the velocity for which the centrifugal force balances the gravitational force) as a solar-Z star has a higher surface rotation velocity due to its smaller radius (one quarter of $Z_{\odot}$ radius for a very low-Z 20 $M_{\odot}$ star). In the models presented below, the initial ratio v/v_{crit} is the same or slightly higher than for solar Z (see Hirschi 2007 for more details). This corresponds to initial surface velocities in the range of $600 - 800 \ \mathrm{km\,s^{-1}}$. These fast initial rotation velocities are supported by chemical evolution models of Chiappini *et al.* (2006b) discussed in the next section. The mass loss prescriptions used in the Geneva stellar evolution code are described in detail in Meynet & Maeder (2005). In particular, the mass loss rates depend on metallicity as $\dot{M} \sim (Z/Z_{\odot})^{0.5}$, where Z is the mass fraction of heavy elements at the surface of the star.

How do rotation induced processes vary with metallicity? The surface layers of massive stars usually accelerate due to internal transport of angular momentum from the core to the envelope. Since at low Z, stellar winds are weak, this angular momentum dredged up by meridional circulation remains in the star, and the star more easily reaches critical rotation. At the critical limit, matter can easily be launched into a keplerian disk which probably dissipates under the action of the strong radiation pressure of the star.

The efficiency of meridional circulation (dominating the transport of angular momentum) decreases towards lower Z because the Gratton-Öpik term of the vertical velocity of the outer cell is proportional to $1/\rho$. On the other hand, shear mixing (dominating the mixing of chemical elements) is more efficient at low Z. Indeed, the star is more compact and therefore the gradients of angular velocity are larger and the mixing timescale (proportional to the square of the radius) is shorter. This leads to stronger internal mixing of chemical elements at low Z (Meynet & Maeder 2002).

The history of convective zones (in particular the convective zones associated with shell H burning and core He burning) is strongly affected by rotation induced mixing (see Hirschi 2007). The most important rotation induced mixing takes place while helium is burning inside a convective core. Primary carbon and oxygen are mixed outside of the convective core into the H-burning shell. Once the enrichment is strong enough, the H-burning shell is boosted (the CNO cycle depends strongly on the carbon and oxygen mixing at such low initial metallicities). The shell then becomes convective and leads to an important primary nitrogen production. In response to the shell boost, the core expands and the convective core mass decreases. At the end of He burning, the CO core is less massive than in the non-rotating model. Additional convective and rotational mixing brings the primary CNO to the surface of the star. This has consequences for the stellar yields. The yield of ^{16}O, being closely correlated with the mass of the CO core, is therefore reduced due to the strong mixing. At the same time the carbon yield is slightly increased. The relatively "low" oxygen yields and "high" carbon yields are produced over a large mass range at $Z = 10^{-8}$ (Hirschi 2007). This is one possible explanation for the possible high [C/O] ratio observed in the most metal-poor halo stars (see Fig. 14 in Spite *et al.* 2005 and Fabbian *et al.* 2008) and in DLAs (Pettini *et al.* 2008).

Models of metal-free stars including the effect of rotation (see contribution by Ekström and Ekström *et al.* 2008) show that stars may lose up to 10 % of their initial mass due to the star rotating at its critical limit (also called break-up limit). The mass loss due to the star reaching the critical limit is non-negligible but not important enough to drastically change the fate of the metal-free stars. The situation is very different at very low but non-zero metallicity (Meynet *et al.* 2006, Hirschi 2007). The total mass of an $85\,M_\odot$ model at $Z = 10^{-8}$ is shown in Fig. 1 (*left*) by the top solid line. This model, like metal-free models, loses around 5% of its initial mass when its surface reaches break-up velocities in the second part of the MS. At the end of core H burning, the core contracts and the envelope expands, thus decreasing the surface velocity and its ratio to the critical velocity. The mass loss rate becomes very low again until the star crosses the HR diagram and reaches the RSG stage. In the cooler part of the H-R diagram, the mass loss becomes very important. This is due to the dredge-up by the convective envelope of CNO elements to the surface increasing its overall metallicity. The total metallicity, Z, is used in this model (including CNO elements) for the metallicity dependence of the mass loss. Therefore depending on how much CNO is brought up to the surface, the mass loss becomes very large again. The CNO brought to the surface comes from primary C and O produced in the He-burning region and from primary N produced in the H-burning one.

The fate of rotating stars at very low Z is therefore probably the following:

• $M < 40\,M_\odot$: mass loss is insignificant and matter is only ejected into the ISM during the SN explosion (see contributions by Nomoto and Tominaga in this volume), which could be very energetic if fast rotation is still present in the core at the core collapse.

• $40\,M_\odot < M < 60\,M_\odot$: mass loss (at critical rotation and in the RSG stage) removes 10–20% of the initial mass of the star. The star probably dies as a black hole without a SN explosion and therefore the feedback into the ISM is only due to stellar winds, which are slow.

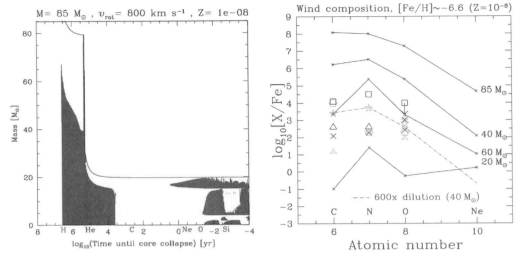

Figure 1. *Left:* Structure evolution diagram for a 85 $M_\odot$ model with $v_{\rm ini} = 800$ km s^{-1} at $Z = 10^{-8}$. Coloured areas correspond to convective zones along the Lagrangian mass coordinate as a function of the time left until the core collapse. The burning stage abbreviations are given below the time axis. The top solid line shows the total mass of the star. Strong mass loss during the RSG stage removes a large fraction of total mass of the star. *Right:* Composition in [X/Fe] of the stellar wind for the $Z = 10^{-8}$ models (solid lines). For HE1327-2326 (*red stars*), the best fit for the CNO elements is obtained by diluting the composition of the wind of the 40 $M_\odot$ model by a factor 600 (see Hirschi 2007 for more details).

- $M > 60$ $M_\odot$: high mass loss removes a significant amount of mass and the stars enter the WR phase. These stars therefore die as type Ib/c SNe and possibly as GRBs.

3. Nucleosynthesis and galactic chemical evolution

Rotation induced mixing leads to the production of primary nitrogen, ^{13}C and ^{22}Ne. In this section, we compare the chemical composition of our models with carbon-rich EMP stars and include our stellar yields in a galactic chemical evolution (GCE) model and compare the GCE model with observations of EMP stars. We also study the weak s process production that can be expected with the primary ^{22}Ne obtained in our models.

3.1. *The most metal-poor star known to date, HE1327-2326*

Significant mass loss in very low-Z massive stars offers an interesting explanation for the strong enrichment in CNO elements of the most metal-poor stars observed in the halo of the galaxy (see Meynet *et al.* 2006, Hirschi 2007). The most metal-poor star known to date, HE1327-2326 (Frebel *et al.* 2006) is characterised by very high N, C and O abundances, high Na, Mg and Al abundances, a weak s-process enrichment and depleted lithium. The star is not evolved so it has not had time to bring self-produced CNO elements to its surface and is most likely a subgiant. By using one or a few SNe and using a very large mass cut, Limongi *et al.* (2003) and Iwamoto *et al.* (2005) are able to reproduce the abundances of most elements. However they are not able to reproduce the nitrogen surface abundance of HE1327-2326 without rotational mixing. The abundance pattern observed at the surface of that star presents many similarities with the abundance pattern obtained in the winds of very metal-poor fast rotating massive star models. HE1327-2326 may therefore have formed from gas, which was mainly enriched by stellar winds of rotating very low metallicity stars. In this scenario, a first generation of stars (PopIII)

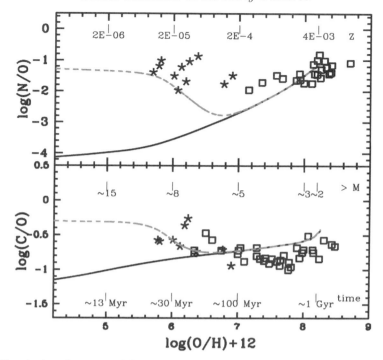

Figure 2. Chemical evolution model predictions of the N/O and C/O evolution, in the galactic halo, for different stellar evolution inputs. The solid curves show the predictions of a model without fast rotators at low metallicities. The dashed lines show the effect of including a population of fast rotators at low metallicities (for the data see Chiappini *et al.* 2006a and references therein).

pollutes the interstellar medium to very low metallicities ([Fe/H]~-6). Then a PopII.5 star (Hirschi 2005), such as the 40 $M_\odot$ model calculated here, pollutes (mainly through its wind) the interstellar medium out of which HE1327-2326 forms. This would mean that HE1327-2326 is a third generation star. The CNO abundances are well reproduced, in particular that of nitrogen, which according to Frebel *et al.* (2006) is 0.9 dex higher in [X/Fe] than oxygen. This is shown in Fig. 1 (*right*) where the abundances of HE1327-2326 are represented by the red stars and the best fit is obtained by diluting the composition of the wind of the 40 $M_\odot$ model by a factor 600. When the SN contribution is added, the [X/Fe] ratio is usually lower for nitrogen than for oxygen. It is interesting to note that the very high CNO yields of the 40 $M_\odot$ stars brings the total metallicity Z above the limit for low mass star formation obtained in Bromm & Loeb (2003).

3.2. *Primary nitrogen and* ^{13}C

The high N/O plateau values observed at the surface of very metal poor halo stars require very efficient sources of primary nitrogen. Rotating massive stars can inject a large amount of primary N on short time scales. They are therefore very good candidates to explain the N/O plateau observed at very low metallicity. According to the heuristic model of Chiappini *et al.* (2005), a primary nitrogen production of about 0.15 $M_\odot$ per star is necessary. Using stellar yield calculations taking into account the effects of rotation at $Z = 10^{-8}$ in a chemical evolution model for the galactic halo with infall and outflow, both high N/O and C/O ratios are obtained in the very metal-poor metallicity range in agreement with observations (see details in Chiappini *et al.* 2006a). This model is shown

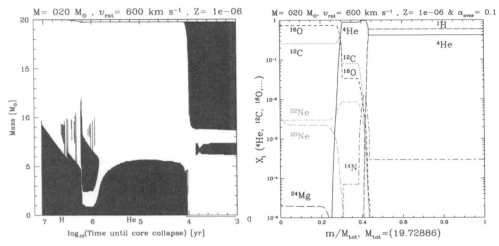

Figure 3. *Left*: Structure evolution diagram (see description of Fig. 1) for a rotating 20 $M_\odot$ model at $Z = 10^{-6}$ during H- and He-burning phases. *Right*: Chemical composition at the end of core He burning. Just above the core, one sees that the maximum abundance of ^{22}Ne is around 1% in mass fraction and at the end of core He burning, around 0.5% is burnt in the core, providing plenty of neutrons for s process.

in Fig. 2 (dashed magenta curve). In the same figure, a model computed without fast rotators (solid black curve) is also shown. Fast rotation enhances the nitrogen production by ∼3 orders of magnitude. These results also offer a natural explanation for the large scatter observed in the N/O abundance ratio of *normal* metal-poor halo stars: given the strong dependency of the nitrogen yields on the rotational velocity of the star, we expect a scatter in the N/O ratio which could be the consequence of the distribution of the stellar rotational velocities as a function of metallicity. As explained above, the strong production of primary nitrogen is linked to a very active H-burning shell and therefore a smaller helium core. As a consequence, less carbon is turned into oxygen, producing high C/O ratios (see Fig. 2). Although the abundance data for C/O is still very uncertain, a C/O upturn at low metallicities is strongly suggested by observations (see Fabbian *et al.* 2008). Note that this upturn is now also observed in very metal poor DLA systems (Pettini *et al.* 2008).

In addition, stellar models of fast rotators have a great impact on the evolution of the ^{12}C/^{13}C ratio at very low metallicities (Chiappini et al. 2008). In this case, we predict that, if fast rotating massive stars were common phenomena in the early Universe, the primordial interstellar medium of galaxies with a star formation history similar to the one inferred for our galactic halo should have ^{12}C/^{13}C ratios between 30-300. Without fast rotators, the predicted ^{12}C/^{13}C ratios would be ∼ 4500 at [Fe/H] = −3.5, increasing to ∼ 31000 at around [Fe/H] = −5.0 (see Fig.2 in Chiappini et al. 2008). Current data on EMP giant normal stars in the galactic halo (Spite *et al.* 2006) agree better with chemical evolution models including fast rotators. The expected difference in the ^{12}C/^{13}C ratios, after accounting for the effects of the first dredge-up, between our predictions with/without fast rotators is of the order of a factor of 2-3. However, larger differences (a factor of ∼ 60 − 90) are expected for giants at [Fe/H]= −5 or turnoff stars already at [Fe/H] = −3.5. To test our predictions, challenging measurements of the ^{12}C/^{13}C in more extremely metal-poor giants and turnoff stars are required.

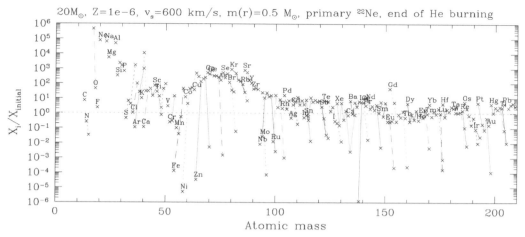

Figure 4. *Left*: Ratio of the abundance at the end of He-burning to the initial abundance for a model including primary ^{22}Ne. We obtain large overproduction factors between 100 and 1000 in the mass range (A) 60 to 90.

3.3. *Primary ^{22}Ne and s process at low Z*

Models at $Z = 10^{-8}$ show a production of primary ^{22}Ne during He burning. We also started calculating models at different Z to determine over which Z range the primary production of ^{22}Ne and also ^{14}N is important. In Fig. 3, we show the properties of a 20 $M_\odot$ model at $Z = 10^{-6}$ up to the end of He burning. Around 0.5% (in mass fraction) of ^{22}Ne is burnt during core He burning and therefore leads to a significant neutron release. We calculated the s process by using the primary nitrogen of the rotating 20 $M_\odot$ model at $Z = 10^{-6}$ inside a one-zone model (based on the Basel network and using an updated version of the *reaclib* reaction rate library) for s-process during He-burning. The first results are shown in Fig. 4. Large overproduction factors (100-1000) are obtained, however the process is not primary (see Pignatari et al. 2008 for more details).

4. Conclusions and outlook

The inclusion of the effects of rotation changes significantly the simple picture in which stellar evolution at low Z is just stellar evolution without mass loss. Strong mixing is induced between the helium and hydrogen burning layers leading to a significant production of primary ^{14}N, ^{13}C and ^{22}Ne. Rotating stellar models also predict strong mass loss during the RSG stage for stars more massive than 60 $M_\odot$. These models predict the formation of WR and type Ib/c SNe down to almost Z=0. The chemical composition of the stellar winds is compatible with the CNO abundance observed in the most metal-poor star known to date, HE1327-2326. GCE models including the stellar yields of these rotating star models are able to better reproduce the early evolution of N/O, C/O and ^{12}C/^{13}C in our galaxy. These new stellar evolution models predict a large neutron release during core He burning and we present here the corresponding s-process production during He burning.

Large surveys of EMP stars (SEGUE, OZ surveys), of GRBs and SNe (Swift and GLAST satellites) and of massive stars (e.g. VLT FLAMES survey) are underway and will bring more information and constraints on the evolution of massive stars at low Z. On the theoretical side, more models are necessary to fully understand and study the complex interplay between rotation, magnetic fields, mass loss and binary interactions at different metallicities. Large grids of models at low Z will have many applications, for example to study the evolution of massive stars and their feedback in high redshift objects like Lyman-break galaxies and damped Ly-alpha systems.

Acknowledgements

R. Hirschi: Royal Society (2008/R1) and the organizers and M. Pignatari: MIRG-CT-2006-046520 MC Grant (EU FP6), and the NSF grant PHY 02-16783 (JINA).

References

Asplund, M. 2005, *ARA&A*, 43, 481
Beers, T. C. & Christlieb, N. 2005, *ARA&A*, 43, 531
Bromm, V. & Loeb, A. 2003, *Nature*, 425, 812
Chiappini, C., Ekström, S., Meynet, G., *et al.* 2008, *A&A*, 479, L9
Chiappini, C., Hirschi, R., Matteucci, F., *et al.* 2006a, in Proceedings of Nuclei in the Cosmos IX, CERN, PoS(NIC-IX)080
Chiappini, C., Hirschi, R., Meynet, G., *et al.* 2006b, *A&A*, 449, L27
Chiappini, C., Matteucci, F., & Ballero, S. K. 2005, *A&A*, 437, 429
Chieffi, A. & Limongi, M. 2004, *ApJ*, 608, 405
Ekström, S., Meynet, G., & Maeder, A. 2007, ArXiv e-prints0709.0202, proc. "First Stars III"
Ekström, S., Meynet, G., Chiappini, C., *et al.* 2008, *A&A* accepted, aph0807.0573
Fabbian, D., Nissen, P. E., Asplund, *et al.* 2008, in Conf. Precision Spectroscopy in Astrophysics, ed. Santos, N. C., *et al.*, 45-46
Frebel, A., Christlieb, N., Norris, J. E., Aoki, W., & Asplund, M. 2006, *ApJ*, 638, L17
Fukuda, I. 1982, *PASP*, 94, 271
Heger, A., Fryer, C. L., Woosley, S. E., Langer, N., & Hartmann, D. H. 2003, *ApJ*, 591, 288
Heger, A. & Langer, N. 2000, *ApJ*, 544, 1016
Heger, A. & Woosley, S. E. 2002, *ApJ*, 567, 532
Hirschi, R. 2005, in IAU Symposium 228, ed. V. Hill, P. François, & F. Primas, 331–332
Hirschi, R. 2007, *A&A*, 461, 571
Hirschi, R., Chiappini, C., Meynet, G., *et al.* 2008, IAU250, 217–230, arXiv:0802.1675
Iwamoto, N., Umeda, H., Tominaga, N., Nomoto, K., & Maeda, K. 2005, *Science*, 309, 451
Limongi, M., Chieffi, A., & Bonifacio, P. 2003, *ApJ*, 594, L123
Maeder, A., Grebel, E. K., & Mermilliod, J.-C. 1999, *A&A*, 346, 459
Maeder, A. & Meynet, G. 2005, *A&A*, 440, 1041
Marigo, P. 2002, *A&A*, 387, 507
Martayan, C., Floquet, M., Hubert, A. M., *et al.* 2007, *A&A*, 472, 577
Meynet, G., Ekström, S., & Maeder, A. 2006, *A&A*, 447, 623
Meynet, G., Ekström, S., & Maeder, A., *et al.* 2008, IAU250, 147-160, arXiv:0802.2805
Meynet, G. & Maeder, A. 2000, *A&A*, 361, 101
—. 2002, *A&A*, 390, 561
—. 2005, *A&A*, 429, 581
Pettini, M., Zych, B. J., Steidel, C. C., *et al.* 2008, *MNRAS*, 385, 2011
Pignatari, M., Gallino, R., Meynet, G., *et al.* 2008, *ApJL* accepted
Pustilnik, S. A., Tepliakova, A. L., Kniazev, A. Y., *et al.* 2008, *MNRAS*, 388, L24
Schneider, R., Omukai, K., Inoue, A. K., & Ferrara, A. 2006, *MNRAS*, 369, 1437
Spite, M., Cayrel, R., Hill, V., *et al.* 2006, *A&A*, 455, 291
Spite, M., Cayrel, R., Plez, B., *et al.* 2005, *A&A*, 430, 655
Spruit, H. C. 2002, *A&A*, 381, 923
Tominaga, N., Umeda, H., & Nomoto, K. 2007, *ApJ*, 660, 516

Low-Metallicity Star Formation:
From the First Stars to Dwarf Galaxies
Proceedings IAU Symposium No. 255, 2008
L.K. Hunt, S. Madden & R. Schneider, eds.

Revision of Star-Formation Measures

Claus Leitherer

Space Telescope Science Institute, 3700 San Martin Drive, Baltimore, MD 21218, USA
email: leitherer@stsci.edu

Abstract. Rotation plays a major role in the evolution of massive stars. A revised grid of stellar evolutionary tracks accounting for rotation has recently been released by the Geneva group and implemented into the Starburst99 evolutionary synthesis code. Massive stars are predicted to be hotter and more luminous than previously thought, and the spectral energy distributions of young populations mirror this trend. The hydrogen ionizing continuum in particular increases by a factor of up to 3 in the presence of rotating massive stars. The effects of rotation generally increase towards shorter wavelengths and with decreasing metallicity. Revised relations between star-formation rates and monochromatic luminosities for the new stellar models are presented.

Keywords. stars: evolution, stars: rotation, galaxies: dwarf, galaxies: evolution, galaxies: irregular, galaxies: starburst, galaxies: stellar content

1. Background

The star-formation rates of galaxies are commonly determined from the integrated light emitted at a certain wavelengths, such as the V band. Comparison with theoretically predicted mass-to-light (M/L) ratios can in principle provide the total stellar mass and, in combination with an appropriate timescale, the star-formation rate. This fundamental methodology goes back to Tinsley's (1980) pioneering work. Although the reasoning is immediately intuitive, two major challenges need to be tackled.

The first challenge is the a priori unknown stellar initial mass function (IMF). Observed values of M/L_V in galaxy centers are around $5 - 100$. At the same time we know that the main contributors to the galaxy light are upper main-sequence (MS) and evolved low-mass stars. These stars have $M/L_V \approx 1$. To put these numbers into perspective, the average M/L_V of all known stars within 20 pc of the Sun is about $1 - 2$ (Faber & Gallagher 1979), and an early-type MS star has $M/L_V \approx 10^{-2}$. Since the mass-to-light ratio in galaxy centers is not too sensitive to dark matter (at least for disk galaxies; cf. E. Brinks' talk at this conference), the apparent discrepancy suggests that most of the stellar mass is hidden from view because most stars have lower luminosity, and therefore have lower mass than indicated by the spectrum. For the purpose of this paper I will assume we can correct for this effect by assuming a known, universal IMF (Kroupa 2007 and this conference).

The second challenge involves the M/L of individual stars of all masses. At the high-mass end, this quantity is not accessible to direct measurements and can only be predicted by stellar evolution models: masses are poorly known because of the scarcity of very massive binaries with mass determinations, and luminosities are elusive because most of the stellar light is emitted in the ionizing ultraviolet. The purpose of this paper is to discuss how the latest generation of stellar evolution models including rotation differs from its predecessor, and how these new models affect the predictions of the evolutionary synthesis code Starburst99 (Leitherer *et al.* 1999; Vázquez & Leitherer 2005). Some of the results presented here can be found in Vázquez *et al.* (2007).

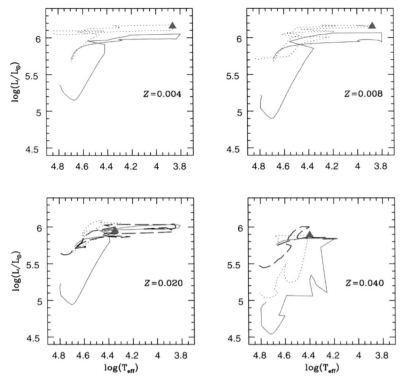

Figure 1. Comparison of the evolution of a 60 $M_\odot$ star without (solid) and with (dotted) rotation for four different metal abundances. $v_{\rm rot} = 300$ km s^{-1}. The line denotes an exploratory model with $v_{\rm rot} = 0$ and all other parameters identical to the rotating model. See Vázquez et al. (2007) for additional details.

2. Stellar Evolution Models with Rotation

Until the late 1990's the evolution of massive stars was thought to be determined by the chemical composition, stellar mass, and mass-loss rate, plus atomic physics and some secondary adjustable parameters. The resulting model grid led to reasonable agreement both with observations of individual stars and of stellar populations. Subsequently it was recognized that stellar rotation can play a key role in the evolution of massive stars (Maeder & Meynet 2000). Evidence of anomalous stellar surface abundances on the MS, lifetimes of certain evolutionary phases, and revised lower mass-loss rates support the concept of rotation. Rotation modifies the hydrostatic structure, induces additional mixing and affects the stellar mass loss (cf. R. Hirschi's conference contribution).

In Fig. 1 I illustrate the effects of rotation on the evolution of a 60 $M_\odot$ star. Rotation leads to generally higher luminosities and higher effective temperatures for massive stars. This is the result of the larger convective core and the lower surface opacity in the presence of rotation. (Recall that hydrogen is the major opacity source and any decrease of its relative abundance by mixing lowers the opacity and therefore increases the temperature.) Fig. 1 suggests a significant luminosity increase of a 60 $M_\odot$ star even on the MS. This trend is present in all massive stars down to $\sim$20 $M_\odot$, depending on metallicity. The lower the metallicity, the more important the influence of rotation, which ultimately becomes the dominant evolution driver for metal-free stars (Hirschi et al. 2008).

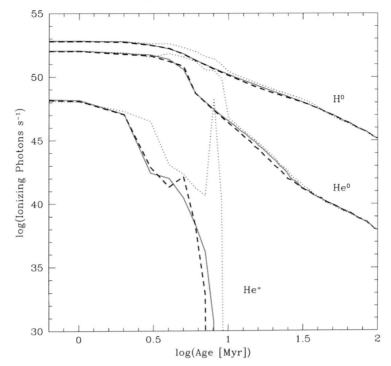

Figure 2. Number of photons in the H^0, He^0, and He^+ continuum (solar composition). Line types as in Fig. 1.

3. Revised Population Models

Vázquez *et al.* (2007) implemented the full grid of rotating evolution models into Starburst99. Stellar atmospheres and/or empirical spectral libraries were attached for each mass and at each time step. The atmospheres that were used for hot stars are those published by Smith, Norris, & Crowther (2002). For an assumed IMF one can then compute the full spectral energy distribution (SED) and its evolution with time. All models quoted here are for a Salpeter IMF with mass cut-offs at 1 and 100 $M_\odot$. The Hawaii group is independently using these SED's as input for photo-ionization modeling with the Mappings code (E. Levesque, these proceedings).

The most dramatic changes with respect to prior models occur at the short-wavelength end of the SED. The ionizing luminosities for a singular burst with mass 10^6 $M_\odot$ are shown in Fig. 2. Since the most massive stars are more luminous and hotter, their ionizing luminosities increase during O-star dominated phases (2 – 10 Myr). The increase reaches a factor of 3 in the hydrogen ionizing continuum and several orders of magnitude in the neutral and ionized helium continua. The predictions for the latter need careful scrutiny, as the photon escape fraction crucially depends on the interplay between the stellar parameters supplied by the evolution models and the radiation-hydrodynamics of the atmospheres. In contrast, the escape of the hydrogen ionizing photons has little dependence on the particulars of the atmospheres and consequently is a relatively safe prediction.

Luminosities for selected wavelengths and passbands are reproduced in Fig. 3. In addition to the previously discussed changes when O stars dominate, the figure suggests significant revisions at epochs when red supergiants are present (10 – 20 Myr). The new

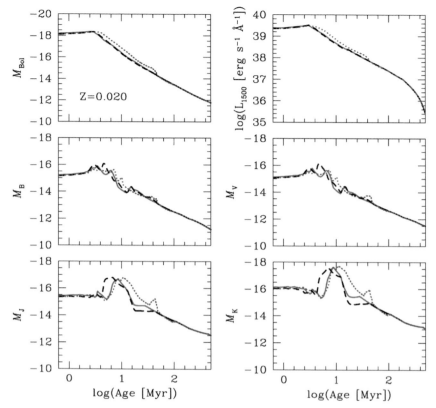

Figure 3. M_{Bol}, L_{1500}, M_{B}, M_{V}, M_{J}, and M_{K} vs. time. The bands most affected by the new models are those in the ultraviolet and infrared. Solar chemical composition. Line types as in Fig. 1.

evolution models with rotation predict an enhanced red supergiant phase which becomes noticeable, e.g., in the higher K band luminosity. The fact that all curves converge after ~50 Myr is an artifact: the models with rotation only reach down to a mass of 9 $M_\odot$, and the traditional tracks were used at masses below that value. However, the effects of rotation are expected to be small at these lower masses.

4. Implications for Star-Formation Indicators

The results discussed so far apply to stellar populations forming quasi-instantaneously. Choosing an instantaneous population makes it easier to identify physical processes in the SED since a particular epoch in time is usually associated with a specific stellar mass interval. While singular bursts are a good approximation for the star-formation history of, e.g., a stellar cluster, galaxies are better described by a star-formation equilibrium when stellar birth and death rates are identical. In this case one can derive relations between the star-formation rate and monochromatic luminosity independent of age.

Star-formation rates as a function of luminosity for several strategic wavelengths were determined for steady-state populations of age 100 Myr. At that epoch massive stars have reached an equilibrium for all wavelengths considered here. The IMF is the same as before. The new relations for stellar models with rotation having solar chemical composition are:

$$SFR\ [M_\odot\ \mathrm{yr}^{-1}] = 3.55 \times 10^{-54} N_{\mathrm{LyC}}\ [\mathrm{s}^{-1}] \tag{4.1}$$

$$SFR \ [M_\odot \ \mathrm{yr}^{-1}] = 3.39 \times 10^{-41} L_{1500} \ [\mathrm{erg \ s}^{-1} \ \mathrm{\mathring{A}}^{-1}] \tag{4.2}$$

$$SFR \ [M_\odot \ \mathrm{yr}^{-1}] = 6.31 \times 10^{-40} L_V \ [\mathrm{erg \ s}^{-1} \ \mathrm{\mathring{A}}^{-1}] \tag{4.3}$$

$$SFR \ [M_\odot \ \mathrm{yr}^{-1}] = 1.48 \times 10^{-44} L_{\mathrm{FIR}} \ [\mathrm{erg \ s}^{-1}]. \tag{4.4}$$

For comparison, if eqs. (4.1), (4.2),(4.3), and (4.4) were derived with the previous tracks (as currently implemented in Starburst99), the conversion factors between luminosity and star-formation rate would be 4.42×10^{-54}, 4.07×10^{-41}, 6.76×10^{-40}, and 1.78×10^{-44}, respectively. The revised relations lead to somewhat lower star-formation rates when applied to the commonly used star-formation measures. The largest effect is for the ionizing photon flux, which can be determined, e.g., from the Hα luminosity. The new rates will be about 25% lower for the same Hα luminosity. In practice, this decrease is hardly significant because other systematic uncertainties, such as the IMF scaling, are more important.

For lower metallicities, the M/L of rotating stars becomes even lower, and this trend is reflected in the M/L of the populations. Consequently the conversion coefficients in eqs. (4.1), (4.2),(4.3), and (4.4) for 20% solar composition become 2.77×10^{-54}, 3.31×10^{-41}, 5.50×10^{-40}, and 1.43×10^{-44}, respectively. The difference between the new conversion at 20% and the previous conversion at solar composition for the ionizing luminosity reaches almost a factor of 2, which is clearly non-negligible.

To summarize, we find noticeable changes in the theoretically predicted M/L ratios of stellar populations computed with the new grid of stellar evolutionary tracks with rotation. Whenever hot, massive stars contribute to the SED, the revised luminosities are higher and the spectrum is harder. The effects are subtle at optical and infrared wavelengths but significant in the ultraviolet. Single stellar populations with ages of several Myr are predicted to have ionizing fluxes that are higher by a factor of up to 3. Steady-state populations are less affected because of the diluting effect of ongoing star formation. Nevertheless, the conversion factor between Hα luminosity and star-formation rate may change by 25% or more.

A prudent Starburst99 user may want to take the new calibrations with care. While there is general consensus that the new evolution models with rotation are a quantum leap over their predecessors, these tracks are still in an exploratory stage and further testing is needed. Ultimately, the new grid and the corresponding revision of the star-formation measures will become the default in Starburst99. It is frustrating from the perspective of the evolutionary synthesis modeler that stellar rotation introduces a new free parameter that reduces some of the deterministic concepts of the previous model generation.

References

Faber, S. M. & Gallagher, J. S. 1979, *ARA&A*, 17, 135

Hirschi, R., Chiappini, C., Meynet, G., Maeder, A., & Ekström, S. 2008, *IAU Symp. 250, Massive Stars as Cosmic Engines*, ed. F. Bresolin, P. A. Crowther, & J. Puls (Cambridge: CUP), 217

Kroupa, P. 2007, in: *IAU Symp. 241, Stellar Populations as Building Blocks of Galaxies*, ed. A. Vazdekis & R. F. Peletier (Cambridge: CUP), 109

Leitherer, C., *et al.* 1999, *ApJS*, 123, 3

Maeder, A. & Meynet, G. 2000, *ARA&A*, 38, 143

Smith, L. J., Norris, R. P. F., & Crowther, P. A. 2002, *MNRAS*, 337, 1309

Tinsley, B. M. 1980, *Fund. Cosm. Phys.*, 5, 287

Vázquez, G. A. & Leitherer, C. 2005, *ApJ*, 621, 695

Vázquez, G. A., Leitherer, C., Schaerer, D., Meynet, G., & Maeder, A. 2007, *ApJ*, 663, 995

Low-Metallicity Star Formation:
From the First Stars to Dwarf Galaxies
Proceedings IAU Symposium No. 255, 2008
L.K. Hunt, S. Madden & R. Schneider, eds.

© 2008 International Astronomical Union
doi:10.1017/S174392130802499X

Stars at Low Metallicity in Dwarf Galaxies

Eline Tolstoy[1], Giuseppina Battaglia[2], and Andrew Cole[3]

[1]Kapteyn Institute, University of Groningen, the Netherlands
email: etolstoy@astro.rug.nl

[2]European Southern Observatory,
Karl-Schwarzschild str. 2, Garching bei München, Germany
email: gbattagl@eso.org

[3]School of Mathematics and Physics, University of Tasmania,
Hobart, Tasmania, Australia
email: andrew.cole@utas.edu.au

Abstract. Dwarf galaxies offer an opportunity to understand the properties of low metallicity star formation both today and at the earliest times at the epoch of the formation of the first stars. Here we concentrate on two galaxies in the Local Group: the dwarf irregular galaxy Leo A, which has been the recent target of deep HST/ACS imaging (Cole *et al.* 2007) and the Sculptor dwarf spheroidal, which has been the target of significant wide field spectroscopy with VLT/FLAMES (Battaglia 2007).

Keywords. galaxies: dwarf, galaxies: evolution, (galaxies:) Local Group

1. Introduction

Studies of individual low mass stars allow us to trace the properties of stars and thus the Interstellar Medium out of which they were formed back to the earliest times. Dwarf galaxies, which often have simple, small and predominantly old stellar populations, in principle, offer us the most straight forward route to the identification and study of these ancient stars. Dwarf galaxies are also presumed to be the most numerous type of object in the early Universe, to an even greater extent than they are today. If individual stars which formed before the Epoch of Reionisation, before the time when their light could reach us directly, can be identified as low mass stars today, they can provide one of the few probes of the "Dark Ages" of cosmic history and potentially offer unique information about which physical processes led to the re-ionisation of the Universe. Dwarf galaxies also offer us the opportunity to study low metallicity star formation much as it may have occurred in the early Universe.

The first challenge is to identify the oldest and least evolved stars in an ancient stellar population and the second is to put them into the context of the formation and evolution of the entire galaxy. This is easiest to attempt in the dwarf galaxies that are close enough for star formation histories (SFHs) to be accurately determined from Main Sequence Turnoff (MSTO) photometry and where follow-up spectroscopic abundances of the brightest ancient stars, the Red Giant Branch (RGB) stars, are possible. It is the combination of Colour-Magnitude Diagram (CMD) analysis and spectroscopic abundances of individual stars that allows us to most accurately determine the star formation and chemical evolution history of a galaxy from the earliest times.

2. Which dwarf galaxy?

In the Local Group there is a large selection of dwarf galaxies which can in principle be used to study low metallicity star formation (e.g., Mateo 1998). The broad range of dwarf galaxy types is usually studied with quite different techniques, because their distances, sizes and surface brightnesses vary by a large amount. The properties and inter-relations of different types of dwarf galaxy are not always easy to understand when we must use different techniques to study them, and often observe different fractions of the stellar populations and over different fractions of the galaxy volume.

Is there a reason to study one dwarf galaxy rather than another? It of course depends on what you are looking for. Some dwarf galaxies still have on-going star formation at very low metallicity, presumably much as it was at the earliest times. But to find an actual star that was formed in the earliest times it is better to target galaxies containing predominantly ancient stars, so as not to have to look through more recent star formation episodes. On the other hand, some galaxies are just easier to study than others (due to distance, size, concentration, location in the sky, heliocentric velocity etc...). Here we give a brief overview of the different groupings of dwarf galaxy types in the Local Universe which can be well resolved into individual stars and of the techniques generally used to study them:

The *ultra compact dwarf galaxies* include actively star forming blue compact dwarf (BCD) galaxies (e.g., I Zw 18) and currently dormant dwarf ellipticals, dEs, (e.g., M 32) that are typically quite distant. The few dEs in the Local Group, are all to be found around M 31, ∼1 Mpc away, and there are no obvious BCDs (with the possible exception of IC 10, hidden behind significant foreground obscuration from the Milky Way). The compactness (and distance) typical of these systems means that resolved studies require the Hubble Space Telescope (HST).

Dwarf irregular galaxies (dIs) are numerous within the Local Group although they are typically at a distance >400 kpc. These are HI-rich systems of varying sizes, masses and luminosities which are currently forming stars with a variety of rates, from extremely low (e.g., Pegasus) to almost zero (e.g., DDO 210) to very high (e.g., NGC 6822). Detailed studies down to the oldest MSTOs of these types of galaxies require HST-like sensitivity and image stability. Ca II triplet (CaT) surveys of their RGB populations are possible (although challenging).

Dwarf spheroidal galaxies (dSphs) are typically associated with large galaxies like our own, and so can be much closer than dIs, with the majority at distances < 130 kpc, although there are also several more distant examples. They look very much like the old extended stellar populations seen in dIs, and this suggests that the major difference is that they lack gas and recent star formation. They have typically not formed stars for at least several 100 Myr (e.g., Fornax), and in several cases much longer (e.g., Sculptor dSph formed the majority of its stars more than 10 Gyrs ago). The proximity of dSphs makes it easier to carry out spectroscopic studies of their resolved stellar populations, although this requires wide field instrumentation to efficiently gain an overview as they are typically > 1 degree across on the sky. These galaxies have been the subject of much recent attention due to new wide-field multi-fibre spectroscopic facilities such as VLT/FLAMES and Magellan/MIKE.

There are also the newly discovered *ultra faint systems*, and it is not yet entirely clear that they are all galaxies. They are so faint and diffuse that they are only picked up by all sky surveys carried out with great care and attention to uniformity, such as SDSS (e.g., Belokurov *et al.* 2007, Martin *et al.* 2008). Often when their individual stars are spectroscopically followed up they are found to only contain a few RGB stars (if

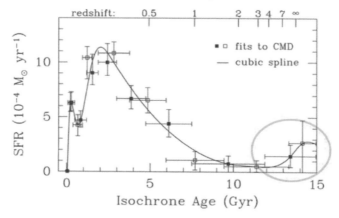

Figure 1. The derived SFH of Leo A, from Cole *et al.* (2007). The data points with 1σ error bars show the best fits to the CMD. The heavy line is a spline fit to the results, and is the best estimate for Leo A's SFH. The circled area represents region of most interest to understand the early evolution of Leo A, and its relation to the Epoch of Reionisation. It will be best probed by studying the RR Lyr variable star population.

any) which makes separating them out from the Galactic foreground and determining their properties quite challenging (e.g., Siegel *et al.* 2008). In several cases it is almost impossible to obtain a statistically meaningful sample. There is evidence to suggest that their average metallicity is lower than any globular cluster, and also than other more luminous dwarf galaxies (Simon & Geha 2007), and that the most metal-poor stars are to be found in these systems (A. Frebel, this volume).

To date, no one has made an equally careful study of the SFH (from CMD analysis) and a spectroscopic survey of the abundances of a large number of RGB stars in the same dwarf galaxy. HST is ideal for making deep and accurate CMDs of dIs (which are a good match to the field of view and sensitivity limits) but it sees only a very small fraction of a much larger dSph. It would require a dedicated survey effort even from wide field imagers on the ground to cover a sizeable fraction of their volume. Spectroscopic studies of individual stars on the other hand are more straightforward for the closer dSphs. Most dIs are too distant for FLAMES spectrosopic studies of their RGB, but there are smaller field multi-object spectrographs (e.g., FORS on VLT and DEIMOS on Keck) which, with a dedicated effort, could produce the medium resolution spectra for the reasonably sized samples of stars required.

Thus in the absence of one well studied galaxy we discuss two different systems, one with deep HST imaging (Leo A) by Cole *et al.* (2007) and one with wide-field spectroscopic VLT/FLAMES study (Sculptor dSph) by Battaglia (2007). Both studies give interesting (but different) insights into the evolution of dwarf galaxies.

3. The Star-Formation History of Leo A

The Leo A dI is still forming stars with a metallicity very similar to that of its oldest stars, and so it is a good probe of how star formation proceeds at very low metallicity, and also how the SFH varies over time in a relatively isolated system most likely depending predominantly upon internal processes.

Leo A is one of the least luminous, most metal-poor gas-rich galaxies in the Local Group (e.g., Mateo 1998). It is at a distance of $\sim$790 kpc and it contains a small population of young, massive stars, with a current star formation rate of $1-2 \times 10^{-4} M_\odot/yr$ (Hunter &

Elmegreen 2004). There are only a few small HII regions, and from spectroscopy of four of these, van Zee *et al.*(2006) determined an oxygen abundance of 12+log(O/H)=7.38±0.10 (assuming solar [O/Fe] this corresponds to [Fe/H]= −1.5), which is one of the lowest ISM metal abundances known, and only slightly more metal rich than I Zw 18 (at 12+log(O/H)=7.20, or [Fe/H]= −1.7 for solar [O/Fe]). Leo A is thus one of the least evolved gas-rich galaxies in the Local Group. There have been several past attempts to determine the SFH of this system (Tolstoy 1996; Tolstoy *et al.* 1998; Schulte-Ladbeck *et al.* 2002), but they all lacked deep MSTO photometry to enable an accurate determination of the SFH going back to the earliest times. Dolphin *et al.* (2002) provided the first definitive proof of truly ancient stars in Leo A with the detection of RR Lyrae variable stars.

As part of a major programme to study isolated Local Group dwarf galaxies, deep images of Leo A were obtained with the ACS on HST over 16 orbits by Cole *et al.* (2007). From the resulting CMDs, reaching down to apparent [absolute] magnitudes of $(M_{475}, M_{814}) \approx (29\ [+4.4], 27.9\ [+3.4])$, an accurate star formation rate as a function of time over the entire history of the galaxy was determined (see Fig. 1). From this we can see that 90% the star formation in Leo A happened during the last 8 Gyr. There was a peak in the star formation rate 1.5 − 3 Gyr ago, when stars were forming at a level 5−10 times the current rate. The CMD analysis of Leo A only requires a very slight metallicity evolution with time. The mean inferred metallicity in the past is consistent with measurements of the present-day gas-phase oxygen abundance.

There appears to have been only a small and uncertain amount of star formation in Leo A at the earliest times (highlighted as a circle in Fig. 1). The level determined from CMD analysis alone is clearly not very reliable as can be seen from the error bars. However, there is no doubt that it exists due to the presence of RR Lyrae variable stars (Dolphin *et al.* 2002). These previously discovered RR Lyrae were at the limit of detection with the WIYN telescope, but with the help of these new ACS data, additional RR Lyrae and more detailed information such as ages and metallicities can be obtained (Fiorentino, Saha *et al.* in prep.). Thus in Leo A it can be seen that some stars were formed around the epoch of re-ionisation, but the bulk were formed much later. Careful modelling and a more detailed look at current observations are required to understand the relation (if any) between the star formation rate at early times in Leo A and the events surrounding the re-ionisation of the Universe.

4. Spectroscopy of Red Giant Branch Stars in Sculptor dSph

From the point of view of picking out the most ancient populations in nearby galaxies, the diffuse Sculptor dSph, with no significant star formation over the last 10 Gyr, is a good target. Sculptor is a close companion of the Milky Way located at a distance of ∼80 kpc. Observations of the stellar population of Sculptor have revealed a sizeable population of RR Lyrae variable stars (e.g., Kaluzny *et al.* 1995), clearly indicating that its stellar population contains a globular cluster age component. Sculptor also shows tantalizing evidence in its extended horizontal branch for a chemical enrichment history consistent with extended early star formation episodes (e.g., Hurley-Keller *et al.* 1999; Majewski *et al.* 1999).

As part of the DART project (Tolstoy *et al.* 2006) a detailed photometric and spectroscopic survey of the Sculptor dSph was carried out (Tolstoy *et al.* 2004; Battaglia 2007; Battaglia *et al.* 2008a, Hill *et al.*, in prep). The ESO/2.2m Wide Field Imager (ESO/WFI) was used to obtain photometry out to and beyond the nominal tidal radius to study the spatial properties of the resolved stellar population. VLT/FLAMES spectra

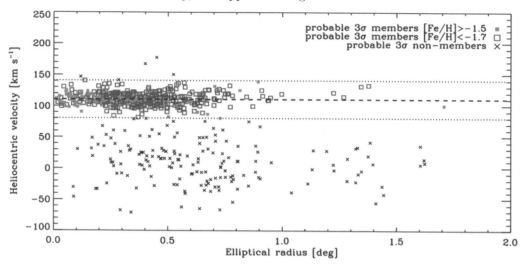

Figure 2. From the FLAMES LR sample of individual RGB stars in the Sculptor dSph (Battaglia 2007). The positions in the galaxy are given as Elliptical radii plotted against the heliocentric velocity of each star. The metal-rich stars ([Fe/H]> −1.5) are the solid (red) squares and the metal-poor stars ([Fe/H]< −1.7) are open (blue) squares, and the probable non-member stars are small (black) crosses, and the limits for membership are given by dotted lines about, $v_{hel} = 110.6$ km/s the heliocentric velocity (dashed line). The colder kinematics and centrally concentrated nature of the metal-rich stars are clearly visible compared to the metal-poor stars which have a larger velocity dispersion and are more uniformly distributed over the galaxy. Colours only available in the electronic version.

were obtained both at high resolution (R ∼ 20 000) for a detailed abundance analysis of ∼80 RGB stars and at low resolution (R ∼6500) in the CaT region for a sample of ∼470 RGB stars.

The high resolution data consist of a single FLAMES pointing (25′ diameter) in the central 25′ of Sculptor which was observed for nearly 3 nights, and from the resulting spectra a detailed abundance analysis was carried out for a range of individual elements, including Na, Mg, Ca, Ti, Fe, Mn, Ni, Y, Ba and Eu (Hill *et al.*, in prep.). This study allows a detailed picture of the chemical evolution of Sculptor from the earliest times.

The low resolution data comprise 15 FLAMES pointings distributed over the entire galaxy out to the nominal tidal radius. These provide accurate velocities and CaT metallicity estimates ([Fe/H]), for a large fraction of the Sculptor RGB population (Battaglia 2007). A sample of 470 probable kinematic members of Sculptor was studied and stars of different metallicities were found to have different spatial distributions and kinematic properties, with the more metal-rich stars being more centrally concentrated and kinematically colder than the more metal-poor stars (see Fig 2). Signs of rotation were also found (Battaglia *et al.* 2008a), and the measured velocity gradient of $7.6^{+3.0}_{-2.2}$ km s^{-1} deg^{-1} makes the shape of Scl consistent with being flattened by rotation. Battaglia *et al.* 2008a also determined the mass of Sculptor within 1.8 kpc using detailed mass modelling and find two separate components in Sculptor distinguished by metallicity, spatial extent and kinematics (Tolstoy *et al.* 2004). The new mass of Sculptor is largely independent of the exact distribution of dark matter in its central region, and with a M/L∼160, this gives a dynamical mass for the system, within 1.8 kpc, of $M_{dyn} = 3 \times 10^8 M_\odot$, which is a factor ∼10 higher than the previous value obtained from a much smaller and more centrally concentrated sample of stars (Queloz *et al.* 1995).

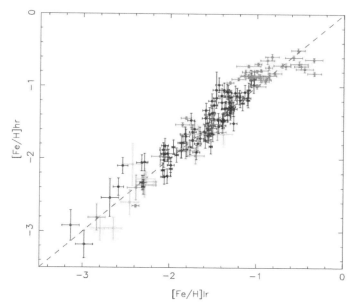

Figure 3. Low resolution (R∼6000) CaT metallicity measurements of [Fe/H] versus high resolution (R⩾ 20 000) direct measurements of [Fe/H] from an average of 50+ individual Fe lines. Included in this plot are HR and LR FLAMES/Giraffe data for Sculptor dSph (93 stars, black) and Fornax dSph (36 stars, red), from Battaglia *et al.* (2008b), and in addition follow-up spectroscopy with HET for 8 stars in Draco (cyan) and 7 stars in Ursa Minor (magenta) and 6 stars (green) in the globular cluster M 3 (Shetrone *et al.*, in prep.) and 3 stars (dark blue) in Carina from MIKE/Magellan (Venn *et al.*, in prep). Colours only available in on-line verison.

One useful advantage of studying a galaxy at both high and low resolution has been that we are able to compare in detail the values of [Fe/H] determined from the CaT with those determined directly from large numbers (50+) of individual Fe lines observed at higher resolution and also to take into account any potential effects due to differing α-element abundances (Battaglia *et al.* 2008b). We find a very good match over the range of metallicities we have sampled (see Fig 3). This is quite a surprising result as it should be expected that the CaT method should stop working, or certainly the slope of the relation between the luminosity weighted equivalent width and [Fe/H] ought to change. So far there has been no clear observational evidence for this, and theoretical studies are under way to try to understand when and how this method should stop working (e.g., Starkenburg, Hill *et al.*, in prep.; see poster by E. Starkenburg, this volume).

In all our observations of the Sculptor dSph so far, we have not found any stars more metal poor than [Fe/H]∼ −3 (Battaglia 2007), and this is also true for the sample of 4 dSph galaxies looked at with VLT/FLAMES (Helmi *et al.* 2006). This lower limit is very similar to the lower limit observed in the Intergalactic Medium metallicities at redshifts, z∼3 (e.g., Cowie & Songaila 1998). This could be interpreted as evidence that dSph galaxies formed out of an IGM at this redshift, however the presence and properties of the blue Horizontal Branch and RR Lyr variable stars, which are as ancient as those in globular clusters, would seem to contradict this. It may instead indicate that dSph galaxies played a significant role in the enrichment of the IGM *before* z∼3. This is broadly consistent with the results of Ferrara & Tolstoy (2000) who found that galaxies with (initial) gas masses of ∼ a few $10^8 M_\odot$ are likely to be the major pollutants of the IGM because they will provide the maximum amount of mass (and metal) loss. This is a balance between limiting the losses so that the galaxy (and hence star formation) is not disrupted in a major way and not making the galaxy so big that little mass can be totally lost to the system. In this scenario galaxies that are so small that one star formation episode destroys the entire galaxy also do not get the opportunity to be significant contributors to the IGM. Sculptor certainly falls into the "blowout" category shown in Fig. 3 of Ferrara & Tolstoy, which means that it loses gas and metals during star formation but not so much that it is immediately destroyed ("blowaway"). The presence of Sculptor in this regime may help to explain its early evolution.

5. What have we learnt?

The lack of very low metallicity stars in dSph suggests a threshold in the metallicity of the oldest stars in these galaxies. Interestingly there is some evidence that there are a couple stars with [Fe/H]< −3 in some of the ultra faint dwarf galaxies (A. Frebel, this volume). This combined with the fact that the low overall metallicities of these systems seem to fall into an overall mass-metallicity relation (Simon & Geha 2007; H. Lee, this volume) suggests a global process which regulates the metallicity of a galaxy depending upon its mass. This mass-metallicity relation makes galactic scale winds the most likely culprit for the regulation of metallicity buildup in a small galaxy. Namely, due to these winds smaller galaxies are able to lose gas and metals more easily and so never form very many stars and those that they form are all of low metallicity.

There are, broadly speaking, three major types of faint dwarf galaxy in the vicinity of the Miky Way: 1) those interacting with our Galaxy, embedded in our halo (ultra faint dwarf galaxies); 2) those in the halo of our galaxy almost certainly affected dynamically by the presence our Galaxy (dSph); 3) more distant galaxies not obviously associated or ever having interacted with a large galaxy (dIs) and of course there are also intermediate "transition" types. All these systems may once have had very similar progenitors but the degree to which they have been influenced by interaction with our Galaxy varies strongly. It is of course very difficult to understand how different processes have affected these galaxies over a Hubble time as there is insufficient information to accurately trace back the history of these small systems in the context of the Milky Way and the Local Group partly because their 3D velocity is not known and partly because their original mass is also not known.

Every galaxy that we have looked at in sufficient detail to detect the Horizontal Branch – or with MSTO photometry (including Leo A and Sculptor) – clearly contains ancient stars. If this is not unambiguous from the CMD analysis it is then confirmed with detections of RR Lyr variable stars. The fraction of stars > 10 Gyr old does vary considerably between different galaxies, but for all galaxies there is at least ∼ 10% of the stellar population that is ancient, and formed at a redshift z> 3. This still leaves some room for ambiguity about whether dwarf galaxies formed their first stars before or just after reionisation. However there is no evidence to suggest that the oldest stars in dwarf galaxies (including Sculptor) are any different (e.g., younger) than those in Galactic halo globular clusters, but it is very difficult to be precise about ages of stars > 10 Gyr old. In the case of Sculptor an analysis of the well populated Horizontal Branch suggests that an age spread of >2 Gyr exists in this ancient population (Tolstoy *et al.* 2004) and this pushes the epoch at which the first stars could have formed in Sculptor back to *at least* 12−13 Gyr ago. Careful well calibrated MSTO analysis over the entire galaxy combined with spectroscopic abundances is required to address this issue.

Thus dwarf galaxies are clearly all ancient objects, well suited to test the properties of the early Universe. However it is difficult to make an accurate determination of the absolute ages of the oldest stars. It appears that we do not find extremely metal-poor stars in these systems and it is still not clear if this is due to pre-enrichment (they formed later than we presently believe) or if this is telling us something about the properties of the first stars to form in these systems (and hence in the Universe).

References

Battaglia, G., Helmi, A., Tolstoy, E., Irwin, M., Hill, V., & Jablonka, P. 2008b, *ApJL*, 681, L13

Battaglia, G., *et al.* 2008a, *MNRAS*, 383, 183

Battaglia, G. 2007 *Phd Thesis*, University of Groningen, the Netherlands.

Belokurov, V., *et al.* 2007, *ApJ*, 654, 897

Cole, A. A., *et al.* 2007, *ApJL*, 659, L17

Cowie, L. L. & Songaila, A. 1998, *Nature*, 394, 44

Dolphin, A. E., *et al.* 2002, *AJ*, 123, 3154

Ferrara, A. & Tolstoy, E. 2000, *MNRAS*, 313, 291

Helmi, A., *et al.* 2006, *ApJL*, 651, L121

Hunter, D. A. & Elmegreen, B. G. 2004, *AJ*, 128, 2170

Hurley-Keller, D., Mateo, M., & Grebel, E. K. 1999, *ApJL*, 523, L25

Kaluzny, J., *et al.* 1995, *ApJS*, 112, 407

Majewski, S. R., Siegel, M. H., Patterson, R. J., & Rood, R. T. 1999, *ApJL*, 520, L33

Martin, N.F., de Jong, J.T.A., & Rix, H-W. 2008, *ApJ, in press* (arXiv:0805.2945)

Mateo, M. L. 1998, *ARAA*, 36, 435

Queloz, D., Dubath, P., & Pasquini, L. 1995, *A&A*, 300, 31

Schulte-Ladbeck, R. E., *et al.* 2002, *AJ*, 124, 896

Siegel, M. H., Shetrone, M. D., & Irwin, M. 2008, *AJ*, 135, 2084

Simon, J. D. & Geha, M. 2007, *ApJ*, 670, 313

Tolstoy, E., *et al.* 2006, *The ESO Messenger*, 123, 33

Tolstoy, E., *et al.* 1998, *AJ*, 116, 1244

Tolstoy, E. 1996, *ApJ*, 462, 684

Tolstoy, E., *et al.* 2004, *ApJL*, 617, L119

van Zee, L., Skillman, E. D., & Haynes, M. P. 2006, *ApJ*, 637, 269

Walker, M. G., *et al.* 2007, *ApJL*, 667, L53

Low-Metallicity Star Formation:
From the First Stars to Dwarf Galaxies
Proceedings IAU Symposium No. 255, 2008
L.K. Hunt, S. Madden & R. Schneider, eds.

© 2008 International Astronomical Union
doi:10.1017/S1743921308025003

Truncated star formation in dwarf spheroidal galaxies and photometric scaling relations

Sven De Rijcke[1], Sander Valcke[1], Christopher J. Conselice[2], and Samantha Penny[2]

[1]Sterrenkundig Observatorium, University of Ghent,
Krijgslaan 281, S9, B-9000 Gent, Belgium
email: `sven.derijcke@UGent.be`, `sander.valcke@UGent.be`

[2]School of Physics and Astronomy, University of Nottingham, University Park, NG9 2RD, UK
email: `conselice@nottingham.ac.uk`, `ppxsp@nottingham.ac.uk`

Abstract. We investigate the global photometric scaling relations traced by early-type galaxies in different environments, ranging from dwarf spheroidals, over dwarf elliptical galaxies, up to giant ellipticals (-8 mag $\gtrsim M_V \gtrsim -24$ mag). These results are based in part on our new HST/ACS F555W and F814W imagery of dwarf spheroidal galaxies in the Perseus Cluster. We show that at $M_V \sim -14$ mag, the slopes of the photometric scaling relations involving the Sérsic parameters change significantly. We argue that these changes in slope reflect the different physical processes that dominate the evolution of early-type galaxies in different mass regimes. We present N-body/SPH simulations of the formation and evolution of dwarf spheroidals that reproduce these slope changes and discuss the underlying physics. As such, these scaling relations contain a wealth of information that can be used to test models for the formation of early-type galaxies.

Keywords. galaxies: dwarf, galaxies: photometry, galaxies: structure

1. Introduction

Bright elliptical galaxies, or Es, and dwarf elliptical galaxies, or dEs, follow the same photometric and kinematic scaling relations (Graham & Guzmán 2003, Matković & Guzmán 2005, De Rijcke *et al.* 2005, Smith Castelli *et al.* 2008). In the luminosity interval -24 mag $\lesssim M_V \lesssim -14$ mag the parameters of the Sérsic profile follow simple power-laws as a function of luminosity and early and late type galaxies trace parallel Tully-Fisher relations (De Rijcke *et al.* 2007). From this wealth of data a picture of (dwarf) galaxy formation emerges that suggests an underlying unity in the physics driving the formation and evolution of stellar systems, with the environment playing a role that is in many situations subordinate to that of internal processes. More specifically, numerical simulations and semi-analytic models of galaxy formation within a ΛCDM cosmology can account for the observed scaling relations when taking into account supernova feedback in galactic gravitational potential wells steepening with galaxy mass (Carraro *et al.* 2001, Nagashima & Yoshii 2004, Ricotti & Gnedin 2005, Marcolini *et al.* 2006, Valcke, De Rijcke, Dejonghe 2008).

In this contribution, we investigate whether the photometric scaling relations traced by dEs and Es persist down to the dSphs ($M_V \gtrsim -14$ mag).

2. Photometric data

2.1. *Perseus Cluster data*

We have obtained high resolution *Hubble Space Telescope (HST)* Advanced Camera for Surveys (ACS) WFC imaging in the F555W and F814W bands of five fields in the Perseus Cluster core, in the immediate vicinity of NGC1275 and NGC1272, the cluster's brightest members, obtained in 2005 (program GO 10201). The scale of the images is $0.05''$ pixel^{-1}, with a field of view of $202'' \times 202''$, providing a total survey area of ~ 57 arcmin^2. Exposure times were 2368 and 2260 seconds for the F555W and F814W bands, respectively. The fields were chosen to cover the most likely cluster dSphs and dEs identified from ground-based imagery by Conselice, Gallagher, Wyse (2003). For some of these, there is spectroscopic confirmation of their cluster membership (Penny & Conselice 2008). For the others, we use morphological criteria to decide cluster membership. The CAS system for quantifying compactness, asymmetry, and clumpiness/smoothness (Conselice 2003b) proves very useful for rejecting e.g. background spiral galaxies based on a smoothness criterion and background bright ellipticals based on a compactness criterion (see Penny *et al.* 2008). The Perseus dataset straddles the dE-dSph transition at $M_V \sim -14$ mag and is therefore essential to the discussion that follows.

2.2. *Data from the literature*

The photometric data, including resolved photometry for surface brightness profiles, of the Local Group dSphs that are identified as Milky Way satellites are collected from Grebel, Gallagher, Harbeck (2003) and Irwin & Hatzidimitriou (1995), adopting the distances listed in Grebel, Gallagher, Harbeck (2003). Data of the M31 dSph satellites is taken from Peletier (1993), Caldwell (1999), Grebel, Gallagher, Harbeck (2003), McConnachie & Irwin (2006), McConnachie, Arimoto, Irwin (2007), and Zucker *et al.* (2007). Data of three Local Group dSphs that are not linked to a giant host galaxy, the Tucana dSph, DDO210, and KKR25, come from Saviane, Held, Piotto (1996), Grebel, Gallagher, Harbeck (2003), and McConnachie & Irwin (2006). De Rijcke *et al.* (2005) (D05) and Mieske *et al.* (2007) provide photometric data on the early-type dwarf galaxy population of the Fornax cluster. Half of the D05 sample consists of dEs from the NGC5044 and NGC5989 groups. The data of the dSphs and dEs in the Antlia cluster are taken from Smith Castelli *et al.* (2008). Data for the giant elliptical and for Coma dEs is taken from Graham & Guzmán (2003) (GG03).

This sample of early-type galaxies comprises dwarf spheroidals, with -14 mag $\lesssim M_V \lesssim -8$ mag, dwarf ellipticals, with -19 mag $\lesssim M_V \lesssim -14$ mag, and bright ellipticals, with $M_V \lesssim -19$ mag. We plot the positions of the sample galaxies in diagrams of V-band absolute magnitude vs. *(i)* half-light radius R_e (in kpc), vs. *(ii)* the Sérsic exponent n of the best fitting Sérsic profile, vs. *(iii)* the central V-band surface brightness of the best fitting Sérsic profile, and vs. *(iv)* V−I colour.

For the Local Group dSphs for which no Sérsic parameters can be found in the literature, we fit Sérsic profiles, with an added constant background density of stars, to the star counts of the dSphs presented in Irwin & Hatzidimitriou (1995).

We now place these early-type galaxies in diagrams correlating the V-band absolute magnitude M_V, the Sérsic exponent n, the extrapolated central surface brightness $\mu_{0,V}$, and the V−I colour. The goal is to investigate the behaviour of the relations between these structural parameters as a function of luminosity in the range -24 mag $< M_V < -8$ mag and of environment, using galaxies from the Local Group; the NGC5044 and NGC5989 groups; and the Fornax, Perseus, and Coma clusters.

3. Photometric scaling relations

For our Perseus dSphs/dEs, we measure the profiles of surface-brightness, position angle, and ellipticity as a function of the geometric mean of major and minor axis distance using our own software. Residual cosmics, background galaxies, and foreground stars are masked and not used in the fit. The shape of an isophote, relative to the best fitting ellipse, is quantified by expanding the surface brightness variation along this ellipse in a fourth order Fourier series. Apparent ABMAG magnitudes in the F555W and F814W bands are calculated using the zero-points given by Sirianni $et\ al.$ (2005). These magnitudes are corrected for interstellar reddening adopting the color excess E(B−V) = 0.171 mag (Schlegel, Finkbeiner, Davis 1998) and using the prescriptions given in Sirianni $et\ al.$ (2005). These reddening-corrected magnitudes are finally converted into Johnson V and I band magnitudes using the transformations of Sirianni $et\ al.$ (2005).

The smooth 2D surface brightness distribution is integrated over circular apertures out to the last isophote we could reliably measure (which is at $\mu_{ABMAG} \approx 27$ mag arcsec^{-2} in both the F555W and F814W images) to derive model independent structural parameters, such as the total apparent magnitude and the half-light radius in each band. For such deep images of galaxies with a roughly exponentially declining surface brightness profile, this truncation results in an insignificant uncertainty on the total luminosity, of the order of a few per cent (see also De Rijcke $et\ al.$ 2005). V−I colors are measured using the V and I-band flux inside the I-band half-light radius. We fit a Sérsic profile to the V-band surface brightness profiles of the program galaxies, expressed in mag arcsec^{-2}.

4. Numerical simulations

The basic SPH scheme of HYDRA was modied to include star formation, supernova feedback, chemical enrichment and gas cooling. The simulations start from a homogeneous gas cloud collapsing onto a Dark Matter (DM) halo. Based on a suite of numerical

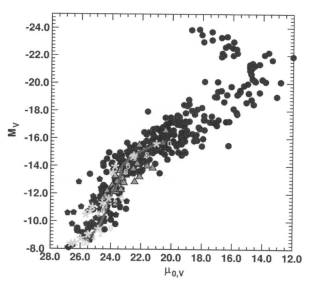

Figure 1. Absolute V-band magnitude versus Sérsic central surface brightness. The dSphs deviate significantly from the dEs and Es (black data points). The models of Valcke, De Rijcke, Dejonghe (2008) nicely reproduce the break at $M_V \sim -14$ mag (dark grey line). Truncating the star-formation histories of the models 0, 2, 4, 6, and 8 Gyrs ago widens the relation considerably in the dSph regime but does not destroy the relation (light grey asterisks).

tests, we decided to use 30 000 gas particles and 15 000 DM particles in the simulations. All model galaxies exhibit episodic star formation. This is a direct consequence of star formation in a shallow potential well. Supernovae heat the gas, causing it to expand. The gas density consequently drops below the threshold for star formation. After the lapse of an adiabatic cooling period, the gas is again allowed to cool and contract. Star formation thus appears to be a self-regulating mechanism.

Since substantial amounts of gas remain in and around the model galaxies throughout the duration of the simulations, one expects tidal interactions and ram-pressure stripping in dense environments to have a significant effect on their SFHs. As a rudimentary way of implementing environmental effects, we simply switch off star formation at some instant in the past and let the stellar population age passively until the present, when the model's observable properties are evaluated. This mimics the effects of ram-pressure stripping or "starvation" (Tully & Trentham 2008). The cluster potential imposes a Roche limit around a galaxy. Gas outside this limit is no longer available for star formation, leaving the galaxy with only a very limited gas supply to form stars.

5. Discussion and conclusions

These results are discussed at length in Valcke, De Rijcke, Dejonghe (2008), De Rijcke *et al.* (submitted), Penny *et al.* (submitted), and De Rijcke *et al.* (in prep.).

There is considerable uniformity in the photometric properties of early-type galaxies, from dwarfs to giants. Photometric parameters quantifying the structure and stellar populations of early-type galaxies, such as the half-light radius, R_e the central surface brightness $\mu_{0,V}$, the Sérsic exponent n, and V−I color all correlate with galaxy luminosity over a range of more than 6 orders of magnitude in luminosity. We have collected photometric data of dSphs/dEs from different environments (galaxy groups and clusters) but find no significant differences between them.

The scaling relations involving the Sérsic parameters, contrary to previous claims, do not keep a constant slope over the whole luminosity range. The Sérsic exponent n varies with luminosity L as $n \propto L^{0.25-0.3}$ for galaxies brighter than $M_V \approx -14$ mag but scatters around a constant value within the range $n \approx 0.5 - 1.0$ for fainter dSphs. This is in agreement with the fact that the surface brightness profiles of dSphs can be well approximated by King profiles with a concentration in the range $c \approx 3 - 10$. Central surface brightness increases with luminosity until the formation of the very brightest, cored ellipticals. At $M_V \approx -14$ mag, the slope of the $M_V - \mu_{0,V}$ changes abruptly. We show that the M_V vs. V−I is essentially a metallicity-luminosity relation of old stellar populations, keeping the same slope over the whole luminosity range investigated here.

Clearly, the absolute magnitude $M_V \approx -14$ mag is not just an arbitrary divide between dSphs and dEs. The rather abrupt changes in the slopes of some of the photometric scaling relations suggest that below and above this luminosity, different physical processes dominate the evolution of early type galaxies. Basically, the balance between the steepness of the gravitational potential, which drives gas inwards, and supernova explosions, which blow gas outwards, tips over at this crucial luminosity. The near-independence of these scaling relations with respect to environment can be appreciated by investigating models with truncated star-formation histories. Early truncation of star formation can widen the photometric relations, especially in the very low-mass regime, but does not destroy them. These results will be discussed in more detail elsewhere (De Rijcke *et al.*, in prep.).

References

Caldwell N., 1999, AJ, 118, 1230

Carraro G., Chiosi C., Girardi L., & Lia C., 2001, MNRAS, 335, 335

Conselice C. J., 2003, ApJS, 147, 1

Conselice C. J. & Gallagher, J. S., III, Wyse R. F. G., 2003, AJ, 125, 66

De Rijcke S., Michielsen D., Dejonghe H., Zeilinger W. W., & Hau G. K. T., 2005, MNRAS, 360, 853 (D05)

De Rijcke S., Zeilinger W. W., Hau G. K. T., Prugniel P., & Dejonghe H., 2007, ApJ, 659, 1172

Graham A. W. & Guzmán R., 2003, AJ, 126, 1787 (GG03)

Grebel E. K. & Gallagher J. S. III, Harbeck D., 2003, AJ, 125, 1926

Irwin M. & Hatzidimitriou D., 1995,

Marcolini A., D'Ercole A., Brighenti F., & Recchi S., 2006, MNRAS, 371, 64

Matković A. & Guzmán R., 2005, MNRAS, 362, 289

McConnachie A. W. & Irwin M. J., 2006, MNRAS, 365, 1263

McConnachie A. W., Arimoto N., & Irwin M., 2007, MNRAS, 379, 379

Mieske S., Hilker M., Infante L, & Mendes de Oliviera C., 2007, A&A, 463, 503

Nagashima M. & Yoshii Y., 2004, ApJ, 610, 23

Peletier R. F., 1993, A&A, 271, 51

Penny S. J. & Conselice C. J., 2008, MNRAS, 383, 247

Penny S. J., Conselice C. J., De Rijcke S., Held E. V., 2008, submitted to MNRAS

Ricotti M. & Gnedin N. Y., 2005, ApJ, 629, 259

Saviane I., Held E. V., Piotto G., 1996, A&A, 315, 40

Schlegel D. J., Finkbeiner D. P., Davis M., 1998, ApJ, 500, 525

Sirianni M., Jee M. J., Benítez N., Blakeslee J. P., Martel A. R., Meurer G., Clampin M., De Marchi G., Ford H. C., Gilliland R., Hartig G. F., Illingworth G. D., Mack J., McCann W. J., 2005, PASP, 117, 1049

Smith Castelli A. V., Bassino L. P., Richtler T., Cellone S. A., Aruta C., Infante L., 2008, MNRAS, 386, 2311

Tully R. B. & Trentham N., 2008, AJ, 135, 1488

Valcke S., De Rijcke S., Dejonghe H., 2008, accepted for publication in MNRAS, astro-ph/0807.0397V

Zucker D. B., Kniazev A. Y., Martínez-Delgado D., Bell E. F., Rix H.-W., et al., 2007, ApJ, 659, L21

Low Metallicity Star Formation:
From the First Stars to Dwarf Galaxies
Proceedings IAU Symposium No. 255, 2008
L. Hunt, S. Madden, & R. Schneider, eds.

© 2008 International Astronomical Union
doi:10.1017/S1743921308025015

Relics of Primordial Star Formation: The Milky Way and Local Dwarfs

Timothy C. Beers[1], Young Sun Lee[1], and Daniela Carollo[2]

[1]Department of Physics & Astronomy, CSCE: Center for the Study of Cosmic Evolution, and
JINA: Joint Institute for Nuclear Astrophysics, Michigan State University,
East Lansing, MI 48824, USA
email: beers@pa.msu.edu, lee@pa.msu.edu

[2]Research School of Astronomy and Astrophysics, Australian National University, Mount
Stromlo Observatory, Cotter Road, Weston, ACT 2611, Australia, and INAF-Osservatorio
Astronomico di Torino, 10025 Pino Torinese, Italy,
email: carollo@mso.anu.edu.au

Abstract. Massive spectroscopic surveys of stars in the thick disk and halo populations of the Galaxy hold the potential to provide strong constraints on the processes involved in (and the timing of) the assembly history of the primary structural components of the Galaxy. In this talk, we explore what has been learned from one of the first such dedicated surveys, SDSS/SEGUE. Over the course of the past three years, SEGUE has obtained spectra for over 200,000 stars, while another hundred thousand stars been added from the calibration star observations of the (primarily extragalactic) SDSS, and other directed programs. A total of well over 10,000 stars with [Fe/H] < −2.0 have been discovered, including several hundred with [Fe/H] < −3.0. Their kinematics have revealed a inner/outer halo structure of the Galaxy.

New determinations of the alpha element ratios for tens of thousands of these stars are reported. Correlations of the alpha-element ratios with kinematics and orbital parameters can be used to test models of the likely formation of the thick-disk and halo components. These new data will (eventually) be considered in connection with possible associations with the present dwarf satellite galaxies of the Milky Way.

Keywords. Astronomical data bases: surveys, techniques: spectroscopic, methods: data analysis, stars: fundamental parameters, Galaxy: disk, Galaxy: halo

1. Introduction

The modern picture of galaxy formation, as described in the context of the ΛCDM paradigm, suggests that large galaxies like the Milky Way are formed from a series of mergers involving the mergers of dark matter dominated sub-haloes with a mass distribution set by the initial conditions that emerged from the Big Bang. These initial mergers are followed by prolonged (continuing to the present) accretion of numerous lower mass sub-haloes, likely in a non-dissipative manner (e.g., Bekki & Chiba 2001). The process is expected to leave behind many hundreds of sub-haloes, most of which should have formed small dwarf-like galaxies in their centers. Such objects should in principle be detectable in the outskirts of the large galaxies they will eventually be accreted by.

The long-standing discrepancy between the numbers of presently observed dwarf galaxies surrounding the Milky Way and this expectation is known as the "missing satellite problem" (Diemand *et al.* 2007, and references therein). The magnitude of this discrepancy has been reduced (but not yet solved) by the discovery of numerous low luminosity dwarf spheroidals (dSphs) by imaging from the Sloan Digital Sky Survey (Belokurov *et al.* 2006; Zucker *et al.* 2006; Walsh *et al.* 2008). Since many of these newly discovered dSphs exhibit surface brightnesses at the very limit of what might be detectable from the ground, it remains possible that the missing satellite problem could be completely resolvable, but just beyond our current reach with ground-based observations.

Another expectation from ΛCDM-based simulations is that many of the satellites, and their attendant stellar populations, may have already been accreted, and mixed (or partially mixed) into the halo populations of the Milky Way. In order to test this idea, we are required to investigate likely signatures of their presence, as reflected in the kinematics and chemistry of the halo populations. Here we report on a current effort to do just this, based on the large (and growing) body of medium-resolution spectroscopy of stars in the halo observed during the course of the SDSS and its first extension (SDSS-II). Below we present a few of the most pertinent highlights, including the recognition of the dual halo structure of the Galaxy (Carollo *et al.* 2007), its decomposition into individual components (Carollo *et al.*, in prep.), and the measurement of alpha abundance ratios for many tens of thousands of stars from SDSS (Lee *et al.*, in prep.).

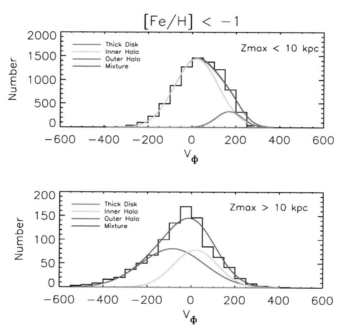

Figure 1. An example maximum-likelihood decomposition of the contribution of thick-disk, inner-halo, and outer-halo populations to the overall observed distribution of V_Φ, the rotation with respect to the Galactic center, based on an analysis of the SDSS/SEGUE calibration stars from DR-6 (Adelman-McCarthy *et al.* 2008). All stars in the figure have [Fe/H] < -1.0. The upper panel applies to stars with $Z_{\mathrm{max}} < 10$ kpc; the lower panel applies to stars with $Z_{\mathrm{max}} > 10$ kpc, where Z_{max} refers to the maximum distance from the Galactic plane reached by an individual star during the course of its orbit. Note that the inner-halo population completely dominates the observed distribution in the upper panel, while a higher dispersion, net retrograde outer-halo population is required to account for the extended low velocity tail in the lower panel.

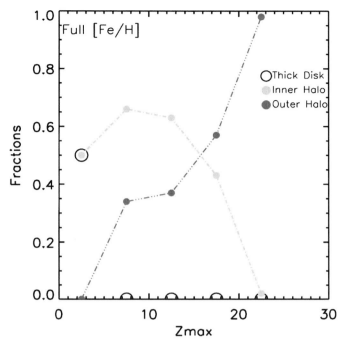

Figure 2. Application of the decomposition models of Carollo *et al.* (in prep.) in order to estimate the contribution of various populations as a function of Z_{max}, over the entire metallicity range of the SDSS/calibration star sample. The inner halo is indicated by the green line (and dots), while the outer halo is represented by the red line (and dots). The thick disk is represented by open black circles.

2. The inner/outer halo structure of the Galaxy

It has been long speculated that the halo of the Milky Way might be structurally complex, and comprise more than a single stellar population. Individual observational signatures that might be explained by a dual halo model have emerged slowly over the past few decades, and have been reported by many authors. This picture has now begun to come into focus. Recently, Carollo *et al.* (2007) used calibration stars from SDSS/SEGUE to demonstrate convincingly that the halo is indeed clearly divisible into two broadly overlapping structural components, an inner and an outer halo, which exhibit different spatial density profiles, stellar orbits, and stellar metallicities. While the inner halo has a modest net prograde rotation (now estimated to be 20 ± 5 km/s), the outer halo exhibits a net retrograde rotation (now estimated to be -85 ± 9 km/s), and a peak metallicity three times lower ([Fe/H] $= -2.2$) than that of the inner halo ([Fe/H] $= -1.6$). These properties indicate that the individual halo components likely formed in fundamentally different ways, possibly through successive dissipational (inner) and dissipationless (outer) mergers, and the tidal disruption of proto-Galactic clumps. Work is now in progress to derive the velocity ellipsoids and metallicity distribution functions of the individual populations (see Fig. 1).

The application of these decomposition models to the DR-6 calibration stars can also be used to derive the approximate fractions of stars contributed to the Galaxy from each stellar population. Fig. 2 indicates that while the thick disk and inner halo dominate within 5 kpc of the plane, the inner halo dominates between 5 and 15 kpc from the plane, and the outer halo dominates beyond 15 kpc.

3. The measurement of alpha abundance ratios

Before estimation of the $[\alpha/\text{Fe}]$ ratios for SDSS/SEGUE stars, one first requires estimates of the fundamental atmospheric parameters T_{eff}, log g, and [Fe/H].

In an effort to provide robust determinations that remain valid over the large range of parameter space and S/N explored by SEGUE, the atmospheric parameters are derived from a set of techniques, based on a number of different calibrations. These approaches, which collectively are applied by the SEGUE Stellar Parameter Pipeline (SSPP), include techniques for finding the minimum distance (parameterized in various ways) between observed spectra and grids of synthetic spectra (e.g., Allende Prieto *et al.* 2006), non-linear regression models (e.g., Re Fiorentin *et al.* 2007, and references therein), correlations between broadband colors and the strength of prominent metallic lines, such as the CaII K line (Beers *et al.* 1999), auto-correlation analysis of a stellar spectrum (Beers *et al.* 1999, and references therein), obtaining fits of spectral lines (or summed line indices) as a function of broadband colors (Wilhelm *et al.* 1999), or the behavior of the CaII triplet lines as a function of broadband color (Cenarro *et al.* 2001). Details of these procedures and tests of the validity of the resulting parameter estimates are presented in a series of three papers, Lee *et al.* (2008a), Lee *et al.* (2008b), and Allende Prieto *et al.* (2008), to which the interested reader is referred.

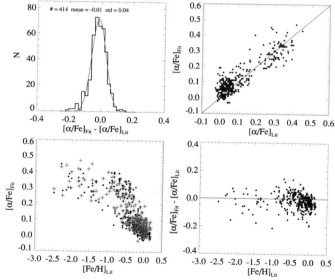

Figure 3. Comparison of $[\alpha/\text{Fe}]$ from our estimates (Fit) with those from the literature (Lit), based on spectra from the ELODIE library (Moultaka *et al.* 2004). The upper left panel is a histogram of the residuals, overplotted with a Gaussian fit. The red crosses in the bottom left panel are our determinations, while the black dots represent the literature values. Residuals in our determinations as a function of $[\alpha/\text{Fe}]_{\text{Lit}}$ and $[\text{Fe}/\text{H}]_{\text{Lit}}$ are shown in the right-hand panels.

The precision of the parameter estimates varies with the S/N of the spectra. At the median S/N of the SEGUE spectra (roughly 20/1), the estimates have typical errors of $\delta T_{\text{eff}} = 150$ K, $\delta \log g = 0.30$ dex, and $\delta[\text{Fe/H}] = 0.25$ dex, respectively. Tests on the accuracy of the atmospheric parameter estimates indicate that there exist negligible zero-point offsets over most of the parameter space.

Lee *et al.* (in prep.) describe a spectral fitting method for derivation of $[\alpha/\text{Fe}]$ ratios from medium-resolution spectroscopy, based on a large grid of synthetic spectra covering a range in atmospheric parameters, and with $[\alpha/\text{Fe}]$ ratios in the range $-0.1 \leqslant \alpha/\text{Fe}$ $\leqslant +0.6$. This approach is validated by comparison with an external library of high-resolution spectra (ELODIE; Moultaka *et al.* 2004), for stars with estimates of alpha-element abundances that have already appeared in the literature.

A total of 414 spectra for stars selected from the ELODIE library were processed in the same way as the synthetic spectra, after degrading them to the SDSS resolving power $R = 2000$. Estimates of their $[\alpha/\text{Fe}]$ were obtained by spectral matching to the synthetic spectra grid. Fig. 3 shows the results of the comparison. A Gaussian fit to the residuals between our values and those from the literature indicate that there is little systematic offset, and further indicates a very small scatter (standard deviation of 0.04 dex). No trends in the estimated alpha abundances as a function of $[\text{Fe/H}]$ are noted for stars with $[\text{Fe/H}] < -0.5$ in the right-hand panels of this figure.

A noise-injection test indicates that, for SDSS spectra with $S/N > 20/1$, $[\alpha/\text{Fe}]$ can be estimated with an rms scatter < 0.1 dex (and negligible zero-point offset), which is sufficiently precise to carry out inspection of the variations of alpha abundances as a function of $[\text{Fe/H}]$ and other quantities.

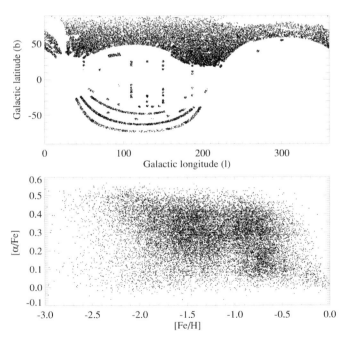

Figure 4. (upper panel) Spatial distribution of the sample (17450 stars) in a Galactic coordinate system. (bottom panel) $[\alpha/\text{Fe}]$ as a function of $[\text{Fe/H}]$.

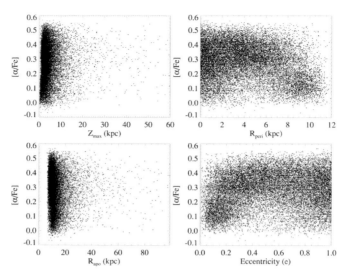

Figure 5. Distribution of $[\alpha/\mathrm{Fe}]$ with respect to $\mathrm{Z_{max}}$, $\mathrm{R_{peri}}$, $\mathrm{R_{apo}}$, and orbital eccentricity.

4. Correlations of alpha abundance ratios with metallicity and orbital parameters

A total of 39,167 spectrophotometric and telluric calibration stars from SDSS/SEGUE were selected and processed by the SSPP in order to derive their stellar atmosphere parameters. Then their $[\alpha/\mathrm{Fe}]$ ratios were estimated as described above. Signal-to-noise cuts were then applied to this sample, leaving 17,450 stars with alpha-element abundance estimates based on spectra with $S/N > 20/1$. The spatial distribution of these stars is shown in the upper panel of Fig. 4. The bottom panel of this figure shows the distribution of $[\alpha/\mathrm{Fe}]$ as a function of $[\mathrm{Fe}/\mathrm{H}]$.

At lower metallicities, inspection of Fig. 4 reveals the existence of a plateau with $[\alpha/\mathrm{Fe}]$ $\sim +0.4$, which is a typical value for halo stars with metallicities below -1.0. Above $[\mathrm{Fe}/\mathrm{H}] = -1.0$, the alpha abundances decrease with metallicity and reach the solar value at $[\mathrm{Fe}/\mathrm{H}] = 0$, as expected. It is also clear that there exists a substantial fraction of low-metallicity stars with $[\alpha/\mathrm{Fe}] < +0.2$. At low metallicities, there is a suggestion of a bifurcation of the alpha abundances into two primary groupings, high- and low-alpha stars, with a dearth of stars having intermediate alpha-element enrichment.

Orbital parameters for a smaller subset of the data (those having well-measured proper motions, and passing additional cuts to ensure they belong to a sufficiently local sample that assumptions involved in the derivation of the orbital parameters are satisfied) can be used to examine the distribution of $[\alpha/\mathrm{Fe}]$ with $\mathrm{Z_{max}}$, $\mathrm{R_{peri}}$, $\mathrm{R_{apo}}$, and eccentricity. Fig. 5 shows these distributions.

It is difficult to discern any obvious trends in the behavior of the alpha element abundances with orbital parameters; there are certainly regions which appear over- or underpopulated, but it is premature to suggest their causes. However, inspection of these data has only just begun.

It is clear that full understanding of the connection between the measured alpha abundance distributions and models for the formation and evolution of the Galaxy will require comparison with more sophisticated numerical simulations of the galaxy assembly process (which take into account chemistry and kinematics), such as those just now becoming available (e.g., Tumlinson *et al.* 2006; Johnston *et al.* 2008). It would also be very desirable to obtain higher resolution spectroscopy of targeted inner- and outer-halo stars, from which more accurate estimates of [α/Fe] ratios could be derived. Several groups have already begun this effort.

Acknowledgements

Funding for the SDSS and SDSS-II has been provided by the Alfred P. Sloan Foundation, the Participating Institutions, the National Science Foundation, the U.S. Department of Energy, the National Aeronautics and Space Administration, the Japanese Monbukagakusho, the Max Planck Society, and the Higher Education Funding Council for England. The SDSS Web Site is http://www.sdss.org/.

This work received partial support from grants AST 07-07776 and PHY 02-15783; Physics Frontier Center / Joint Institute for Nuclear Astrophysics (JINA), awarded by the US National Science Foundation.

References

Adelman-McCarthy, J. K, Agüeros, M. A., Allah, S. S, Allende Prieto, C., & Anderson, K. S. J., *et al.* 2008, *ApJS*, 175, 297

Allende Prieto, C., Beers, T. C., Wilhelm, R., Newberg, H. J., & Rockosi, C. M., *et al.* 2006, *ApJ*, 636, 804

Beers, T. C., Rossi, S., Norris, J. E., Ryan, S. G., & Shefler, T. 1999, *AJ*, 117, 981

Bekki, K. & Chiba, M. 2001, *ApJ*, 558, 666

Allende Prieto, C., Sivarani, T., Beers, T. C., Lee, Y. S., & Koesterke, L. *et al.* 2008, *AJ*, in press (arXiv:0710.5780)

Belokurov, V., Zucker, D. B., Evans, N. W., Gilmore, G., & Vidrih, S., *et al.* 2006, *ApJ*, 642, L137

Carollo, D., Beers, T. C., Lee, Y. S., Chiba, M., & Norris, J. E., *et al.* 2007, *Nature*, 450, 1020

Cenarro, A. J., Cardiel, N, Gorgas, J., Peletier, R. F., Vazdekis, A., & Prada, F. 2001, *MNRAS*, 326, 959

Diemand, J., Kuhlen, M., & Madau, P. 2007 2007, *ApJ*, 667, 859

Johnston, K. V., Bullock, J. S., Sharma, S., Font, A., Robertson, B. E., & Leitner, S. N. 2008, *ApJ*, in press (arXiv:0807.3911)

Lee, Y. S., Beers, T.C., Sivarani, T., Allende Prieto, C., & Koesterke, L., *et al.* 2008, *AJ*, in press (arXiv:0710.5645)

Lee, Y. S., Beers, T. C., Sivarani, T., Johnson, J. A., & An, D., *et al.* 2008, *AJ*, in press (arXiv:0710.5778)

Moultaka, J., Ilovaisky, S. A., Prugniel, P., & Soubiran, C. 2004, *PASP*, 116, 693

Re Fiorentin, P., Bailer-Jones, C. A. L., Lee, Y. S., Beers, T. C., & Sivarani, T., *et al.* 2007, *A&A*, 467, 1373

Tumlinson, J. 2006, *ApJ*, 641, 1

Walsh, S. M., Willman, B., & Jerjen, H. 2008, *AJ*, submitted (arXiv:0807.3345)

Wilhelm, R., Beers, T. C., & Gray, R. O. 1999, *AJ*, 117, 2308

Zucker, D. B., Belokurov, V., Evans, N. W., Wilkinson, M. I., & Irwin, M. J., *et al.* 2006, *ApJ*, 643, L103

Low-Metallicity Star Formation:
From the First Stars to Dwarf Galaxies
Proceedings IAU Symposium No. 255, 2008
L.K. Hunt, S. Madden & R. Schneider, eds.

© 2008 International Astronomical Union
doi:10.1017/S1743921308025027

Galactic archeology
with extremely metal-poor stars

Yutaka Komiya[1], Takuma Suda[2], Asao Habe[2],
and Masayuki Fujimoto[2]

[1] Department of Astronomy, Faculty of Science, Tohoku University,
Sendai, Miyagi Prefecture 980-8578, Japan
email: komiya@astro.tohoku.ac.jp
[2] Department of Cosmosciences, Hokkaido University,
Sapporo, Hokkaido 060-0810, Japan

Abstract. Extremely metal-poor (EMP) stars are thought to be formed in the low-mass protogalaxies as building blocks of the Milky Way and can be probes to investigate the early stage of galaxy formation and star formation in the early universe. We study the formation history of EMP stars in the Milky Way halo using a new model of chemical evolution based on the hierarchical theory of the galaxy formation. We construct the merging history of the Milky Way halo based on the extended Press-Schechter formalism, and follow the star formation and chemical evolution along the merger tree. The abundance trends and number of low-mass stars predicted in our model are compared with those of observed EMP stars. Additionally, in order to clarify the origin of hyper metal poor stars, we investigate the change of the surface metal abundances of stars by accretion of interstellar matter. We also investigate the pre-enrichment of intergalactic matter by the first supernovae.

Keywords. stars: abundances, (stars:) binaries: general, stars: chemically peculiar, stars: formation, stars: luminosity function, stars: mass function

1. Introduction

Recent surveys detected thousands of EMP stars in the Milky Way halo (Beers *et al.* 1992, Christlieb *et al.* 2001), hundreds of which are observed by high dispersion spectroscopy (e.g. Aoki *et al.* 2007, Spite *et al.* 2005). Thanks to these observations, we can discuss the statistical features of EMP stars. These EMP stars are formed in the ancient Galaxy and are useful probes of star formation in the early universe and metal-poor environments. Such a study of the early universe and formation of the Galaxy with metal-poor stars is called Galactic Archaeology. However, we can observe only low-mass survivors among stars formed in the early universe, while massive stars should have ended their lives leaving their nucleosynthetic signatures. Hereafter, we call all the stellar population of $[\mathrm{Fe/H}] \lesssim -2.5$ "EMP population" and low mass subset "EMP survivors".

One prominent observational feature of EMP survivor is that the fraction of carbon-enhanced EMP (CEMP) stars is very large ($\gtrsim 20\%$; see e.g., Suda *et al.* 2008) compared with more metal-rich counterparts (a few %). In our previous study (Komiya *et al.* 2007), we showed that CEMP stars are formed as secondary components of binary systems where matter enriched with carbon is transferred from intermediate-mass primary stars. Based on the stellar evolution model of Fujimoto *et al.* (2000), we can constrain the IMF of EMP stars by requiring the statistical features of CEMP stars to be consistent with model predictions. Observed large fractions of CEMP stars in EMP stars, number ratio between CEMP stars with and without s-process element enhancement, total number of EMP survivors, and metallicity distribution function (MDF) of stars with $-2.5 \gtrsim [\mathrm{Fe/H}] \gtrsim -4$, imply a massive IMF with medium mass $M_{md} \sim 10 \, M_\odot$ (Komiya *et al.* 2007, Komiya *et al.* 2008a). This result suggests that most of EMP survivors are secondary companions of massive or intermediate-mass stars when they were formed.

Two intriguing problems of EMP stars are the scarcity of stars of $[Fe/H] \lesssim -4$ and the existence of three hyper metal-poor (HMP) stars, having metallicities $[Fe/H] < -4.5$ Christlieb *et al.* (2002), Frebel *et al.* (2005), Norris *et al.* (2007). Karlsson (2005) and Karlsson (2006) insist that this metallicity gap can be interpreted as a result of a metal diffusion process in the stochastic chemical evolution model. On the other hand, some studies advocate a change in the IMF around $[Fe/H] = -4$ (Tumlinson 2006, Salvadori *et al.* 2006). For the origin of HMP stars, Suda *et al.* (2004) propose the possibility that they are remnant Pop. III stars formed without metals and polluted by the accretion of interstellar matter (ISM) and binary mass transfer. Umeda & Nomoto (2003) proposed that these stars are formed from the ejecta of the first supernovae with peculiar abundance patterns. Karlsson (2006) point out that the small amount of iron in HMP stars is explained by pre-enrichment by Pop. III stars. In this paper, we investigate the MDF of EMP stars using the model of chemical evolution with a semi-analytic merger tree and explain the origin of the metallicity gap, taking into account hierarchical galaxy formation. We also study two hypotheses for the origin of HMP stars: surface pollution and pre-enrichment by the first supernovae.

2. Hierarchical Chemical Evolution

In order to investigate the effects of structure formation processes in the early stages of chemical evolution, we plant a merger tree using the extended Press-Schechter approach (Bond *et al.* 1991, Lacey & Cole 1993). We calculate the merger tree of halo with a mass of $10^{12} \, M_\odot$ in a ΛCDM universe ($\Omega_M = 1 - \Omega_\Lambda = 0.3, \Omega_b = 0.045, h = 0.7, \sigma_8 = 0.9$) with the method constructed in Somerville & Kolatt (1999). Mass resolution of mini-halos, $M_{h,l}$, is determined by the halo mass when the virial temperature becomes $T_{vir}(M_{h,l}, z) = 10^3$K (Tegmark *et al.* 1997, Nishi & Susa 1999). We use the result of Lacey & Cole (1993) to calculate the time for merged mini-halos orbit as satellites. We use simple assumptions about star formation and chemical evolution. Star formation efficiency is constant: 10^{10} /yr and the binary fraction is 50%. For the EMP population, we assume a log-normal IMF with $M_{md} \sim 10 \, M_\odot$, derived by Komiya *et al.* (2007) and flat mass ratio distribution, $n(q) = 1$. The stellar yields and lifetimes are taken from Tominaga *et al.* (2007) and Schaerer (2002), respectively. Metals ejected by the supernovae spread instantaneously and homogeneously in their host halos. For the first stars in the minihalos, we assume that $M_{md} = 50 \, M_\odot$ and the binary fraction is set to 0, to suppress star formation in the host halo.

Figure 1 shows the resultant MDF. The slope of the MDF as well as the total observed number are consistent with theoretical results in the metallicity range $-2.5 \gtrsim [Fe/H] \gtrsim -4.5$. The metallicity cut-off at $[Fe/H] \sim -4$ is also well reproduced by the hierarchical scenario. In the current model, typical Type II supernovae (SN) eject $\sim 0.07 \, M_\odot$ of iron into the ISM. The metallicity of the primordial halo of typical mass ($\sim 10^6 \, M_\odot$) becomes

$$[Fe/H] = \log \frac{0.07 \, M_\odot}{10^6 \, M_\odot (\Omega_b/\Omega_m)} - \log Z_\odot \sim -3.5. \qquad (2.1)$$

On average, the next generation of stars is expected to have typical metallicities of $[Fe/H] \gtrsim -3.5$ in the mini-halo. If mini-halos have larger mass, stars with metallicities of $[Fe/H] \sim -4$ can be formed. Our result suggests that HMP stars with metallicity $[Fe/H] \lesssim -4.5$ are likely to be born before the first supernova in the host cloud.

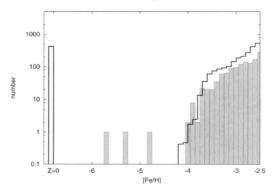

Figure 1. Comparison of the theoretical and observational MDF of EMP stars. The shaded histogram shows the observed number of stars with the HK and HES surveys Beers (2005). The solid line shows the estimated number of stars in the detectable survey areas assuming a uniform distribution of halo stars.

In the lowest metallicity range, however, our numerical results do not agree with the observations. Our model predicts a large number of Pop. III stars. They are second generation stars in the halos with the first stars of mass $m > 50\,M_\odot$. These stars of $M > 50\,M_\odot$ become black holes without iron ejection, and second generation stars without iron are formed. The predicted number of second generation Pop. III stars becomes large because we assume that the IMF of second generation stars with $Z = 0$ is the same as EMP stars. In contrast to this result, the observed number of stars with [Fe/H] < -4.5 is very small. It suggests that the change of IMF from the first stars ($M_{\rm md} \gtrsim 50$) to the EMP population ($M_{\rm md} \sim 10$) is caused by metals, not by radiation from the first stars. But the critical metallicity to enhance the low mass star formation is not necessarily [Fe/H] ~ -4.

3. Origin of HMP stars

3.1. *Accretion of ISM*

EMP stars are likely to modify their surface abundances due to the possible accretion processes during their long lives. In particular, it is pointed out that the small amount of metals in HMP stars comes from interstellar accretion and that they are possibly Pop. III stars polluted by accretion of the ISM (Suda *et al.* 2004). We trace the modifications of surface abundances in individual low-mass Pop. III or EMP stars which survive to date.

In considering the effect of accretion, we adopt the Bondi-Hoyle accretion rate

$$\dot{M} = 4\pi(GM)^2\rho(V_r^2 + c_s^2)^{-3/2}, \tag{3.1}$$

where M, ρ, c_s, and V_r are the stellar mass, the density and the sound velocity of the ISM, and the relative velocity between stars and ambient gas, respectively. We use the following assumptions for dynamics of gas and stars: (1) gas is cooled down to 200K and concentrated around the center of the mini-halo; (2) V_r is negligible in the mother cloud in which EMP stars are formed; (3) once the mother cloud accretes onto a larger halo, stars move with virial velocity, $V_{\rm vir}$, in an ISM of average density $\rho_{\rm av} = \bar{\rho}\Delta_c\Omega_b$. (4) All the ISM accreted on binary systems settles onto the EMP survivors, to give the upper limit of the accretion rate.

Figure 2 shows the MDF taking into account surface pollution by accretion of the ISM. Accreted matter is mixed in the surface convective zone of mass of $M_{\rm SCZ} = 0.2\,M_\odot$ and $0.003\,M_\odot$ for giant and main sequence stars, respectively (Fujimoto *et al.* 1995). Surface

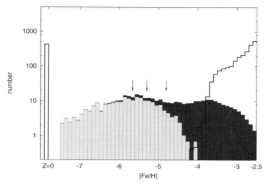

Figure 2. Effect of surface pollution on Pop. III stars and EMP stars. Solid line shows MDF with surface pollution of ISM taken into account. Dark and light gray histograms show the distribution of the polluted main sequence Pop. III stars and giant Pop. III stars, respectively. The different degree of pollution is due to the depth of surface convective zones. Three arrows denote the locations of three HMP stars.

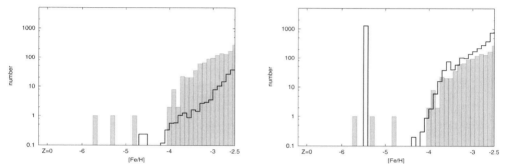

Figure 3. Effect of PISN. *left:* First stars in all mini-halos assumed to become PISN. *right:* First stars in mini-halos formed with $[Fe/H] < [Fe/H]_c - 5.5$ become PISN.

metallicity of main-sequence Pop. III stars is distributed around $[Fe/H] \sim -3$ to -4, and is diluted to $[Fe/H] \sim -5$ to -6 when they evolve to giants. This result implies that HMP stars are possibly polluted Pop. III stars that accrete iron group elements onto their surface from the ISM.

3.2. *Pre-enrichment by the first supernova*

We considered the effect of pair-instability supernovae (PISNe) for Pop. III stars that produce huge explosion energies sufficient to blow off the host mini-halo. We assume that $10\,M_\odot$ of iron is ejected by PISNe and is mixed into the galactic matter instantaneously and homogeneously. Figure 3 shows the resultant MDF with the outflow of gas and metals by PISNe taken into account. At first, we assume that a single PISNe event occurs in all mini-halos (see left panel). In this case, overproduction of metals by PISNe reduces the number of EMP stars and the resultant MDF is inconsistent with observations.

We also assume the case that the PISNe of first stars occur in the mini-halo with $Z \leqslant Z_{cr} = 10^{-5.5}\,Z_\odot$ and $T_{vir} < 10^4$K (see right panel). In this model, the theoretical MDF with $[Fe/H] > -5$ is quite similar to Figure 1 and consistent with observations. However, too many stars with $Z \sim Z_{cr}$ are formed instead of stars with $Z = 0$. In any case, if the PISNe contribute to the pre-enrichment of the Galaxy, the observed abundances of HMP stars should reflect the yields of PISNe, although the observations do not support this.

These results are inconsistent with observations but we cannot reject the possibility that some PISNe occur in the early universe because their yields may be masked by accreted ISM in the currently observed HMP stars. Our results will provide some constraints on the understanding of the first supernovae and pre-enrichment history of the Galaxy.

4. Conclusions

We modelled the formation history of EMP stars with realistic merging history and simple chemical evolution. Theoretical MDFs of EMP stars are calculated and compared with observations. Our hierarchical model naturally explains the metallicity cut-off at [Fe/H] ~ -4. On the other hand, the theoretical MDF with [Fe/H] $\lesssim -5$ is inconsistent with observations. This suggests that the formation process and mass distribution of stars with [Fe/H] $\lesssim -5$ may differ from more metal-rich populations.

Our estimate of chemical enrichment implies that HMP stars are formed before the first supernovae in their host halos. Considering the effect of surface pollution in the hierarchical model, we show that they are possibly polluted Pop. III stars. We also consider the effect of pre-enrichment by PISNe, but these models fail in reproducing the observed MDF.

References

Abel, T, Bryan, G. L., & Norman, M. L. 2002, *Science*, 295, 93

Aoki, W., Beers, T. C., Christlieb, N., Norris, J. E., Ryan, S. G., & Tsangarides, S.2007 *ApJ*, 655, 492

Beers, T. C., Preston, G. W., & Shectman, S. A. 1992, *AJ*, 103, 1987

Beers, T. C., Christlieb, N., Norris, J. E., Bessell, M. S., Wilhelm, R., Allende P. A., Yanny, B., Rockosi, C., Newberg, H. J., Rossi, S., & Lee, Y. S. 2005, Proc. IAUS 228, 175

Bond, J. R., Cole, S., Efstathiou, G., & Kaiser, N. 1991, *ApJ*, 379, 440

Christlieb, N., Green, P. J., Wisotzki, L., & Reimers, D. 2001, *A&A*, 375, 366

Christlieb, N., Bessell, M. S., Beers, T. C., Gustafsson, B., Korn, A., Barklem, P. S., Karlsson, T., Mizuno-Wiedner, M., Rossi, S. 2002, *Nature*, 419, 904-906

Frebel, A., Aoki, W., Christlieb, N., *et al.* 2005, *Nature* 434, 871-873

Fujimoto, M. Y., Sugiyama, K., Iben, I. Jr., & Hollowell, D. 1995, *ApJ* , 444, 175

Fujimoto, M. Y., Ikeda, Y., & Iben, I. Jr. 2000, *ApJ*, 529, L25

Karlsson 2006, *A&A*,

Karlsson 2006, *ApJl*, 641, L41

Komiya, Y., Suda, T., Minaguchi, H., Shigeyama, T., Aoki, W., & Fujimoto, Y. M. 2006 *ApJ* 658, 367

Komiya, Y., Suda, T., & Fujimoto, Y. M. 2008 *ApJ* submitted.

Komiya, Y., Suda, T., Asao, H., & Fujimoto, Y. M. 2008, in preparation.

Lacey, C., & Cole, S. 1993, *MNRAS*, 262, 627

Nishi, R., & Susa, H. 1999, *ApJ*, 523, L103

Norris, John E., Christlieb, N., Korn, A. J., Eriksson, K., Bessell, M. S., Beers, Timothy C., Wisotzki, L., Reimers, D. 2007 *ApJ* 670, 774

Salvadori, S., Schneider, R., & Ferrara, A. 2006, *MNRAS*, 381, 647

Schaerer, D 2002, *A&A*, 382, 28

Somerville, R. S., & Kolatt, T. S. 1999, *MNRAS*, 305, 1

Spite, M. *et al.* 2005, *A&A* , 430, 655

Suda, T., Aikawa, M., Machida, M. N., Fujimoto, M. Y., & Iben, I. Jr. 2004, *ApJ*, 611, 476

Suda, T. *et al.* 2008, *PASJ*, in press

Tegmark, M., Silk, J., Rees, M. J., Blanchard, A., Abel, T., & Palla, F. 1997, *ApJ*, 474, 1

Tumlinson, J. 2006, *ApJ*, 641, 1

Tominaga, N., Umeda, H., & Nomoto, K. 2007, *ApJ*, 660, 516

Umeda, H. & Nomoto, K. 2003, *Nature*, 422, 871

Low-Metallicity Star Formation:
From the First Stars to Dwarf Galaxies
Proceedings IAU Symposium No. IAU255, 2008
L. Hunt, S. Madden & R. Schneider, eds.

Stellar Archaeology: Using Metal-Poor Stars to Test Theories of the Early Universe

Anna Frebel, Jarrett L. Johnson and Volker Bromm

McDonald Observatory and Department of Astronomy, University of Texas,
1 University Station, C1402, Austin TX, 78712
email: anna, jljohnson, vbromm@astro.as.utexas.edu

Abstract. Constraints on the chemical yields of the first stars and supernova can be derived by examining the abundance patterns of different types of metal-poor stars. We show how metal-poor stars are employed to derive constraints of the formation of the first low-mass stars by testing a fine-structure line cooling theory. The concept of stellar archaeology, that stellar abundances truly reflect the chemical composition of the earliest times, is then addressed. The accretion history of a sample of metal-poor stars is examined in detail in a cosmological context, and found to have no impact on the observed abundances. Predictions are made for the lowest possible Fe and Mg abundances observable in the Galaxy, $[\text{Fe/H}]_{min} = -7.5$ and $[\text{Mg/H}]_{min} = -5.5$. The absence of stars below these values is so far consistent with a top-heavy IMF. These predictions are directly relevant for future surveys and the next generation of telescopes.

Keywords. stars: abundances, stars: Population II, Galaxy: halo, early Universe

1. Introduction

The first stars, the so-called Population III (Pop III), were the key drivers of early cosmic evolution. They produced large amounts of hydrogen-ionizing radiation which initiated the reionization of the universe. The first supernova (SN) explosions then provide the pristine intergalactic medium (IGM) with the first heavy elements. Based on numerical simulations, the current theoretical models of Pop III star formation suggests that the Pop III initial mass function (IMF) was top-heavy (Bromm, Coppi & Larson 2002). Testing this crucial prediction is one of the main goals of the upcoming James Webb Space Telescope (JWST), but it is important to apply complementary probes that are accessible already now.

We here consider a "near-field cosmology" by assessing low-mass, metal-poor Galactic halo stars as probes of the nucleosynthetic signature of the first stars. The most metal-poor Galactic halo stars are now frequently used in an attempt to reconstruct the onset of the chemical and dynamical formation processes of the Galaxy. These stars are an easily-accessible local equivalent of the high-redshift Universe, and are used to address near-field cosmological questions.

2. Testing a Low-Mass Star Formation Theory

Large samples of these old, metal-poor objects are now being employed to test theoretical predictions about the early Universe. Compared to the first stars that presumably were very massive ($100 \, M_\odot$), the most metal-poor stars are of low mass ($M < 1 \, M_\odot$). Currently, two competing scenarios (fine-structure line cooling and dust cooling) responsible for cooling of the early interstellar medium are being debated in the literature.

Fine-structure line cooling through C I and O II may be responsible for sufficient fragmentation of near-primordial gas which led to the formation of the first low-mass stars (Bromm & Loeb 2003). We have developed an "observer-friendly" formulation of the fine-structure line cooling theory (Frebel, Johnson & Bromm (2007)) that incorporates the observed C and/or O abundances of metal-poor stars; $D_{trans} = \log(10^{[C/H]} + 0.3 \times 10^{[O/H]}) \geqslant -3.5$. If this theory is correct, a low-mass star that is still observable today, thus has to lay above the critical metallicity limit of $D_{trans} = -3.5$. The observations thus far support this theory, and furthermore we predict that future stars to be found with $[Fe/H] < -4.0$ will show significant C and/or O overabundances.

We have further populated the Figure 1 of Frebel, Johnson & Bromm (2007) with lower limits for D_{trans}, based on new $[Fe/H]$ and $[C/Fe]$ data of various stars from the literature. Among them is HE 0558−4840 with $[Fe/H] = -4.8$. Its C abundance provides a lower limit which lies somewhat below the D_{trans} limit. The O abundance, which is not yet available, will determine if the star will remain below the limit. Within the theoretical uncertainties, however, HE 0557−4840 still does not violate the fine-structure line cooling theory. More such "borderline" examples are crucial for a successful test of this theory. We also note that this is one of the few theoretical works that actually can be confirmed or refuted with observational data. We therefore encourage observers to add their C and/or O measurements of metal-poor stars to such a plot in order to fully map the observable regions, and to test the fine-structure line theory of Bromm & Loeb (2003) with the most solid observational material. We are currently studying new carbon-poor stars from the bright metal-poor sample of Hamburg/ESO survey which will then be added to our diagram.

3. Validating Stellar Archaeology

Abundances of metal-poor stars are primarily used to infer details of the chemical composition of the early Universe and the onset of Galactic chemical evolution. It is hereby assumed that the observed abundances reflect the conditions of the interstellar medium at the time and place of their formation. Accretion of material from the interstellar medium while a star orbits in the Galaxy for $\sim 10\,$Gyr has long been suggested to affect the observed abundance patterns, although only simplified calculations have been carried out. Iben (1983) calculated a basic "pollution limit" of $[Fe/H] = -5.7$. Based on Bondi-Hoyle accretion, he predicted that no stars could be found with Fe abundances below this value because they would have accreted too much material from the ISM.

To assess the potential accretion history of individual halo stars in more detail, we carried out a kinematic analysis of metal-poor stars with $[Fe/H] < -2.5$ selected from the Sloan Digital Sky Survey (DR6). The necessary input data (e.g., abundances, radial velocities, proper motions, distances) are available on the DR6 website†. All stars were followed for $10\,$Gyr in a three component potential adopted from Johnston (1998). Accretion was assumed to take place only during disk crossings, whereby the disk height was set to $100\,$pc with a density of $n = 5\,\mathrm{cm}^{-3}$. The amount of accreted Fe and Mg are calculated based on the total accreted amount for every crossing. A simple model for chemical evolution is taken into account in which the interstellar medium has scaled solar abundances. In Figure 1, we compare the total accreted Mg amount for every star with the observed abundances. The stars all have lower "accreted abundances" than observed abundances, sometimes by several dex. This generally confirms that accretion does not

† http://www.sdss.org/dr6/

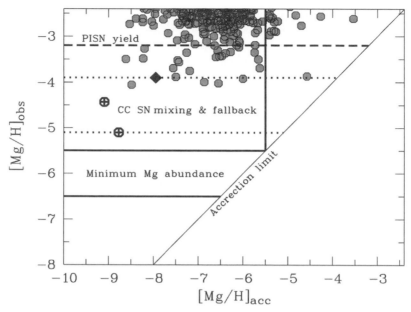

Figure 1. Observed vs. "accreted" Mg abundances for the sample of metal-poor stars (*red circles*). The three most Fe-poor stars are indicated (*crossed circles, diamond*). All stars have accreted less metals (> one order of magnitude) than what is observed, thus validating stellar archaeology. The minimum Mg abundance range, calculated under the assumption of a top-heavy Pop III IMF, is indicated. The Mg abundances arising from a PISN event in an atomic cooling halo is also shown (*dashed line*), Mg levels of pre-enrichment from a 25 $M_\odot$ mixing and fallback SN (*dotted lines*). The stars left of the vertical line that can be used to place constraints on the Pop III IMF, where accretion cannot affect whether they lie above or below the minimum abundance range, predicted for a top-heavy IMF.

dominate the abundance pattern of old metal-poor stars. The same is found in the case of Fe.

We also investigated the special case where each star is moving once through a dense molecular cloud. In this case, the observed abundance pattern might be dominated by the signature of the interstellar medium. We find, however, that the stars still have slightly lower "accreted abundances" than observed abundances. Whether every star indeed runs through such a cloud remains unclear, but even if it does, the accretion process heavily depends on the space velocity of a star. We thus consider this a "maximum accretion" case which serves as a very robust upper limit to the total accreted amount for each star. This leads to the conclusion that the concept of stellar archaeology is accurate and that the observed abundances of metal-poor stars truly reflect the chemical compositions of the time and place of their formation. Furthermore, it becomes obvious that kinematic information is vital for the identification of the lowest-metallicity stars in the Milky Way.

4. The Nucleosynthetic Yields of the First Stars

The metal enrichment provided by the first stars, assuming a top-heavy IMF, that may still be observable in the most metal-poor stars is schematically illustrated in Fig. 2. There are several progenitor mass ranges of SNe that would have provided different levels

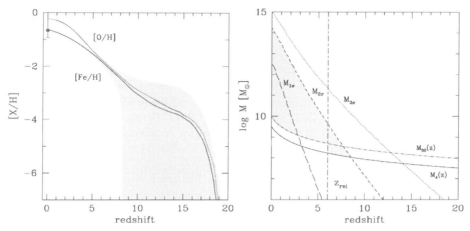

Figure 2. *Left panel*: Evolution of GM iron and oxygen (solid and dotted lines) abundance. Lines are the average values over 100 realizations of the merger tree and the shaded area the $\pm 1\sigma$ dispersion region for [Fe/H]. The point is the measured [O/H] in high-velocity clouds (Ganguly *et al.* 2005). *Right panel*: Evolution of the mass corresponding to $1/2/3\sigma(M, z)$ density peaks (long-dashed/short-dashed/dotted lines); the solid and dotted-dashed lines show the evolution of $M_4(z)$ and $M_{30}(z)$ respectively. The selected reionization redshift $z_{rei} = 6$ is also shown.

halo MDF at [Fe/H]< -2.5 form in unpolluted haloes accreting Fe-enhanced gas from the GM. The initial [Fe/H] abundance of the gas is then fixed by the corresponding GM abundance at the virialization redshift (z_{vir}). The evolution of the GM iron and oxygen abundance is shown in Fig. 2 (left panel).

4. The dSph evolution

We can now study the formation and evolution of dSphs inside our cosmological picture by assuming that dSph candidates belong to MW building blocks.

4.1. *Birth environment*

The first point to address is the selection criteria to identify dSphs among various MW progenitors. We select dSph candidates on the basis of their host halo mass using two criteria: one founded on reionization and the other on dynamical arguments. We already discussed (Sec. 2) that $M_4(z)$ is the minimum halo mass to trigger the SF; thus dSph candidates must have $M > M_4(z)$. During reionization, however, the increase of the Inter Galactic Medium (IGM) temperature causes the growth of the Jeans mass and consequent suppression of gas infall in low-mass objects. Essentially, below a characteristic halo mass-scale, the gas fraction is drastically reduced compared to the cosmic value (Gnedin 2000) and the SF consequently quenched. We adopt the simplest prescription assuming that, after reionization, SF is suppressed in haloes with circular velocity $v_c < 30$ km/s. Thus, when $z < z_{rei}$ (here fixed at $z_{rei} = 6$), dSph candidates must have a mass $M_{30}(z) > M(v_c = 30$ km/s $, z)$. Second, we want to select dSphs among virializing haloes which could become MW satellites. Following the results of N-body cosmological simulations by Diemand, Madau & Moore (2005) we require that dSph candidates correspond to low-σ density fluctuations at the virialization epoch (z_{vir}). In particular we select those objects with masses $M < M_{2\sigma}$. Figure 2 (right panel) provides a summary of our selection criteria. Since the probability to form newly virializing haloes with $M > M_{30}$ is very low at each redshift, the first criterion implies that the formation of dSphs is unlikely to occur below z_{rei}. Thus, dSph candidates are selected in a very narrow redshift range $6 < z < 8.5$. In this interval the mean GM iron abundance is [Fe/H] ~ -3: the dSph birth environment has been naturally pre-enriched up to [Fe/H] values consistent with those implied by observations of the MDFs by Helmi *et al.* (2006).

4.2. *Star formation history*

The dSph evolution is highly influenced by mechanical feedback effects, which are more intense in low mass objects (McLow & Ferrara 2000, Ferrara & Tolstoy 2002). The intermittent SFR (see Fig. 3 of SFS08) we derive for a typical dSphs is a consequence of such strong mechanical feedback processes. After $\sim$100 Myr from the dSph formation the mass of gas lost by SN-driven winds becomes larger than the remaining gas into the galaxy, causing a complete blow-away of the gas. The dSph becomes gas free and the SF is stopped. At subsequent times, however, gas returned by evolved stars can be gradually collected: the dSph enters in a rejuvenation phase and the SF can start again. From the beginning of the rejuvenation phase the subsequent evolution of the galaxy proceeds like in the first 100 Myr of its life although the SFR is more than 2 orders of magnitude lower due to the paucity of returned gas. As a consequence we found that the 99% of today living stars is formed during the first 100 Myr of the dSph life, implying that dSphs are dominated by an ancient stellar population.

5. Some observable properties of dSph

In the following Sections we will compare our numerical results with some of the most relevant observations. Given the amount of available data, we take Sculptor as a reference dSph to compare with.

5.1. *The Color-Magnitude Diagram*

The Color-Magnitude Diagram (CMD) represents one of the best tools to study the SFH of a galaxy. Starting from our numerical results for a typical dSph (SFR, ISM metallicity evolution, IMF) we have computed the corresponding synthetic CMD using the publicly available IAC-STAR code by Aparicio & Gallart (2004). A simple randomization procedure has been adopted in order to simulate observational errors into the synthetic CMD and compare numerical results with data (see SFS08 for details). Despite the simple procedure used and the IAC-STAR limitations ($Z_{min} > 0.005 Z_\odot$) we can observe that the match between theoretical and experimental points is quite good (Fig. 3 left panel).

5.2. *The Metallicity Distribution Function*

We can finally focus on the Sculptor MDF. In Fig. 3 (right panel) the observed (Helmi *et al.* 2006) and simulated MDFs are compared, the latter being normalized to the total number of observed stars (513). As can be inferred from the Figure, the model shows a good agreement with the observed MDF, particularly for [Fe/H]< -1.5; a marginally significant deviation is present at larger [Fe/H] values (see SFS08 for the discussion). From our analysis we found that infall rate controls the shape of the low-[Fe/H] MDF tail while mechanical feedback (and so differential winds) strongly affects the high-[Fe/H] region. The two remaining free parameters of the models (infall and differential winds) are fixed in order to match the Sculptor MDF.

6. Conclusions

We proposed a cosmological approach to investigate the formation and evolution of dSph galaxies. In this scenario dSphs represent fossil objects that virialize at $z = 7.2 \pm 0.7$

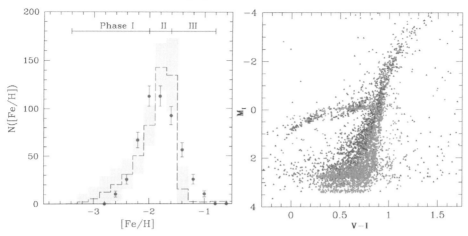

Figure 3. *Left panel*: Comparison between the observed (Helmi *et al.* 2006) (points) and simulated Sculptor MDF (histogram). Error bars are Poissonian errors. The histogram is averaged MDF over the surviving number of satellites (~20) in 100 realizations of the merger tree (~2000 objects) and the shaded area the $\pm 1\sigma$ dispersion region among different realizations. *Right Panel*: Comparison between the CMD of the Sculptor stellar population observed by Tolstoy *et al.* (2004) (triangles) and the synthetic CMD (open points) derived for a typical dSph galaxy with total mass $M = 1.6 \times 10^8 M_\odot$ which virialize at redshift $z_{vir} = 7.2$.

(i.e. in the pre-reionization era $z > z_{rei} = 6$) in the MW environment which at that epoch had already been pre-enriched up to [Fe/H]~ −3. Their dynamical masses are in the narrow range $M = (1.6 \pm 0.7) \times 10^8 M_\odot$. Our results match several observed properties of Sculptor, used as a template of dSphs: (i) the Metallicity Distribution Function; (ii) the Color Magnitude Diagram; (iii) the decrement of the stellar [O/Fe] abundance ratio for [Fe/H]> −1.5; (iv) the dark matter content and the light-to-mass ratio; (v) the HI gas mass content.

References

Bromm, V., Ferrara, A., Coppi, P. S., & Larson, R. B., 2001, MNRAS, 328, 969

Christlieb, N., *et al.* 2002, *Nature*, 419, 904

Frebel, A., Christlieb, N., Norris, J. E., Aoki, W., & Asplund, M. 2005, *Nature*, 434, 871

Fujita, A., Mac Low, M. M., Ferrara, A., & Meiksin, A., 2004, ApJ, 613, 159

Ganguly, R., Sembach, K. R., Todd, T. M., & Savage B. D. 2005, *ApJ*, 157, 251

Helmi, A., *et al.* 2006, *ApJ*, 651, L121

Kereš, D., Katz, N., Weinberg, D. H., & Davé, R., 2005, MNRAS, 363, 2

Mac Low, M.-M. & Ferrara, A. 1999, *ApJ*, 513, 1421

Omukai, K., 2000, ApJ, 534, 809

Omukai, K., Tsuribe, T., Schneider, R., & Ferrara, A., 2005, ApJ, 626, 627

Salvadori, S., Schneider, R., & Ferrara, A. 2007, *MNRAS*, 381, 647

Salvadori, S., Ferrara, A., & Schneider, R. 2008, *MNRAS*, 381, 647

Schneider, R., Ferrara, A., Natarajan, P., & Omukai, K., 2002, ApJ, 571, 30

Schneider, R., Ferrara, A., Salvaterra, R., Omukai, K., & Bromm V., 2003, Nat, 422, 869

Schneider, R., Omukai, K., Inoue, A. K., & Ferrara, A., 2006, MNRAS, 369, 1437

Tolstoy, E., *et al.* 2004, *ApJ*, 617, L119

Vader, J. P., 1986, ApJ, 305, 669

Session VI

Low-metallicity star formation in the local Universe

Low-Metallicity Star Formation:
From the First Stars to Dwarf Galaxies
Proceedings IAU Symposium No. 255, 2008
L.K. Hunt, S. Madden & R. Schneider, eds.

Blue Compact Dwarf Galaxies: Laboratories for probing the Primordial Universe

Trinh X. Thuan

University of Virginia, Astronomy Department
P.O. Box 400325, Charlottesville, VA 22904-4325, USA
email: txt@virginia.edu

Abstract. Blue Compact Dwarf (BCD) galaxies are the most metal-deficient star-forming galaxies known in the universe, with metallicities ranging from 1/40 to 1/3 that of the Sun. I review how they constitute excellent nearby laboratories for studying big bang nucleosynthesis and star formation and galaxy evolution processes in a nearly primordial environment.

Keywords. galaxies: dwarf, galaxies: abundances, galaxies: ISM, galaxies: star clusters, galaxies: individual (I Zw 18), galaxies: individual (SBS 0335−052), stars: winds, stars: outflows, stars: Wolf-Rayet

1. Blue Compact Dwarfs and Galaxy Formation

Galaxy formation is one of the most fundamental problems in astrophysics. To understand how galaxies form, we need to unravel how stars form from the primordial gas and how the first stars interact with their surrounding environments. While much progress has been made in finding large populations of galaxies at high redshifts ($z \geqslant 2$), truly young galaxies in the process of forming remain elusive in the distant universe. The spectra of those far-away galaxies generally indicate the presence of a substantial amount of heavy elements, implying previous star formation and metal enrichment (Shapley *et al.* 2004). Instead of focussing on high-redshift galaxies, another approach is to study the properties of the massive stellar populations and their interaction with the ambient interstellar medium (ISM) in a class of nearby metal-deficient dwarf galaxies, called Blue Compact Dwarf (BCD) galaxies, which are the least chemically evolved star-forming galaxies known in the universe. These galaxies have an oxygen abundance in the range $12+\log(\mathrm{O/H}) = 7.1 - 8.3$, i.e. $1/40 - 1/3$ that of the Sun if the solar abundance of Asplund *et al.* (2005), $12+\log(\mathrm{O/H})_\odot = 8.7$, is adopted. Thus, the massive stellar populations of BCDs have properties intermediate between those of massive stars in solar-metallicity galaxies such as the Milky Way and those of the first stars. BCDs constitute then excellent nearby laboratories for studying physical processes of galaxy and star formation and chemical enrichment processes in environments that are sometimes much more pristine than those in known high-redshift galaxies. The proximity of BCDs allows studies of their structure, metal content, and stellar populations with a sensitivity, precision, and spatial resolution that faint distant high-redshift galaxies do not allow. In the hierarchical model of galaxy formation, large galaxies result from the merging of dwarf galaxies which are the first structures to collapse and form stars. These building-block galaxies are too faint and small to be studied at high redshifts, while we stand a much better chance of understanding them with local BCDs.

Studies of these very chemically unevolved galaxies will also shed light on galaxy formation theories. Cold Dark Matter (CDM) models predict that low-mass dwarf galaxies could still be forming at the present epoch because they originate from density fluctuations considerably smaller and less dense than those giving rise to the giant ones. Thus, if it could be shown that the most chemically unevolved BCDs are also young, i.e. their first stars did not form until $\leqslant$ 1 Gyr ago, then the existence of young dwarf galaxies in the local universe will put strong constraints on the primordial density fluctuation spectrum. A key issue in galaxy formation studies is the observed anti-correlation between the stellar mass of a galaxy and the formation epoch of the stars in it, which is often referred to as "downsizing" (Neistein *et al.* 2006). In the downsizing scenario, massive galaxies are the first to form stars at a high rate, while dwarf galaxies can remain without forming stars and be chemically unevolved for nearly a Hubble time.

2. A Brief History

Extragalactic very metal-deficient systems were first discussed by Sargent & Searle (1970) and Searle & Sargent (1972). Carrying out a spectroscopic survey of a sample of galaxies selected by Zwicky (1971) on the Palomar Sky Survey photographic plates to have a very compact appearance, the pair came across two very interesting objects, I Zw 18 and II Zw 40. Because the spectra of the two galaxies are "strikingly different from those of galaxies of the Hubble sequence" and resemble those of H II regions in the Galaxy and the Large Magellanic Cloud, Sargent & Searle (1970) dubbed them "Isolated extragalactic H II regions". Abundance measurements show both systems to be metal-deficient (about 1/40 solar for I Zw 18 and 1/4 solar for II Zw 40). Because of their low metallicities (more extreme than those of H II regions in the outskirts of spiral galaxies), Searle & Sargent (1972) argued that these galaxies could be either young, in the sense that the bulk of their star formation has occurred in recent times, or that they could be older systems with star formation in them occurring in intense bursts separated by long quiescent periods. Nearly a decade later, Thuan & Martin (1981) showed that I Zw 18 and II Zw 40 are part of a general class of dwarf extragalactic systems undergoing intense bursts of star formation, producing young massive blue stars in a localized compact region, which they called Blue Compact Dwarf (BCD) galaxies. Thuan & Martin (1981) defined a BCD by the following criteria: 1) it has a low luminosity ($M_B \geqslant -18$; 2) its optical spectrum exhibits strong narrow emission lines superposed on a blue continuum, similar to that of an H II region; and 3) it has a compact appearance ($D_{25} = 1$–2 Kpc) and a high mean surface brightness ($S_B \leqslant 22$ mag arcsec^{-2}). Criterion 1) ensures that the object is a dwarf galaxy. It excludes more massive emission-line galaxies with density waves such as Seyfert galaxies. The dwarf criterion also ensures that the BCD has a low heavy-element abundance. Later, Sandage & Binggeli (1984) extended their dwarf classification system to include BCDs, but their connection with the other types of dwarf galaxies (the dwarf ellipticals and the dwarf magellanic irregulars) is still not yet clear. BCDs have also been sometimes called "H II galaxies" although this term refers to a larger class of emission-line galaxies that include also luminous high-metallicity objects with spiral structure. While on the Palomar Sky Survey plates, the BCDs are "almost stellar in appearance, with no obvious underlying galaxy", later CCD surveys showed them to have nearly always an underlying more extended low-surface-brightness (LSB) component on which are superposed the high-surface brightness compact star-forming regions. The LSB component is redder than the star-forming regions, indicative of an older stellar population. Loose & Thuan (1986) have proposed a morphological classification scheme for BCDs based on the morphology of both the star-forming regions and of the underlying component. They distinguish four main types of BCDs: 1) the iE BCDs which show a complex structure with several centers of star formation and irregular (i) isophotes in the central regions superposed on a LSB component with elliptical (E) isophotes. They are by far the most common type of BCDs; 2) the nE galaxies with a nuclear (n) star-forming region at the center of a LSB elliptical (E) component; 3) the iI BCDs with irregular (i) star-forming regions superposed on a LSB component with irregular (I) outer isophotes. An interesting subset of iI galaxies are "cometary" galaxies where the star-forming region constitutes the head of the comet and the LSB component the tail; and 4) the i0 galaxies which have irregular (i) star-forming regions but no (0) evident extended underlying older stellar population. The two extremely metal-deficient BCDs discussed below as the prototypes of the two main modes of star formation in BCDs, I Zw 18 and SBS 0335–052, are both of type i0. Fig. 1a shows a HST/ACS picture of I Zw 18 (Izotov & Thuan 2004). It possesses two star-forming regions in the main body, the brighter northwest (NW) and the fainter southeast (SE) components, separated by 8". There is another blue irregular star-forming region 22" NW of the main body, called the C component. An interferometric H I map of I Zw 18 by van Zee *et al.* (1998) shows that the C component is embedded within a common H I envelope with the main body (Fig. 3a). Fig. 1b shows a HST/ACS UV picture of SBS 0335–052E (Thuan & Izotov 2005). The star formation in SBS 0335–052E occurs mainly in six superstar clusters (SSCs) with ages $\leqslant 25$ Myr, within a region of 2" or 520 pc in size. Dust is clearly present and mixed spatially with the SSCs. The SSCs are roughly aligned in the SE-NW direction, and there is a systematic increase in reddening of the clusters away from the brightest one. The observed color dependence on position may be the combined effects of differential extinction by dust and color evolution with time due to sequential propagating star formation. There is a supershell of radius 380 pc, delineating a large supernova cavity (Thuan *et al.* 1997). In contrast to other BCDs, both IZw 18 and SBS 0335–052E do not show evidence of an extended low-surface-brightness underlying component of red old stars.

3. The most metal-deficient BCDs known and the metallicity floor

Extremely metal-deficient emission-line galaxies at low redshift are very rare. For more than three decades, I Zw 18 held the record as the most metal-deficient emission-line

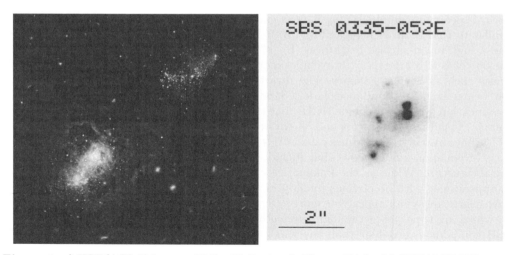

Figure 1. a) HST/ACS V image of I Zw 18 (Izotov & Thuan 2004). (b) HST/ACS UV image of SBS 0335–052 (Thuan & Izotov 2005). North is up and east is to the left. Large supershells of ionized gas are seen delineating supernova cavities in both I Zw 18 and SBS 0335–052. However, in both BCDs, no extended low surface brightness underlying component of red old stars is visible.

galaxy known, with an oxygen abundance 12+log(O/H) = 7.17 ± 0.01 in its NW component and 7.22 ± 0.02 in its SE component (Thuan & Izotov 2005). That it was one of the first BCDs discovered (Sargent & Searle 1970) was just sheer beginner's luck. During the two decades that followed, many surveys have been carried out to search for other I Zw 18-like galaxies, without much success. Only in 1990, did Izotov *et al.* (1990) discover the BCD SBS 0335–052E with an oxygen abundance comparable to that of I Zw 18, 12+log(O/H)= 7.31 ± 0.01 (Thuan & Izotov 2005). Only very recently has I Zw 18 been displaced by the BCD SBS 0335–052W. The latter was discovered when SBS 0335–052E was mapped in the H I line with the VLA (Pustilnik *et al.* 2001). The VLA map showed that a huge H I complex is associated with SBS 0335–052E, having an overall size of about 66 by 22 kpc (Fig. 3b). Two prominent, slightly resolved H I peaks are seen in the integrated H I map, separated in the east-west direction by 22 kpc (84"). While the eastern peak is coincident with SBS 0335–052E, the western peak is associated with another faint BCD, SBS 0335–052W. Izotov *et al.* (2005) have determined its oxygen abundance to be 12+log (O/H) = 7.12 ± 0.03, making it the emission-line galaxy with the lowest metallicity now known in the local universe. More recently, Pustilnik *et al.* (2005) have drawn attention to another extremely metal-deficient dwarf galaxy, DDO 68 with 12+log(O/H) = 7.14 ± 0.03 (Izotov & Thuan 2007).

Can we find more extremely low-metallicity galaxies? We stand a better chance of finding them in very large spectroscopic surveys. One of the best surveys suitable for such a search is the Sloan Digital Sky Survey (SDSS) which include the spectra of a million galaxies. The first three releases yielded no extremely metal-deficient objects. The lowest-metallicity emission-line galaxies found in these early releases all have 12+log(O/H)⩾7.4 (e.g. Kniazev *et al.* 2003). Recently, Izotov *et al.* (2006) and Izotov & Thuan (2007) have started a spectroscopic survey of SDSS galaxies selected to have [O III]λ4959/Hβ ⩽ 1 and [N II]λ6583/Hβ ⩽ 0.05. These spectral properties uniquely select out low-metallicity dwarfs, since no other type of galaxy possesses them. Those authors discovered two

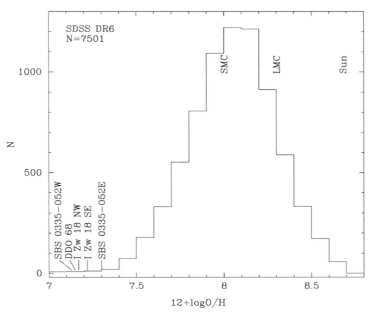

Figure 2. The metallicity distribution of all galaxies in the SDSS Data Release 6 with a H II region-like spectrum and a detected [O III]λ4363 line.

new extremely metal-deficient galaxies, J0113+0052 (UGC 772) with 12+log(O/H) = 7.24 ± 0.05, and J2104–0035 with 12+log(O/H) = 7.26 ± 0.03. All other objects have 12+log(O/H) between 7.3 and 7.8. Thus, at the present time, the 7 most metal-deficient objects known in the local universe, in order of increasing metallicity (see also Guseva in these proceedings), are: SBS 0335-052W, DDO 68, I Zw 18 NW, I Zw 18 SE, J0113+0052, J2104–0035 and SBS 0335-052E. I Zw 18 has lost its title of "most metal-deficient star-forming dwarf" it held for some 35 years, to be relegated to third place. Fig. 2 shows the metallicity distribution of the 7501 emission-line galaxies selected from the SDSS Data Release 6 to have a H II region-like spectrum and a detected [O III]λ4363 line (for accurate abundance determination). It is clear that the low-metallicity end of the distribution is filling in (compare with the list of metal-poor galaxies given 8 years ago by Kunth & Östlin 2000). However, despite much effort, no H II region with 12+log(O/H) ≤ 7.1 has been found. The existing data appear to suggest the existence of an oxygen abundance floor for the ionized gas of BCDs.

4. Is the neutral gas in BCDs primordial?

How about the neutral gas component? BCDs all possess an important neutral gas component to feed the star formation (e.g. Thuan & Martin 1981, Thuan *et al.* 1999b). One may suppose that, if the BCD is young, i.e. undergoing now one of its first bursts of star formation, then the H I gas in BCDs may be primordial, devoid of any heavy element. Is this the case? The advent of *FUSE* has allowed to attack that problem. Using the H II regions in the BCD as sources of UV light shining through the H I envelope (Fig. 3), *FUSE* can obtain absorption spectra which can be used to determine the heavy element abundances in the neutral gas of the BCD. Thus far some 8 BCDs and dwarf irregular galaxies have been observed by *FUSE* (Lebouteiller, these proceedings). The general result is that there appears to be also a metallicity floor for the H I gas of BCDs, at about the same level as for the ionized gas. The oxygen abundance of the neutral gas appears to be constant at $12+\log(O/H)_{neutral} \sim 7.0$, for those BCDs with $12+\log(O/H)_{ionized} \leqslant 8$, increasing with the abundance of the ionized gas in BCDs with higher metallicities (Lebouteiller). This supports the suggestion of Thuan *et al.* (2005) that the matter from which dwarf emission-line galaxies formed was preenriched to a common level of about 2% of the abundance of the Sun, possibly by Population III stars. This idea appears to be also supported by the oxygen abundances 12+log(O/H)

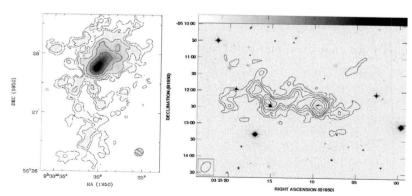

Figure 3. VLA H I maps of: a) I Zw 18 (van Zee *et al.* 1998). The main body and the C component are located at the H I peak; b) the SBS0335–052 complex (Pustilnik *et al.* 2001). SBS 0335-052E and SBS 0335-052W are located at the two H I peaks.

between 6.7 and 7.6 derived by e.g. Telfer *et al.* (2002) for the intergalactic medium, using ultraviolet absorption lines in Lyα absorbers.

5. The hot gas component

The generation of large amounts of hot (a few times 10^6 K) and rarefied gas, resulting from the injection of energy and momentum into the cold ambient interstellar medium by stellar winds from massive stars and supernovae (SNe), has important consequences for the subsequent dynamical and chemical evolution of BCDs. The starburst events that are responsible for the hot X-ray emitting gas last about 10^7 yr each, and are separated by long quiescent periods of more than 1–2 Gyr (Fanelli *et al.* 1988). Tenorio-Tagle (1996) has proposed a "galactic fountain" scenario in which the SNe nucleosynthesis products are driven to high galactic latitudes in a hot phase before cooling down and falling back onto the stellar body as molecular droplets. This naturally explains the uniform abundances that are observed in separate H II regions within the same BCD. To test this galactic fountain scenario, Thuan *et al.* (2004) have obtained Chandra X-ray observations of three of the most metal-deficient BCDs known, SBS 0335-052E, SBS 0335-052W, and I Zw 18. X-ray emission is indeed detected from all three BCDs. The 0.5–10.0 keV luminosities of these objects are in the range 1.3–$8.5{\times}10^{39}$ ergs s^{-1}. But more than 80% of the X-ray emission comes from point sources, which Thuan *et al.* (2004) attribute to ultraluminous high-mass X-ray binaries. There are hints of faint extended diffuse X-ray emission in SBS 0335052E and I Zw 18, probably associated with the superbubbles visible in both BCDs. However, the X-ray images do not show the hot gas breaking out from the stellar body, as predicted by the galactic fountain scenario. These observations also constrain chemical evolution models of dwarf galaxies which rule out closed-box models and require gas infall or loss of metals by galactic winds to account for the observed chemical abundances (Matteucci, these proceedings). There is no evident sign of galactic winds in the X-ray maps of SBS 0335-052E, SBS 0335-052W, and I Zw 18.

6. The "active" versus "passive" mode of star formation in BCDs

We now discuss star formation in BCDs. Based on the detailed investigation of two of the most metal-deficient and most studied BCDs, SBS 0335052E and I Zw 18, star formation in BCDs appears to occur in two very different modes: an "active" mode (we follow the terminology of Hirashita & Hunt 2004) of which SBS 0335052E is the prototype, and a "passive" mode, of which I Zw 18 is the prototype. The active mode is characterized by super-star cluster (SSC) formation, a high star formation rate (SFR), a very compact size, hot dust, and significant amounts of molecules such as H_2. On the other hand, the passive mode is characterized by an absence of SSCs, a low SFR, a larger size, cooler dust and no significant amount of H_2. The observed characteristics of SBS 0335052E and I Zw 18 are summarized in Table 1. Clearly, metallicity is not the distinguishing factor between these two modes of star formation since both BCDs have similarly low heavy element abundance. Hirashita & Hunt (2004) have suggested that the difference between the two modes can be understood through a difference in size and density of the star-forming regions. The active mode occurs in regions that are compact and dense. On the other hand, the passive mode occurs in regions that are diffuse (with radius $\geqslant 100$ pc) and less dense.

Indeed, the SSCs in SBS 0335-052E are compact ($\leqslant$25 pc) and dense ($\sim$100 pc^{-2}) (Thuan *et al.* 1997), while the star clusters in I Zw 18 are instead ordinary young clusters, $\geqslant$300 pc in size and $\leqslant$0.02 pc^{-2}, comparable to those in M 101, and not as extreme as

Table 1. Active versus Passive Star Formation in SBS 0335–052E and I Zw 18

SBS 0335-052E	I Zw 18
Super star clusters (SSCs)	No SSCs
n_e (S II) $= 390$ cm^{-3}	n_e (S II) $\leqslant 100$ cm^{-3}
n_e (radio) $= 2000$ cm^{-3}	
Compact radio source: 17 pc	less compact
Total SFR(radio, IR)$= 0.7$-0.8 M$_\odot$ yr^{-1}	Total SFR (radio) $= 0.1$ M$_\odot$ yr^{-1}
Thermal emission: 9000 O7V stars	Thermal emission: 1200 O7V stars
High-ionization lines: [Ne v] (7 Ryd)	No high-ionization lines
H$_2$	No H$_2$
Very dusty : M$_{dust}= 1.5$-6×10^3 M$_\odot$	Not so dusty
Continuum MIR peak: $\sim 28 \mu$m	Continuum peaks longwards of 70μm

those in 30 Doradus. The electron density of SBS 0335–052E as derived from the [S II] $\lambda\lambda 6717$, 6731 optical emission line ratio is equal to 390 cm^{-3}, while that derived for I Zw 18 is $\leqslant 100$ cm^{-3}. The electron density as derived from the radio emission in SBS 0335-052E, which probes more extincted star-forming regions, is even higher. The radio emission of SBS 0335-052E shows significant free-free absorption at 1.4 GHz (Hunt et al. 2004), while there is no evidence for such a spectral turnover in I Zw 18 (Hunt et al. 2005). The inferred density for SBS 0335–052E from the radio free-free absorption is ~ 2000 cm^{-3}. The radio emission of SBS 0335-052E comes from an extremely compact region (its diameter is 17 pc), while the radio emission in I Zw 18 is diffuse and extended, despite a similar global thermal/non-thermal mix (at 4.8 GHz: 40/60 in I Zw 18 and 30/70 in SBS 0335-052E). The radio luminosity of I Zw 18 is considerably smaller than that of SBS 0335-052E; the non-thermal radio luminosity of I Zw 18 is only $\sim 5\%$, and its thermal component only about 25% of that of SBS 0335-052E. The total SFR in I Zw 18 as derived from the radio emission is ~ 0.1 M$_\odot$ yr^{-1}, nearly one order of magnitude lower than that the one in SBS 0335-052E.

Before the mid-infrared (MIR) *ISO* observations of SBS 0335-052E by Thuan et al. (1999), it was generally thought that low-metallicity BCDs would not contain much dust. Thuan et al. (1999) found that although SBS 0335-052E is one of the most metal-deficient BCDs known (4% solar), it is unexpectedly bright in the MIR range: as much as 75% of the total luminosity of SBS 0335-052E comes as MIR radiation. Later *Spitzer* observations by Houck et al. (2004) have shown that the total dust mass in SBS 0335–052E is $\sim 1.5 \times 10^3$ M$_\odot$. The MIR spectrum peaks around 28 μm, which indicates that there is very little cold dust in SBS 0335-052E. Contrary to the spectra of typical solar metallicity starburst galaxies, the spectrum of SBS 0335-052E does not show polycyclic aromatic hydrocarbon (PAH) emission features. This is due to the very low metallicity of the BCD (the PAH features are weaker at lower metallicities, see e.g. Wu et al. 2006) but also to its very high UV luminosity density which destroys PAHs. On the other hand, the continuum MIR emission of I Zw 18 has a much steeper slope, more like that of a starburst galaxy. It peaks longward of 70 μm, indicating a significant amount of cold dust in I Zw 18. As for SBS 0335-052E, the MIR spectrum of I Zw 18 does not show any PAH feature, because of its very low metallicity (Wu et al. 2007).

Concerning the molecular content, *FUSE* did not detect any diffuse H$_2$ in either SBS0335–052E (Thuan et al. 2005) or I Zw 18 (Vidal-Madjar et al. 2001). The absence of diffuse H$_2$ is due to the combined effects of a low H I density in the neutral gas envelope, a large UV flux that destroys H$_2$ molecules and a low metallicity that makes grains on which to form H$_2$ molecules scarce. However Vanzi et al. (2000) did detect several H$_2$ emission lines in their near-infrared spectrum of SBS0335–052E. These lines observed in the K band are generally consistent, within the errors, with both thermal and fluorescent excitation by the strong UV field. This implies that the detected H_2 must be clumpy and associated only with the dense star-forming regions. These clumps should be denser than 1000 cm^{-3} and have a temperature greater than 1000 K (Thuan *et al.* 2005).

7. High-ionization radiation

There exists also a difference between the active and passsive modes concerning the amount of very hard ionizing radiation in the star-forming regions. SBS 0335–052E contains very hard radiation, with energies in excess of 95 eV, while the radiation in I Zw 18 is not as hard. The hardness of the ionizing radiation in BCDs has long been known to increase with decreasing metallicity. This is supported by the fact that the strong nebular emission line He IIλ4686 is often seen in the spectra of BCDs, with a flux that increases with decreasing metallicity of the ionized gas (Guseva *et al.* 2000). Besides He II emission, high-ionization emission lines of heavy elements ions are also seen in the spectra of some BCDs, such as the high-ionization emission line [Fe v]λ4227 in the BCDs Tol 1214-277 (12+logO/H = 7.55) and SBS 0335-052E. The presence of this line, just as that of the He IIλ4686 line, requires ionizing radiation with photon energies in excess of 4 Rydberg. Later, the [Ne v]$\lambda\lambda$ 3346, 3426 emission lines were detected in the spectra of the BCDs SBS0335-052E, HS0837+4717 (12+logO/H = 7.60), Tol 1214–277, and perhaps of Mrk 209 (12+logO/H = 7.82) (Fig. 4a) (Thuan & Izotov 2005). The existence of these lines requires the presence of hard radiation with photon energies above 7.1 Rydberg, i.e., in the range of soft X-rays. Such very hard ionizing radiation is unexpected: it has been seen before in AGN, but *not* in star-forming galaxies. By contrast, while the spectrum of I Zw 18NW does show the He IIλ4686 line and a hint of the [Fe v]λ4227 line, the [Ne v]$\lambda\lambda$ 3346, 3426 lines are conspicuously absent (Thuan & Izotov 2005). This means

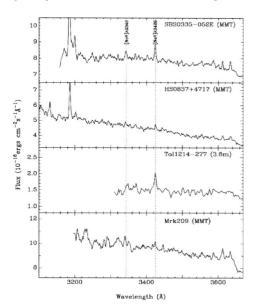
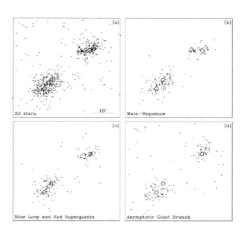

Figure 4. a) Spectra of all known BCDs with detected [Ne v]$\lambda\lambda$ 3346, 3426 lines. The location of these lines are indicated by dotted vertical lines (Thuan & Izotov 2005); b) Spatial distributions of (a) stars of all types, (b) MS stars, (c) BL and RSG stars, and (d) AGB stars in I Zw 18 (dots). Open circles show the locations of the NW and SE components in the main body and of the central cluster in the C component (Izotov & Thuan 2004).

that I Zw 18NW contains radiation as hard as 4 Rydberg, but not as hard as 7.1 Rydberg, as in SBS 0335–052E. There is a general trend of higher [Ne v], [Fe v] and He II emission at lower metallicities. However metallicity cannot be the only factor which controls the hardness of the radiation, since SBS 0335-052 and I Zw 18 have about the same metallicity.

It is interesting to note that *Spitzer* IRS spectra of some higher-metallicity BCDs such as Haro 3 (12+logO/H = 8.34; Hunt *et al.* 2006, these proceedings) and Mrk 996 (12+logO/H = 8.0; Thuan *et al.* 2008) show the presence of the [O IV]$\lambda 25.89\mu$m line, with an ionization potential of 54.9 eV, just beyond the He II edge at 54.4 eV. However, the optical spectra of these objects do not show the presence of the high-ionization [Fe v]$\lambda 4227$ and He II$\lambda 4686$ nebular emission lines which both also have an ionization potential of 54.4 eV. This implies that the hard ionizing radiation in these objects comes entirely from a region that is hidden in the optical but seen in the MIR.

What is the origin of the hard ionizing radiation beyond 4 Rydberg, and especially beyond 7 Rydberg? Thuan & Izotov (2005) found that the hardness of the ionizing radiation with photon energies greater than 4 Rydberg does not generally depend on the burst age as measured by the equivalent width of Hβ. But there is a clear absence of He II emission in very young 3–4 Myr starbursts with EW(Hβ) $\geqslant 300$ Å, implying that the source of ionization with photon energy greater than 4 Rydberg is not massive stars in their main-sequence phase, but in their post-main-sequence phase. Strong He II emission is seen in starbursts with EW(Hβ) $\sim$50–300 Å, suggesting that less massive stars and their descendants also contribute radiation to ionize helium. However present non-rotating photoionization models of main-sequence stars (such as the Costar models of Schaerer & de Koter 1997) and of Wolf-Rayet stars (Schaerer & Vacca 1998) cannot reproduce the observed intensity of the He II emission in very low metallicity BCDs. It would be interesting to see whether the new rotating stellar models by the Geneva group can do a better job. The X-ray luminosity in luminous high-mass X-ray binaries (HMXBs) observed in a few very metal-deficient BCDs, such as SBS 0335–052 (Thuan *et al.* 2004), can account for their [Ne v] emission. However, the scarce X-ray data do not show a one-to-one correlation between the presence of a HMXB and [Ne v] emission (SBS 0335–052 shows [Ne v] emission, but I Zw 18 does not, although both have HMXBs) so that HMXBs are probably ruled out as the main source of hard ionizing radiation with energy above 7.1 Ryd. Thuan & Izotov (2005) found that the most likely source of [Ne v] emission is probably fast radiative shocks moving with velocities $\sim$ 450 km s^{-1} through a dense ISM with electron number densities of several hundreds cm^{-3} (Dopita & Sutherland 1996). These shocks can probably be produced via the evolution of 50–100 M$_\odot$ stars, formed in very compact and dense SSCs in the "active" mode of star formation. These fast radiative shocks are evidenced by broad components in the line profiles of the strong emission lines (Izotov *et al.* 2007b). However the presence of compact SSCs and broad components is not necessarily accompanied by high-ionization emission. Metallicity appears to play an important role. [Ne v] emission is detected only in low- metallicity galaxies with 12+logO/H $\leqslant 7.8$. In higher-metallicity galaxies, the postshock regions are cooled more efficiently. Therefore, their high-ionization regions are smaller and their emission lower.

8. The age of I Zw 18

We now turn to the question posed more than 3 decades ago by Searle & Sargent (1972): are the most metal-deficient BCDs young galaxies, i.e. they formed their first stars at a relatively recent time ($\leqslant 1$ Gyr ago), or are they old galaxies which formed most of their stars several Gyr ago, in which case the present starburst is just the last episode of a series of such events? For the vast majority of BCDs (more than 99%), the first alternative is true: their CCD images show clearly an extended low-surface-brightness component indicative of an older stellar population. Photometric studies combined with spectral population synthesis models of a few BCDs give ages of a few Gyr for the older stellar population, considerably smaller than the age of the universe. For example, the study of the two "cometary" BCDs Mrk 59 and Mrk 71 by Noeske *et al.* (2000) gives an age of $\sim$ 4 Gyr for the older stellar population, consistent with the age found by Thuan & Izotov (2005b) from a *HST/WFPC2* color-magnitude diagram study of Mrk 71 (NGC 2366).

However, the most metal-deficient BCDs, those with 12+logO/H $\leqslant$ 7.3 discussed in Section 3, do not appear to have an extended lower surface brightness component. Of the objects listed in Section 3, only I Zw 18 and SBS0335-052E have been extensively photometrically studied. Photometry combined with spectrophotometric data give ages $\leqslant$ 1 Gyr for the compact underlying population of these two objects (Papaderos *et al.* 1998 for SBS 0335-052E; Papaderos *et al.* 2002 and Hunt *et al.* 2003 for I Zw 18). The most direct way to determine the age of a galaxy is by constructing its color-magnitude diagram (CMD). Among the galaxies discussed in Section 3, only I Zw 18 and DDO 68 are near enough to be resolved into stars by *HST*. Izotov & Thuan (2004) (hereafter IT04) obtained deep V and I *HST/ACS* images to construct a CMD of I Zw 18. They did not detect a well developed red giant branch (RGB), which led them to conclude that the age of I Zw 18 is $\sim$ 500 Myr. They also derive a distance to I Zw 18 in the range 12.6–15 Mpc. Later, Aloisi *et al.* (2007)(hereafter A07) (see also these proceedings) obtained more ACS data for I Zw 18 and combining with the data of IT04 constructed a deeper CMD of I Zw 18. According to these authors, there is a red giant branch (RGB) in I Zw 18, the tip of which (TRGB) gives a distance of 18.2 Mpc, in agreement with the distance given by one classical Cepheid they found in the galaxy. Because of the presence of RGB stars, A07 concluded that the age of the oldest stars in I Zw 18 is $\geqslant$ 1 Gyr, an age at least twice as large as the age of 0.5 Gyr found by IT04. Despite the work of A07, we feel that the verdict is not in yet concerning the age of I Zw 18. Derived ages depend on theoretical isochrones, the location of which in the CMD depends on the adopted distance to the galaxy. A distance as large as 18.2 Mpc for I Zw 18 poses several problems which need to be understood before we can accept it. First, the distance determination rests on a single classical Cepheid, and the period-luminosity of Cepheids at low metallicities is not well-known. Second, at the distance of 18.2 Mpc, the *I* absolute magnitude of the Asymptotic Giant Branch (AGB) stars in I Zw 18 would be brighter by $\sim$ 0.8 mag than those in other low-metallicity BCDs. The magnitude difference between the TRGB and the AGB stars in I Zw 18 would be some 1.8 mag in the *I* band instead of $\leqslant$ 1 mag as in other galaxies. To reconcile the properties of the AGB stars in I Zw 18 with those in other dwarf galaxies, IT04 argued that the distance of I Zw 18 should be close to 15 Mpc. Furthermore, the determination of the TRGB by A07 relies on a very small number of stars. Even if I Zw 18 contains RGB stars, their number is unusually small as compared to other BCDs with 12+log O/H $\geqslant$ 7.3, which possess well-populated RGBs and well-defined TRGBs.

Finally, the spatial distribution of stars with different ages in I Zw 18 is not easily understood if the latter is an old galaxy. Fig. 4b shows that the older AGB stars are not more spread out spatially than the younger main-sequence (MS), blue loop (BL) and red supergiant (RSG) stars. Such a distribution is drastically different from the situation in other BCDs, where the old AGB and RGB stars are distributed over a considerably larger area as compared to younger stars because of diffusion and relaxation processes of stellar ensembles. If anything, the reverse appears to be true in I Zw 18: the MS and BL + RSG stars are distributed over a larger area around the main body as compared to the AGB stars. This suggests that the star formation in the past responsible for the AGB stars was more concentrated in the main body, while recent star formation responsible for the MS and BL + RSG stars is more spread out. I Zw 18 appears to be a young galaxy in the process of forming from the inside out.

9. The Primordial ^{4}He Abundance

Because they are the least chemically evolved star-forming objects known in the universe, BCDs contain very little helium manufactured by stars after the big bang. They are thus excellent objects in which to measure the primordial ^{4}He abundance. In the standard theory of big bang nucleosynthesis (SBBN), given the number of light neutrino species, the abundances of light elements depend only on one cosmological parameter, the baryon density parameter, the present ratio of the baryon mass density to the critical density of the universe. This means that accurate measurements of the primordial abundances of each of the four light elements can provide, in principle, a direct measurement of the baryonic mass density. Because of the strong dependence of its abundance on baryonic mass density, deuterium has become the baryometer of choice ever since accurate measurement of D/H in high-redshift low-metallicity QSO Lyα absorption systems became possible. While a single good baryometer like D is sufficient to derive the baryonic mass density from big bang nucleosynthesis, accurate measurements of the primordial abundances of at least two different relic elements are required to check the consistency of SBBN. Among the remaining relic elements, ^{3}He has undergone significant chemical evolution after the big bang, making it difficult to derive its primordial abundance while the derivation of the primordial ^{7}Li abundance in metal-poor halo stars in the Galaxy is beset by difficulties such as the uncertain stellar temperature scale and the temperature structures of the atmospheres of these very cool stars. Thus ^{4}He remains the light element of choice to test SBBN.

For the last decade and a half, I have been involved with my colleague Yuri Izotov in a sustained effort to obtain a precise determination of the primordial He mass fraction Y_p. However, to detect small deviations from SBBN (such as deviations from the standard rate of Hubble expansion or a possible asymmetry between the numbers of neutrinos and antineutrinos in the early universe), and make cosmological inferences, Y_p has to be determined to a level of accuracy of less than 1%. Attaining that precision requires many conditions to be met. First, the observational data have to be of excellent quality. This has been the concern of our group. Over the years, we have been obtaining high signal-to-noise ratio spectroscopic data of low-metallicity extragalactic H II regions, and our sample now includes a total of 86 H II regions in 77 galaxies (Izotov *et al.* 2007). This constitutes by far the largest sample of high-quality data reduced in a homogeneous way to investigate the problem of the primordial helium abundance. Second, all known systematic effects that may affect the Y_p determination must be taken into account. They include different sets of He I line emissivities and reddening laws, collisional and fluorescent enhancements of He I recombination lines, underlying He I stellar absorption lines, collisional excitation of hydrogen lines, temperature and ionization structure of the H II region, and deviation of He I and H emission-line intensities from case B. Using Monte Carlo methods to solve simultaneously for the above systematic effects, Izotov *et al.* (2007) find that most of the present uncertainty in Y_p comes from the He I emissivities. They find values of Y_p equal to 0.2472 ± 0.0012 and 0.2516 ± 0.0011 for the 2 sets of He I emissivities currently in the literature. The first value agrees well with the value given by SBBN theory, while the second value would imply slight deviations from SBBN.

Acknowledgements

I wish to thank L. Hunt, S. Madden and R. Schneider for organizing a wonderful conference in an enchanting setting.

References

Aloisi, A., Clementini, M., & Tosi, M. *et al.* 2007, *ApJ*, 667, L151
Asplund, M., Grevesse, N., & Sauval, A.J. 2005, in: T.G. Barnes III & F. N. Bash (eds.), *Cosmic Abundances as Records of Stellar Evolution and Nucleosynthesis*, ASP Ser. 336, p. 25
Dopita, M. A. & Sutherland, R. S. 1996, *ApJS*, 102, 161
Fanelli, M. N., O'Connell, R. W., & Thuan, T. X. 1988, *ApJ*, 334, 665
Guseva, N. G., Izotov, Y. I., & Thuan, T. X. 2000, *ApJ*, 531, 776
Hirashita, H. & Hunt, L. K. 2004, *A&A*, 421, 555
Houck, J. R. *et al.* 2004, *ApJS*, 154, 211
Hunt, L. K., Thuan, T. X., & Izotov, Y. I. 2003, *ApJ*, 588, 281
Hunt, L. K., Dyer, K. K., Thuan, T. X., & Ulvestad, J. S. 2004, *ApJ*, 606, 853
Hunt, L. K., Dyer, K. K., & Thuan, T. X. 2005, *A&A*, 436, 837
Hunt, L. K., Thuan, T. X., Sauvage, M., & Izotov, Y. I. 2006, *ApJ*, 653, 222
Izotov, Y. I. & Thuan, T. X. 2004, *ApJ*, 616, 768
Izotov, Y. I. & Thuan, T. X. 2007, *ApJ*, 665, 1115
Izotov, Y. I., Thuan, T. X., & Guseva, N. G. 2005, *ApJ*, 632, 210
Izotov, Y. I., Thuan, T. X., & Stasińska, G. 2007, *ApJ*, 662, 15
Izotov, Y. I., Thuan, T. X., & Guseva, N. G. 2007b, *ApJ*, 671, 1297
Izotov, Y. I., Guseva, N.G., Lipovetsky, V.A., *et al.* 1990, *Nature*, 343, 238
Izotov, Y. I., Papaderos, P., Guseva, N. G., Fricke, K. J., & Thuan, T. X. 2006, *A&A*, 454, 137
Kniazev, A. Y., Grebel, E. K., Hao, L., *et al.* 2003, *ApJ*, 593, L73
Kunth, S. & Östlin, G. 2000, *A&AR*, 10, 1
Loose, H.-H. & Thuan, T. X. 1986, in: D. Kunth, T.X. Thuan & J.T.T. Van (eds.), *Star-forming dwarf galaxies and related objects*, (Gif-sur-Yvette: Editions Frontières), p. 73
Neistein, E., van den Bosch, F. C., & Dekel, A. 2007, *MNRAS*, 372, 933
Noeske, K. G., Guseva, N. G., Fricke, K. J. *et al.* 2000, *A&A*, 361, 33
Papaderos, P., Izotov, Y. I., Fricke, K. J. *et al.* 1998, *A&A*, 338, 43
Papaderos, P., Izotov, Y. I., Thuan, T. X. *et al.* 2002, *A&A*, 393, 461
Pustilnik, S. A., Kniazev, A. Y., & Pramskij, A. G. 2005, *A&A*, 443, 91
Pustilnik, S. A., Brinks, E., Thuan, T. X., Lipovetsky, V. A., & Izotov, Y. I. 2001, *AJ*, 121, 1413
Sargent, W. L. W. & Searle, L. 1970, *ApJ*, 162, L155
Sandage, A. & Binggeli, B. 1984, *AJ*, 89, 919
Schaerer, D. & de Koter, A. 1997, *A&A*, 322, 598
Schaerer, D. & Vacca, W. D. 1998, *ApJ*, 497, 618 '
Searle, L. & Sargent, W.L.W. 1972, *ApJ*, 173, 25
Shapley, A. E., Erb, D. K.; Pettini, M, Steidel, C. C., & Adelberger, K. L. 2004, *ApJ*, 612, 108
Telfer, R. C., Kriss, G. A., Zheng, W., Davidsen, A. F., & Tytler, D. 2002, *ApJ*, 579, 500
Tenorio-Tagle, G. 1996, *AJ*, 111, 1641
Thuan, T. X. & Martin, G. E. 1981, *ApJ*, 247, 823
Thuan, T. X. & Izotov, Y. I. 2005, *ApJS*, 161, 240
Thuan, T. X. & Izotov, Y. I. 2005b, *ApJ*, 627, 739
Thuan, T. X., Izotov, Y. I., & Lipovetsky, V.A. 1997, *ApJ*, 477, 661
Thuan, T. X., Sauvage, M., & Madden, S. 1999, *ApJ*, 516, 783
Thuan, T. X., Lecavelier des Etangs, A., & Izotov, Y. I. 2005, *ApJ*, 621, 269
Thuan, T. X., Hunt, L. K., & Izotov, Y. I. 2008, *ApJ*, 689,
Thuan, T. X., Lipovetsky, V. A., Martin, J.-M., & Pustilnik, S. A. 1999b, *A&AS*, 139, 1

Thuan, T. X., Bauer, F. E, Papaderos, P., & Izotov, Y. I. 2004, *ApJ*, 606, 213

van Zee, L., Westpfahl, D., Haynes, M., & Salzer, J. 1998, *AJ*, 115, 1000

Vanzi, L., Hunt, L. K., Thuan, T. X., & Izotov, Y. I. 2000, *A&A*, 363, 493

Vidal-Madjar, A., *et al.* 2000, *ApJ*, 538, L77

Wu, Y., Charmandaris, V., Hao, L., *et al.* 2006, *ApJ*, 639, 157

Wu, Y., Charmandaris, V., Hunt, L. K., *et al.* 2007, *ApJ*, 662, 952

Zwicky, F. 1971, *Catalogue of Selected Compact Galaxies and of Post Eruptive Galaxies*, Bern, Switzerland

Low-Metallicity Star Formation:
From the First Stars to Dwarf Galaxies
Proceedings IAU Symposium No. 255, 2008
L.K. Hunt, S. Madden & R. Schneider, eds.

The size-density relation of H II regions in blue compact dwarf galaxies

Hiroyuki Hirashita[1] and Leslie K. Hunt[2]

[1]Institute of Astronomy and Astrophysics, Academia Sinica,
P.O. Box 23-141, Taipei 10617, Taiwan, R.O.C.
email: hirashita@asiaa.sinica.edu.tw

[2]INAF - Istituto di Radioastronomia/Sezione Firenze,
Largo E. Fermi, 5, 50125 Firenze, Italy
email: hunt@arcetri.astro.it

Abstract. We investigate the size-density relation of H II regions in blue compact dwarf galaxies (BCDs) by compiling observational data of their size (D_i) and electron density (n_e). We find that the size-density relation follows a relation with constant column density ($n_e \propto D_i^{-1}$) rather than with constant luminosity ($n_e \propto D_i^{-1.5}$). Such behavior resembles that of Galactic H II regions, and may imply an underlying "scale-free" connection. Because this size-density relation cannot be explained by static models, we model and examine the evolution of the size-density relation of H II regions by considering the star formation history and pressure-driven expansion of H II regions. We find that the size-density relation of the entire BCD sample does not result from an evolutionary sequence of H II regions but rather reflects a sequence with different initial gas densities (or "hierarchy" of density). We also find that the dust extinction of ionizing photons is significant for the BCD sample, despite their blue optical colors. This means that as long as the emission from H II regions is used to trace massive star formation, we would miss the star formation activity in dense environments even in low-metallicity galaxies such as BCDs.

Keywords. dust, extinction, galaxies: dwarf, galaxies: evolution, galaxies: ISM, galaxies: star clusters, HII regions

1. Introduction

H II regions are a class of objects important in studying star formation activity; they are ionized by young massive stars, which trace recent star formation. However, our knowledge of extragalactic H II regions is still lacking compared with that of Galactic H II regions because of relative observational difficulties. Moreover, physical properties of extragalactic H II regions have a wider spread than those of Galactic H II regions.

The total mass of massive stars (sources of ionizing photons) can be observationally quantified by estimating the number of ionizing photons. A simple argument of a Strömgren sphere indicates that $\dot{N}_{ion} \propto D_i^3 n_e^2$, where $\dot{N}_{ion}$ is the number of ionizing photons emitted per unit time, D_i is the diameter of the H II region, and n_e is the electron number density in the H II region. Thus, in principle, by examining n_e and D_i of an H II region, we can estimate the ionizing photon luminosity. Observationally, the diameter (D_i) and the electron density (n_e) of Galactic H II regions are known to have a relation roughly fitted by $n_e \propto D_i^{-1}$ (e.g., Garay & Lizano 1999), while the above argument of the Strömgren sphere indicates that $n_e \propto D_i^{-1.5}$ if $\dot{N}_{ion}$ is constant.

It is surprising that the size-density relation of extragalactic H II regions also follows $n_e \propto D^{-1}$ as shown by Hunt *et al.* (2003) for blue compact dwarf galaxies (BCDs), although the relation is *scaled-up* compared with the Galactic H II regions. The relation $n_e \propto D^{-1}$ is interpreted as a *constant ionized-gas column density*. However, the physical mechanism that reproduces a constant column density has never been clarified. Since some BCDs have intense star formation activity represented by super star clusters (e.g., Thuan *et al.* 1997), which are never seen in the Galactic disk, extragalactic H II regions, especially those in BCDs, enable us to examine the properties of H II regions over a wide range of the ionizing photon luminosity, spatial extent, etc. Since BCDs have generally low metallicity, giant H II regions in BCDs may be used as a "test bench" of high-redshift intense star formation at the early stage of chemical evolution.

2. Data

We adopt the sample in Hunt *et al.* (2003), who measure the sizes (D_i) of BCDs by using the archival data of the *Hubble Space Telescope* (*HST*). The electron number density n_e is measured optically, and taken from the literature. This sample is called *HST* sample. In order to examine the compact extremes of our models, we also incorporate some high-resolution radio data of BCDs from the literature, with sizes and densities inferred from the high-frequency thermal radio continuum (see Hirashita & Hunt 2008, hereafter HH08). Finally, in addition to BCDs, we include H II region data of other types of galaxies from the literature.

Fig. 1 (HH08 for more details) shows the trends of the data. The Galactic sample of Kim & Koo (2001), as noted by them, can be fitted by $n_e \propto D_i^{-1}$. The H II regions in BCDs follow a similar size-density relation with the same slope (power-law index), but scaled-up relative to the Galactic one.

The power-law size-density relation of H II regions indicates that the star formation is self-similar, that is, there is no characteristic scale of star formation. This scale-free

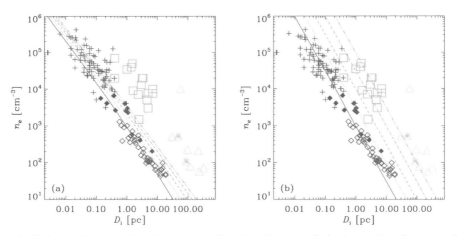

Figure 1. Relation between the electron number density n_e and the ionization diameter D_i for various samples. The open triangles and open squares show the data of the *HST* sample and the radio sample of blue compact dwarf galaxies, respectively. The two data points with asterisk correspond to SBS 0335−052 (the upper) and I Zw 18 (the lower). The Galactic H II regions are shown by the crosses (Garay & Lizano 1999) and the open diamonds (Kim & Koo 2001). The filled diamonds correspond to the sample of the Small Magellanic Cloud. In addition to the observational data, some theoretical predictions in the static models are shown in each panel: (a) The solid, dotted, dashed, dot-dashed, and dot-dot-dot-dashed lines present the results with $\dot{N}_{ion} = 3 \times 10^{49}$, 3×10^{50}, 3×10^{51}, 3×10^{52}, and 3×10^{53} s^{-1}, respectively. The Galactic dust-to-gas ratio ($\kappa = 1$) is assumed. (b) Same as Panel (a) but with a lower dust-to-gas ratio ($\kappa = 0.1$).

nature of H II regions is supported by the data for He 2−10 and SBS 0335−052, two of the HST sample; they both host compact dense H II regions when probed by radio high-resolution observations, but appear larger and more tenuous with HST. This means that the same regions when traced with higher spatial resolution turn out to be denser and smaller, but as shown above they follow the same size-density relation as that of larger complexes.

3. Static models

We interpret the size-density relation of the H II regions compiled in the previous section. Here, we basically relate the size and density of H II regions by using the Strömgren sphere, where an ionizing source in a uniform and static medium is assumed (Spitzer 1978). We also include the effect of dust extinction, which is suggested to be important in determining the size of H II regions (e.g., Inoue *et al.* 2001). Since we treat static H II regions in this section, the models described here are called static models. The detailed formulation can be found in HH08 and here we only review some important features.

The most important parameter for the model is the extinction over the Strömgren radius, τ_{Sd}, estimated as (Hirashita *et al.* 2001)

$$\tau_{Sd} = 0.87 \left(\frac{\mathcal{D}}{6 \times 10^{-3}} \right) \left(\frac{n_H}{10^2 \ \mathrm{cm}^{-3}} \right)^{1/3} \left(\frac{\dot{N}_{ion}}{10^{48} \ \mathrm{s}^{-1}} \right)^{1/3}, \tag{3.1}$$

where $\mathcal{D}$ is the dust-to-gas mass ratio, n_H is the number density of hydrogen nuclei, and $\dot{N}_{ion}$ is the number of ionizing photons emitted per unit time. We assume the dust-to-gas ratio of the solar neighborhood to be $\mathcal{D}_\odot = 6 \times 10^{-3}$ (Spitzer 1978). We adopt various constant values for $\mathcal{D}$ and do not follow the time evolution of $\mathcal{D}$ in order to avoid uncertainty concerning the chemical evolution models. For convenience, we define the dust-to-gas ratio normalized to the solar neighborhood value, κ, as

$$\kappa \equiv \mathcal{D}/\mathcal{D}_\odot . \tag{3.2}$$

In Fig. 1, the results of the static models are plotted over the observational samples for various $\dot{N}_{ion}$ (D_i is the diameter of the ionized region, and n_e is the electron number density). In Fig. 1a, the dust-to-gas ratio is assumed to be Galactic ($\kappa = 1$), while in Fig. 1b, $\kappa = 0.1$ is adopted to take into account the low metal content of the BCD sample. As shown in Fig. 1a, the size-density relation is relatively insensitive to the change of $\dot{N}_{ion}$, since most of the ionizing photons are absorbed by dust grains. Because of this weak dependence of D_i on $\dot{N}_{ion}$, it is extremely difficult to explain the data of the extragalactic sample, unless we assume an extremely large $\dot{N}_{ion}$. However, the size-density relation of the BCDs, is explained quite easily if we assume a lower dust-to-gas ratio typical of the BCD sample ($\kappa = 0.1$). For this value of dust-to-gas ratio, D_i increases almost in proportion to $\dot{N}_{ion}^{1/3}$.

If dust extinction were absent, the size-density relation under a constant $\dot{N}_{ion}$ would follow $n_e \propto D_i^{-3/2}$. However, both Galactic and extragalactic H II regions have a shallower slope than $n_e \propto D_i^{-3/2}$ in the size-density relation. Both samples show $n_e \propto D_i^{-1}$ rather than $n_e \propto D_i^{-3/2}$, and cannot be explained by a constant $\dot{N}_{ion}$. This implies that the size-density relation is almost certainly not a sequence with a constant $\dot{N}_{ion}$. Rather it is probable that the relation should be considered with varying $\dot{N}_{ion}$, and we examine this in the following.

4. Evolution

The evolution of $\dot{N}_{\rm ion}$ is calculated based on Hirashita & Hunt (2006) under a given star formation history (SFH). We extend our models to include the effect of dust extinction according to Arthur *et al.* (2004). Indeed, dust extinction significantly reduces the size of H II regions especially for compact ones. The details of the formulation are found in HH08.

We choose the initial hydrogen number density, $n_{\rm H0}$, and the gas mass, $M_{\rm gas}$, so that the results are consistent with the sizes, densities, and star formation rates of SBS 0335−052 and I Zw 18. The models calculated with these initial conditions are called the SBS 0335−052 model and the I Zw 18 model, and the selected values are $n_{\rm H0} = 7 \times 10^3$ cm^{-3} and $M_{\rm gas} = 8.1 \times 10^6$ $M_\odot$ for the former and $n_{\rm H0} = 100$ cm^{-3} and $M_{\rm gas} = 1.4 \times 10^7$ $M_\odot$ for the latter. We also examine a case with $n_{\rm H0} = 10^5$ cm^{-3} and $M_{\rm gas} = 10^7$ $M_\odot$ to investigate the dense gas of the radio sample.

In Figs. 2a and b, we show the time evolution of the ionized region on the size-density diagram for the constant SFH and the exponentially decaying SFH, respectively. The evolutionary track in the size-density diagram explains the upper locus of *HST* data including SBS 0335−052 with an age of ∼ 3–10 Myr. Such dense and compact BCDs are classified as "active" in Hunt *et al.* (2003). The I Zw 18 model reproduces the lower part of the BCD sample, and such relatively low-density and diffuse BCDs are called "passive". Therefore, those two classes should have different initial conditions. This picture is consistent with Hirashita & Hunt (2006).

We also observe that the results from the highest initial density reproduce the data points of the radio sample. This means that the radio sample can be understood as an extension of the "active" BCDs toward higher density. Since we assume a similar gas mass among compact sources, SBS 0335−052, and I Zw 18, the observational data

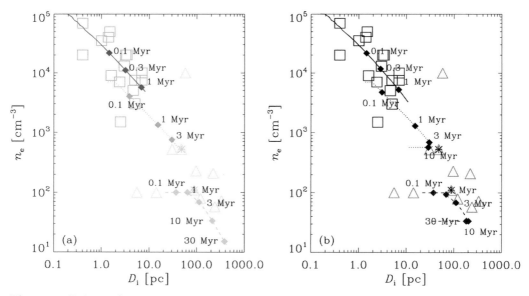

Figure 2. Relation between the electron number density $n_{\rm e}$ and the ionization diameter $D_{\rm i}$ for (a) the constant SFH and (b) the exponentially decaying SFH. The models for compact sources, SBS 0335−052, and I Zw 18 are shown by the solid, dotted, and dashed lines, respectively. The ages are indicated on the corresponding points. The open triangles and open squares are the observational data of the *HST* sample and the radio sample, respectively. The two data points with asterisk correspond to SBS 0335−052 (the upper) and I Zw 18 (the lower).

implies that all the sample is converting 10^7 $M_\odot$ of gas to stars. This may indicate that the final stellar mass formed as a result of the current star formation in the radio sample is comparable to the star formation activity traced in the optical. Thus, we should recognize the importance of embedded compact H II regions such as the radio sample.

5. Summary and discussion

We have investigated the size-density relation of H II regions in BCDs. Motivated by the observed size-density relation, $n_e \propto D_i^{-1}$, we have modeled and examined the size-density relation of ionized regions by considering the star formation history and pressure-driven expansion of H II regions. By comparing the results with the observed size-density relation of BCDs, we have shown that the entire sample cannot be understood as an evolutionary sequence with a single initial condition. But rather, the size-density relation reflects a sequence with different initial gas densities.

We have also found that the extinction of extragalactic H II regions is also significant. Thus, the H II regions can be regarded as "dust-extinction limited", in the sense that the dust optical depth of the ionizing photons is roughly of order unity. This is consistent with the observed size-density relation of H II regions which follows a constant column density of ionized gas. The dust extinction of ionizing photons is particularly severe over the entire lifetime of the compact radio sample with typical densities of $\gtrsim 10^4$ cm^{-3}. This means that the compact radio sample constitutes a different population from the *HST* one and that we would tend to miss the star formation activity in such dense regions if we use the emission from H II regions (hydrogen recombination lines, free-free continuum) as indicators of star formation rate.

Our results may have a great impact in the cosmological context. Our sample has a similar mass to the building blocks in the Universe. The escape of ionizing photons from them is considered to have contributed to reionize the Universe at high redshift. We have shown that even in low-metallicity galaxies, the dust extinction of ionizing photons is significant. If the same situation holds for high-redshift building blocks, dust may significantly affect the escape fraction of ionizing photons. This highlights the importance of studying dust enrichment in the early Universe. ALMA will be a useful tool to detect the light reprocessed by dust grains. Hirashita *et al.* (2008) have already analyzed data of far-infrared dust emission for a sample of BCDs. Such far-infrared data of nearby metal-poor objects are useful to construct strategies for ALMA observations of high-redshift galaxies.

References

Arthur, S. J., Kurtz, S. E., Franco, J., & Albarrán, M. Y. 2004, *ApJ*, 608, 282
Garay, G., & Lizano, S. 1999, *PASP*, 111, 1049
Hirashita, H., & Hunt, L. K. 2006, *A&A*, 460, 67
Hirashita, H., & Hunt, L. K. 2008, *A&A*, to be submitted (HH08)
Hirashita, H., Inoue, A. K., Kamaya, H., & Shibai, H. 2001, *A&A*, 366, 83
Hirashita, H., Kaneda, H., Onaka, T., & Suzuki, T. 2008, *PASJ*, accepted
Hunt, L. K., Hirashita, H., Thuan, T. X., Izotov, Y. I., & Vanzi, L. 2003, in: V. Avila-Reese, C. Firmani, C. Frenk, & C. Allen (eds.), *Galaxy Evolution: Theory and Observations*, RevMexAA SC (*astro-ph/0310865*)
Inoue, A. K., Hirashita, H., & Kamaya, H. 2001, *ApJ*, 555, 613
Kim, K.-T., & Koo, B.-C. 2001, *ApJ*, 549, 979
Spitzer, L., Jr. 1978, *Physical Processes in the Interstellar Medium*, (New York: Wiley)
Thuan, T. X., Izotov, Y. I., & Lipovetsky, V. A. 1997, *ApJ*, 463, 120

Low-Metallicity Star Formation:
From the First Stars to Dwarf Galaxies
Proceedings IAU Symposium No. 255, 2008
L.K. Hunt, S. Madden & R. Schneider, eds.

Probing Globular Cluster Formation in Low Metallicity Dwarf Galaxies

Kelsey E. Johnson[1,2], Leslie K. Hunt[3], and Amy E. Reines[1]

[1]Astronomy Department, University of Virginia, P.O. Box 400325,
Charlottesville, VA, 22904, USA
email: kej7a@virginia.edu

[2]The National Radio Astronomy Observatory, 520 Edgemont Road,
Charlottesville, VA 22903, USA

[3]INAF-Istituto di Radioastronomia-Sez. Firenze, L.go, Fermi 5, I-50125 Firenze, Italy

Abstract. The ubiquitous presence of globular clusters around massive galaxies today suggests that these extreme star clusters must have been formed prolifically in the earlier universe in low-metallicity galaxies. Numerous adolescent and massive star clusters are already known to be present in a variety of galaxies in the local universe; however most of these systems have metallicities of $12 + \log(O/H) > 8$, and are thus not representative of the galaxies in which today's ancient globular clusters were formed. In order to better understand the formation and evolution of these massive clusters in environments with few heavy elements, we have targeted several low-metallicity dwarf galaxies with radio observations, searching for newly-formed massive star clusters still embedded in their birth material. The galaxies in this initial study are HS 0822+3542, UGC 4483, Pox 186, and SBS 0335-052, all of which have metallicities of $12 + \log(O/H) < 7.75$. While no thermal radio sources, indicative of natal massive star clusters, are found in three of the four galaxies, SBS 0335-052 hosts two such objects, which are incredibly luminous. The radio spectral energy distributions of these intense star-forming regions in SBS 0335-052 suggest the presence of $\sim 12,000$ equivalent O-type stars, and the implied star formation rate is nearing the maximum starburst intensity limit.

Keywords. galaxies: star clusters, galaxies: starburst, HII regions, stars: formation

1. Introduction

Ancient globular clusters are ubiquitous around massive galaxies in the relatively nearby universe (e.g. Harris 1991; Brodie & Strader 2006), providing an important fossil record of conditions in the earlier universe. In fact, given the likely rate of mortality for these massive clusters is $\gtrsim 90\%$ (Fall & Zhang 2001; Whitmore *et al.* 2007), globular clusters must have been formed prolifically during this epoch. However, until the launch of the *Hubble Space Telescope*, little was known about the formation and early evolution of these extreme clusters.

With the availability of high spatial resolution imaging, large numbers of massive and dense young blue clusters were found in a variety of galaxies. These adolescent massive clusters were dubbed "super star clusters" (SSCs), and studying them became a bit of a cottage industry. Over the following years, strong lines of evidence suggested that many SSCs will, in fact, evolve into globular clusters (e.g. Whitmore 2003). However, a critical issue linking the present-day SSCs and the ancient globular clusters remains: we have not observationally constrained how the low metal abundance in the early universe affected the formation of massive star clusters. As discussed elsewhere in these proceedings, metallicity can affect star formation and evolution in a variety of ways, including the hardness of the stellar spectra, cooling and pressure in the natal material, and dust formation (Schaerer 2002; Smith, Norris, & Crowther 2002; Tumlinson, Venkatesan, & Shull 2004; Bate 2005).

One way to approach this problem and derive observational constraints on the effect that low metallicity has on the formation of massive star clusters is to study SSC formation that is taking place in low-metallicity environments in the relatively nearby universe.

2. Embedded Massive Star Clusters Known in Dwarf Galaxies

By using sensitive radio observations, a number of dwarf galaxies have been found to host natal SSCs, including He 2-10, NGC 5253, IC 4662, II ZW 40, NGC 4490, NGC 4449, NGC 3125, NGC 2573, NGC 4214, and Haro 3 (Turner *et al.* 2000; Beck *et al.* 2000; Johnson & Kobulnicky 2003; Johnson *et al.* 2004; Johnson *et al.* 2003; Reines, Johnson, & Goss 2008; Johnson *et al.* in prep.; Aversa *et al.* in prep.). A few of these galaxies have truly impressive populations of natal clusters; in particular, He 2-10 has at least four massive embedded star clusters that (in sum) account for nearly all of thermal-infrared emission from that galaxy (Vacca, Johnson, & Conti 2002). However, the problem is that despite being dwarfs, these galaxies do not have particularly low metallicities; none of the objects listed above has a metallicity lower than $12 + \log(\text{O/H}) \sim 8.0$.

3. The Low-Metallicity Sample

In order to investigate massive star cluster formation in galaxies with even lower metallicities, we have used the Very Large Array to obtain radio observations of four dwarf galaxies with metallicities of $12 + \log(\text{O/H}) < 7.75$. The galaxies in this sample are HS 0822+3542, UGC 4483, Pox 186, and SBS 0335-052. The observations were carried out at several wavelengths ranging from 1.3 cm to 6 cm in order to detect the characteristic thermal free-free emission that originates in the dense regions of ionized gas surrounding newly formed clusters. Down to relatively stringent detection thresholds, no embedded clusters were detected in HS 0822+3542, UGC 4483, or Pox 186. Because of the sensitivity of these observations, the upper limits on possible natal clusters that could be present but not detected is significant. In these cases, the maximum masses of natal clusters that the three galaxies with non-detections could host ranges from $\sim 8 \times 10^3 M_\odot$ to $\sim 4 \times 10^4 M_\odot$, all of which are smaller than the stellar mass of the R136 cluster in the 30 Dor region of $\sim 6 \times 10^4 M_\odot$ (Hunter *et al.* 1995, extrapolated down to 0.1 $M_\odot$). In other words, if natal clusters with masses similar to R136 existed in these three galaxies, they would have been detected. Fortunately, the story does not end there.

4. The Incredible Natal Clusters in SBS 0335-052

In contrast to the other three galaxies in the sample, SBS 0335-0552 hosts extremely luminous thermal radio sources in the southern region of the galaxy (Figure 1). The ionizing flux inferred for the entire radio region suggests the equivalent of $\sim 12,000$ O-type stars are present in the combined radio sources, and the associated instantaneous star formation rate is a remarkable $\sim 23 M_\odot \text{yr}^{-1} \text{Mpc}^{-1}$. A comparison between the radio emission from the compact sources and the entire region suggests that up to $\sim 50\%$ of the ionizing flux could be leaking from the compact HII regions. This high percentage of escaping flux is consistent with the results of Reines *et al.* (2008), which suggest that the interstellar medium surrounding the natal clusters in SBS 0335-052 is porous and clumpy. Model HII region fits to the radio data for these sources indicate that the average density

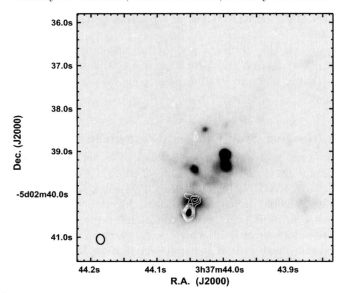

Figure 1. VLA 3.6 cm contours overlaid on an HST ACS F550M image of SBS 0335-053. The radio sources clearly correspond to previously known super star clusters in this galaxy. The synthesized beam is shown in the lower left corner.

of the star-forming regions is $n_e \gtrsim 10^3 - 10^4 \mathrm{cm}^{-3}$. However, given the evidence that the interstellar medium is inhomogeneous, peak densities are likely to reach significantly higher values.

5. Concluding Remarks

From this small sample of four galaxies with a detection rate of 25%, one might be tempted to conclude that the conditions required for the formation of massive star clusters in low-metallicity dwarf galaxies are rare. Such a conclusion would present a problem for understanding globular cluster formation in the very early universe. First, the low rate of detections of thermal radio sources, at least in part, also reflects the very short time that clusters spend in the embedded phase; similar non-detection rates have been recovered from larger surveys not restricted to dwarf galaxies (Aversa *et al.* in prep); in other words, trying to catch these massive clusters while they are still embedded in their birth material virtually guarantees a low detection rate. Perhaps more importantly, we must also keep in mind that today's dwarf galaxies may not be the same beasts as the dwarf galaxies of ~10 Gyr ago. In particular, the density of galaxies was higher and interactions more frequent in the past. In this light, SBS 0335-052 is truly a gem in the relatively nearby universe for enabling this type of study, offering both very low metallicity and also extremely vigorous star formation. Unfortunately, at a distance of ~ 55 Mpc, sensitivity and spatial resolution are already issues in obtaining and interpreting observations of SBS 0335-052, and this galaxy is the nearest currently known example of a such a low-metallicity galaxy undergoing a starburst.

While this single galaxy SBS 0335-052 is a fascinating case study, it is important to keep an eye on the big picture. In this case, the main focus of this line of research is trying to constrain the physical conditions in the early universe that prodigiously gave rise to the ancient globular clusters we see today. There are a number of ways in which low metallicity might affect the formation and early evolution of massive star clusters, however the impact of metallicity in these intense star-forming environments remains an open issue. Both more extensive and more detailed studies are called for, and we are still a long way from that goal. However, one thing does seem quite clear; if the conditions are right, low-metallicity dwarf galaxies are certainly capable of producing globular clusters. The coming decade should bring about rapid progress in this area; the future is bright with a number of sensitive, high spatial resolution, and long wavelength observatories scheduled to become available over the next several years.

References

Bate, M. R. 2005, MSRAS, 363, 363

Beck, S. C., Turner, J. L., & Kovo, O. 2000, AJ, 120, 244

Brodie, J. P. & Strader, J. 2006, ARA&A, 44, 193

Fall, S.M. & Zhang, Q. 2001, ApJ, 561, 751

Harris, W. E. 1991, ARA&A, 29, 543

Johnson, K.E. & Kobulnicky, H.A. 2003, ApJ, 597, 923

Johnson, K. E., Indebetouw, R., & Pisano, D. J. 2003, AJ, 126, 101

Johnson, K.E., Indebetouw, R.I., Watson, C., & Kobulnicky, H.A. 2004, AJ, 128, 610

Reines, A.E., Johnson, K.E, & Hunt, L.K. 2008, AJ, 136, 141

Schaerer, D. 2002, AAP, 382, 28

Smith, L.J., Norris, R.P.F., Crowther, P.A. 2002, MNRAS, 337, 1309

Tumlinson, J., Venkatesan, A., Shull, & J.M. 2004, ApJ, 612, 602

Turner, J. L., Beck, S. C., & Ho, P. T. P. 2000, ApJL, 532, L109

Vacca, W.D., Johnson, K.E., & Conti, P.S. 2002 AJ, 123, 772

Whitmore, B. C. 2003, in A Decade of Hubble Space Telescope Science, 153

Whitmore, B. C., Chandar, R., & Fall, S.M. 2007, 133, 1067

Low-Metallicity Star Formation:
From the First Stars to Dwarf Galaxies
Proceedings IAU Symposium No. 255, 2008
L.K. Hunt, S. Madden & R. Schneider, eds.

Broad line emission in dwarf galaxies: the first detection of low-metallicity AGN

Yuri I. Izotov

Main Astronomical Observatory,
27 Zabolotnoho str., Kyiv, 03680, Ukraine
email: `izotov@mao.kiev.ua`

Abstract. Observations of AGN show that they generally possess a high metallicity, varying from solar to supersolar metallicities. This is the case since AGN are usually found in massive, bulge-dominated galaxies that have converted most of their gas into stars by the present epoch. Since AGN metallicity is strongly correlated with stellar mass, low-metallicity AGN are expected to be in low-mass dwarf galaxies. However, until now, searches in low-mass galaxies have only turned up AGN with metallicities around half that of typical AGN, i.e. with solar or slightly subsolar values. We report the discovery of four low-metallicity dwarf galaxies in the Data Release 6 of the Sloan Digital Sky survey, with $12 + \log O/H$ in the range 7.4–8.0, and that appear to harbor an AGN. In the course of a long-range program to search for extremely metal-deficient emission-line dwarf galaxies, we have come across four galaxies with very unusual spectra: the strong permitted emission lines, mainly the Hα line, show very prominent broad components, with full widths at zero intensity corresponding to velocities varying between 2200 and 3500 km s^{-1}, and extraordinarily large broad Hα luminosities, varying from 3×10^{41} to 2×10^{42} erg s^{-1}. The Balmer lines show a very steep decrement, suggesting collisional excitation and that the broad emission comes from very dense gas ($N_e \gg 10^4$ cm^{-3}). Only the presence of an accretion disk around an intermediate-mass black hole in the dwarf galaxies appears to account for these properties.

Keywords. galaxies: active, galaxies: abundances, galaxies: dwarf, galaxies: nuclei

1. Introduction

Active galactic nuclei (AGN) are usually found in massive, bulge-dominated galaxies that have converted most of their gas into stars by the present epoch, therefore their gas metallicities are generally high (Storchi-Bergmann *et al.* 1998, Hamann *et al.* 2002). A question then arises: do low-metallicity AGN exist? If so, can we find them in low-mass galaxies? To address these questions, Groves *et al.* (2006) have searched the Sloan Digital Sky Survey (SDSS) Data Release 4 (DR4) spectroscopic galaxy sample. Using diagnostic line ratios and imposing an upper mass limit of 10^{10} $M_\odot$ to restrict themselves to low-mass galaxies, they are left with a sample of only ~ 40 AGN, which they found to appear to have metallicities around half that of typical AGN, i.e. solar or slightly subsolar values. The same high metallicity range is found in the sample of low-mass AGN of Greene & Ho (2007). Assessing their findings, Groves *et al.* (2006) are led to another question: "Why are there no AGN with even lower metallicities?" Here, following Izotov & Thuan (2008), we suggest that these low-metallicity AGN do exist although they are extremely rare.

In the course of a long-range program to search for extremely metal-deficient emission-line dwarf galaxies, Izotov *et al.* (2007) have used the SDSS DR5 database. While studying that sample to look for emission-line galaxies (ELGs) with broad components in their strong emission lines, Izotov *et al.* (2007) came across four galaxies with very unusual spectra. Their spectra shown in Fig. 1, resemble those of moderately to very low-metallicity high-excitation H II regions: their oxygen abundances are in the range $12+\log$ O/H ~ 7.4–8.0. Izotov *et al.* (2007) found that there is however a striking difference: the strong permitted emission lines, mainly the Hα $\lambda 6863$ line, show very prominent broad components. These are characterized by somewhat unusual properties: 1) their Hα full widths at zero intensity $FWZI$ vary from 102 to 158 Å, corresponding to expansion velocities between 2200 and 3500 km s^{-1}; 2) the broad Hα luminosities L_{br} are extraordinarily large, varying from 3×10^{41} to 2×10^{42} erg s^{-1}. This is to be compared with the range 10^{37}–10^{40} erg s^{-1} found by Izotov *et al.* (2007) for the other ELGs with broad-line emission; 3) the Balmer lines show a very steep decrement, suggesting collisional excitation and that the broad emission comes from very dense gas ($N_e \gg 10^4$ cm^{-3}). The very large Hα luminosities are most likely associated with SN shocks or AGN. Izotov *et al.* (2007) have considered type IIn SNe because their Hα luminosities are larger ($\sim 10^{38}$–10^{41} erg s^{-1}) than those of the other SN types and they decrease less rapidly.

2. Broad emission and diagnostic diagrams

To decide whether type IIn SNe or AGN are responsible for the broad emission in these galaxies, monitoring of their spectral features on the relatively long time scale of several years is necessary. If broad features are produced by IIn type SNe, then we would expect a decrease in the broad line luminosities. No significant temporal evolution would be expected in the case of an AGN. In order to check for temporal evolution, Izotov & Thuan (2008) obtained second-epoch spectra of the above four galaxies with broad emission, using the 3.5 m Apache Point Observatory (APO) telescope.

Comparison of SDSS and APO broad Hα fluxes shows that they have remained nearly constant (with variations $\leqslant 20\%$) over a period of ~ 3–7 years. This likely rules out the hypothesis that the broad line fluxes are due to type IIn SN because their Hα fluxes should have decreased significantly over this time interval.

There remains the AGN scenario. Can accretion disks around black holes in these low-metallicity dwarf galaxies account for their spectral properties? The spectra of the four objects do not show clear evidence for the presence of an intense source of hard nonthermal radiation: the [Ne V] $\lambda 3426$, [O II] $\lambda 3727$, He II $\lambda 4686$, [O I] $\lambda 6300$, [N II] $\lambda 6583$, and [S II] $\lambda\lambda 6717, 6731$ emission lines, which are usually found in the spectra of AGN, are weak or not detected. Aside from He II $\lambda 4686$, the apparent weakness of such emission lines, however, may be accounted for by the low metallicities of our galaxies. Another way to check for the presence of an AGN in a galaxy is to check for its location in the emission-line diagnostic diagram of Baldwin *et al.* (1981) (BPT). It can be seen in Fig. 2a that all four objects lie in the region corresponding to star-forming galaxies (SFG), to the left of the region occupied by AGN with low-mass black holes and with metallicities ranging from 2 to 1/4 that of the Sun (Greene & Ho 2007). However, their locations in the SFG region do not necessarily disqualify them as AGN candidates. Photoionization models of AGN show that lowering their metallicity moves them to the left of the BPT diagram, so that they end up in the SFG region (Groves *et al.* 2006, Stasińska *et al.* 2006). Thus the BPT diagram is unable to distinguish between SFGs and low-metallicity AGN. Admitting that there is an AGN in our dwarf galaxies, can we account for the weakness of the high-ionization lines? Photoionization models with only AGN nonthermal ionizing radiation do predict detectable He II $\lambda 4686$ and [Ne V] $\lambda 3426$ emission lines. To make the observed spectra agree with the models, one solution is to dilute the nonthermal ionizing radiation from the AGN by thermal radiation from surrounding hot massive stars. In Fig. 2a, we show the results of Izotov & Thuan (2008) CLOUDY calculations (Ferland *et al.* 1998) of H II regions ionized by a composite radiation consisting of different proportions of stellar and nonthermal radiation. Two curves, characterized by different metallicities, are shown by solid lines: the lower one is for $12 + \log O/H = 7.3$ and the upper one is for $12 + \log O/H = 7.8$, typical of the metallicities of our objects. Each model point is labeled by the ratio R of nonthermal-to-thermal ionizing radiation. A slope $\alpha = -1$ has been adopted for the non-thermal power-law spectrum over the whole wavelength range under consideration ($f_\nu \propto \nu^\alpha$). The calculations have been done with a number of ionizing photons $Q_{th} = 10^{53}$ s^{-1} for stellar radiation, $Q_{nonth} = RQ_{th}$ for nonthermal radiation and $N_e = 10^4$ cm^{-3}. Higher densities would move the curves to the right. The dotted lines in Fig. 2a show the corresponding models with $\alpha = -2$. They are very similar to the models with $\alpha = -1$ when $R \leqslant 1$, but fall below for $R \geqslant 1$. It is seen that models with $12 + \log O/H = 7.8$ and in which the nonthermal ionizing radiation contributes $\leqslant 10\%$ of the total ionizing radiation can account well for the location of all four galaxies in the BPT diagram, independently of the slope of the power-law spectrum.

How about the high-ionization lines? In Fig. 2b, we show the diagnostic diagram for [Ne v] $\lambda 3426$/Hβ vs. [N ii] $\lambda 6583$/Hα (thick lines) and He ii $\lambda 4686$/Hβ vs. [N ii] $\lambda 6583$/Hα (thin lines). As in Fig. 2a, CLOUDY models with $\alpha = -1$ and -2 are shown by solid and dotted lines. The vertical dashed line separates models with $12 + \log O/H = 7.3$ (Fig. 2b, left) from those with $12 + \log O/H = 7.8$ (Fig. 2b, right). The shaded rectangle shows the region of the upper limits of $\sim 1\%$–2% of the Hβ flux, set for [Ne v] $\lambda 3426$/Hβ and He ii $\lambda 4686$/Hβ in our objects, in the observed range of their [N ii] $\lambda 6583$/Hα ratio. If we adopt $12 + \log O/H = 7.8$ as typical for our galaxies, then Fig. 2b shows that models that satisfy the non-detectability limit of the high-ionization lines (i.e. that fall within the shaded box) are characterized either by a steep slope and a not excessively small R ($\alpha = -2$ and $R \sim 0.1$) or by a shallower slope and a very low R ($\alpha = -1$ and $R \sim 0.03$). It is also possible that the absence of strong high-ionization lines is caused by a high covering factor of the accretion disk. In this case the hard radiation would be absorbed inside the dense accretion disk and no high-ionization forbidden lines would be formed.

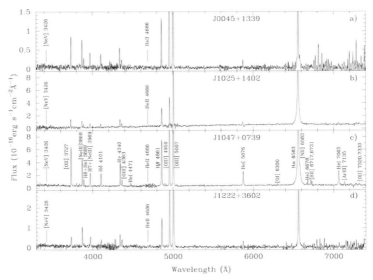

Figure 1. Redshift-corrected 3.5 m Apache Observatory second-epoch spectra of four low-metallicy emission-line dwarf galaxies thought to contain AGN. The locations of the non-detected [Ne v] $\lambda 3426$ and He ii $\lambda 4686$ high-ionization emission lines are shown in all panels. Other emission lines are labeled in panel c).

3. Black hole virial masses

It has been shown (see e.g. Kaspi *et al.* 2000) that continuum and broad line luminosities in AGN can be used to determine the size and geometry of the broad emission-line region and the mass of the central black hole. Examining a large sample of broad-line AGN, Greene & Ho (2005) have found that the Hα luminosity scales almost linearly with the optical continuum luminosity and that a strong correlation exists between the Hα and Hβ line widths. On the basis of these two empirical correlations, those authors have derived the relations for the central black hole mass.

In Table 1 we list the extinction-corrected broad Hα luminosities $L(\mathrm{H}\alpha)$ and continuum luminosities $\lambda L_\lambda(5100)$ for the four galaxies, as derived from the SDSS spectra. The extinction coefficient was set equal to the one derived for the narrow Balmer hydrogen lines. Since the reddening due to dust extinction in dense regions may be larger than that derived from the narrow hydrogen emission lines, the derived $L(\mathrm{H}\alpha)$ should be considered as lower limits. The $L(\mathrm{H}\alpha)$ and $\lambda L_\lambda(5100)$ of our galaxies follow closely the correlation between Hα and continuum luminosities found by Greene & Ho (2005). This implies that our galaxies are very likely the same type of objects as those considered by Greene & Ho (2005). Therefore, we can use equations from Greene & Ho (2005) for the determination of the central black hole masses. The masses $M_{\mathrm{BH}}(\mathrm{H}\alpha)$ and $M_{\mathrm{BH}}(5100)$ derived from the broad Hα and continuum luminosities are shown in Table 1. These masses are in the range $\sim 5 \times 10^5~M_\odot - 3 \times 10^6~M_\odot$, lower or similar to the mean black hole mass of $1.3 \times 10^6~M_\odot$ found by Greene & Ho (2007) for their sample of low-mass black holes. Since the luminosities used to derive the masses of the central black holes are lower limits, the derived masses should also be considered as lower limits.

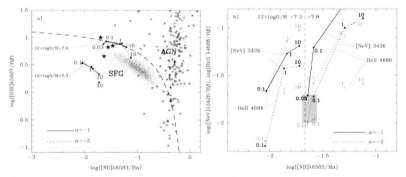

Figure 2. a) The BPT diagram (Baldwin *et al.* 1981) for low-ionization emission lines. Plotted are the ∼10,000 ELGs from Izotov *et al.* (2007) (cloud of points), the low-mass black hole sample of Greene & Ho (2007) (asterisks) and the four low-metallicity AGN in Table 1 (stars). The dashed line separates star-forming galaxies (SFG) from active galactic nuclei (AGN) (Kauffmann *et al.* 2003). The solid and dotted lines connect CLOUDY photoinization models computed for H II regions ionized by a composite radiation consisting of different proportions of stellar and nonthermal radiation. The two upper curves are characterized by 12+log O/H = 7.8 and the two lower ones by 12+log O/H = 7.3. Each model point is labeled by the ratio of nonthermal–to-thermal ionizing radiation. All curves have been calculated adopting a number of ionizing photons $Q = 10^{53}$ s^{-1} for stellar radiation, different slopes of the nonthermal spectral energy distributions $f_\nu \propto \nu^\alpha$ (solid lines are for $\alpha = -1$ and dotted lines are for $\alpha = -2$). A density $N_e = 10^4$ cm^{-3} is adopted. Higher densities would shift the curves to the right. b) The diagnostic diagram for high-ionization emission lines: [Ne V] $\lambda3426$/Hβ vs. [N II] $\lambda6583$/Hα (thick lines) and He II $\lambda4686$/Hβ vs. [N II] $\lambda6583$/Hα (thin lines). The same CLOUDY models as in a) are shown. The shaded region shows the upper intensity limits of the high-ionization lines [Ne V] $\lambda3426$ and He II $\lambda4686$ relative to Hβ in the four galaxies considered here. The dashed vertical line separates models with 12+logO/H = 7.30 (left) from those with 12+logO/H = 7.80 (right).

Table 1. Hα and continuum luminosities and masses of the black holes[1]

Object	L_{br} (**H**α)[2]	**FWHM**$_{br}$ (**H**α)[3]	λL_λ (**5100**)[2]	M_{BH} (**H**α)[6]	M_{BH} (**5100**)[4]
J0045+1339	2.74×10^{41}	1540	7.59×10^{42}	2.43×10^{6}	2.00×10^{6}
J1025+1402	3.21×10^{41}	680	3.62×10^{42}	5.07×10^{5}	2.50×10^{5}
J1047+0739	1.57×10^{42}	1050	2.32×10^{43}	3.05×10^{6}	1.91×10^{6}
J1222+3602	2.80×10^{41}	790	7.17×10^{42}	6.34×10^{5}	5.10×10^{5}

Notes:
[1] Parameters are derived from the 2.5 m SDSS spectra (Izotov *et al.* 2007).
[2] In units erg s^{-1}.
[3] In units km s^{-1}.
[4] In solar masses.

4. Conclusion

We study here the broad line emission in four low-metallicity star-forming dwarf galaxies with 12+logO/H $\sim$ 7.4–8.0. Our main conclusions are following:

1. The steep Balmer decrements of the broad hydrogen lines and the very high luminosities of the broad Hα line in all four galaxies (3×10^{41} to 2×10^{42} erg s^{-1}) suggest that the broad emission arises from very dense and high luminosity regions such as those associated with accretion disks around black holes. If so, these four objects would harbor a new class of AGN residing in low-metallicity dwarf galaxies, with an oxygen abundance that is considerably lower than the solar or super-solar metallicity of a typical AGN.

2. There is no obvious spectroscopic evidence for the presence of a source of a non-thermal hard ionizing radiation in all four galaxies: high-ionization emission lines such as He II λ4686 and [Ne V] λ3426 emission lines were not detected at the level $\leqslant 1-2$ percent of the Hβ flux. The predicted fluxes of the high-ionization lines are below the detectability level if the spectral energy distribution $f_\nu \propto \nu^\alpha$ of the ionizing nonthermal radiation has $\alpha \sim -1$ and the nonthermal ionizing radiation is significantly diluted by the thermal stellar ionizing radiation contributing $\leqslant 3-10$ percent of the total ionizing radiation.

3. The lower limits of the masses of the central black holes M_{BH} of $\sim 5 \times 10^5$ $M_\odot$ – 3×10^6 $M_\odot$ in our galaxies are among the lowest found thus far for AGN.

References

Baldwin, J. A., Phillips, M. M., & Terlevich, R. 1981, *PASP*, 93, 5
Ferland, G. J., Korista, K. T., Verner, D. A., Ferguson, J. W., Kingdon, J. B., & Verner, E. M. 1998, *PASP*, 110, 761
Greene, J. E. & Ho, L. C. 2005, *ApJ*, 630, 122
Greene, J. E. & Ho, L. C. 2007, *ApJ*, 670, 92
Groves, B. A., Heckman, T. M., & Kauffmann, G. 2006, *MNRAS*, 371, 1559
Hamann, F. *et al.* 2002, *ApJ*, 564, 592
Izotov, Y. I. & Thuan, T. X. 2008, *ApJ*, in press; preprint arXiv:0807.2029
Izotov, Y. I., Thuan, T. X., & Guseva, N. G. 2007, *ApJ*, 671, 1297
Kauffmann, G., *et al.* 2003, *MNRAS*, 346, 1055
Kaspi, S., Smith, P. S., Netzer, H., Maoz, D., Jannuzi, B. T., & Giveon, U. 2000, *ApJ*, 533, 631
Stasińska, G., Cid Fernandes, R., Mateus, A., Sodré, L., Jr., & Asari, N. V. 2006, *MNRAS*, 371, 972
Storchi-Bergmann, T., Smitt, H. R., Calzetti, D., & Kinney, A. L. 1998, *AJ*, 115, 909

Low-Metallicity Star Formation:
From the First Stars to Dwarf Galaxies
Proceedings IAU Symposium No. 255, 2008
L.K. Hunt, S. Madden & R. Schneider, eds.

Ionized gas in dwarf galaxies:
Abundance indicators

Grażyna Stasińska

[1]LUTH, Observatoire de Paris, CNRS, Université Paris Diderot; Place Jules Janssen 92190
Meudon, France
email: grazyna.stasinsa@obspm.fr

Abstract. We discuss the four basic methods to derive HII region abundances in metal-poor galaxies by presenting a few recent results obtained with these methods. We end up by commenting on the yet unsolved problem of temperature fluctuations in HII regions, which may plague abundance determinations, as well as the discrepancy between abundances derived from recombination lines and collisionally excited lines, to which inhomogeneous chemical composition might be the explanation.

Keywords. (ISM:) HII regions, galaxies: abundances, galaxies: dwarf

1. Introduction

The ionized gas provides the best way to determine elemental abundances in dwarf galaxies. We present and discuss the four basic methods to derive abundances from HII regions, using examples drawn from the recent literature.

2. Te-based methods

When the data allow the determination of the electron temperature (Te) *directly* from the spectra (generally using the [O III] $\lambda4363$/[O III] $\lambda5007$ ratio), ionic abundances are readily obtained from observed emission line intensities, since these are proportional to the abundances and to the Te-dependent line emissivities. Generally, the lines that are used for abundance determinations are the strongest ones in the spectrum, which are collisionally excited forbidden lines for the heavy elements, and recombination lines for H and He. The abundances of the elements are then obtained by applying ionization correction factors, derived from simple considerations or from photoionization model grids. Izotov *et al.* (2006) have provided a series of analytical formulae to derive abundances with such methods, based on photoionization models for giant HII regions using the stellar energy distributions computed using Starburst 99 (Leitherer *et al.* 1999) with the updated model atmospheres by Smith *et al.* (2002).

In Figure 1, we show an example of a result obtained with Te-based abundance determinations in HII regions and planetary nebulae belonging to two Magellanic irregular galaxies: NGC 3109 and the Small Magellanic Clouds (Peña *et al.* 2007). The figure shows that the oxygen abundances are remarkably similar in all of the HII regions of each galaxy. The scatter is 0.07 dex and 0.09 dex, respectively. This implies that mixing has been very strong in the interstellar medium of these galaxies. Such a result is obtained due to the quality of the observations and the accuracy of the method. One can also see that the range in oxygen abundances for the planetary nebulae is significantly larger. This is due to a combination of nucleosynthesis processes in the planetary nebula progenitors and to chemical evolution of the galaxies.

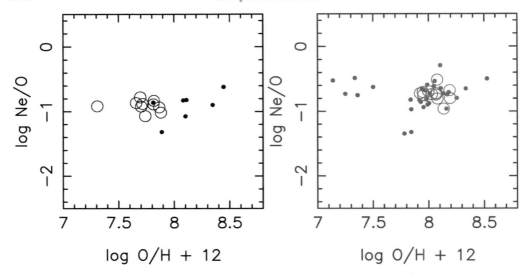

Figure 1. HII regions (large circles) and planetary nebulae (small points) in two Magellanic irregular galaxies: NGC 3109 (left) and the Small Magellanic Cloud (right).

3. Strong line methods

When the [O III] $\lambda4363$ line is not observed, the metallicity can still be estimated using *statistical* methods, based on strong lines only. These methods take advantage of the fact that the metallicity appears to be strongly linked to the mean effective temperature of the stellar radiation field and to the ionization parameter, U, of the nebula. As a result, in first approximation, one can consider that the strong line spectrum of a giant HII region is essentially determined by the metallicity.

Strong line methods have to be calibrated. In the low metallicity regime, which is the one of interest for dwarf galaxies, calibration is relatively easy, since one can use giant HII regions in which the weak [O III] $\lambda4363$ line has been measured (in the high metallicity regime, the calibration is more difficult, and less reliable).

Strong line methods are expected to be less accurate than Te-methods. Moreover, they can be biased, if applied to a category of objects that do not share the same structural properties as the sample which was used to calibrate the method.

The various panels of Fig. 2 show the values of different line ratios used as metallicity indicators as a function of $12 + \log$ O/H. These are: [Ne III] $\lambda3869$/[O III] $\lambda5007$, proposed by Nagao *et al.* (2006), ([O II] $\lambda3727$ + [O III] $\lambda5007$)/Hβ, introduced by Pagel *et al.* (1979), [O III] $\lambda5007$/[N II] $\lambda6584$, first proposed by Alloin *et al.* (1979), [S III] $\lambda9069$/[O III] $\lambda5007$, proposed by Stasińska (2006), [O III] $\lambda5007$/[O II] $\lambda3727$, considered by Nagao *et al.* (2006), and [N II] $\lambda6584$/Hα, first used by Storchi Bergmann *et al.* (1994). The curves drawn with symbols correspond to sequences of photoionization models ionized by blackbody radiation (for simplicity) with varying metallicities (i.e. the abundances of all the heavy elements vary in step with that of oxygen, except nitrogen whose abundance increases more rapidly with O/H to mimic secondary nitrogen production). The curves drawn with circles correspond to models of different ionization parameters, with larger symbols representing larger ionization parameters. The curves drawn with triangles correspond to an effective temperature, $T_{\rm eff}$ of 60,000 K, the curves drawn with circles correspond to $T_{\rm eff} = 50,000$ K and the curves drawn with squares to 40,000 K. Clearly, some of the indicators vary very little – if at all – with O/H. Rather, they depend on $T_{\rm eff}$ and on U. The thick curves correspond to the calibrations of these abundance indicators as given by Nagao *et al.* (2006) (except for [S III] $\lambda 9069$/[O III] $\lambda 5007$ which is given by Stasińska 2006). Figure 2 shows that the metallicity dependence of these metallicity indicators is largely due to the variation of effective temperature and ionization parameter with metallicity. If then such metallicity indicators are used to compare the abundances of two samples that are expected to be characterized by systematically different values of effective temperature and/or ionization parameter, the interpretation in terms of abundances may be wrong.

This is likely the case in a recent study by Sanchez *et al.* (2008) aiming at the characterization of two groups of galaxies: blue compact dwarf (BCD) galaxies selected from the Sloan Digital Sky Survey (SDSS), and quiescent blue compact dwarf (QBCD) galaxies, also selected from the SDSS on the basis of photometric criteria mainly. Using the [N II] $\lambda 6584$/Hα index, these authors find that QBCDs have larger HII region-based oxygen abundances than BCDs, a property difficult to explain in terms of galactic chemical evolution. However, by definition, QBCDs are expected to have cooler exciting stars than BCDs, since the most massive stars have already disappeared. It is also likely that the ionization parameter in QBCD HII regions is smaller than those in BCDs. Indeed, the total number of ionizing photons from an ageing star burst decreases in time and, in addition, the HII region is expected to be more extended at larger ages due to expansion. Both factors increase the value of [N II] $\lambda 6584$/Hα at a given metallicity: thus the O/H abundance derived using the same calibration line as for BCDs will result in overestimated abundances, as seen in Fig. 2.

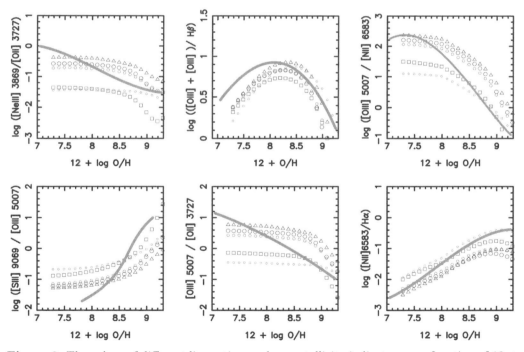

Figure 2. The values of different line ratios used as metallicity indicators as a function of $12 + \log$ O/H for sequences of photoionization models. Triangles: $T_{\mathrm{eff}} = 60{,}000$ K; circles: $T_{\mathrm{eff}} = 50{,}000$ K; squares $T_{\mathrm{eff}} = 40{,}000$ K. The symbol sizes indicate the ionization parameters, with larger symbols representing larger ionization parameters. The thick curves correspond to the calibrations of these abundance indicators as given by Nagao *et al.* (2006).

4. Comparison to grids of photoionization models

It is often considered that, in the cases where the electron temperature cannot be obtained directly from the spectra, abundances derived by comparison with grids of photoionization models are more reliable than abundances obtained by strong line methods. The idea behind such an opinion is that photoionization models take into account the physics of interaction of the photons with nebular matter and predict trustworthy emission line intensities.

There are, however, a number of conditions to be fulfilled before abundances derived in such a way can be considered reliable.

First of all, as is obvious, the grid must be built by varying all the important independent parameters that determine the spectrum of an HII region. Second, the grid must be adequately meshed. It is preferable that the grid contains the entire range of observed line ratios, so that extrapolations are not needed. The possibility of multiple solutions must be clearly identified and conveniently dealt with.

Another important aspect is the criteria chosen for fitting a given object. The common procedure is to use a χ^2 method on (a subset of) line ratios. This is potentially dangerous and may lead to important biases in the derived abundances. A good example of such biases is the determination of oxygen abundances in emission line galaxies from the Sloan Digital Sky Survey by Tremonti et al. (2004). These authors compared the intensities of the strongest emission lines in these objects (Hα, Hβ, [O II] λ3727, [O III] λ5007, [N II] λ6584, [S II] λ6716, λ6731) with a grid of 2×10^5 photoionization models computed by Charlot & Longhetti (2001) and corresponding to different assumptions about the effective gas parameters (metallicity, ionization parameter, dust-to-gas ratio, star formation histories etc...). Yin et al. (2007) showed the presence of an important offset between the values of O/H derived by Tremonti et al. (2004) and those derived directly with the classical Te-based method. In addition, they found that the offset strongly correlated with N/O. This led them to point out that, since in Tremonti et al. (2004), all the lines are used simultaneously, any difference in an abundance ratio with the ratio used in the Charlot & Longhetti (2001) models has the potential of causing offsets. They attribute the difference between the O/H values determined by Tremonti et al. (2004) and those obtained with the Te-method to the crude way in which secondary nitrogen enrichment is treated in the Charlot & Longhetti (2001) grid.

5. Tailored model-fitting

Tailored-model fitting is a better way to obtain abundances, but it is obviously a more complicated process. As explained in Stasińska (2007), however, here also a number of conditions must be fulfilled. One is that the model must reproduce *all* of the emission lines that bear information on the physical conditions in the ionized gas, and not only the strongest lines. The other is that the observed lines must allow a full abundance diagnostic (which is not always the case): very different solutions are possible if there is no diagnostic of the electron temperature.

It is sometimes difficult to fit detailed observations with a photoionization model. Then the abundances are not obtained with the required accuracy. One such example is the case of the extremely metal-poor blue compact galaxy, I Zw 18. Stasińska & Schaerer (1999) found it impossible to reproduce the observed [O III] λ4363/5007 ratio with a model that had a geometry reflecting the gross features in the Hα image, and that was powered by a radiation obtained from stellar population synthesis models fitting the observed stellar data. The best models yielded a [O III] λ4363/5007 ratio too low by about 30%, leading to an uncertainty in the O/H ratio by about 20% in this case. In view of uncertainties generally quoted for abundance determinations, this may seem unimportant. However, I Zw 18 is an exemplary test case, where one should be able to fit the data perfectly. Péquignot (2008) reconsidered the problem, and introduced the presence of diffuse matter between ionized filaments, in a way similar to the model adopted by Jamet *et al.* (2005). It turns out that such a model can explain the observed spectrum in most of its details. A propitious circumstance is that, in the meantime, collisional strengths for the excitation of Lyman lines of hydrogen have been recomputed and turn out to be lower than the ones used by Stasińska & Schaerer, making cooling less efficient. It is however not clear whether the model by Péquignot (2008) can reproduce the high electron temperatures observed even at large distances from the exciting cluster by Vílchez & Iglesias-Páramo (1998). The comparison of the works by Stasińska & Schaerer (1999) and Péquignot (2008) leads to an interesting remark: prior to modelling, the observational data must be analyzed critically; this is a difficult task, and the views adopted by several authors may differ in details that turn out to be important to constrain the models.

6. Old, unsolved problems in ionized nebulae affecting abundance determinations

It has long been known that nebular temperatures derived from various indicators are different. While nebulae are not expected to be exactly isothermal, the observed differences are larger than expected from photoionization models. Peimbert (1967) postulated the existence of temperature fluctuations to account for these differences, and developed a formalism to estimate the magnitude of these fluctuations from observations using his famous parameter t^2. Peimbert & Costero (1969) showed that ignoring t^2 in abundance derivations led to an underestimate of the abundances and proposed a formalism to account for this t^2. Since then, the nebular astronomical community is divided as to the reality of temperature fluctuations in HII regions. While numerous studies point towards a value of t^2 typically of 0.03-0.04, little direct evidence is seen. Perhaps the most convincing direct observational argument is provided by the high spatial resolution imaging of the Orion nebula by O'Dell *et al.* (2003).

Recently, the derivation of the electron temperatures from recombination lines of O^{++} allowed an independent determination of t^2, by comparison with the temperature derived from [O III] λ4363/5007 (see García-Rojas & Esteban 2007 and references therein). However, it has been argued that the different temperatures derived for O^{++} could in fact be due to an inhomogeneous chemical composition, with oxygen-rich clumps embedded in a medium of "normal" chemical composition (Tsamis *et al.* 2003, Tsamis & Péquignot 2005). Such clumps could be produced by the scenario of Tenorio-Tagle (1996) for the enrichment of the interstellar medium by supernova ejecta. One may ask what is the meaning of the derived abundances in this case. Stasińska *et al.* (2007) have examined this question and found that, at least in the cases they considered, optical recombination lines strongly overestimate the average oxygen abundance, while collisionally excited lines overestimate them only slightly. Peimbert *et al.* (2007) do not share this point of view, and argue that the correct oxygen abundances in HII regions are those derived from recombination lines which, in the case of metal-rich objects, lead to values about twice as large than when derived from collisionally excited lines.

References

Alloin, D., Collin-Souffrin, S., Joly, M., & Vigroux, L., 1979, A&A 78, 200

Charlot, S. & Longhetti, M., 2001, MNRAS, 323, 887

García-Rojas, J. & Esteban, C., 2007, ApJ, 670, 457

Izotov, Y. I., Stasińska, G., & Meynet, G. *et al.*, 2006, A&A, 448, 955

Jamet, L., Stasińska, G., Pérez, E., González Delgado, R. M., & Vílchez, J. M., 2005, A&A, 444, 723

Leitherer, C., *et al.*, 1999, ApJS, 123, 3

Nagao, T., Maiolino, R., & Marconi, A., 2006, A&A, 459, 85

O'Dell, C. R., Peimbert, M., & Peimbert, A., 2003, AJ, 125, 2590

Pagel, B. E. J., Edmunds, M. G., & Blackwell, D. E. *et al.*, 1979, MNRAS, 189, 95

Peimbert, M. & Costero, R., 1969, BOTT, 5, 3

Peimbert, M., 1967, ApJ, 150, 825

Peimbert, M., Peimbert, A., Esteban, C., García-Rojas, J., Bresolin, F., Carigi, L., Ruiz, M. T., & López-Sánchez, A. R., 2007, RMxAC, 29, 72

Peña, M., Stasińska, G., & Richer, M. G., 2007, A&A, 476, 745

Péquignot, D., 2008, A&A, 478, 371

Smith, L. J., Norris, R. P. F., & Crowther, P. A., 2002, MNRAS, 337, 1309

Stasińska G., Tenorio-Tagle, G., Rodríguez, M., & Henney, W. J., 2007, A&A, 471, 193

Stasińska, G., 2007, arXiv, 704, arXiv:0704.0348

Stasińska, G., 2006, A&A, 454, L127

Stasińska, G., 2004, cmpe.conf, 115

Stasińska, G. & Schaerer, D., 1999, A&A, 351, 72

Stasińska, G., Tenorio-Tagle, G., Rodríguez, M., & Henney, W. J., 2007, A&A, 471, 193

Storchi-Bergmann, T., Calzetti, D., & Kinney, A. L., 1994, ApJ, 429, 572

Tenorio-Tagle, G., 1996, AJ, 111, 1641

Tremonti, C. A., *et al.*, 2004, ApJ, 613, 898

Tsamis, Y. G. & Péquignot, D., 2005, MNRAS, 364, 687

Tsamis, Y. G., Barlow, M. J., Liu, X.-W., Danziger, I. J., & Storey P. J., 2003, MNRAS, 338, 687

Vílchez, J. M. & Iglesias-Páramo, J., 1998, ApJ, 508, 248

Yin, S. Y., Liang, Y. C., Hammer, F., Brinchmann, J., Zhang, B., Deng, L. C., & Flores, H., 2007, A&A, 462, 535

Low-Metallicity Star Formation:
From the First Stars to Dwarf Galaxies
Proceedings IAU Symposium No. 255, 2008
L.K. Hunt, S. Madden & R. Schneider, eds.

© 2008 International Astronomical Union
doi:10.1017/S1743921308025118

SMC in space and time: a project to study the evolution of the prototype interacting late-type dwarf galaxy

M. Tosi[1], J. Gallagher[2], E. Sabbi[3], K. Glatt[4,5], E. K. Grebel[4], C. Christian[3], M. Cignoni[6,1], G. Clementini[1], A. Cole[7], G. Da Costa[8], D. Harbeck[2], M. Marconi[9], M. Meixner[3], A. Nota[3], M. Sirianni[3] and T. Smecker-Hane[10]

[1] INAF - Osservatorio Astronomico di Bologna
Via Ranzani 1, I-40127, Bologna, Italy
email: monica.tosi@oabo.inaf.it

[2] University of Wisconsin, Madison, WI, USA

[3] STScI, Baltimore, MD, USA

[4] Heidelberg University, Heidelberg, D

[5] Basel University, Basel, CH

[6] Bologna University, Bologna, I

[7] University of Tasmania, AU

[8] Research School of Astronomy & Astrophysics, ANU, AU

[9] INAF - OA Capodimonte, Napoli, I

[10] University of California, Irvine, CA, USA

Abstract. We introduce the *SMC in space and time*, a large coordinated space and ground-based program to study star formation processes and history, as well as variable stars, structure, kinematics and chemical evolution of the whole SMC. Here, we present the Colour-Magnitude Diagrams (CMDs) resulting from HST/ACS photometry, aimed at deriving the star formation history (SFH) in six fields of the SMC. The fields are located in the central regions, in the stellar halo, and in the wing toward the LMC. The CMDs are very deep, well beyond the oldest Main Sequence Turn-Off, and will allow us to derive the SFH over the entire Hubble time.

Keywords. (galaxies:) Magellanic Clouds, galaxies: dwarf, galaxies: evolution, galaxies: stellar content

1. Introduction

The Small Magellanic Cloud (SMC) is the closest late-type dwarf and has many properties similar to those of the vast majority of this common class of galaxies. Its current metallicity ($Z \simeq 0.004$ in mass fraction, as derived from HII regions and young stars) is typical of dwarf irregular and Blue Compact Dwarf (BCD) galaxies, the least evolved systems, hence the most similar to primeval galaxies. Its mass (between 1 and $5 \times 10^9 M_\odot$, e.g. Kallivayalil *et al.* 2006 and references therein) is at the upper limit of the range of masses typical of late-type dwarfs. These characteristics, combined with its proximity, make the SMC the natural benchmark to study the evolution of late-type dwarf galaxies. Moreover, its membership to a triple system allows detailed studies of interaction-driven modulations of the star formation activity.

A wealth of data on the SMC are available in the literature, although not as much as for its bigger companion, the LMC. Yet, much more are needed for a better understanding of how the SMC has formed and evolved. We have thus embarked on a long-term project to study the evolution of the SMC in space and time. Our project plans to exploit the high performances in depth, resolving power or large field of view of current and forthcoming, space and ground based, telescopes, such as HST, VLT, Spitzer, SALT and VST.

2. The SMC in Space and Time

Primary goals of our long-term project are the derivation of the star formation history (SFH) in the whole SMC from deep and accurate photometry and of stellar chemical abundances from high and intermediate resolution spectra. These data will allow us to infer the age-metallicity relation (or lack thereof) of stars resolved in different regions of the galaxy and to better constrain numerical models for the chemical evolution of the various SMC regions as well as for the galaxy as a whole. Since the SFH and the age-metallicity relation are key parameters in chemical evolution modeling, these new models will be of unprecedented accuracy for an external galaxy, reaching the level of reliability currently attained only for the solar neighbourhood.

Part of the project is devoted to the study of the SMC variable stars of all types, to classify them and use the unique aspects of variability to get their physical properties. We will study the spatial distribution of the various types of variables and this will provide unique information on the space and time confinement of the formation of their parent stellar populations. Standard candles such as the RR Lyraes will also provide information on the 3D structure of the galaxy and on its reddening distribution.

The SFH will be derived from Colour-Magnitude diagrams (CMD) using the synthetic CMD technique. This kind of study has already been performed by other authors (e.g. Dolphin 2001, Harris & Zaritsky 2004, Chiosi *et al.* 2006, Noel *et al.* 2007). Our plan, however, is to have CMDs several magnitudes fainter than the oldest main-sequence (MS) turn-off (TO) for the entire galaxy, including its halo and the wing in the direction of the LMC, allowing us to infer for the first time the SFH of the whole SMC over the entire Hubble time.

Time has already been awarded to this project on HST (PIs A. Nota and J. Gallagher) and on VLT (PI E.K. Grebel) and is guaranteed on VST (PI V. Ripepi). HST/ACS photometry was acquired in Cycle 13 for 4 young star clusters, 7 older clusters and 6 fields. Fig. 1 shows the location of our HST/ACS targets (except for the two outermost old clusters). Results from these data have already been published on the young clusters NGC 346 (Nota *et al.* 2006 and Sabbi *et al.* 2007) and NGC 602 (Carlson *et al.* 2007) and on the seven old clusters (Glatt *et al.* 2008a for NGC 121, and Glatt *et al.* 2008b for NGC 339, NGC 416, NGC 419, Lindsay 1, Lindsay 38 and Kron 3). More details on the young clusters are presented by Sabbi *et al.* in this volume, while the derivation of the SFH in the region of NGC 602 is described by Cignoni *et al.* (2008).

3. The HST/ACS fields

The six SMC fields have been observed in Cycle 13 (GO 10396, PI J. Gallagher) with the Wide Field Channel of the HST/ACS in the F555W (V) and F814 (I) bands for a total of 12 orbits. The target fields (indicated by white circles in Fig. 1) have been chosen to maximize possible stellar population differences between different SMC regions. Three fields are in the central region: one (SF4) close to the barycenter of the young population, one (SF1) close to the barycenter of the old population, and one (SF5) in an intermediate zone. Two fields are located in the wing, the SMC extension towards the LMC: one (SF9) in the wing outer part, and one (SF10) in its inner part. The last field (SF8) is in the opposite side, in what can be considered the SMC halo.

The observations were performed with the ACS/WFC, following a standard dithering pattern to improve PSF sampling, allow for hot pixel and cosmic ray removal and fill the gap between the two WFC detectors. The photometric analysis has been performed independently with two packages suited for PSF fitting in crowded fields: Stetson's Daophot and Anderson's imgxy-WFC.01x10. Extensive artificial star tests have been performed on the images to assess photometric errors, incompleteness and blending factors.

The CMDs resulting from the application of Anderson's photometric package are shown in Fig.2, where a 7 Gyr isochrone with metallicity Z=0.001 (from Angeretti *et al.* 2007) is also plotted for reference. As in all resolved galaxies, there is a clear stellar density gradient from the SMC center to the periphery, the field containing the largest number of stars (25300) being SF5. All regions turn out to contain old stellar populations whose evolutionary phases are visible in the CMDs: main-sequence (MS), subgiant branch (SGB), red giant branch (RGB), clump and asymptotic giant branch (AGB). The age of these old populations appears to be of several Gyr, mostly around 7 Gyr.

All fields, but the halo one, also show the blue plume typical of late-type dwarf galaxies, populated by high and intermediate mass stars in the main-sequence phase or at the blue edge of the blue loops (corresponding to the central He-burning phase). It is interesting to notice the contrast between the outer wing, where in spite of a relatively low number

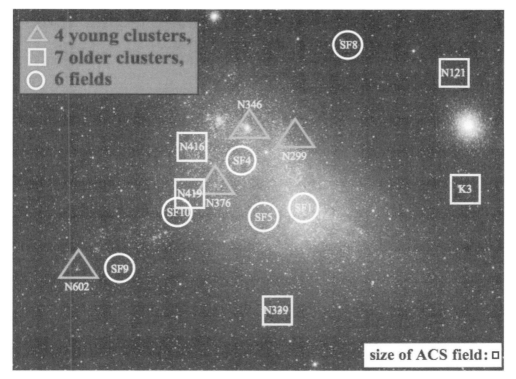

Figure 1. Location of our HST/ACS targets, overimposed on an SMC image (courtesy Stéphane Guisard). The young clusters are indicated by (orange) triangles, the older clusters by (yellow) squares, and the fields by the white circles. The name of the target is shown, with N standing for NGC and K for Kron. Lindsay 1 and Lindsay 38 fall out of the shown sky area. The actual size of the ACS field is shown in the bottom-right corner.

of measured stars the blue plume is well populated with young stars, and the much older halo in the SF8 field.

An interesting feature of the six CMDs is the apparent homogeneity of the old populations: in all panels of Fig.2 a) the old SGBs have roughly the same magnitude and the old MS TOs have roughly the same colour, and b) the clumps have roughly the same magnitude and colour. This circumstance suggests that no large differences in age and metallicity exist among the SMC old stars, irrespectively of their spatial location. By inspecting the CMDs in more detail (for instance with the aid of the 7 Gyr isochrone), we do see that the stars in the halo field (SF8) are bluer than the others, presumably because they are metal poorer and/or less reddened, whilst those at the barycenter of

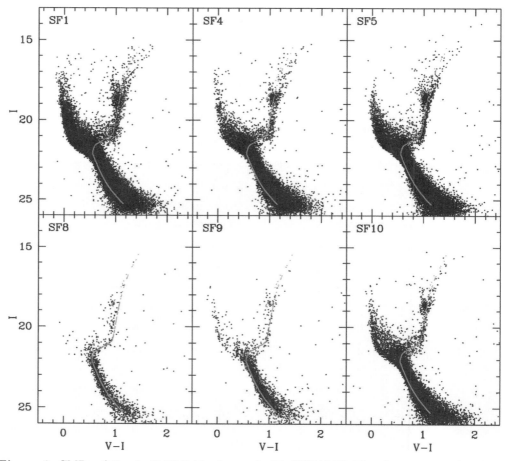

Figure 2. CMDs of the six SMC fields observed with HST/ACS. The three top panels refer to the central regions: SF1, with 28000 measured stars, SF4, with 16000 stars, and SF5, with 25300 stars. The bottom-left panel refer to the halo field, SF8, with 1550 stars. The central bottom panel shows the CMD of the outer wing field SF9, with 2440 stars, while the CMD of the inner wing field, SF10, with 13700 stars, is in the bottom-right panel. The 7 Gyr isochrone with Z = 0.001, interpolated by Angeretti *et al.* (2007) from the Padova stellar evolution models (Fagotto *et al.* 1994a and Fagotto *et al.* 1994b) is also shown in all panels for reference. The assumed intrinsic distance modulus and reddening are $(m-M)_0 = 18.9$ and $E(B-V) = 0.08$.

the SMC young population (SF4) are the reddest of all, probably because they are metal richer or more reddened. Also the widths of the various evolutionary phases appear somewhat different from one field to the other, which could be due either to actual age or metallicity spreads or to differences in the distribution of stars in distance within the SMC. At any rate, the CMDs are too similar to each other in the phases relative to old stars to allow for macroscopic differences in the early evolution of the six fields. The six regions seem to share a relatively late onset of the bulk of star formation activity, not much earlier than 7 Gyr ago. This result is in agreement with the findings by Dolphin *et al.* (2001) and Noel *et al.* (2007) for other SMC regions, and at variance with Harris & Zaritsky (2004) conclusions, based, however, on shallower photometry not reaching the old MS TO.

To infer the details of the SFH in the six regions, fully exploiting all the photometric information, we will apply the synthetic CMD method taking into account photometric errors, incompleteness and crowding effects as estimated with the artificial star tests. For a better assessment of the uncertainties involved in the SFH derivation, we will apply three different and independent approaches of the synthetic CMD method: Cignoni's (see e.g. Cignoni *et al.* 2006), Cole's (see e.g. Cole *et al.* 2007) and Tosi's (see e.g. Tosi *et al.* 1991 and Angeretti *et al.* 2005). The CMDs presented here already allow to forecast interesting results. When the VST will be operating and we will cover the whole SMC, wing included, with B, V, I photometry reaching several magnitudes below the oldest MS TO, we will be able to assess precisely how the star formation activity has evolved both in space and in time.

Acknowledgements

This work has been partially supported through NASA-HST funding to JSG, PRIN-INAF-2005, and Swiss NSF 200020-113697.

References

Angeretti, L., Fiorentino, G., & Greggio, L. 2007, in A. Vazdekis & R. F. Peletier eds *IAU Symp.*, 241, (CUP), p. 41

Angeretti, L., Tosi, M., Greggio, L., Sabbi, E., Aloisi, A., & Leitherer, C. 2005, *AJ* 129, 2203

Carlson, L. R., Sabbi, E., Sirianni, M., Hora, J. L., Nota, A., Meixner, M., Gallagher, J. S., Oey, M. S., Pasquali, A., Smith, L. J., Tosi, M., & Walterbos, R. 2007, *ApJ*, 665, L109

Chiosi, E., Vallenari, A., Held, E. V., Rizzi, L., & Moretti, A. 2006, *A&A*, 452, 179

Cignoni, M., Degl'Innocenti, S., Prada Moroni, P. G., & Shore, S. N. 2008, *A&A*, 459, 783

Cignoni, M., Sabbi, E., Nota, A., Tosi, M., Degl'Innocenti, S., Prada Moroni, P., Angeretti, L., Carlson, L., Gallagher, J., Meixner, M., Sirianni, M., & Smith, L.J. 2008, *AJ*, submitted

Dolphin, A. E., Walker, A. R., Hodge, P. W., Mateo, M., Olszewski, W. W., Schommer, R. A., & Suntzeff, N. B. 2001, *ApJ*, 562, 303

Fagotto, F., Bressan, A., Bertelli, G., & Chiosi, C. 1994a, *A&AS*, 104, 365

Fagotto, F., Bressan, A., Bertelli, G., & Chiosi, C. 1994b, *A&AS*, 105, 29

Glatt, K., Gallagher, J. S., Grebel, E. K., Nota, A., Sabbi, E., Sirianni, M., Clementini, G., Tosi, M., Harbeck, D., Koch, A., & Cracraft, M. 2008, *AJ*, 135, 1106

Glatt, K., Grebel, E. K., Sabbi, E.,Gallagher, J. S., Nota, A., Sirianni, M., Clementini, G., Tosi, M., Harbeck, D., Koch, A., Kayser, A., & Da Costa, G. 2008, *AJ*, in press, arXiv:0807.3744

Harris, J. & Zaritsky, D. 2004, *ApJ*, 127, 1531

Kallivayalil, N., van der Marel, R. P., & Alcock, C. 2006, *ApJ*, 652, 1213

Noel, N. E. D., Gallart, C., Costa, E., & Mendez, R. A., 2007, *AJ*, 133, 2037

Nota, A., Sirianni, M., Sabbi, E., Tosi, M., Meixner, M., Gallagher, J, Clampin, M., Oey, M. S.,
    Smith, L. J., Walterbos, R., & Mack, J. 2006, *ApJ*, 640, L29
Sabbi, E., Sirianni, M., Nota, A., Tosi, M., Gallagher, J., Meixner, M., Oey, M. S., Walterbos,
    R., Pasquali, A., Smith, L. J., & Angeretti, L. 2007, *AJ*, 133, 44
Sabbi, E., Sirianni, M., Nota, A., Tosi, M., Gallagher, J., Smith, L., Angeretti, L., Meixner, M.,
    Oey, M. S., Walterbos, R., & Pasquali, A. 2008, *AJ*, 135, 173
Tosi, M., Greggio, L., Marconi, G., & Focardi, P. 1991, *AJ*, 102, 951

Low-Metallicity Star Formation:
From the First Stars to Dwarf Galaxies
Proceedings IAU Symposium No. 255, 2008
L.K. Hunt, S. Madden & R. Schneider, eds.

A New Age and Distance for I Zw 18, the Most Metal-Poor Galaxy in the Nearby Universe

A. Aloisi[1,2], **G. Clementini**[3], **M. Tosi**[3], **F. Annibali**[1], **R. Contreras**[3], **G. Fiorentino**[3], **J. Mack**[1], **M. Marconi**[4], **I. Musella**[4], **A. Saha**[5], **M. Sirianni**[1,2], and **R. P. van der Marel**[1]

[1]Space Telescope Science Institute,
3700 San Martin Drive, Baltimore, MD 21218, USA
email: aloisi@stsci.edu

[2]On assignment from the Space Telescope Division of the European Space Agency

[3]INAF-Osservatorio Astronomico di Bologna,
Via Ranzani 1, I-40127 Bologna, Italy

[4]INAF-Osservatorio Astronomico di Capodimonte,
Via Moiariello 16, I-80131 Napoli, Italy

[5]National Optical Astronomy Observatory,
P.O. Box 26732, Tucson, AZ 85726

Abstract. The blue compact dwarf galaxy I Zw 18 holds the record of the lowest metallicity ever observed in the local universe. As such, it represents the closest analog to primordial galaxies in the early universe. More interestingly, it has recurrently been regarded as a genuinely young galaxy caught in the process of forming in the nearby universe. However, stars of increasingly older ages are found within I Zw 18 every time deeper high-resolution photometric observations are performed with the Hubble Space Telescope (HST): from the original few tens of Myrs to, possibly, several Gyrs. Here we summarize the history of I Zw 18 age and present an ongoing HST/ACS project which allowed us to precisely derive the galaxy distance by studying its Cepheid variables, and to firmly establish the age of its faintest resolved populations.

Keywords. galaxies: dwarf, galaxies: starburst, galaxies: evolution, galaxies: stellar content, galaxies: distances, galaxies: redshifts, galaxies: individual (I Zw 18), (stars: variables:) Cepheids

1. Introduction

The blue compact dwarf (BCD) galaxy I Zw 18 is one of the most intriguing nearby objects. Since its discovery by Zwicky (1966), it holds the record of the lowest nebular oxygen abundance of all known star-forming galaxies in the nearby Universe (12 + $\log(O/H) = 7.2$, corresponding to 1/50 $Z_\odot$; Skillman & Kennicutt 1993). This is even more intriguing considering the many efforts devoted over the past 50 years to the search for very metal-poor galaxies at redshift zero. I Zw 18 also has a high gas fraction (e.g., van Zee *et al.* 1998) and an extremely high star formation (SF) rate per unit mass (Searle & Sargent 1972) resulting in a blue young stellar population that dominates the integrated luminosity and color. All this observational evidence makes I Zw 18 a chemically unevolved stellar system. At a distance of about 12-15 Mpc, it actually represents the closest analog to primordial galaxies in the early universe and it has long been regarded as a possible example of a galaxy undergoing its first burst of star formation.

Many Hubble Space Telescope (HST) studies have focused on the evolutionary state of I Zw 18. The first HST/WFPC2 studies (Hunter & Thronson 1995; Dufour *et al.* 1996)

seemed to confirm that the stars in I Zw 18 are only a few tens of Myr old. From a new improved photometric reduction of the same HST/WFPC2 archival images, our group managed to detect fainter asymptotic giant branch (AGB) stars with ages of at least several hundreds of Myr (Aloisi, Tosi, & Greggio 1999). These results were confirmed by Östlin (2000) through deep HST/NICMOS imaging. More recently, Izotov & Thuan (2004) presented new deep HST/ACS imaging observations. Their I vs. $V - I$ CMD shows no sign of an RGB (i.e., low-mass stars with ages $\sim$1-13 Gyr that are burning H in a shell around a He core) at an assumed distance $D \lesssim 15$ Mpc. Their conclusion is that the most evolved (AGB) stars are not older than 500 Myr and that I Zw 18 is a *bona-fide* young galaxy. This result was subsequently challenged by Momany *et al.* (2005) and our group (Tosi *et al.* 2006) based on a better photometric analysis of the same data. This showed that many red sources do exist at the expected position of an RGB, and that their density in the CMD drops exactly where a RGB tip (TRGB, at the luminosity of the He flash) would be expected. However, the small number statistics, large photometric errors, and incompleteness, did not allow a more conclusive statement about the possible existence of an RGB in I Zw 18.

The actual nature of I Zw 18 has important cosmological implications. According to hierarchical formation scenarios, dwarf ($M \lesssim 10^9$ M$_\odot$) galaxies should have been the first systems to collapse and start forming stars. Indeed an RGB has been detected in all metal-poor dwarf irregular galaxies of the Local Group and BCDs within $D \lesssim 15$ Mpc that have been imaged with HST (e.g., SBS 1415+437, Aloisi *et al.* 2005, and references therein). I Zw 18 has remained the only elusive case so far.

The lack of RGB evidence has also made it impossible to pinpoint the distance of I Zw 18 via the TRGB method. Its distance therefore continues to be debated. With a recession velocity of 745 ± 3 km s^{-1}, I Zw 18 has often been assumed to be at a distance of $\sim$10 Mpc ($H_0 = 75$ km s^{-1}Mpc^{-1}). Correction for Virgocentric infall implies a slightly larger distance between 10 and 14.5 Mpc ($30.0 \lesssim m - M \lesssim 30.8$; Östlin 2000). Izotov *et al.* (2000) argued that I Zw 18 should be as distant as 20 Mpc to provide consistency between the CMD, the presence of Wolf-Rayet stars and the ionization state of the H II regions. But they suggested a shorter distance $D \lesssim 15$ Mpc from the brightness of AGB stars in Izotov & Thuan (2004).

2. New HST/ACS Observations

We were awarded 24 additional orbits with ACS over a three-month period starting in October 2005 (GO program 10586, PI Aloisi) to better understand the evolutionary status of I Zw 18. The observations were obtained in 12 different epochs in F606W and F814W to: (1) build a deeper CMD to search for RGB stars; (2) detect and characterize Cepheids at the lowest metallicity available in the local universe; and (3) use both the Cepheids and a possible TRGB detection to determine an accurate distance to I Zw 18 (see Aloisi *et al.* 2007).

PSF-fitting photometry was performed on deep images that were obtained by combining the exposures in each filter with MultiDrizzle. After application of CTE and aperture corrections, the count rates were transformed to Johnson-Cousins V and I magnitudes. Values shown and discussed hereafter are corrected for $E(B-V) = 0.032$ mag of Galactic foreground extinction, but not for any extinction intrinsic to I Zw 18. The archival ACS data in F555W and F814W (GO program 9400, PI Thuan) were also re-processed in a similar manner. The two ACS datasets were then combined using several different approaches and rejection schemes. The results discussed here were obtained by demanding that stars should be detected in all the four deep images (V and I for both datasets). At

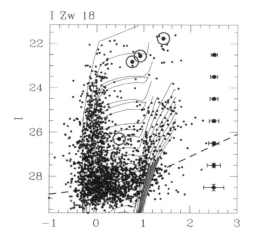
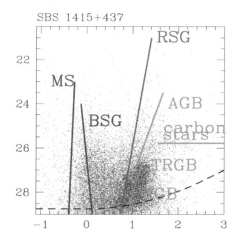

Figure 1. (a) HST/ACS CMD for I Zw 18 (Aloisi *et al.* 2007). Median photometric errors at $V - I = 1$ (determined by comparison of measurements from GO-9400 and GO-10586) are also shown as function of I. Padua isochrones from 5.5 Myr to 10 Gyr are overlaid, with the RGB phase for isochrones from 1.7 to 10 Gyr colored red. The isochrones have metallicity $Z = 0.0004$ (as inferred from the H II regions of I Zw 18) and are shown for the distance $D = 18.2$ Mpc ($m - M = 31.30$). The CMD includes stars in both the main and secondary bodies of I Zw 18. Blue open circles highlight the four confirmed variables, which are plotted according to their intensity-averaged magnitudes. (b) HST/ACS CMD for SBS 1415 + 437 (Aloisi *et al.* 2005). The main evolutionary sequences seen in the data are indicated in approximate sense as colored straight lines: main sequence (MS), blue supergiants (BSG), red supergiants (RSG), the red giant branch (RGB) with its tip (TRGB), the asymptotic giant branch (AGB), and carbon stars. Both CMDs are corrected for Galactic foreground extinction. Dashed lines are estimates of the 50% completeness level. The vertical axes of the panels are offset from each other by 0.61 mag, i.e. the difference in distance modulus between the galaxies (see Fig. 2). Some ~10 times more stars were detected in SBS 1415+437, owing to its smaller distance; stars for this galaxy are shown with smaller symbols. The faint red stars in both galaxies indicate that these metal-poor BCD galaxies started forming stars $\gtrsim 1$ Gyr ago.

the expense of some depth, this approach has the advantage of minimizing the number of false detections and therefore providing relatively "clean" CMDs.

3. Results and Interpretation

Fig. 1a shows the resulting I vs. $V - I$ CMD of I Zw 18. The CMD shows faint red stars exactly at the position where an RGB would be expected (see the Padua isochrones overplotted in the figure). Figure 2 shows the luminosity function (LF) of the red stars. It shows a sharp drop towards brighter magnitudes, exactly as would be expected from a TRGB. The magnitude of the discontinuity, $I = 27.27 \pm 0.14$ mag, implies a distance modulus $m - M = 31.30 \pm 0.17$ mag (e.g., Bellazzini *et al.* 2001), i.e., $D = 18.2 \pm 1.5$ Mpc. This assumes that the evolved RGB stars have negligible intrinsic extinction. The TRGB distance is consistent with the distance as inferred from the analysis of the Cepheid variables identified by our program (Aloisi *et al.* 2007). This agreement further supports our interpretation of the LF drop in Figure 2 as a TRGB feature.

The evidence for an RGB in I Zw 18 is further strengthened by comparison to another BCD, SBS 1415+437, observed by us with a similar HST/ACS set-up (Aloisi *et al.* 2005). This galaxy is not quite as metal poor as I Zw 18 ($12 + \log(O/H) = 7.6$) and is somewhat nearer at $D \approx 13.6$ Mpc. But taking into account the differences in distance and

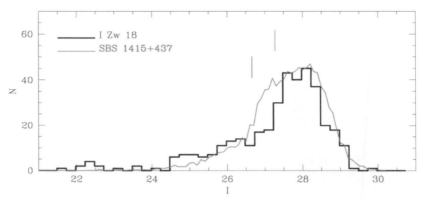

Figure 2. *I*-band LFs for stars with red colors in the range $V - I = 0.75$–1.5 mag, inferred from the CMDs in Fig. 1. Normalizations are arbitrary. Vertical marks indicate the positions of the TRGB, as determined using a Savitzky-Golay filtering technique developed by one of us (see Cioni *et al.* 2000). At these magnitudes there is a steep LF drop towards brighter magnitudes, due to the end of the RGB sequence. By contrast, the LF drop towards fainter magnitudes at $I > 28$ mag is due to incompleteness in both cases. Apart from a shift $\Delta(m - M) \approx 0.61$, these metal-poor BCD galaxies have very similar LFs.

completeness, the CMDs of these galaxies look very similar. Since SBS 1415+437 has an unmistakable RGB sequence, this suggests that such an RGB sequence exists in I Zw 18.

It has been previously suggested that I Zw 18 has no stars with ages $\gtrsim 500$ Myr. It is therefore interesting to ask the question whether there exist Star Formation Histories (SFHs) which do not have stars older than ~ 1 Gyr (and by extension, do not have an RGB), yet which can still fit: (1) the observed number of faint red stars in I Zw 18 ($V - I \gtrsim 1.0$, $I \gtrsim 27.3$); and (2) the observed LF drop-off at $I \approx 27.3$. Preliminary results of SFH modeling that we have performed indicate that such SFHs do not exist.

4. Conclusions

We have obtained new deep HST/ACS observations of I Zw 18 that provide improved insight into the evolutionary status of this benchmark metal-poor BCD. Our results indicate that this galaxy contains RGB stars, in agreement with findings for other local metal-poor BCDs studied with HST. Underlying old ($\gtrsim 1$ Gyr) populations are therefore present in even the most metal-poor systems. The coherent picture that emerges is that these galaxies did not form recently ($z < 0.1$) and may well be as old as the first systems that collapsed in the early universe. Deeper studies (well below the TRGB) will be needed to pinpoint the exact onset of the star formation in these extreme objects.

We find that at $D = 18.2 \pm 1.5$ Mpc, I Zw 18 is more distant than the values ~ 15 Mpc that have often been assumed in previous work. This may explain why it has remained difficult for so long to unambiguously detect or rule out the presence of old resolved (RGB) stars in this object. The data that we have compiled on Cepheid stars in I Zw 18 will be unique for probing the properties of variable stars at metallicities that have never before been probed.

Acknowledgements

Support for proposals #9361 and #10586 was provided by NASA through a grant from STScI, which is operated by AURA, Inc., under NASA contract NAS 5-26555.

References

Aloisi, A., Clementini, G., Tosi, M., Annibali, F., Contreras, R., Fiorentino, G., Mack, J., Marconi, M., Musella, I., Saha, A., Sirianni, M., & van der Marel, R. P. 2007, *ApJ* 667, L151

Aloisi, A., Tosi, M., & Greggio, L. 1999, *AJ* 118, 302

Aloisi, A., van der Marel, Mack, J., Leitherer, C., Sirianni, M., & Tosi, M. 2005, *ApJ* 631, L45

Bellazzini, M., Ferraro, F. R., & Pancino, E. 2001, ApJ, 556, 635

Cioni, M.-R. L., van der Marel, R. P., Loup, C., & Habing, H. J. 2000, A&A, 359, 601

Dufour, R. J., Garnett, D. R., Skillman, E. D., & Shields, G. A. 1996, in: C. Leitherer, U. Fritze-von Alvensleben, & J. Huchra (eds.), *From Stars to Galaxies: The Impact of Stellar Physics on Galaxy Evolution*, Proc. (San Francisco: ASP), p. 358

Hunter, D. A., & Thronson, H. A. 1995, *ApJ* 452, 238

Izotov, Y. I., Papaderos, P., Thuan, T. X., Fricke, K. J., Foltz, C., & Guseva, N. G. 2000, in: A. Weiss, T. G. Abel, & V. Hill (eds.), *The First Stars*, Proc. (Berlin: Springer), p. 303

Izotov, Y. I., & Thuan, T. X. 2004, *ApJ* 616, 768

Momany, Y., Held, E. V., Saviane, I., Bedin, L. R., Gullieuszik, M., Clemens, M., Rizzi, L., Rich, M. R., & Kuijken, K. 2005, *A&A* 439, 111

Östlin, G. 2000, *ApJ* 535, L99

Searle, L., & Sargent, W. L. W. 1972, *ApJ* 173, 25

Skillman, E. D., & Kennicutt, R. C. 1993, *ApJ* 411, 655

Tosi, M., Aloisi, A., Mack, J., & Maio, M. 2007, in: F. Combes & J. Palous (eds.), *Galaxy Evolution Across the Hubble Time*, Proc. IAU Symposium No. 235 (San Francisco: ASP), p. 65

van Zee, L., Westpfahl, D., Haynes, M. P., & Salzer, J. J. 1998, *AJ* 115,1000

Zwicky, F. 1966, *ApJ* 143, 192

Low-Metallicity Star Formation:
From the First Stars to Dwarf Galaxies
Proceedings IAU Symposium No. 255, 2008
L.K. Hunt, S. Madden & R. Schneider, eds.

© 2008 International Astronomical Union
doi:10.1017/S1743921308025131

The ACS LCID project: accurate measurements of the full star formation history in low metallicity, isolated, Local Group dwarf galaxies

Matteo Monelli[1]† and the LCID team

[1]Instituto de Astrofísica de Canarias,
C/ Vía Láctea, 38205, La Laguna, Tenerife, Spain
email: monelli@iac.es

Abstract. We present here the latest results of the *LCID project (Local Cosmology from Isolated Dwarfs)*, aimed at recovering the full star formation history (SFH) of six isolated dwarf galaxies of the Local Group (LG). Our method of analysis is based on the IAC-pop code, which derives the SFH of a resolved stellar system by comparing the observed and a model color-magnitude diagram (CMD). We summarize here basic technical issues and the main results concerning our sample of galaxies. We show that LeoA is the only object showing a clear delay in the onset of the major SF event, while all the other galaxies present a dominant component older than 10 Gyrs.

Keywords. galaxies: evolution, galaxies: dwarf, (galaxies:) Local Group

1. Introduction

The study of nearby resolved galaxies is a powerful tool to shed light on the formation and evolution of galaxies. In particular, the importance of isolated galaxies relies on the fact that they are likely approaching the Local Group (LG) for the first time, and therefore they spent most of their life time free from the influence of giant galaxies. These objects are ideal candidates to study the first stages of their formation and evolution, and can provide key insight in the cold dark matter scenery. Our approach to recover the SFH of resolved stellar systems is based on sophisticated modelling of their CMD. In particular, the most reliable constraints can be derived from the main sequence turn-offs (MSTO, Gallart *et al.* 1999). Therefore, the galaxies of our sample were selected balancing two opposing requirements. Isolated, and therefore distant, objects were required, but deep and accurate photometry was mandatory to detect the TO stars with good signal-to-noise ratio (>10). The final sample includes galaxies of different morphological type: Cetus and Tucana (dSph), LGS3 and Phoenix (transition dIrr/dSph), IC1613 and LeoA (dIrr). With the exception of Phoenix, for which we used WFPC2 data, all of the galaxies were observed with the ACS camera on *Hubble Space Telescope* (HST). Our method is based on the application of the IAC-star and IAC-pop codes. The latter makes use of a genetic algorithm to derive the star formation rate (SFR) and the chemical evolution law of a stellar system, comparing the observed and a model CMDs. A few technical aspects of the method are discussed (a complete description can be found in Aparicio & Gallart 2004 and in Aparicio & Hidalgo 2008).

† Local Cosmology from Isolated Dwarfs, http://www.iac.es/project/LCID

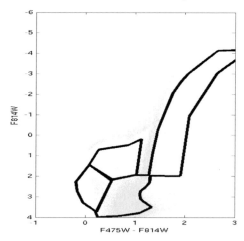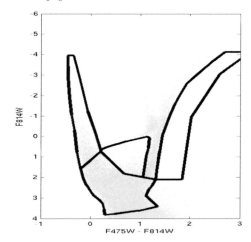

Figure 1. Two examples of parametrization of a CMD: Tucana (*left*) and IC1613 (*right*). The four regions define the *bundles* which are sub-divided into boxes for the star counting. We adopt the finest sampling in the MS and MSTO regions.

2. The method: IAC-pop

The basic idea of IAC-pop is that any SFH is mainly characterized by two parameters, the age and the metallicity of the stars. The SFH, $\Psi(t, Z)$, is therefore a function described as the combination of *single populations*, defined as the ensembles of stars formed per unit of time and metallicity, and can be expressed as:

$$\Psi(t, Z) = A \sum_i \alpha_i \psi_i \qquad (2.1)$$

where $\Psi(t, Z)$ is the global SFH, ψ_i are the contributions of the single populations, and A is the normalization constant. In this representation, the weights α_i give a quantitative estimate of the mass of gas transformed into stars at that particular age and with that particular metallicity. From a practical point of view, the single populations are identified using a discrete set of time and metallicity bins. The values adopted in this work are the following: age bins of 1 Gyr in the range 0 to 15 Gyr, except the last bin between 13 and 15 Gyr, Z = [0.0001, 0.0003, 0.0005, 0.0007, 0.001, 0.0015, 0.002], being this last value extended to 0.003 in the case of LGS3 and LeoA, and to 0.005 for IC1613.

The model CMD is generated using the IAC-star simulator, and requires a number of different ingredients:

• a set of theoretical stellar evolution libraries. Here we used the BaSTI (Pietrinferni *et al.* 2004) models, because it covers the full metallicity range expected in the observed galaxies;

• the SFR, which we set constant at any age;

• the metallicity law, Z(t), which we define such that the metallicity of stars of any age is uniformly distributed in the range 0.0001 < Z < 0.005;

• the initial mass function (IMF), taken from Kroupa (2001);

• the binary fraction, β, and the relative mass distribution q, the default values being $\beta = 60\%$, $q > 0.5$.

Other auxiliary but fundamental information is necessary to compare the model with the observations. First, an independent estimate of the distance modulus (DM) and the extinction E(B-V) are mandatory to shift the theoretical and observed diagrams in the same magnitude system. Second, the observational errors (crowding, blending,

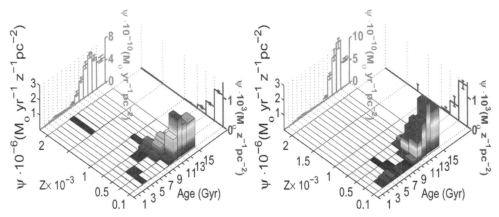

Figure 2. 3-D representation of the SFH of the two dSph galaxies in our sample: Tucana and Cetus.

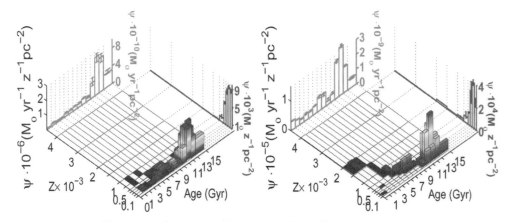

Figure 3. Same as for Fig. 2, but for LGS3 and Phoenix.

incompleteness, photometric errors, . . .) have to be properly simulated in the model CMD (Aparicio & Gallart 1995). This is done through extensive tests adding synthetic stars to the images, and repeating the full process of the photometric analysis. On average, 700,000 synthetic stars were used for each galaxy.

The observable that IAC-pop uses to compare the data with the model is the number of stars, using a χ^2 merit function (Mighell 1999). A key point of the method is that the CMDs have to be parametrized in order to perform the star count. This is made by defining a set of boxes covering the CMDs (see Fig. 1). It is important to notice that different features of the CMD are sampled in different ways. Fig. 1 identifies various regions (*bundles*); each of them is sub-divided into boxes. Extensive tests disclosed that one has to rely the most on the features where most of the information resides, and where the theoretical uncertainities of the models are the smallest. Thefore, we adopt the finest sampling on the MS, the TO and the sub-giant branch regions. Note that we don't use at all the red giant and horizontal branches, given their small temporal resolution and theroretical uncertainities, respectively.

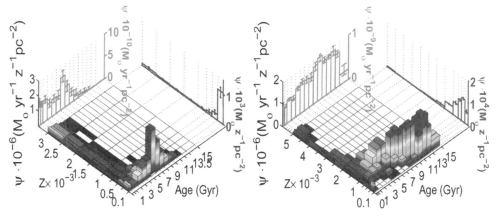

Figure 4. Same as for Fig. 3, but for LeoA and IC1613.

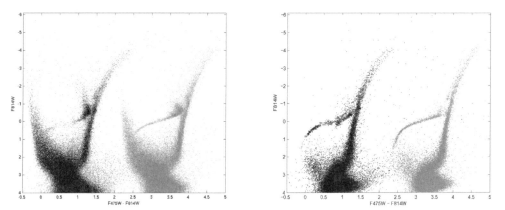

Figure 5. In each panel are shown the observed CMD (*left*) and the CMD corresponding to the derived SFH. The two galaxies represented are IC1613 and Tucana.

3. On the solutions and their stability

Figs. 2 to 4 present the SFH of the six sample galaxies.

i) Cetus and Tucana: these two dSph galaxies are predominantly old and metal-poor systems. The bulk of the SF occurred at old ages, the peak being around 12 Gyrs ago. Both present declining SFR, with negligible star formation at epochs more recent than 10 Gyrs;

ii) LGS3 and Phoenix: the two transition galaxies in our sample present a continuous SF all over their lifetime. Again, the most important episode of SF occurred at old epochs (>12 Gyr), followed by a steady decline until ≈7 Gyr ago. However, both systems were able to form stars until present ages, with a mild but continuous process.

iii) LeoA and IC1613: LeoA presents the most striking example of delayed SF (Cole *et al.* 2007). Although the presence of an old population is testified by the presence of RR Lyrae stars (Dolphin *et al.* 2002), this galaxy has experienced a dominating event of SF ≈6-8 Gyrs ago. IC1613 seems to show a similar trend, even if to a much less dramatic extent. The right panel of Fig. 4 shows that the SF of this galaxy became more and more intense for ages from ≈13 to ≈10 Gyrs, reached a peak and then started declining.

Fig. 5 shows the CMD corresponding to the solution for IC1613 and Tucana. The overall agreement is satisfactory, and it is interesting to note that the RGB is very well reproduced, even if no information at all from this feature has been used to derive these solutions.

To study the uncertainities and the stability of these solutions, we first have to identify the possible sources of error. For the sake of clarity, we can classify them in three groups:

- *observational errors*: crowding, incompleteness, ...;
- *theoretical errors*: uncertainties in the stellar evolution models;
- *external errors*: distance, extinction, binary fraction, IMF, ...

We assume that the observational errors are properly modelled by our reduction procedure, and a full discussion concerning the use of different stellar evolution libraries (BaSTI and the Padova library, Bertelli *et al.* 1994) will be presented in a forthcoming paper (Hidalgo *et al.*, in prep.).

Our work-in-progress is mainly focused at analysing the external errors, because this will allow us to estimate the impact of our assumptions on the derived SFH. This is done by changing the input parameters of the model CMD. In particular, we are testing different: *i)* fraction of binary stars (from 0 to 100%); *ii)* values of the distance (± 0.15 mag); *iii)* values of the extinction; *iv)* IMF (changing one or both the IMF indices).

The in-depth analysis and results will be presented in the papers in preparation.

References

Aparicio, A. & Gallart, C. 1995, *AJ*, 110, 2105

Aparicio, A. & Gallart, C. 2004, *AJ*, 128, 1465

Aparicio, A. & Hidalgo, S. 2008, *AJ*, Submitted

Bertelli, G., Bressan, A., Chiosi, C., Fagotto, F., & Nasi, E. 1994 *A&AS*, 106, 275

Cole, A. A., Skillman, E. D., Tolstoy, E., Gallagher, J. S., III, Aparicio, A., Dolphin, A. E., Gallart, C., Hidalgo, S. L., Saha, A. t, Stetson, P. B., & Weisz, D. R. 2007 *ApJ* (Letters), 659, 17

Dolphin, A. E., Saha, A., Claver, J., Skillman, E. D., Cole, A. A., Gallagher, J. S., Tolstoy, E., Dohm-Palmer, R. C., & Mateo, M. 2002 *AJ*, 123, 3154

Gallart, C., Freedman, W. L., Aparicio, A., Bertelli, G., & Chiosi, C. 1999, *AJ*, 1181, 2245

Kroupa, P. 2001, *MNRAS*, 322, 231

Mighell, K. J., 1999, *ApJ*, 518, 380

Pietrinferni, A., Cassisi, S., Salaris, M., & Castelli, F. 2004, *ApJ*, 612, 168

Low-Metallicity Star Formation:
From the First Stars to Dwarf Galaxies
Proceedings IAU Symposium No. 255, 2008
L.K. Hunt, S. Madden & R. Schneider, eds.

© 2008 International Astronomical Union
doi:10.1017/S1743921308025143

Low Metallicity Galaxies at $z \sim 0.7$: Keys to the Origins of Metallicity Scaling Laws

David J. Rosario[1], Carlos Hoyos[2], David Koo[1] and Andrew Phillips[1]

[1] Dept. of Astronomy and Astrophysics, University of California – Santa Cruz,
Santa Cruz, California, USA, 95064
email: rosario@ucolick.org, koo@ucolick.org, phillips@ucolick.org

[2] Departmento de Fisica Teórica, Universidad Autónoma de Madrid, Carretera de Colmenar
Viejo kn 15.600 28049, Madrid, Spain
email: charly.hoyos@uam.es

Abstract. We present a study of remarkably luminous and unique dwarf galaxies at redshifts of $0.5 < z < 0.7$, selected from the DEEP2 Galaxy Redshift survey by the presence of the temperature sensitive [OIII]λ4363 emission line. Measurements of this important auroral line, as well as other strong oxygen lines, allow us to estimate the integrated oxygen abundances of these galaxies accurately without being subject to the degeneracy inherent in the standard R_{23} system used by most studies. [O/H] estimates range between 1/5–1/10 of the solar value. Not surprisingly, these systems are exceedingly rare and hence represent a population that is not typically present in local surveys such as SDSS, or smaller volume deep surveys such as GOODS.

Our low-metallicity galaxies exhibit many unprecedented characteristics. With B-band luminosities close to L$_*$, thse dwarfs lie significantly away from the luminosity-metallicity relationships of both local and intermediate redshift star-forming galaxies. Using stellar masses determined from optical and NIR photometry, we show that they also deviate strongly from corresponding mass-metallicity relationships. Their specific star formation rates are high, implying a significant burst of recent star formation. A campaign of high resolution spectroscopic follow-up shows that our galaxies have dynamical properties similar to local HII and compact emission line galaxies, but mass-to-light ratios that are much higher than average star-forming dwarfs.

The low metallicities, high specific star formation rates, and small halo masses of our galaxies mark them as lower redshift analogs of Lyman-Break galaxies, which, at $z \sim 2$ are evolving onto the metallicity sequence that we observe in the galaxy population of today. In this sense, these systems offer fundamental insights into the physical processes and regulatory mechanisms that drive galaxy evolution in that epoch of major star formation and stellar mass assembly.

Keywords. surveys, galaxies: abundances, galaxies: kinematics, galaxies: starburst

1. Introduction

The evolution of metal abundances in galaxies is inextricably linked to a host of physical processes and properties which determine their baryonic distribution, such as star formation regulation, stellar mass functions, ISM mixing, outflows and gaseous infall, supernova enrichment, stellar winds, and grain formation and depletion. A broad brush panorama of the complex interplay of these different forces is empirically manifested in the form of the Stellar Mass-Metallicity Relationship (MZR), apparent in both gaseous and stellar abundances. Work with local galaxy samples (e.g., Tremonti *et al.* 2004, Lee *et al.* 2006) show that the expected metallicity of a galaxy increases with increasing stellar content, over 5 orders of magnitude in stellar mass with roughly constant scatter. Various models exist to explain the form of this trend, though current observational constraints are unable to easily distinguish between them. A number of authors in these proceedings

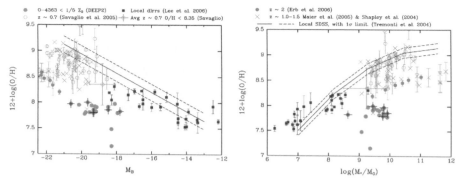

Figure 1. Luminosity-Metallicity Relation (LZR – left) and Mass-Metallicity Relation (MZR – right) for the 4363-selected DEEP2 galaxies (red points), compared to other local, intermediate and high-redshift samples. The solid and dashed black lines are the mean trend and 1σ scatter for SDSS galaxies from Tremonti *et al.* (2004). The green error bars are the mean for $z \sim 0.7$ galaxies of Savaglio *et al.* (2005) in the same luminosity and mass range of our sample.

explore the theoretical ramifications of various classes of models, such as those dominated by stellar feedback, or those with mass dependent star-formation efficiencies.

In this contribution, we discuss a new sample of intermediate redshift dwarf galaxies which have remarkably low metallicities for their measured luminosities and stellar content. More details of this study can be found in Hoyos *et al.* (2005) and Hoyos *et al.* (2008).

2. The Sample: Dataset, Selection and Measurements

The DEEP2 Redshift Survey is currently the deepest and most complete large-volume extragalactic spectroscopic survey at $z \sim 1$. The survey dataset consists of Keck/DEIMOS slit spectra of galaxies brigher than $R_{AB} = 24.1$ in four widely separated fields, over a wavelength range of approximately 6500–9000 Å. Compared to previous galaxy redshift surveys of similar depth, DEEP2 is quite efficient, yielding redshifts accurate to better than 2σ in roughly 70% of the slits. Details of the DEEP2 survey can be found in Coil *et al.* (2004) (photometry), Davis *et al.* (2003) (survey design) and Davis *et al.* (2005) (mask-making strategy and first results).

Low-metallicity galaxies were selected by searching for the [OIII]λ4363 auroral emission line in the DEEP2 spectral dataset, through a combination of automated line-fitting and visual inspection of candidates. This yielded a final sample of 25 galaxies with bonafide [OIII]λ4363 detections between the redshifts $0.55 < z < 0.7$. This choice of redshift range ensures that the [OIII]$\lambda\lambda$4959, 5007 doublet and Hβ are also in the spectrum. Gaussian fits were made to the emission lines of the [OIII] triplet, Hβ, and where possible, the [OII]$\lambda\lambda$3726, 3729 doublet, from which line fluxes were estimated after applying a standardized flux calibration determined for DEIMOS spectra. For galaxies in the lower redshift range of our sample, the [OII] line lies beyond the blue end of the spectra, so its strength was estimated from Hβ using the strong correlation between the equivalent widths (EWs) of the two lines found for the other sample galaxies.

3. Metallicities, Luminosities and Stellar Masses

The ratio [OIII]$\lambda\lambda$4363/5007 is a sensitive indicator of the electron temperature in the [OIII]-emitting gas. Through the use of photoionization models of HII regions, the temperature of the [OII]-emitting zone can then be evaluated, enabling an unbiased estimate of the light-weighted oxygen abundance of the gas in the star-forming region. Following the procedure of Pérez-Montero & Díaz (2003) and using calibrations from Pagel *et al.* (1992), we estimate oxygen abundances for our galaxies in the range of $7.4 < 12 + [O/H] < 8.3$, with typical uncertainties around 0.1 dex, comparable to that of local BCDs.

Accurate stellar masses are the key to understanding the inter-relationship between stellar processes and metal enrichment in our galaxies and, consequently, must be determined as accurately as possible. We combine optical BRI and NIR K_s-band photometry in two independent ways to estimate the stellar masses of our sample. Our principal method is as follows: after correcting the rest-frame colors for the substantial emission line and nebular continuum contamination (0.16 mag in $U - B$ and 0.2 mag in $B - V$), we fit the BRI (and, where possible, K_s) colors of our galaxies with low-metallicity two-burst BC03 models (Bruzual & Charlot 2003), optimizing the relative fractions of the two SSP components, as well as the age of the older component. The age of the youngest burst was fixed at 2 Myr, consistent with the average EW of Hβ in the spectra of the 4363-selected sample. The resultant stellar masses have statistical uncertainties of about 0.2 dex.

These were compared to masses derived from the prescription of Lin *et al.* (2007), which are based on BRIK$_s$ SED fits to the main set of DEEP2 galaxies. Using this approach gives masses that are very comparable to the more detailed method described above. In the rest of this proceedings, including in all figures, we will adopt the stellar masses determined by the two-burst BC03 fits.

Restframe B-band luminosities of our galaxies were determined using k-corrections calculated from SED fits to the overall DEEP2 galaxy sample, following the method of Willmer *et al.* (2006).

In addition to oxygen abundances, luminosities and stellar masses, we derive star-formation rates (SFRs) from the Hβ flux, along the lines of Kennicutt *et al.* (1994). Our galaxies have intermediate to high Specific SFRs (i.e., the SFR per unit stellar mass – a measure of recent to past star-formation). Roughly half have among the highest SSFRs of all DEEP2 star-forming systems. Clearly, our galaxies are going through a burst of substantial current star-formation, which supports the choice of a two-burst framework in the estimation of their stellar masses.

4. Metallicity Scaling Relations

In Fig. 1, we plot the Luminosity-Metallicity and Mass-Metallicity relations (LZR and MZR) for our sample, compared to local SDSS galaxies (Tremonti *et al.* 2004) and local dIrrs (Lee *et al.* 2006), as well as normal star-forming galaxies at intermediate redshifts (Shapley *et al.* 2004, Savaglio *et al.* 2005, Maier *et al.* 2005) and high redshifts (Erb *et al.* 2006). Remarkably, our objects deviate strongly from both local scaling laws and those at comparable redshifts, by as much as 0.8 dex.

The abundances for all the comparison samples in Fig. 1 are derived from strong line methods, such as R_{23} and the N/O calibration. In particular, the use of the R_{23} assumes in all cases that the galaxies lie along a high metallicity branch, which places a lower limit on the abundances that can be derived by this method. Clearly, the galaxies in our sample have luminosities and stellar masses comparable to some normal star-forming systems, but with much lower metallicities. The unchecked use of R_{23} can alter the form and scatter of metallicity scaling relations, as well as result in incorrect abundances for a fraction of star-forming systems. Care must be taken when applying this method to

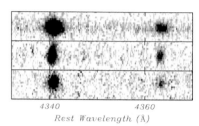

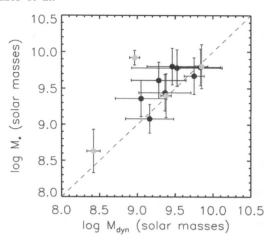

Figure 2. Sections of DEEP2 two-dimensional spectra of three 4363-selected galaxies, showing the Hγ and [OIII]λ4363 lines.

Figure 3. Dynamical vs. Stellar Masses for 4363-selected galaxies with Keck/HIRES spectra. Four galaxies with resolved HST/ACS images in AEGIS are plotted as yellow points. For the black points, M_{dyn} is estimated assuming $R_{em} = 2$ kpc, with a 0.3 dex uncertainty in the radius reflected in the errors. The dashed line has a slope of unity.

derive abundances in large-volume extragalactic spectral datasets or those with complex selection functions.

The locations of our sample galaxies in the LZR and MZR imply that they are analogs of normal star-forming systems at $z \gtrsim 2$, such as Lyman-Break Galaxies (LBGs). They promise to serve as testbeds for the physical conditions and processes which evolve such systems to the galaxy population of today.

5. Dynamical Masses and Mass-to-Light Ratios

The emission line properties of the 4363-selected sample are very similar to local HII galaxies, as well as Compact Narrow-Emission Line Galaxies (CNELGs) found at intermediate redshifts (Koo *et al.* 1995). To compare their dynamical properties as well, we embarked on a program on Keck/HIRES spectroscopy to get resolved line widths and line profiles for most of our sample objects. Despite the optical faintness of these galaxies, their high emission line EWs ensured clean line measurements in all cases. From our measured line widths σ (in km/s), we estimate a dynamical mass using the following relation (from Guzmán *et al.* 2003): $M_{dyn}/M_{\odot} = 1.1 \times 10^6 \, R_{em} \, \sigma^2$. Here, R_{em} is the half-light radius of the line emitting distribution of the galaxy, in kpc. Since we lack resolved imaging for most of our sample, we assume $R_{em} \sim 2$ kpc (but see below).

In Fig. 3, we plot the stellar mass of the galaxies with HIRES spectra against their dynamical masses. Quite unexpectedly, they are roughly the same, within the degree of uncertainty of the stellar and dynamical masses (about 0.2 dex). Implied total mass-to-light ratios (M/L) are around unity, very low compared to normal dwarf galaxies with similar kinematic properties, which typically have $M/L \sim 10 - 100$ or more. This result is fairly insensitive to the method of dynamical mass estimates or the half-light radius of the galaxy, which may account for factors of a few in dynamical mass.

Atek, Hakim *"Lyman-alpha emission regulation in star-forming galaxies"*
(poster/atek_h.pdf)

Aykutalp, Aycin *"The impact of metallicity on early star formation"*
(poster/aykutalp_a.pdf)

Barzdis, Arturs *"High-resolution spectroscopy of metal-poor star HD 187216"*
(poster/barzdis_a.pdf)

Bernard, Edouard *"The ACS LCID project: Short-period variables"*
(poster/bernard_e.pdf)

Bianchi, Simone *"Dust formation and survival in supernova ejecta"*
(poster/bianchi_s.pdf)

Brito de Freitas, Daniel *"Lithium abundances in evolved members of Galatic open clusters"*
(poster/britodefreitas_d.pdf)

Brott, Ines *"Constraints on rotational mixing from Magellanic Clouds B-stars"*
(poster/brott_i.pdf)

Carlson, Lynn *"A panchromatic view of Magellanic star formation: From optical to infrared"*
(poster/carlson_l.pdf)

Cescutti, Gabriele *"Inhomogeneous chemical evolution models: From the Galactic halo to the dwarf spheroidal galaxies"*
(poster/cescutti_g.pdf)

Champavert, Nicolas *"Chemical evolution of galaxies using a new multiphase chemodynamical code"*
(poster/champavert_n.pdf)

Choi, Ena *"The chemical enrichment of the early building blocks"*
(poster/choi_e.pdf)

Cignoni, Michele *"Star formation history in the SMC: NGC 602 and NGC 346"*
(poster/cignoni_m.pdf)

Cristallo, Sergio *"AGB nucleosynthesis at very low metallicities"*
(poster/cristallo_s.pdf)

Cumming, Robert *"Stellar kinematics in blue compact galaxies"*
(poster/cumming_r.pdf)

de la Rosa, Ignacio *"Galaxy assembly in the densest environments"*
(poster/delarosa_i.pdf)

Ekta *"HI in the most metal-deficient galaxies"*
(poster/ekta_-.pdf)

For, Bi-Qing *"Searching for α-poor stars in the Galactic halo"*
(poster/for_b.pdf)

Galametz, Maud *"Studying the dust properties of low-metallicity dwarf galaxies at submillimeter wavelengths with LABOCA"*
(poster/galametz_m.pdf)

Gall, Christa *"Dust formation modelling in supernovae ejecta"*
(poster/gall_c.pdf)

Goessl, Claus *""Blue" pulsating variable stars in local metal poor dwarf galaxies"*
(poster/goessl_c.pdf)

Goncalves, Denise *"Planetary Nebulae: Enlightening the luminosity-metallicity relation of the Local Group dwarf galaxies"*
(poster/goncalves_d.pdf)

Gouliermis, Dimitrios *"The Magellanic clouds as templates of star formation at low metallicities: The census of pre-main sequence stars"*
(poster/gouliermis_d.pdf)

Gratier, Pierre *"Large scale CO(2-1) mapping of Local Group galaxies: The case of the dwarf galaxy NGC 6822"*
(poster/gratier_p.pdf)

Guseva, Natalia *"Most metal-deficient emission-line galaxies: New discoveries"*
(poster/guseva_n.pdf)

Hasegawa, Kenji *"Secondary star formation in a Pop III object: Dependence of UV feedback on the mass of source star"*
(poster/hasegawa_k.pdf)

Hibbard, John *"GBT HI observations of low-metallicity galaxies from the SDSS"*
(poster/hibbard_j.pdf)

Hidalgo-Gamez, Ana *"LFs and SFRs for a sample of dwarf spiral galaxies"*
(poster/hidalgo-gamez_a.pdf)

Hocuk, Seyit *"Thermodynamic properties of molecular clouds and the IMF in dwarf galaxies"*
(poster/hocuk_s.pdf)

Hood, Michael *"Kinematics of the stellar populations of M 33"*
(poster/hood_m.pdf)

Hunt, Leslie *"The Spitzer view of low-metallicity star formation: Haro 3 and Mrk 996"*
(poster/hunt_l.pdf)

Husti, Laura *"Theoretical interpretation of the metal-poor barium giant HD 123396"*
(poster/husti_l.pdf)

James, Bethan *"Spectral mapping of the anomalous blue compact dwarf galaxy: Mrk 996"*
(poster/james_b.pdf)

Johnson, Jarrett *"The occurrence of metal-free galaxies in the early Universe"*
(poster/johnson_l.pdf)

Kehrig, Carolina *"A study of very low metallicity HII galaxies using integral field spectroscopy"*
(poster/kehrig_c_1.pdf)

Kehrig, Carolina *"The star-forming dwarf galaxy population in the local universe and beyond: The first 3D spectroscopic analysis of blue compact dwarf galaxies"*
(poster/kehrig_c_2.pdf)

Klapp, Jaime *"Evolution and nucleosynthesis of the first stars"*
(poster/klapp_j.pdf)

Koleva, Mina *"When did dwarf elliptical galaxies start to form stars?"*
(poster/koleva_m.pdf)

Krticka, Jiri *"CNO driven winds of hot iňĄrst stars"*
(poster/krticka_j.pdf)

Lanfranchi, Gustavo *"The evolution of [α/Fe] in Carina dwarf spheroidal galaxy: Constraints from new data"*
(poster/lanfranchi_g.pdf)

Leaman, Ryan *"First Metallicity Distribution From VLT FORS2 CaT: Spectroscopy of RGB Stars in the dwarf irregular galaxy WLM"*
(poster/leaman_r.pdf)

Lebouteiller, Vianney *"Mid-infrared study of giant HII regions"*
(poster/lebouteiller_v.pdf)

Lucatello, Sara *"Looking for binaries at low metallicity"*
(poster/lucatello_s.pdf)

Maio, Umberto *"Onset of star formation and impact on the surroundings"*
(poster/maio_u.pdf)

Makarov, Dmitry *"Star formation history reconstruction of nearby dwarf galaxies"*
(poster/makarov_d.pdf)

Makarova, Lidia *"Ancient and recent epoch of star formation in nearby dwarf galaxies"*
(poster/makarova_l.pdf)

Martin-Manjon, Mariluz *"POPStar, a new grid of evolutionary synthesis models: Emission line diagnostics for evolving HII regions"*
(poster/martin-manjon_m_1.pdf)

Martin-Manjon, Mariluz *"Modelling starbursts in HII galaxies: From chemical to spectro-photometric evolutionary self-consistent models"*
(poster/martin-manjon_m_2.pdf)

Masseron, Thomas *"Very metal-poor AGB witnesses"*
(poster/masseron_t.pdf)

Matsuoka, Kenta *"Cosmic metallicity evolution traced by radio galaxies"*
(poster/matsuoka_k.pdf)

Milone, Andre *"Single-aged stellar population models with empirical variable Mg-enhancement"*
(poster/milone_a.pdf)

Oliveira, Joana *"Ice chemistry in young stellar objects in the Magellanic clouds"*
(poster/oliveira_j.pdf)

Papaderos, Polychronis *"Stellar populations and extended ionized gas emission in blue compact dwarf galaxies"*
(poster/papaderos_p.pdf)

Placco, Vinicius *"A search for metal-poor stars based on carbon overabundance"*
(poster/placco_v.pdf)

Pustilnik, Simon *"Galaxy DDO 68 as a testbed for models of very low metallicity SF, massive stars and cosmological mergers"*
(poster/pustilnik_s.pdf)

Roederer, Ian *"Subtle nucleo-kinematic differences in the stellar halos of the Milky Way galaxy"*
(poster/roederer_i.pdf)

Schleicher, Dominik *"Primordial magnetic fields in the dark ages of the universe"*
(poster/schleicher_d.pdf)

Schlesinger, Katharine *"Determining the low-mass end of the initial mass function using metal-poor stars"*
(poster/schlesinger_k.pdf)

Schuster, William *"The San Pedro Martir survey of high-velocity and metal-poor stars"*
(poster/schuster_w.pdf)

Shao, Zhengyi *"Color gradients along discs of spiral galaxies"*
(poster/shao_z.pdf)

Simpson, Caroline *"VII Zw 403: A blue compact dwarf case study"*
(poster/simpson_c.pdf)

Smecker-Hane, Tammy *"The star formation history of the Leo I dSph galaxy"*
(poster/smecker-hane_t.pdf)

Snigula, Jan *"AGB stars in metal-poor dwarf galaxies: LPVs and the fuel consumption theorem as tracers of stellar populations"*
(poster/snigula_j.pdf)

Starkenburg, Else *"How accurate is the NIR CaII triplet at very low metallicity?"*
(poster/starkenburg_e.pdf)

Suda, Takuma *"The star formation history of the Milky Way halo explored with stellar evolution and nucleoeosynthesis of EMP stars, and with their stellar abundances database"*
(poster/suda_t.pdf)

Telles, Eduardo *"Kinematics and the ISM phases in II Zw 40: Optical and infrared IFU spectroscopy"*
(poster/telles_e.pdf)

Tesileanu, Ovidiu *"Jets from young stars: Radiative MHD simulations"*
(poster/tesileanu_o.pdf)

Tsujimoto, Takuji *"New insights into the first stars in the Galactic halo and dwarf spheroidal galaxies"*
(poster/tsujimoto_t.pdf)

Umemura, Masayuki *"First star simulations down to CDM damping scales"*
(poster/umemura_m.pdf)

Valcke, Sander *"Simulations of the formation and evolution of isolated dwarf galaxies"*
(poster/valcke_s.pdf)

Wallerstein, George *"A comparison of the globular clusters associated with the Sagittarius and Fornax galaxies"*
(poster/wallerstein_g.pdf)

Yakobchuk, Taras *"Relatively young age of the red stellar populations in extremely metal-deficient dwarf galaxies"*
(poster/yakobchuk_t.pdf)

Yi, Sukyoung *"First stars as a possible origin for the helium-rich populations in globular clusters"*
(poster/yi_s.pdf)

Zacs, Laimons *"Abundance analysis of extremely metal-poor star HD 112869"*
(poster/zacs_l.pdf)

Author Index

Subject Index

413